T0213602

Lecture Notes in Computer Science 9826

Commenced Publication in 1973
Founding and Former Series Editors:
Gerhard Goos, Juris Hartmanis, and Jan van Leeuwen

Editorial Board

More information about this series at http://www.springer.com/series/7407

Gul Agha · Benny Van Houdt (Eds.)

Quantitative Evaluation of Systems

13th International Conference, QEST 2016
Quebec City, QC, Canada, August 23–25, 2016
Proceedings

 Springer

Editors
Gul Agha
University of Illinois
Urbana, IL
USA

Benny Van Houdt
University of Antwerp
Antwerp
Belgium

ISSN 0302-9743   ISSN 1611-3349   (electronic)
Lecture Notes in Computer Science
ISBN 978-3-319-43424-7   ISBN 978-3-319-43425-4   (eBook)
DOI 10.1007/978-3-319-43425-4

Library of Congress Control Number: 2015944718

LNCS Sublibrary: SL1 – Theoretical Computer Science and General Issues

Printed on acid-free paper

This Springer imprint is published by Springer Nature
The registered company is Springer International Publishing AG Switzerland

Preface

Welcome to the proceedings of QEST 2016, the 13th International Conference on Quantitative Evaluation of Systems. QEST is a leading forum on quantitative evaluation and verification of computer systems and networks, through stochastic models and measurements. QEST was first held in Enschede, The Netherlands (2004), followed by meetings in Turin, Italy (2005), Riverside, USA (2006), Edinburgh, UK (2007), St. Malo, France (2008), Budapest, Hungary (2009), Williamsburg, USA (2010), Aachen, Germany (2011), London, UK (2012), Buenos Aires, Argentina (2013), Florence, Italy (2014) and, most recently, in Madrid, Spain (2015).

This year's QEST was held in Quebec City, Canada, and colocated with the 27th International Conference on Concurrency Theory (CONCUR 2016) and the 14th International Conference on Formal Modeling and Analysis of Timed Systems (FORMATS 2016).

As one of the premier fora for research on quantitative system evaluation and verification of computer systems and networks, QEST covers topics including classic measures involving performance and reliability, as well as quantification of properties that are classically qualitative, such as safety, correctness, and security. QEST welcomes measurement-based studies and analytic studies, diversity in the model formalisms and methodologies employed, as well as development of new formalisms and methodologies. QEST also has a tradition in presenting case studies, highlighting the role of quantitative evaluation in the design of systems, where the notion of system is broad. Systems of interest include computer hardware and software architectures, communication systems, embedded systems, infrastructural systems, and biological systems. Moreover, tools for supporting the practical application of research results in all of the aforementioned areas are also of interest to QEST. In short, QEST aims to encourage all aspects of work centered around creating a sound methodological basis for assessing and designing systems using quantitative means.

The Program Committee (PC) consisted of 30 experts and we received a total of 46 submissions. Each submission was reviewed by three reviewers, either PC members or external reviewers. The review process included a one-week PC discussion phase. In the end, 21 full papers and three tool demonstration papers were selected for the conference program. The program was greatly enriched by the QEST keynote talk of Carey Williamson (University of Calgary, Canada), the joint keynote talk with FORMATS 2016 of Ufuk Topcu (University of Texas at Austin, USA), and the joint FORMATS 2016 and CONCUR 2016 keynote of Scott A. Smolka (Stony Brook University, USA). We believe the overall result is a high-quality conference program of interest to QEST 2016 attendees and other researchers in the field.

We would like to thank a number of people. Firstly, thanks to all the authors who submitted papers, as without them there simply would not be a conference. In addition, we would like to thank the PC members and the additional reviewers for their hard work and for sharing their valued expertise with the rest of the community, as well as

EasyChair for supporting the electronic submission and reviewing process. We are also indebted to our proceedings chair, Karl Palmskog, and to Alfred Hofmann and Anna Kramer for their help in the preparation of this volume. Thanks also to the Web manager, Andrew Bedford, the local organization chair, and general chair, Josée Desharnais, for their dedication and excellent work. Finally, we would like to thank Joost-Pieter Katoen, chair of the QEST Steering Committee, for his guidance throughout the past year, as well as the members of the QEST Steering Committee.

We hope that you find the conference proceedings rewarding and will consider submitting papers to QEST 2017.

August 2016   Gul Agha
                                                                        Benny Van Houdt

Organization

General Chair

Josée Desharnais   Université Laval, Canada

Program Committee Co-chairs

Gul Agha   University of Illinois, USA
Benny Van Houdt   University of Antwerp, Belgium

Local Organization Chair

Josée Desharnais   Université Laval, Canada

Proceedings and Publications Chair

Karl Palmskog   University of Illinois, USA

Steering Committee

Alessandro Abate   University of Oxford, UK
Luca Bortolussi   University of Trieste, Italy
Javier Campos   University of Zaragoza, Spain
Pedro D'Argenio   Universidad Nacional de Córdoba, Argentina
Boudewijn Haverkort   University of Twente, The Netherlands
Jane Hillston   University of Edinburgh, UK
Andras Horvath   University of Turin, Italy
Joost-Pieter Katoen   RWTH Aachen University, Germany
William Knottenbelt   Imperial College London, UK
Gethin Norman   University of Glasgow, UK
Anne Remke   University of Twente, The Netherlands
Enrico Vicario   University of Florence, Italy

Program Committee

Alessandro Abate   University of Oxford, UK
Nail Akar   Bilkent University, Turkey
Christel Baier   Technical University of Dresden, Germany
Nathalie Bertrand   Inria Rennes, France
Luca Bortolussi   University of Trieste, Italy
Peter Buchholz   Technical University of Dortmund, Germany

Ana Bušic	Inria Paris, France
Javier Campos	University of Zaragoza, Spain
Rohit Chadha	University of Missouri, USA
Florin Ciucu	University of Warwick, UK
Andres Ferragut	Universidad ORT, Uruguay
Dieter Fiems	Ghent University, Belgium
Anshul Gandhi	Stony Brook University, USA
Tingting Han	Birkbeck, University of London, UK
John Hasenbein	University of Texas, USA
Jane Hillston	University of Edinburgh, UK
William Knottenbelt	Imperial College London, UK
Sasa Misailovic	MIT, USA
Pavithra Prabhakar	Kansas State University, USA
Sriram Sankanarayanayan	University of Colorado Boulder, USA
M. Zubair Shariq	University of Iowa, USA
Evgenia Smirni	College of William and Mary, USA
Mark Squillante	IBM, USA
Tetsuya Takine	Osaka University, Japan
Peter Taylor	University of Melbourne, Australia
Miklós Telek	Technical University of Budapest, Hungary
Enrico Vicario	University of Florence, Italy
Mahesh Viswanathan	University of Illinois, USA

Additional Reviewers

Alexander Andreychenko	Elena Gómez-Martínez	Laura Nenzi
Benoît Barbot	Illes Horvath	Marco Paolieri
Simona Bernardi	Jean-Michel Ilié	Elizabeth Polgreen
Laura Carnevali	Nadeem Jamali	Daniël Reijsbergen
Nathalie Cauchi	Jorge Julvez	Ricardo J. Rodríguez
Diego Cazorla	Charalampos	Andreas Rogge-Solti
Milan Ceska	Kyriakopulous	Dimitri Scheftelowitsch
Taolue Chen	Wenchao Li	Sadegh Soudjani
Daniel Gburek	Andras Meszaros	Max Tschaikowski
Blaise Genest	Dimitrios Milios	Feng Yan

Abstracts of Invited Talks

A Stroll Down Speed-Scaling Lane

Carey Williamson

Department of Computer Science, University of Calgary, Calgary, AB, Canada
carey@cpsc.ucalgary.ca

Abstract. This talk provides a retrospective look at the past, present, and future of speed scaling systems. Such systems have the ability to auto-scale their service capacity based on demand, which introduces many interesting tradeoffs between response time (a classic performance metric) and energy efficiency (a relatively recent performance metric of growing interest).

The talk highlights key results and observations from the past two decades of speed scaling research, which appears in both the theory and systems research communities. One theme in the talk is the dichotomy between the assumptions, approaches, and results in these two research communities. Another theme is that modern processors support surprisingly sophisticated speed scaling functionality, which is not yet well-harnessed by current algorithms or operating systems.

During the stroll, I will also share some insights and observations from our own work on speed scaling designs, including coupled, decoupled, and turbocharged systems. This work includes analytical and simulation modeling, as well as empirical system measurements. The talk closes with thoughts about future opportunities in speed scaling research.

V-Formation as Optimal Control

Scott A. Smolka

Department of Computer Science, Stony Brook University,
Stony Brook, NY, USA
sas@cs.stonybrook.edu

Abstract. In this talk, I will present a new formulation of the V-formation problem for migrating birds in terms of model predictive control (MPC). In this approach, to drive a flock towards a desired formation, an optimal *velocity adjustment* (acceleration) is performed at each time-step on each bird's current velocity using a model-based prediction window of T time-steps. I will present both centralized and distributed versions of this approach. The optimization criteria used is based on fitness metrics of candidate accelerations that V-formations are known to exhibit. These include *velocity matching*, *clear view*, and *upwash benefit*. This MPC-based approach is validated by showing that for a significant majority of simulation runs, the flock succeeds in forming the desired formation. These results help to better understand the emergent behavior of formation flight, and provide a control strategy for flocks of autonomous aerial vehicles. This talk represents joint work with Radu Grosu, Ashish Tiwari, and Junxing Yang.

Adaptable Yet Provably Correct Autonomous Systems

Ufuk Topcu

Department of Aerospace Engineering and Engineering Mechanics,
University of Texas at Austin, Austin, TX, USA
utopcu@utexas.edu

Abstract. Acceptance of autonomous systems at scales at which they can make societal and economical impact hinges on factors including how capable they are in delivering complicated missions in uncertain and dynamic environments and how much we can trust that they will operate safely and correctly. In this talk, we present a series of algorithms recently developed to address this need. In particular, these algorithms are for the synthesis of control protocols that enable agents to learn from interactions with their environment and/or humans while verifiably satisfying given formal safety and other high-level mission specifications in nondeterministic and stochastic environments.

We take two complementing approaches. The first approach merges data efficiency notions from learning (e.g., so-called probably approximate correctness) with probabilistic temporal logic specifications. The second one leverages permissiveness in temporal-logic-constrained strategy synthesis with reinforcement learning.

Contents

Performance Modeling

Markov Processes

Property-Driven State-Space Coarsening for Continuous Time Markov Chains

Michalis Michaelides[1]([⊠]), Dimitrios Milios[1], Jane Hillston[1], and Guido Sanguinetti[1,2]

[1] School of Informatics, University of Edinburgh, Edinburgh, UK
mic.michaelides@ed.ac.uk
[2] SynthSys, Centre for Synthetic and Systems Biology, University of Edinburgh, Edinburgh, UK

Abstract. Dynamical systems with large state-spaces are often expensive to thoroughly explore experimentally. Coarse-graining methods aim to define simpler systems which are more amenable to analysis and exploration; most current methods, however, focus on a priori state aggregation based on similarities in transition rates, which is not necessarily reflected in similar behaviours at the level of trajectories. We propose a way to coarsen the state-space of a system which optimally preserves the satisfaction of a set of logical specifications about the system's trajectories. Our approach is based on Gaussian Process emulation and Multi-Dimensional Scaling, a dimensionality reduction technique which optimally preserves distances in non-Euclidean spaces. We show how to obtain low-dimensional visualisations of the system's state-space from the perspective of properties' satisfaction, and how to define macro-states which behave coherently with respect to the specifications. Our approach is illustrated on a non-trivial running example, showing promising performance and high computational efficiency.

1 Introduction

Reasoning about behavioural properties of dynamical systems is a central goal of formal modelling. Recent years have witnessed considerable progress in this direction, with the definition of formal languages [9,10] and logics [12] which enable compact representations of dynamical systems, and mature reasoning tools to model-check properties in an exact [15] or statistical way [14,20].

While such advances are indubitably improving our understanding of dynamical systems, the applicability of these techniques in practical scenarios is still largely hindered by computational issues. In particular, systems with large state-spaces quickly become infeasible to analyse via exact methods due to the phenomenon of state-space explosion; even statistical methods may require computationally expensive and extensive simulations. State-space reduction methodologies aim to construct more compact representations for complex systems. Such

M. Michaelides, D. Milios and G. Sanguinetti are supported by the European Research Council under grant MLCS 306999. J. Hillston is supported by the EU project, QUANTICOL 600708.

G. Agha and B. Van Houdt (Eds.): QEST 2016, LNCS 9826, pp. 3–18, 2016.
DOI: 10.1007/978-3-319-43425-4_1

reduced-state systems are generally amenable to more effective analysis and may yield deeper insights into the structure and dynamics of the system.

Broadly speaking, state-space reduction can be achieved by either model simplification, usually by abstracting some system behaviours into a simpler system, or state aggregation, often by exploiting symmetries or approximate invariances. A prime example of model simplification is the technique of time-scale separation, which replaces a large system with multiple weakly dependent sub-systems [5]. Most aggregation methods, instead, are based on grouping different states with similar behaviour with respect to their transition probabilities. This idea is at the core of the concept of *approximate lumpability*, which extends the exact lumpability relationship by aggregating states based on a pre-defined metric on the outgoing exit rates [1,7,11,17,19].

In this paper we propose a novel state-space reduction paradigm by shifting the focus from the infinitesimal properties of states (i.e. their transition rates) to the global properties of trajectories. Namely, we seek to aggregate states that yield *behaviourally similar* trajectories according to a set of pre-defined logical specifications. Intuitively, two states will be aggregated if trajectories starting from either state exhibit similar probabilities of satisfying the logical specifications. We define a statistical algorithm based on statistical model checking and Gaussian Process emulation to define this behavioural similarity across the whole state-space of the system. We then propose a dimensionality reduction and clustering pipeline to aggregate states and define reduced (non-Markovian) dynamics. To illustrate our approach, we give a running example of model reduction for the Susceptible-Infected-Recovered-Susceptible (SIRS) model, a non-trivial, non-linear stochastic system widely used in epidemiology. Our results show that property-driven aggregation can yield an effective tool to reduce the complexity of stochastic dynamical systems, leading to non-trivial insights in the structure of their state-space.

2   Background

2.1   Population Continuous Time Markov Chains

A Continuous Time Markov Chain (CTMC) is a continuous-time Markovian stochastic process over a discrete state-space $\mathcal{S}$. We will consider only *population* models, where the state-space is organised along populations: in this case, the state-space is indexed by the counts of each population $n_i \in \mathbb{N} \cup \{0\}$. Population CTMCs (pCTMCs) are frequently used in many scientific and engineering domains; we will use here the notation of chemical reactions as it is widespread and intuitively appealing. Transitions in a pCTMC are denoted as

$$r_1 X_1 + \ldots r_n X_n \xrightarrow{\tau(\boldsymbol{X})} s_1 X_1 + \ldots s_n X_n$$

meaning that r_i particles of type X_i are consumed and s_j particles of type X_j are produced when the specific transition takes place. $\tau(\boldsymbol{X})$ is a transition rate which depends on the current state of the system.

It is easy to show that waiting times between transitions are exponentially distributed random variables; this observation is the basis of exact simulation algorithms for pCTMCs, such as the celebrated Gillespie algorithm [13]. The Gillespie algorithm generates trajectories of a pCTMC by randomly choosing the next reaction to occur and the time to elapse until the reaction occurs.

Example 1.1. We introduce here our running example, the Susceptible-Infected-Recovered-Susceptible (SIRS) model of epidemic spreading. The SIRS model is a discrete stochastic model of disease spread in a population, where individuals in the population can be in one of three states, Susceptible, Infected and Recovered. There are different variations of the model, some open (individuals can enter and exit the system), others with individuals relapsing to a susceptible state after having recovered. Here, we consider a relapsing, closed system, which evolves in a discrete, 2-dimensional state-space, where dimensions are the number of Susceptible and Infected individuals in the population (Recovered numbers are uniquely determined since the total population is constant). We also introduce a spontaneous infection of a susceptible individual with constant rate, independent of the number of infected individuals, to eliminate absorbing states.

With a population size of N, states in the 2D space can be represented by $x = (S, I), S \in \{0, \cdots, N\}, I \in \{0, \cdots, N - S\}$ for a total of $(N + 1)(N + 2)/2$ states. The chemical reactions for this system are:

infection $S + I \xrightarrow{\alpha} 2I$;

spontaneous infection $S \xrightarrow{\beta/5} I$;

recovery $I \xrightarrow{\beta} R$;

relapsing $R \xrightarrow{\beta} S$.

We set the infection rate $\alpha = 0.005$, recovery rate $\beta = 0.01$, and population size $N = S + I + R = 100$, for a total of 5151 states in this SIRS system. Sample trajectories of the system were simulated using the Gillespie algorithm.

2.2   Temporal Logic and Model Checking

We formally specify trajectory behaviours by using temporal logic properties. We are particularly interested in properties that can be verified on single trajectories, and assume metric bounds on the trajectories, so that they are observed only for a finite amount of time. Metric Interval Temporal logic (MITL) offers a convenient way to formalise such specifications.

Formally, MITL has the following grammar:

$$\phi ::= \mathbf{tt} \mid \mu \mid \neg\phi \mid \phi_1 \wedge \phi_2 \mid \phi_1 \mathbf{U}_{[T_1, T_2]} \phi_2,$$

where $\mathbf{tt}$ is the true formula, conjunction and negation are the standard boolean connectives, and the time-bounded until $\mathbf{U}_{[T_1, T_2]}$ is the only temporal modality. Atomic propositions μ are (non-linear) inequalities on population variables.

A MITL formula is interpreted over a function of time x, and its satisfaction relation is given as in [16]. More temporal modalities, such as the time-bounded eventually and always, can be defined in terms of the until operator: $\mathbf{F}_{[T_1,T_2]}\phi \equiv \mathtt{tt}\mathbf{U}_{[T_1,T_2]}\phi$ and $\mathbf{G}_{[T_1,T_2]}\phi \equiv \neg\mathbf{F}_{[T_1,T_2]}\neg\phi$.

MITL formulae evaluate as true or false on individual trajectories; when trajectories are sampled from a stochastic process, the truth value of a MITL formula is a Bernoulli random variable. Computing the probability of such a random variable is a model checking problem. Model checking for MITL properties evaluated on trajectories from a CTMC requires the computation of transient probabilities; despite major computational efforts [15], this is seldom possible exactly due to state-space explosion. Statistical model checking (SMC) methods circumvent such problems by adopting a Monte Carlo perspective: by drawing repeatedly and independently sample trajectories, one may obtain an unbiased estimate of the truth probability, and statistical error bounds can be obtained by employing either frequentist or Bayesian statistical approaches [14,20]. It should be pointed out that such bounds do not carry the same guarantees as numerical results obtained say by transient analysis; however, simply by drawing more samples one may reduce the uncertainty in the bounds arbitrarily.

Example 1.2. MITL formulae can be used effectively to obtain behavioural characterisations of the system's trajectory. We turn again to the SIRS model to illustrate this concept.

Assume one may want to express a global bound on the virulence of the infection, so that the fraction of infected population never exceeds λ. This can be done by considering the formula ϕ_1, defined as

$$\phi_1 :: = \mathbf{G}_{[0,100]}(I < \lambda N) \tag{1}$$

which translates to:

$$\phi_1(x) = \begin{cases} \mathtt{tt} & \text{if } I_t < \lambda N \ \forall t \in [0,100], \\ \neg\mathtt{tt} & \text{otherwise.} \end{cases}$$

Statistical model checking of this formula is trivial: one simply draws a trajectory using Gillespie's algorithm, and monitors that the maximal number of infected does not exceed the specified threshold in the $[0,100]$ interval.

3   Methodology

3.1   High Level Method Description

We first present a high-level description of the proposed methodology; the technical ingredients will be introduced in the following subsections. Figure 1 provides an intuitive roadmap of the approach. The overarching idea is to provide a state-space aggregation algorithm which uses behavioural similarities as an aggregation criterion.

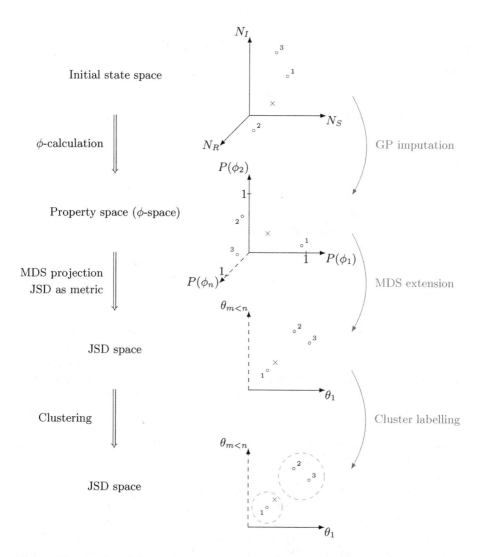

Fig. 1. The sequence of transformations from space to space are shown in the figure. States from the original state-space (blue circles 1–3) are projected to ϕ-space according to satisfaction rate of set properties (found via simulation of the system). MDS is used to project from ϕ-space to a space where JSD of ϕ satisfaction probability distributions between states is preserved as Euclidean distance (in the figure, $\mathrm{JSD}[P_\phi(2) \parallel P_\phi(3)] < \mathrm{JSD}[P_\phi(1) \parallel P_\phi(2)], \mathrm{JSD}[P_\phi(1) \parallel P_\phi(3)]$ so states 2, 3 are placed closer together than 1). The states are then clustered to produce macro-states. Out-of-sample states (red cross) can be projected to ϕ-space, using GP imputation to estimate satisfaction probabilities. MDS extension allows projecting from ϕ-space to JSD space without moving the sampled states. The most likely cluster for the state to belong to (nearest centroid) is the macro-state it belongs to. (Color figure online)

The input to the approach is a CTMC model and a set of MITL formulae $\phi_1, \ldots, \phi_n$ which define the behavioural traits we are interested in. We formalise some of the key concepts through the following definitions.

Definition 1. *A* coarsening map $\mathcal{C}$ *for a CTMC* $\mathcal{M}$ *is a surjective map*

$$\mathcal{M} : \mathcal{S} \to \mathcal{R}, \tag{2}$$

from the state-space $\mathcal{S}$ *of* $\mathcal{M}$ *to a finite set* $\mathcal{R}$*, such that* $\mathrm{card}(\mathcal{S}) \geq \mathrm{card}(\mathcal{R})$.

Definition 2. *The* macro-states *of the coarsened system are the elements of the image of the coarsening map* $\mathcal{C}$.

Therefore, the set of all macro-states is a partition of the set of initial states $\mathcal{S}$, where each element in the partition is a macro-state. In general, there is no way to retrieve the initial state configuration of the system only from information of the macro-state configuration, i.e., the coarsening entails an information loss.

We illustrate the various steps of the proposed procedure in Fig. 1. The first step is to take a sample of possible initial states; we then evaluate the joint satisfaction of the n formulae, given a particular state as initial condition. This implicitly defines a map

$$\Phi \colon \mathcal{S} \to [0, 1]^{2^n} \tag{3}$$

which associates each initial state with the probability of each possible satisfaction pattern of the n formulae. Notice that all of the 2^n possible truth values are needed to ensure correlations between properties are captured. Constructing such a *property map* by exhaustive exploration of the state-space is clearly computationally infeasible; we therefore evaluate it (by SMC) on a subset of possible initial states, and then extend it using a statistical surrogate, a Gaussian Process (Fig. 1 top).

The property representation contains the full information over the dependence of the properties of interest on the initial state. It can be endowed with an information-theoretic metric by using the JSD between the resulting probability distributions. However, the high dimensionality and likely very non-trivial structure of the property representation may make this unwieldy. We therefore propose a dimensionality reduction strategy which maintains approximately the metric structure of the property representation using Multi-Dimensional Scaling (MDS; Fig. 1 middle). MDS will also have the advantage of automatically identifying potentially redundant characterisations, as implied for example by logically dependent formulae.

The low-dimensional output of the MDS projection can then be visually inspected for groups of initial states (*macro-states*) with similar behaviours with respect to the properties. This operation is a *coarsening map*, which can also be automated by using a variety of clustering algorithms.

The model dynamics induce, in principle, a dynamics on this reduced space $\mathcal{R}$. In practice, such dynamics will be non-Markovian and not easily expressible in a compact form; we propose a simple, simulation-based alternative definition which re-uses some of the computation performed in the previous steps to define an empirical, coarse-grained dynamics on the macro-states.

3.2   Satisfaction Probability as a Function of Initial Conditions

The starting point for our approach consists of embedding the initial state-space into the property space, ϕ-space. This is achieved by computing satisfaction probabilities for the 2^n possible truth patterns of the n properties we consider. As in general these satisfaction probabilities can only be computed via SMC, this is potentially a tremendous computational bottleneck. To obviate this problem, we turn the computation of the property map into a machine learning problem: we evaluate the 2^n functions on a (sparse) subset of initial states, and predict their values on the remaining initial states using a Gaussian Process (GP).

GPs have extensively been used in machine learning for regression purposes and it is in this context they are used here. A GP is a generalisation of the multivariate normal distribution to function spaces with infinitely many dimensions; within a regression context, GPs are used to provide a flexible prior distribution over the set of candidate functions underpinning the hypothesised input-output relationship. Given a number of input-output observations (training set), one can use Bayes's rule to condition the GP on the training set, obtaining a posterior distribution over the regression function at other input points. For a review of GPs and their uses in machine learning, we refer the reader to [18].

In our setting, the input-output relationship is the property map from initial states to satisfaction probabilities of the properties. This function is defined over a discrete space, but we can use the population structure of the pCTMC to embed the state-space $\mathcal{S}$ in a (subset) of $\mathbb{R}^D$ for some D. We can then treat the problem as a standard regression problem, learning a function $f_\phi : \mathbb{R}^D \to \mathbb{R}^{2^n}$.

Remark. GPs have already been used to explore the dependence of the satisfaction probability of a formula on model parameters in the so-called Smoothed Model Checking approach [6]. There, the authors proved a smoothness result which justified the use of smoothness-inducing GPs for the problem. It is easy to see that such smoothness does not hold in general for the function f_ϕ; for example, the probability of satisfying the formula $x(0) > N$ has a discontinuity at $x = N$. However, since we only ever evaluate f_ϕ on a discrete set of points, the lack of smoothness is not an issue, as a continuous function can approximate arbitrarily well a discontinuous function when restricted to a discrete set.

Example 1.3. We exemplify this procedure on the SIRS example. We consider here three properties of interest: the global bound encoded in formula ϕ_1 defined in equation (1), and two further properties encoded as

$$\phi_2 :: = \mathbf{F}_{[0,60]} \mathbf{G}_{[0,40]} (0.05N \leq I \leq 0.2N), \tag{4}$$
$$\phi_3 :: = \mathbf{F}_{[30,50]} (I > 0.3N). \tag{5}$$

Satisfaction of ϕ_2 requires that the infection has remained within 5 to 20 % of the total population for 40 consecutive time units, starting anytime in the first 60 time units; satisfaction of ϕ_3 requires that the infection peaks at above 30 % between time 30 and time 50.

The property map in this case would have an 8-dimensional co-domain, representing the probability of satisfaction for each of the 2^3 possible truth values of the three formulae. Figure 2 plots the probability of satisfaction for the three formulae individually, as we vary the initial state. In this case, 10 % of all possible initial states were randomly selected and numerically mapped to the property space via SMC, while the satisfaction probabilities for the remaining 90 % were imputed using GPs. We see that throughout most of the state-space the second property has low probability. Also it is of interest to observe the strong anti-correlation between the first and third properties: intuitively, if there is very high probability that the infection will be globally bounded below 40 % of individuals, it becomes more difficult to reach a peak at above 30 %.

3.3 Dimensionality Reduction of Behaviours

Once states are mapped onto ϕ-space, reducing dimensionality of this space is useful to remove correlations and redundancies in the properties tracked. Properties may often capture similar behaviour, leading to strong correlations in their satisfaction probability. Reducing the dimensionality of the property space mostly retains the information of how behaviour differs from state to state, eliminating redundancies. Moreover, reduced dimensional mappings can aid practitioners to visually identify structures within the state-space of the system.

In order to quantify the similarity of different initial states with respect to property satisfaction, the Jensen-Shannon Divergence (JSD) between the probability distributions of property satisfaction is used as a metric. JSD is an information theoretic symmetric distance between probability distributions — the higher the difference between the distributions, the higher JSD is. Between two distributions, P, Q, JSD is defined as

$$\mathrm{JSD}[P \parallel Q] = \frac{1}{2}(\mathrm{KL}[P \parallel M] + \mathrm{KL}[Q \parallel M]),$$

where $M = 0.5(P + Q)$ the average of the distributions, and $\mathrm{KL}[P \parallel Q] = \sum_i P(i) \log \frac{P(i)}{Q(i)}$, the Kullback-Leibler divergence.

The JSD enables us to derive a matrix of pairwise distances in property space between different initial states. Such a distance is not Euclidean, and is defined in the high-dimensional property space. To map the initial states in a more convenient, low-dimensional space, we employ a dimensionality reduction technique known as Multi-Dimensional Scaling (MDS) [4].

MDS has its roots in the social science literature; it is a valuable and widely used tool in psychology and similar fields where data is collected by assessing similarity between pairs.

Given some points X in an m-dimensional space, MDS finds the position of corresponding points Z in an n-dimensional space, where usually $n < m$, such that a given metric between points is optimally preserved. In the most common case, (also known as Torgerson–Gower scaling or Principal Component Analysis), the metric to be preserved is the Euclidean distance, and is preserved by minimisation of a loss function. This function is generally known as *stress* for metric MDS, but specifically for classical MDS as *strain*.

For the classical MDS case, the projection is achieved by eigenvalue decomposition of a distance matrix of the (normalised) points $XX^\top$, and subsequently reconstructing the points from the n largest (eigenvector, eigenvalue) pairs. This results in Z, a projection of the points to an n-dimensional space, where Euclidean distance is optimally preserved.

In the classical MDS definition, the MDS projection is defined statically for the available data points, and needs ab initio re-computation if new points become available. In [2], the method is extended to new points by constructing a new dissimilarity matrix of new points to old ones, by which the projection of new points will be consistent to that of the old points. The kernel for this new matrix achieves this by replacing the means required for centring with expectations over the old points; such that for points $x, y \in X$

$$\tilde{K}(x,y) = -\frac{1}{2}\left(d^2(x,y) - \frac{1}{n}\sum_{x'} d^2(x', y) - \frac{1}{n}\sum_{y'} d^2(x, y') + \frac{1}{n^2}\sum_{x',y'} d^2(x', y')]\right),$$

where $\tilde{K}(x, y)$ is the kernel used for the dissimilarity matrix, is replaced by

$$\tilde{K}(a,b) = -\frac{1}{2}\left(d^2(a,b) - E_x[d^2(x, a)] - E_{x'}[d^2(b, x')] + E_{x,x'}[d^2(x, x')]\right),$$

where a can be an out-of-sample point ($a \notin X, b \in X$).

This reconstructs the dissimilarity matrix for the original points exactly, and allows us to generalise to out-of-sample points and find their positions in the embedding learned, as described in [2]. Extending MDS allows us to create macro-states based on samples of points, and then project new points on the space created by MDS to find in which clusters they belong.

Example 1.4. We have introduced three properties in Eqs. (1), (4) and (5), and the associated property map. This has an eight-dimensional co-domain, but already some of its properties can be gleaned by the three-dimensional plot of the single-formula probabilities shown in Fig. 2. Particularly, these reveal strong negative correlations, indicating that MDS may prove fruitful.

Figure 3 shows the states projected to a 2D space were proximity implies similar probability distribution over property satisfaction. This was achieved using MDS to project the states, with JSD used as the metric to be preserved as Euclidean distance in the new 2D space. Elements of the square-shaped structure visible in ϕ-space (Fig. 2) are preserved, with the subset of states giving rise to higher probabilities for property ϕ_2 (top of Fig. 2) appearing further from the connected outline (bottom left group in Fig. 3).

3.4   Clustering and Structure Discovery

The MDS projection enables us to visually appreciate the existence of non-trivial structures within the state-space, such as clusters of initial states that produce

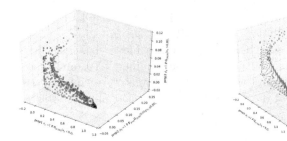

Fig. 2. Left: Projection of states in ϕ-space via SMC (trajectory simulations for each initial state). Notice the non-trivial state distribution structure. Right: Projection of states in ϕ-space using SMC for 10 % of the states, and GP regression to estimate $P(\phi)$ for the rest 90 % of states (red crosses). (Color figure online)

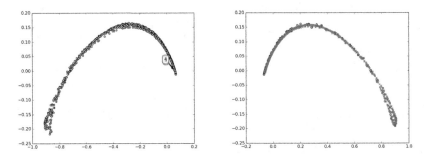

Fig. 3. Left: $P(\phi_1, \phi_2, \phi_3)$ estimated via SMC for each state. MDS was then used to project them from an 8D to a 2D space. Right: GP estimates of $P(\phi_1, \phi_2, \phi_3)$ for 90 % of states (red crosses) produce an almost identical MDS projection. (Color figure online)

similar behaviours with respect to the property specification. Our intuition is that such structures should form the basis to define macro-states of the system, groups of states that will exhibit similar satisfaction probabilities for the properties defined. To automate this process, we propose to use a clustering algorithm to define macro-states. Since our goal is to group states with similar behaviours, we adopt k-means clustering [3], which is based on the Euclidean distance of the states in the MDS space (representative of the JSD between the probability satisfaction distributions). k-means requires specification of the desired number of clusters (the k parameter); this allows the user to select the level of coarsening required. Figure 4 shows the clusters produced in the reduced MDS space for the running SIRS model example, where we set the number of clusters $k = 10$.

3.5   Constructing Coarse Dynamics

Once states have been grouped into macro-states, a major question is how to construct dynamics for the now coarsened system. The coarsened system naturally inherits dynamics from the original (fine-grained) system; however,

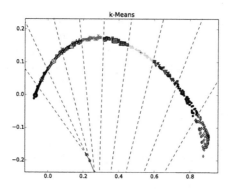

Fig. 4. The states were clustered in the space created by the MDS projection and coloured accordingly, using k-means (10 clusters). Since the Euclidean distance in this space is representative of distance in probability distributions over properties, states with different behaviour should be in different clusters. (Color figure online)

such dynamics are non-Markovian, and in general fully history dependent so that transition probabilities would have the form

$$p(k'|k,t,h) = p(k'|k,t,h)p(t|k,h), \tag{6}$$

where h denotes the history of the process. Simulating such a non-Markovian system is very difficult and likely to be much more computationally expensive than simulating the original system.

We therefore seek to define approximate dynamics which are amenable to efficient simulation, but still capture aspects of the non-Markovian dynamics. The most natural approximation is to replace the system with a semi-Markov system: transitions are still history-independent, but the distribution of sojourn times is non-exponential. To evaluate the sojourn-time distribution, we resort to an empirical strategy, and construct an empirical distribution of sojourn times by re-using the simulated trajectories of the fine system that were drawn to define the coarsening. In other words, once a clustering is defined, we retrospectively inspect the trajectories to construct a histogram distribution of sojourn times, approximating $p(t|k)$.

A possible drawback of this semi-Markov approximation is that it may introduce transitions which are actually impossible in the original state-space. This is because states were clustered based on behaviour rather than transition rates, and therefore states that are actually quite far in the original state-space may end up being clustered together. Since the identity of the original states is lost after the coarse graining, impossible transitions may be introduced.

Retrospectively inspecting whole system trajectories, rather than agnostically examining cluster transitions of the original system with a uniform initial state distribution within the cluster, ameliorates this problem. Similarly, estimates of $p(k'|t,k)$ are produced from the same trajectories; these are the macro-state transition frequencies in each bin of the sojourn time probability histogram. This method avoids a lot of impossible trajectories one might generate, if the above probabilities were estimated by sampling randomly from initial states in a macro-state and looking at when the macro-state is exited and to which macro-state the system transitions. Assuming the original system has a steady state, the empirical dynamics constructed here capture this steady state macro-state distribution; however, accuracy of transient dynamics suffers, and the coarsened system enters the steady state faster than the original system.

Example 1.5. We illustrate and evaluate the quality of the coarsened trajectories with respect to the original ones on the SIRS example. In particular, we examine the probability distribution over the macro-states at different times in the evolution of the system. The macro-state distribution has been estimated empirically by sampling trajectories using the Gillespie algorithm for the fine system, and our coarse simulation scheme for the coarsened system. We have then constructed histograms to capture the distribution of the categorical random variables that represent the macro-state. Finally, we measure the histogram distance between histograms obtained from the fine and the coarse systems. Figure 5 depicts the evolution of the macro-state histograms over time.

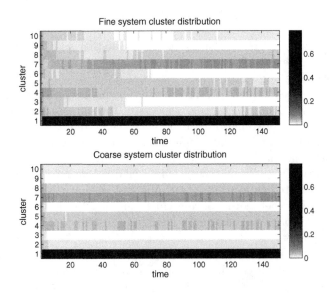

Fig. 5. Evolution of the macro-state histograms over time

Quality of Approximation. In order to put any distance between empirical distributions into context, this has to be compared with the corresponding average self-distance, which is the expected distance value when we compare two samples from the same distribution. In this work, we estimate the self-distance using the result of [8]: given N samples and K bins in the histogram, an upper bound for the average histogram self-distance is given by $\sqrt{(4K)/(\pi N)}$. In our example, we have $K = 10$ histogram bins, which are as many as the macro-states. In practice, a distance value smaller than the self-distance implies that the distributions compared are virtually identical for a given number of samples. In Fig. 6, we see the estimated distances for $N = 10000$ simulation runs for times $t \in [0, 150]$. It can be seen that the steady-state behaviour of the system is captured accurately, as the majority of the distances recorded after time $t = 60$ lie below the self-distance threshold. However, the transient behaviour of the system is not captured as accurately. Upon a more careful inspection of the shape of the histograms in Fig. 5, we see that the coarsened system simply converges more quickly to steady-state. To conclude, we think that the the approximation quality of the steady-state dynamics is a promising result, but a more accurate approximation of the transient behaviour is subject of future work.

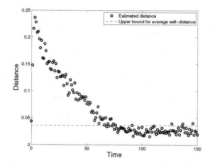

Fig. 6. Evolution of the macro-state histogram distances over time

Computational Savings. State-space coarsening results in a more efficient simulation process, since the coarse system is characterised by lower complexity as opposed to the fine system. We demonstrate these computational savings empirically in terms of the average number of state transitions invoked during simulation. More specifically, we consider a sample of 5000 trajectories of the fine and the coarse system. We have recorded 320 ± 25 initial state transitions on average in each trajectory of the fine system, compared to 56 ± 31 macro-state transitions in trajectories of the coarsened system. The number of transitions in the coarse system is an order of magnitude lower than in the fine one, owing to the reduction of states in the system from a total of 5151 to 10 (the number of macro-states). Clearly, our procedure, particularly the GP imputation, incurs some computational overheads. Table 1 presents the computational savings of using GPs to estimate satisfaction probability distributions for most states, instead of exhaustively exploring the state-space. All simulations were performed using a Gillespie algorithm implementation, taking 1000 trajectories starting at each examined state, running on 10 cores.

Table 1. Real running times for simulations of varying sample size (percentage of state-space) and GP estimation of remaining states.

Sample size	GP &MDS time (s)	Simulation time (s)	Total time (s)	Percentage of exhaustive total time (total time/8516 s)
100 %	1616*	6900	8516	100 %
50 %	1133	3450	4583	54 %
40 %	884	2760	3644	43 %
30 %	595	2070	2665	31 %
20 %	354	1380	1734	20 %
10 %	170	690	860	10 %

* No GP was performed here, just the MDS.

4 Discussion

We presented a novel approach to the coarsening of a CTMC, in order to gain a stochastic process with a much smaller state-space. Unlike previous approaches to CTMC aggregation, which are based on structural properties of the state-space, our approach is based on property satisfaction, allowing the coarse-grained system to focus on abstracting the dynamics in terms of aspects of behaviour that are important in the modelling study. The further steps are to identify key clusters of states in property space, or a lower-dimensional representation of it, and approximate the transition dynamics between them. For example, this approach might be used within multi-scale modelling to reduce the state-space of a lower level model before embedding in a higher-level representation.

Common aggregation techniques, such as exact or approximate lumpability, often impose stringent conditions on the symmetries and transition rates within the original state-space. Moreover, the macro-states produced can be difficult to interpret when the reduction is applied directly at the state-space level (i.e. without a corresponding bisimulation over transition labels). In contrast, the property-based approach allows macro-states to be defined by high-level behaviour, rather than them emerging from an algorithm applied to low-level structure.

The GP regression we employed for estimating satisfaction probability of properties for out-of-sample states proved quite accurate; simulation estimates for 10 % of the states were sufficient to reconstruct the state distribution in the space defined by the probability of property satisfaction, ϕ-space, without substantial loss of structure. Therefore, the proposed approach may be helpful in effectively understanding the behavioural structure of large and complex Markovian systems, with implications for design and verification.

Initial experiments on a simple system show that our approach can be practically deployed, with considerable computational savings. The approach induces coarsened dynamics which empirically match the original system's dynamics in terms of steady-state behaviour. However, the recovery of transient coarse-grained dynamics poses more of a challenge and this will provide a focus for future work. In particular, we will seek to explore the possibility of quantifying the information lost through the coarsening approach, at least asymptotically, for systems which admit a steady state. Exploring the scalability of the approach on more complex, higher dimensional examples will also be an important priority. In general, we expect our approach to be beneficial when simulation costs dominate the overheads incurred by the GP regression approach. This condition will be mostly met for systems with moderately large state spaces but complex (e.g. stiff) dynamics. For extremely large state spaces, the cubic complexity (in the number of retained states) of GP regression may force users to adopt excessively sparse sub-sampling schemes, and it may be preferable to replace the GP regression step with alternative schemes with better scalability. Exploration of these computational trade-offs would likely prove insightful for the methodology.

References

1. Abate, A., Brim, L., Češka, M., Kwiatkowska, M.Z.: Adaptive aggregation of Markov chains: quantitative analysis of chemical reaction networks. In: Kroening, D., Păsăreanu, C.S. (eds.) CAV 2015. LNCS, vol. 9206, pp. 195–213. Springer, Heidelberg (2015)
2. Bengio, Y., Paiement, J.-F., Vincent, P., Delalleau, O., Le Roux, N., Ouimet, M.: Out-of-sample extensions for LLE, Isomap, MDS, eigenmaps, and spectral clustering. In: Proceedings of NIPS, pp. 177–184 (2004)
3. Bishop, C.M.: Pattern Recognition and Machine Learning. Springer-Verlag New York, Inc., Secaucus (2006)
4. Borg, I., Groenen, P.: Modern Multidimensional Scaling: Theory and Applications. Springer, New York (2005)
5. Bortolussi, L., Milios, D., Sanguinetti, G.: Efficient stochastic simulation of systems with multiple time scales via statistical abstraction. In: Roux, O., Bourdon, J. (eds.) CMSB 2015. LNCS, vol. 9308, pp. 40–51. Springer, Heidelberg (2015)
6. Bortolussi, L., Milios, D., Sanguinetti, G.: Smoothed model checking for uncertain continuous-time Markov chains. Inf. Comput. **247**, 235–253 (2016)
7. Buchholz, P., Kriege, J.: Approximate aggregation of Markovian models using alternating least squares. Perform. Eval. **73**, 73–90 (2014)
8. Cao, Y., Petzold, L.: Accuracy limitations and the measurement of errors in the stochastic simulation of chemically reacting systems. J. Comput. Phys. **212**(1), 6–24 (2006)
9. Ciocchetta, F., Hillston, J.: Bio-PEPA: a framework for the modelling and analysis of biological systems. Theor. Comput. Sci. **410**(33), 3065–3084 (2009)
10. Danos, V., Feret, J., Fontana, W., Harmer, R., Krivine, J.: Rule-based modelling of cellular signalling. In: Caires, L., Vasconcelos, V.T. (eds.) CONCUR 2007. LNCS, vol. 4703, pp. 17–41. Springer, Heidelberg (2007)
11. Deng, K., Mehta, P.G., Meyn, S.P.: Optimal Kullback-Leibler aggregation via spectral theory of Markov chains. IEEE Trans. Autom. Control **56**(12), 2793–2808 (2011)
12. Donzé, A., Maler, O.: Robust satisfaction of temporal logic over real-valued signals. In: Chatterjee, K., Henzinger, T.A. (eds.) FORMATS 2010. LNCS, vol. 6246, pp. 92–106. Springer, Heidelberg (2010)
13. Gillespie, D.: Exact stochastic simulation of coupled chemical reactions. J. Phys. Chem. **81**(25), 2340–2361 (1977)

14. Jha, S.K., Clarke, E.M., Langmead, C.J., Legay, A., Platzer, A., Zuliani, P.: A Bayesian approach to model checking biological systems. In: Degano, P., Gorrieri, R. (eds.) CMSB 2009. LNCS, vol. 5688, pp. 218–234. Springer, Heidelberg (2009)
15. Kwiatkowska, M., Norman, G., Parker, D.: PRISM 4.0: verification of probabilistic real-time systems. In: Gopalakrishnan, G., Qadeer, S. (eds.) CAV 2011. LNCS, vol. 6806, pp. 585–591. Springer, Heidelberg (2011)
16. Maler, O., Nickovic, D.: Monitoring temporal properties of continuous signals. In: Lakhnech, Y., Yovine, S. (eds.) FORMATS 2004 and FTRTFT 2004. LNCS, vol. 3253, pp. 152–166. Springer, Heidelberg (2004)
17. Milios, D., Gilmore, S.: Component aggregation for pepa models: an approach based on approximate strong equivalence. Perform. Eval. **94**, 43–71 (2015)
18. Rasmussen, C.E., Williams, C.K.I.: Gaussian Processes for Machine Learning. MIT Press, Cambridge (2006)
19. Tschaikowski, M., Tribastone, M.: A unified framework for differential aggregations in Markovian process algebra. J. Log. Algebr. Meth. Program. **84**(2), 238–258 (2015)
20. Younes, H.L.S., Simmons, R.G.: Statistical probabilistic model checking with a focus on time-bounded properties. Inf. Comput. **204**(9), 1368–1409 (2006)

Optimal Aggregation of Components for the Solution of Markov Regenerative Processes

Elvio Gilberto Amparore[✉] and Susanna Donatelli

University of Torino, Corso Svizzera 187, Torino, Italy
{amparore,susi}@di.unito.it

Abstract. The solution of non-ergodic Markov Renewal Processes may be reduced to the solution of multiple smaller sub-processes (components), as proposed in [4]. This technique exhibits a good saving in time in many practical cases, since components solution may reduce to the transient solution of a Markov chain. Indeed the choice of the components might significantly influence the solution time, and this choice is demanded in [4] to a greedy algorithm. This paper presents a computation of an optimal set of components through a translation into an *integer linear programming* problem (ILP). A comparison of the optimal method with the greedy one is then presented.

1 Introduction

A *Markov Regenerative Process* (MRP) is a stochastic process defined by a sequence of time instants called *renewal times* in which the process loses its memory, i.e. the age of non-exponential (general) events is 0. The behaviour between these points is then described by a time-limited stochastic process. MRPs have been studied extensively in the past [13,16], and many solid analysis techniques exist. MRPs are considered the richest class of stochastic processes for which it is still possible to compute an exact numerical solution, and have therefore attracted a significant interest in the performance and performability community.

This paper considers the subclass of MRP in which the time limited stochastic process is a CTMC, general events are restricted to deterministic ones, and at most one deterministic event is enabled in each state. This type of MRPs arise for example in the solution of Deterministic Stochastic Petri nets (DSPN), in the model-checking of a one-clock CSL^{TA} formula [12] and in Phased-Mission Systems (PMS) as in [8,15].

The steady-state solution of an MRP involves the computation and the solution of its discrete time *embedded Markov chain*, of probability matrix $\mathbf{P}$. The construction of $\mathbf{P}$ is expensive, both in time and memory, because this matrix is usually dense even if the MRP is not. The work in [13] introduces an alternative *matrix-free* technique (actually $\mathbf{P}$-free), based on the idea that $\mathbf{P}$ can be substituted by a function of the basic (sparse) matrices of the MRP.

When the MRP is non-ergodic it is possible to distinguish transient and recurrent states, and specialized solution methods can be devised. The work in [2,4] introduces an efficient steady-state solution for non-ergodic MRPs,

© Springer International Publishing Switzerland 2016
G. Agha and B. Van Houdt (Eds.): QEST 2016, LNCS 9826, pp. 19–34, 2016.
DOI: 10.1007/978-3-319-43425-4_2

in matrix-free form, called *Component Method*. To the best of our knowledge, this is the best available technique for non-ergodic DSPN and for CSL^{TA}, as well as for non-ergodic MRPs in general.

The work in [2,4] and its application to CSL^{TA} in [5] identify a need for aggregating components into bigger ones, and observe that the performance of the algorithm may depend on the number, size, and solution complexity of the components. The aggregation is defined through a set of rules, to decide which components can be aggregated together, and through a greedy-heuristic algorithm that performs aggregations as much as it can. In this paper we observe that the greedy algorithm of [4] may actually find a number of components that is not minimal. The greedy solution seems to work quite well on the reported example, but the lack of optimality makes it hard to determine if it is convenient.

This paper formalizes the optimality criteria used in [4] and defines an ILP for the computation of an optimal set of components: to do so, the component identification problem is first mapped into a graph problem.

The paper develops as follows: Sect. 2 defines the necessary background. Section 3 defines the component identification problem in terms of the MRP graph. Section 4 defines the ILP that computes the optimal set of components. Section 5 discusses the performance of the ILP method and how it compares to the greedy method and concludes the paper.

2   Background and Previous Work

We assume that the reader has familiarity with MRPs. We use the definitions of [13]. Let $\{\langle Y_n, T_n \rangle \mid n \in \mathbb{N}\}$ be the *Markov renewal sequence* (MRS), with *regeneration points* $Y_n \in \mathcal{S}$ on the state space $\mathcal{S}$ encountered at *renewal time instants* T_n. An MRP can be represented as a discrete event system (like in [11]) where in each state a general event g is taken from a set G. As the time flows, the age of g being enabled is kept, until either g fires ($\mathbf{\Delta}$ event), or a Markovian transition, concurrent with g, fires. Markovian events may actually disable g (preemptive event, or $\bar{\mathbf{Q}}$ event), clearing its age, or keep g running with its accumulated age (non-preemptive event, or $\mathbf{Q}$ event).

Definition 1 (MRP Representation). *A representation of an MRP is a tuple* $\mathcal{R} = \langle \mathcal{S}, G, \delta_g, \Gamma, \mathbf{Q}, \bar{\mathbf{Q}}, \mathbf{\Delta} \rangle$ *where* $\mathcal{S}$ *is a finite set of states,* $G = \{g_1 \ldots g_n\}$ *is a set of general events,* δ_g *is the duration of event* g, $\Gamma : \mathcal{S} \to G \cup \{E\}$ *is a function that assigns to each state a general event enabled in that state, or the symbol* E *if no general event is enabled,* $\mathbf{Q} : \mathcal{S} \times \mathcal{S} \to \mathbb{R}_{\geq 0}$ *is the non-preemptive transition rates function (rates of non-preemptive Markovian events),* $\bar{\mathbf{Q}} : \mathcal{S} \times \mathcal{S} \to \mathbb{R}_{\geq 0}$ *is the preemptive transition rates function (rates of preemptive Markovian events),* $\mathbf{\Delta} : \mathcal{S} \times \mathcal{S} \to \mathbb{R}_{\geq 0}$ *is the branching probability distribution (probability of states reached after the firing of the general event enabled in the source state). Let* $\boldsymbol{\alpha}$ *be the initial distribution vector of* $\mathcal{R}$.

Given a subset of states $A \in \mathcal{S}$, let $\Gamma(A) = \{\Gamma(s) \mid s \in A\}$ be the set of events enabled in A. Let the *augmented set* $\widehat{A}$ be defined as set of states A plus the states of $\mathcal{S} \setminus A$ that can be reached from A with one or more non-preemptive Markovian events (**Q** events). To formulate MRP matrices, we use the *matrix filter* notation of [13]. Let $\mathbf{I}^g$ be the matrix derived from the identity matrix of size $|\mathcal{S}|$ where each row corresponding to a state s with $\Gamma(s) \neq \{g\}$ is set to zero. Let $\mathbf{I}^E$ be the same for $\Gamma(s) \neq \{E\}$.

By assuming $\{Y_n, T_n\}$ to be time-homogeneous, it is possible to define the *embedded Markov chain* (EMC) of the MRP. The EMC is a matrix $\mathbf{P}$ of size $|\mathcal{S}| \times |\mathcal{S}|$ defined on the MRS as $\mathbf{P}_{i,j} \stackrel{\text{def}}{=} \Pr\{Y_n = j \mid Y_{n-1} = i\}$. A full discussion on the EMC matrix can be found in [13, Chap. 12]. Matrix $\mathbf{P}$ is usually dense and slow to compute. To avoid this drawback, a matrix-free approach [14] is commonly followed. We now recall briefly the matrix-free method for non-ergodic MRP in reducible normal form.

Definition 2 (RNF). *The* reducible normal form *of an EMC* $\mathbf{P}$ *is obtained by rearranging the states s.t.* $\mathbf{P}$ *is in upper-triangular form:*

$$\mathbf{P} = \begin{bmatrix} \mathbf{T}_1 & \boxed{\mathbf{F}_1} & & \\ & \ddots & & \ddots \\ & & \mathbf{T}_k & \boxed{\mathbf{F}_k} \\ & & & \mathbf{R}_{k+1} \\ & & & & \ddots \\ & & & & & \mathbf{R}_m \end{bmatrix} \begin{matrix} \left. \vphantom{\begin{matrix}a\\b\\c\end{matrix}} \right\} k \geq 0 \ transient \ subsets. \\ \\ \left. \vphantom{\begin{matrix}a\\b\\c\end{matrix}} \right\} (m-k) \ recurrent \ subsets, \\ with \ m > k. \end{matrix} \tag{1}$$

The RNF of $\mathbf{P}$ *induces a directed acyclic graph, where each node is a subset of states* $\mathcal{S}_i$ *(called* component i*). Let* $\mathbf{I}_i$ *be the* filtering identity matrix, *which is the identity matrix where rows of states not in* $\mathcal{S}_i$ *are zeroed.*

When $\mathbf{P}$ is in RNF, the steady-state probability distribution can be computed using the *outgoing probability* vectors $\boldsymbol{\mu}_i$. The vector $\boldsymbol{\mu}_i$ gives for each state $s \in (\mathcal{S} \setminus \mathcal{S}_i)$ the probability of reaching s in one jump while leaving $\mathcal{S}_i$:

$$\boldsymbol{\mu}_i = \left(\mathbf{I}_i \cdot \boldsymbol{\alpha} + \sum_{j<i} (\mathbf{I}_i \cdot \boldsymbol{\mu}_j) \right) \cdot (\mathbf{I} - \mathbf{T}_i)^{-1} \cdot \mathbf{F}_i, \qquad i \leq k \tag{2}$$

Since matrix inversion is usually expensive, a product of a generic vector $\mathbf{u}$ with $(\mathbf{I} - \mathbf{T}_i)^{-1}$ can be reformulated as a linear equations system $\mathbf{x} \cdot (\mathbf{I} - \mathbf{T}_i) = \mathbf{u}$. This system can be computed iteratively using vector × matrix products with $\mathbf{u}\mathbf{T}_i$. The steady state probability of the i-th recurrent subset is given by:

$$\boldsymbol{\pi}_i = \left(\mathbf{I}_i \cdot \boldsymbol{\alpha} + \sum_{j=1}^{k} (\mathbf{I}_i \cdot \boldsymbol{\mu}_j) \right) \cdot \lim_{n \to \infty} (\mathbf{R}_i)^n, \qquad k < i \leq m \tag{3}$$

The Component Method computes first Eq. (2) for all transient components, taken in an order that respects the condition $j < i$ of the formula, and then computes the probability for the recurrent subsets based on Eq. (3).

Since the construction of $\mathbf{P}$ is not always feasible, a matrix-free method has been devised in [4] for the computations of $\mathbf{u}\mathbf{T}_i$ and $\mathbf{u}\mathbf{F}_i$. This generalisation provides: (1) a derivation of the m subsets $\mathcal{S}_i$ which is based only on $\mathbf{Q}$, $\bar{\mathbf{Q}}$ and $\boldsymbol{\Delta}$; (2) the matrix-free form of the sub-terms $\mathbf{T}_i$, $\mathbf{F}_i$ and $\mathbf{R}_i$, to be used in Eqs. (2) and (3). Observing that solution costs may differ depending on the structure of the subterms, it is convenient to distinguish three different matrix-free formulations.

[**Class** C_E] Condition: $\Gamma(\mathcal{S}_i) = \{E\}$. No general event is enabled in the $\mathcal{S}_i$ states. The matrix-free products are defined as $\mathbf{uT}_i = \mathbf{I}_i \cdot \mathbf{a}_i(\mathbf{u})$ and $\mathbf{uF}_i = (\mathbf{I} - \mathbf{I}_i) \cdot \mathbf{a}_i(\mathbf{u})$, with the term $\mathbf{a}_i(\mathbf{u})$ defined as follow, given $\mathbf{I}_i^E = \mathbf{I}_i \cdot \mathbf{I}^E$ and $\mathbf{Q}_i^E = \mathbf{I}_i^E \cdot \mathbf{Q}$:

$$\mathbf{a}_i(\mathbf{u}) = \mathbf{u} \cdot (\mathbf{I}_i^E - \mathrm{diag}^{-1}(\mathbf{Q}_i^E)\mathbf{Q}_i^E)$$

Time cost of a product with $\mathbf{T}_i$ or $\mathbf{F}_i$ is of $O(|\mathbf{Q}_i^E|)$.

[**Class** C_M] Either $|\Gamma(\mathcal{S}_i)| > 1$ or $\Gamma(\mathcal{S}_i) = \{g\} \wedge (\bar{\mathbf{Q}}_i \cdot \mathbf{I}_i \neq \mathbf{0} \vee \boldsymbol{\Delta}_i \cdot \mathbf{I}_i \neq \mathbf{0})$ Let $\mathbf{b}_i(\mathbf{u})$ be defined as:

$$\mathbf{b}_i(\mathbf{u}) = \left(\mathbf{u} \cdot \sum_{g \in G} \mathbf{I}_i^g \cdot e^{\mathbf{Q}_i \delta^g} \right) \cdot \boldsymbol{\Delta} + \left(\mathbf{u} \cdot \sum_{g \in G} \mathbf{I}_i^g \cdot \int_0^{\delta^g} e^{\mathbf{Q}_i x} \, dx \right) \cdot \bar{\mathbf{Q}}$$

The term $\mathbf{b}_i(\mathbf{u})$ gives the probability distribution of the next regeneration state reached with the firing of the general event ($\boldsymbol{\Delta}$ event) or with the preemption of the general event enabled ($\bar{\mathbf{Q}}$ event). Note that the computation of $\mathbf{b}_i(\mathbf{u})$ on a subset $\mathcal{S}_i$ of states has to consider all the states in the augmented set $\widehat{\mathcal{S}}_i$, since we have to consider all states, also outside of the component, in which the system can be found at the next regeneration state. The products with $\mathbf{T}_i$ and $\mathbf{F}_i$ are defined as:

$$\mathbf{uT}_i = \mathbf{I}_i \cdot \big(\mathbf{a}_i(\mathbf{u}) + \mathbf{b}_{\widehat{i}}(\mathbf{u}) \big), \qquad \mathbf{uF}_i = (\mathbf{I} - \mathbf{I}_i) \cdot \big(\mathbf{a}_i(\mathbf{u}) + \mathbf{b}_{\widehat{i}}(\mathbf{u}) \big)$$

The term $(\mathbf{I} - \mathbf{T}_i)^{-1}$ in Eq. (2) requires a fixed-point iteration to be evaluated. The time cost of $\mathbf{b}_i(\mathbf{u})$ is that of the uniformization, which is roughly $O(|\mathbf{Q}_i| \times R^g)$, with R^g the right truncation point [14, Chap. 5] of a Poisson process of rate $\delta_g \cdot \max_{s \in \mathcal{S}_i}(-\mathbf{Q}(s,s))$.

[**Class** C_g] Condition: $\Gamma(\mathcal{S}_i) = \{g\} \wedge \bar{\mathbf{Q}}_i \cdot \mathbf{I}_i = \mathbf{0} \wedge \boldsymbol{\Delta}_i \cdot \mathbf{I}_i = \mathbf{0}$. A single general event g is enabled in $\mathcal{S}_i$, and all the $\boldsymbol{\Delta}$ and $\bar{\mathbf{Q}}$ transitions exits from $\mathcal{S}_i$ in one step. The matrix-free products with $\mathbf{T}_i$ and $\mathbf{F}_i$ are then:

$$\mathbf{uT}_i = 0, \qquad \mathbf{uF}_i = (\mathbf{I} - \mathbf{I}_i) \cdot \mathbf{b}_{\widehat{i}}(\mathbf{u})$$

which means that the term $(\mathbf{I} - \mathbf{T}_i)^{-1}$ in (2) reduces to the identity.

3   Identification of an Optimal Set of Components

As observed in [2, 4], the performance of the Component Method may vary significantly depending on the number, size and class of the considered components. There are two main factors to consider. The first one is that the complexity of the computation of the outgoing probability vector $\boldsymbol{\mu}_i$ in Eq. (2) depends on the class of component $\mathcal{S}_i$, and a desirable goal is to use the most convenient method for each component. The second one is that the presence of many small components, possibly with overlapping augmented sets, increases the solution time, as observed in [4], where it was also experimentally observed that the number of SCCs of a non-ergodic MRP can be very high (tens of thousands is not uncommon) also in non artificial MRPs. Therefore multiple components should be joined into a single one, as far as this does not lead to solving components of a higher complexity class.

In [4] a greedy method was proposed that aggregates components to reduce their number, while keeping the component classes separated. The identification of the components starts from the observation that the finest partition of the states that produces an acyclic set of components are the strongly connected components (SCC), where the bottom SCCs (BSCC) represent the recurrent components of the MRP. The greedy algorithm aggregates components when *feasible and convenient*. Two components can be aggregated if acyclicity is preserved (*feasibility*), thus ensuring that the MRP has a reducible normal form, and if the resulting component has a solution complexity which is not greater than that of the two components (*convenience*). The objective is then to *find the feasible and convenient aggregation with the least number of components*. The greedy algorithm works as follows:

1. Let Z be the set of SCCs of $\mathcal{S}$, and let $F_Z \subseteq Z$ be the frontier of Z, i.e. the set of SCC with *in-degree* of 0 (no incoming edges).
2. Take an SCC s from F_Z and remove it from Z.
3. Aggregate s with as many SCCs from F_Z as possible, ensuring that the class of the aggregate remains the same of the class of s.
4. Repeat the aggregation until Z is empty.

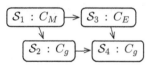

Fig. 1. Counter-example.

The main limitation of this method is that it depends on the visit order, since the aggregation of step 3 only visits the frontier. This limitation is necessary to ensure the acyclicity, but it may lead to sub-optimal aggregations. Indeed Fig. 1 shows the SCCs of an MRP, along with their classes, where the greedy algorithm may fail to provide the minimal aggregation. If the visit order is $\mathcal{S}_1, \mathcal{S}_2, \mathcal{S}_3, \mathcal{S}_4$, at the time of visiting $\mathcal{S}_2$ the *in-degree* of $\mathcal{S}_4$ will still be 1, since $\mathcal{S}_3$ is yet to visit. Therefore the method will not merge $\mathcal{S}_2$ with $\mathcal{S}_4$, resulting in a sub-optimal aggregation. Viceversa, the visit order $\mathcal{S}_1, \mathcal{S}_3, \mathcal{S}_2, \mathcal{S}_4$ allows the greedy algorithm to aggregate $\mathcal{S}_2$ and $\mathcal{S}_4$ together.

The goal of this paper is indeed to propose a method that identifies the optimal set of valid partitions (feasible and convenient).

Definition 3 (MRP Valid Partition). *A set of components of an MRP state space is a valid partition iff (1) the components are acyclic; and (2) each component, which belongs to one of the three classes (C_E, C_g and C_M), should not be decomposable into an acyclic group of sub-components of different classes.*

Acyclicity ensures that the partition is feasible and can be used for the Component Method. Condition (2) ensures convenience, i.e. by aggregating we do not increase the complexity of the solution method required for the component.

Definition 4 (MRP Component Optimization Problem). *The MRP component optimization problem consists in finding a valid partition of the MRP with the smallest number of components.*

It should be clear that this problem does not necessarily result in the fastest numerical solution of the MRP, since other factors, like rates of the components and numerical stability, may come into play: as usual the optimization is only as good as the optimality criteria defined, but results reported in [4] show that the component method is always equal or better, usually much better, than the best MRP solution method that considers the whole MRP. We transform the component optimization into a graph optimization problem for graphs with two types of edges: joinable (for pair of vertices that can stay in the same component) and non-joinable (for pair of vertices that have to be in different components).

3.1   Reformulation as a Graph Problem

We use standard notation for graphs. Let $G = \langle V, E \rangle$ be a directed graph, with V the set of vertices and $E \subseteq V \times V$ the set of edges. Notation $v \rightsquigarrow w$ indicates that vertex w is reachable from vertex v.

Definition 5 (DAG-LJ).   *A labelled directed acyclic graph with joinable edges is a graph $\mathcal{G} = \langle V, \Sigma, Lab, E, E_N \rangle$, where:*

- *$\langle V, E \rangle$ is an acyclic (direct) graph;*
- *Σ is a finite set of labels and $Lab : V \to \Sigma$ is a vertex labelling function;*
- *$E_N \subseteq E$ is the set of non-joinable edges; For ease of reference we also define $E_J = E \setminus E_N$ as the set of joinable edges;*
- *$\forall v, v' \in V, \langle v, v' \rangle \in E_J \Rightarrow Lab(v) = Lab(v')$;*

Notations $v \dashrightarrow^{J} v'$ and $v \xrightarrow{N} v'$ are shorthands for a joinable and a non-joinable edge from v to v', respectively. Given a label $l \in \Sigma$, the *section* D_l of $\mathcal{G}$ is the set of vertices of equal label: $\{v \in V \mid Lab(v) = l\}$. Let $\mathcal{D} = \{D_l \mid l \in \Sigma\}$ be the *set of sections* of $\mathcal{G}$. Let $sect(v)$ be the section of vertex v.

We now define the concept of valid and optimal partition of a DAG-LJ, to later how how an optimal valid partition of $\mathcal{G}$ induces a set of optimal components of the MRP for the component method.

Definition 6. *A valid partition of the vertices V of DAG-LJ $\mathcal{G}$ is a partitioning $\mathcal{P} = \{P_1, \ldots, P_m\}$ of the set of vertices V such that:*

1. *$\forall P \in \mathcal{P}$ and $\forall v, v' \in P : Lab(v) = Lab(v')$;*
2. *$\forall P \in \mathcal{P}: E_N \cap (P \times P) = \varnothing$;*
3. *Partition elements P are in acyclic relation.*

and we indicate with $Parts(\mathcal{G})$ the set of all valid partitions of $\mathcal{G}$.

Note that the presence of a non-joinable edge $v \xrightarrow{N} v'$ implies that v and v' cannot stay in the same partition element, in any valid partition. A joinable edge $v \dashrightarrow^{J} v'$ means that v and v' are allowed to be in the same partition element (and they are, unless other constraints are violated). From a valid partition we can build a graph which is a condensation graph, the standard construction in which all the vertices belonging to the same partition are replaced with a single vertex, from which we can easily check acyclicity.

An optimal partition (not necessarily unique) is then defined as:

Definition 7 (Optimal Partition of $\mathcal{G}$). *A valid partition $\mathcal{P}^* \in Parts(\mathcal{G})$ is optimal if the number of partition elements m is minimal over $Parts(\mathcal{G})$.*

3.2   Partitioning an MRP

MRPs have a natural representation as directed graphs: MRP states are mapped onto vertices and non-zero elements in $\mathbf{Q}, \bar{\mathbf{Q}}$, and Δ are mapped onto edges. Figure 2, upper part shows the graph of an MRP $\mathcal{R}$ of 10 states, s_1 to s_{10}, and one general event g_1. For each state we list the $\Gamma(s_i)$, which is either g_1 or E, if no general event is enabled. Transition rates are omitted. The mapping to DAG-LJ cannot be done at the MRP state level, since this results in general in a non-acyclic directed graph. Since our objective is to find an *acyclic* set of components we can map SCC of the MRP (instead of MRP states) to vertices and connection among SCCs to edges, since SCCs are the finest partition that satisfies acyclicity. When mapping to DAG-LJ, labels are used to account for the class of the SCCs, and non-joinable edges are used to identify connections that violates the convenience of component aggregation.

Definition 8. *Given an MRP $\mathcal{R} = \langle \mathcal{S}, G, \delta_g, \Gamma, \mathbf{Q}, \bar{\mathbf{Q}}, \Delta \rangle$, its corresponding DAG-LJ $\mathcal{G}(\mathcal{R}) = \langle V, \Sigma, Lab, E, E_N \rangle$ is defined as:*

- *$V = SCC(\mathcal{S})$. Each vertex is a strongly connected component of MRP states. Let $states(v)$ be the set of states in the strongly connected component $v \in V$.*
- *The set Σ of labels is $\{ C_E, C_M \} \cup \{ C_g \mid g \in G \}$ and $Lab(v)$ is defined as:*
  - *$Lab(v) = C_E$ iff $\Gamma(states(v)) = E$;*
  - *$Lab(v) = C_g$ with $g \in G$ iff $\Gamma(states(v)) = \{g\}$ and $\forall s, s' \in states(v)$: $\bar{\mathbf{Q}}(s, s') = 0 \wedge \Delta(s, s') = 0$; ($g$ is enabled continuously, no firing that disables and immediately re-enables g is allowed)*
  - *otherwise $Lab(v) = C_M$.*
- *$E = \{ \langle v, v' \rangle : \exists s \in states(v) \text{ and } s' \in states(v') \text{ such that } \mathbf{Q}(s, s') \neq 0 \text{ or } \bar{\mathbf{Q}}(s, s') \neq 0 \text{ or } \Delta(s, s') \neq 0 \}$.*
- *Edge $\langle v, v' \rangle$ is a joinable edge iff $Lab(v) = Lab(v')$ and: (1) either $Lab(v) = M$ or (2) all MRP transitions from the states of v to the states of v' are $\mathbf{Q}$ transitions. All other edges are non-joinable. Note that if there is a joinable and a non-joinable edge between v and v', the former is ignored, since E_J is defined as $E \setminus E_N$.*

$\mathcal{G}(\mathcal{R})$ has $|G|+2$ distinct labels that induce $|G|+2$ distinct sections: (D_E) if the SCC is of class C_E; (D_g) if the SCC is of class C_g, for the general event $g \in G$; (D_M) if the SCC is of class C_M.

Example 1. The MRP of Fig. 2 (upper part) has only two SCCs with more than one state: $\{s_2, s_3, s_4\}$ and $\{s_6, s_7\}$. The bottom-left part shows the DAG-LJ $\mathcal{G}$ built from the SCCs of $\mathcal{R}$. The DAG has three sections: D_E for SCCs of class C_E (all the states of the SCC enables only exponential transitions), D_{g_1} for SCCs in which all states enable g_1 and D_M for the remaining ones. Vertices v_3 and v_5 are connected by a joinable edge, since only $\mathbf{Q}$ transitions connect states of v_3 with states of v_5, while the edge $\langle v_3, v_4 \rangle$ is non-joinable because $\mathbf{\Delta}(s_5, s_8) \neq 0$. the condensation graph of a valid partition of the DAG-LJ is shown on the right of Fig. 2. The partition satisfies the requirements of Definition 6: all vertices in the same partition element have the same label, and it is not possible to go from one vertex to another vertex in the same partition elements through a non-joinable edge. Since the condensation graph is acyclic this is a valid partition.   □

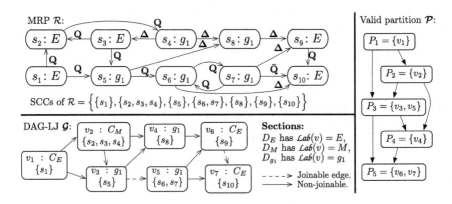

Fig. 2. Example of an MRP, its DAG-LJ, and a valid partition.

We now prove that an optimal partitioning of the DAG-LJ generated from an MRP is a solution of the MRP component optimization problem.

Property 1. If $\mathcal{G}(\mathcal{R})$ is the DAG-LJ of an MRP $\mathcal{R}$, and $\mathcal{P}^* = \{P_1, \ldots, P_m\}$ is an optimal partition of $\mathcal{G}$, then $\mathcal{P}_{\mathcal{R}}^* = \{\mathcal{S}_1, \ldots, \mathcal{S}_m\}$, with $\mathcal{S}_i = \bigcup_{v \in P_i}(states(v))$ is a solution of the MRP component optimization problem of $\mathcal{R}$ according to Definition 4.

Proof. Recall that each partition element $\mathcal{S}_i$ is a set of SCCs of $\mathcal{R}$ and each SCC belongs to one of the three classes (C_E, C_g and C_M). We need to prove that $\mathcal{P}_{\mathcal{R}}^*$ is a solution of the component optimization problem of Definition 4, which requires to prove that $\mathcal{P}_{\mathcal{R}}^*$ is an MRP valid partition and that m is minimal.

A valid MRP partition is characterized by (1) acyclicity and (2) each component should not be decomposable into an acyclic group of sub-components of different classes. Acyclicity of the set of $\mathcal{S}_i$ trivially descends from the acyclicity of $\mathcal{P}^*$. For point (2) we can observe that all SCCs in the same partition element, by Definition 6 condition 1, have the same label and therefore have the same complexity class. Therefore point (2) can be proven by showing that it is never the case that the union of two SCCs of the same class results in a component of a different class if the two SCCs are in the same partition element. If the two SCCs are classified as C_E, then all states are exponential, and the union is still in C_E. If the two SCCs are classified as C_g, then we know that there is no $\bar{Q}$ nor Δ transitions inside the single SCC, so that the classification of their union into an higher class (C_M) can only be originated by an arc between the states of the two SCCs, but the presence of such an arc, by definition of the non-joinable edges in $\mathcal{G}(\mathcal{R})$, produces a non-joinable arc between the DAG-LJ vertices of the two SCCs, and this violates point 2 of Definition 6 (there can be no non-joinable edges between vertices of the same partition element). If the two SCCs are classified as C_M, then all arcs between them, if any, are joinable, and the two SCCs can end up in the same partition element, which is also of class C_M.

Optimality of $\mathcal{P}_\mathcal{R}^* = \{\mathcal{S}_1, \ldots, \mathcal{S}_m\}$ trivially descends from optimality of $\mathcal{P}^* = \{P_1, \ldots, P_m\}$, as it is never the case that two SCCs that can be joined together result in a pair of vertices with a non-joinable edge between them, which is true by definition of $\mathcal{G}(\mathcal{R})$. □

4   Formulation of the ILP

This section defines an optimization problem with integer variables whose solution allows to build $\mathcal{P}^*$. For each vertex $v \in V$ the ILP considers $|\mathcal{D}|$ integer variables : a variable x_v and one variable y_v^D for each section $D \in \mathcal{D} \setminus sect(v)$ (each section excluded that of v). We shall refer to these two types of variables simply as x and y variables. The optimal partition of $\mathcal{G}$ is then built as: $\mathcal{P}^*(\mathcal{G}) = \bigcup_{D \in \mathcal{D}} \left(\bigcup_{i=1}^{N_D} (P_i^D) \right)$ where $P_i^D = \{v \mid v \in D \wedge x_v = i\}$, and N_D is the number of partition elements of section D (optimization target).

Definition 9. *The optimization problem is:*
    Minimize $\sum_{D \in \mathcal{D}} N_D$ subject to:

Rule 1. $\forall v \in V$: $x_v \geq 1$ *and* $\forall D \neq sect(v)$: $y_v^D \geq 0$
Rule 2. $\forall v \in V$: $x_v \leq N_D$
Rule 3. $\forall v, v' \in V$ *with* $sect(v) = sect(v')$ *and* $v \xrightarrow{J} v'$: $x_v \leq x_{v'}$
Rule 4. $\forall v, v' \in V$ *with* $sect(v) = sect(v')$ *and* $v \xrightarrow{N} v'$: $x_v < x_{v'}$
Rule 5. $\forall v \in D, v' \notin D$ *if* $v \xrightarrow{N} v'$ *then:* $x_v \leq y_{v'}^D$
Rule 6. $\forall v \in D, v' \notin D$ *if* $v' \xrightarrow{N} v$ *then:* $y_{v'}^D < x_v$
Rule 7. if $v \xrightarrow{J} v'$ *or* $v \xrightarrow{N} v'$ *then* $\forall D \notin \{sect(v), sect(v')\}$ *add:* $y_v^D \leq y_{v'}^D$

Rule 8. $\forall\, v, w \in D$ *such that* $\neg(v \leadsto w) \wedge \neg(w \leadsto v)$ *add*[1] *the constraint:* $x_v \leq x_w \Rightarrow \forall D' \neq D : y_v^{D'} \leq y_w^{D'}$.

Rule 1 sets a minimum value for the x and y variables. Rule 2 defines the N_D value as the maximum of all x variables of the same section. This value is part of the ILP goal function. Rules 3 and 4 constrains the x variables of the same section: if there is a non-joinable edge the order must be strict. Note that the relative order of the x variables follows the arc sense. No constraint is present if there is no direct edge between v and v'.

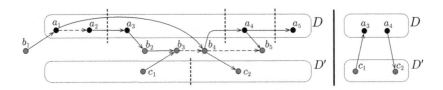

Fig. 3. The use of y variables to respect acyclicity

The remaining constraints take into account the requirement of acyclicity. Observe the portion of DAG-LJ reported in Fig. 3, left. a_i vertices are in section D, c_i are in section D' and b_i are in some other unspecified section(s). Since there is no arc between a_3 and a_4 the first 4 rules do not generate any constraint between the x variables of the two vertices, but if a_3 and a_4 end up in the same partition element acyclicity will be violated. The y variables are then defined as:

$$y_v^D = \max(0, x_w \mid w \in D \wedge w \leadsto v) \qquad (4)$$

For each vertex v, variables y_v^D is the maximum over the x values of the vertices in D that can reach v. The value of y_v^D is used for the definition of the x variables of those vertices $w \in D$ that can be reached from v, if any. If there is an edge $v \rightarrow w$, then x_w has to be strictly greater than y_v^D. Back to Fig. 3, left, $y_{b_4}^D$ stores the maximum value among x_{a_1} and x_{a_3}, therefore $y_{b_4}^D = x_{a_3}$, while $y_{b_4}^{D'}$ has the same value of x_{c_1}. Indeed Rules 5 to 7 of the ILP ensure that the optimal solution of the ILP assigns to each y the correct value, as we shall prove in Theorem 1. In the example Rules 5 to 7 insert the following constraints: $x_{a_3} \leq y_{b_2}^D \leq y_{b_3}^D \leq y_{b_4}^D < x_{a_4}$, therefore $x_{a_3} \neq x_{a_4}$, so x_{a_3} and x_{a_4} end up in different elements of the partition and acyclicty is preserved.

The above rules are effective in generating a constraints between x_v and x_w of the same section only if the two vertices are connected through a path (possibly passing through different sections). Consider the DAG-LJ of Fig. 3, right: Rules 9 to 7 produce four constraints: $x_{a_4} \leq y_{c_2}^D$, $y_{c_1}^D < x_{a_3}$, $x_{c_1} \leq y_{a_3}^{D'}$, and $y_{a_4}^{D'} < x_{c_2}$,

---

[1] This logic implication is not in standard ILP form. It can be transformed [10] in ILP form as follows. Let U be a constant greater than $|V|$. Add a new variable $k_{v,w}$ subject to these constraints: $0 \leq k_{v,w} \leq 1$, $Uk_{v,w} - U < x_v - x_w \leq Uk_{v,w}$ and $\forall D' \in \mathcal{D} \setminus \{D\}$ add $y_v^{D'} \leq y_w^{D'} + Uk_{v,w}$.

that allows for a ILP solution with $x_{a_3} = x_{a_4} = 1$, $x_{c_1} = x_{c_2} = 1$, $y_{c_1}^D = y_{a_4}^{D'} = 0$ and $y_{c_2}^D = y_{a_3}^{D'} = 1$. The final partition will be $\mathcal{P}^* = \{a_3, a_4\} \cup \{c_1, c_2\}$, which clearly violates acyclicity. Rule 8 accounts for these situations for pairs of unconnected (in the $\leadsto$ sense) vertices of the same section, stating that the values of the x and y variables in the ILP solution should respect the property that $x_v \neq x_w \Rightarrow y_v^{D'} \leq y_w^{D'}$ (the $\leq$ relation among x variables should be reflected in the order of the corresponding y variables).

Back to Fig. 3, right, four constraints are inserted by Rule 8: $x_{a_3} \leq x_{a_4} \Rightarrow y_{a_3}^{D'} \leq y_{a_4}^{D'}$, $x_{a_4} \leq x_{a_3} \Rightarrow y_{a_4}^{D'} \leq y_{a_3}^{D'}$, $x_{c_2} \leq x_{c_1} \Rightarrow y_{c_2}^D \leq y_{c_1}^D$, and $x_{c_1} \leq x_{c_2} \Rightarrow y_{c_1}^D \leq y_{c_2}^D$. And the assignment of x and y above does not satisfy the constraint. In this case a feasible solution is either $x_{a_3} > x_{a_4}$, with $x_{c_2} = x_{c_1}$ or $x_{c_2} > x_{c_1}$, with $x_{a_3} > x_{a_4}$. The final partition has then three components and is acyclic.

Rule 8 modifies the constraint on the y variables, and their definition should now be based on a different notion of reachability. Let $v \xrightarrow{*} v'$ be the *one-step extended reachability relation*, which is true if either $\langle v, v' \rangle \in E$ or $sect(v) = sect(v') \wedge x_v \leq x_{v'}$. Let $v \leadsto v'$ be the *extended reachability relation*, defined as the reachability of v' from v using the $\xrightarrow{*}$ relation. The y variables are now:

$$y_v^D = \max\left(0, x_w \mid w \in D \wedge w \leadsto v\right) \tag{5}$$

Theorem 1. *The partition $\mathcal{P}^*$ of $\mathcal{G}$ built on the solution of the ILP of Definition 9 for graph $\mathcal{G}$, is an optimal partition of $\mathcal{G}$ according to Definition 7.*

Proof. We need to show that the ILP solution provides a partition which is valid (as in Definition 6) and which has a minimum number of elements.
Validity is articulated in three conditions, the first two are trivial, as partition elements are built from vertices of the same section (and therefore of equal label) and Rule 4 states that $x_v < x_w$ whenever there is a non-joinable edge between v and w. Acyclicity is also rather straightforward. There is a cycle among two partition elements if it exists a pair of partition elements P_i^D and $P_j^{D'}$ and vertices $v, w \in P_i^D$ and $v', w' \in P_j^{D'}$ such that $v \leadsto v'$ and $w' \leadsto w$. Obviously $x_v = x_w$ and $x_{v'} = x_{w'}$. We show that if such paths exists, then at least one constrain of the ILP is violated. We consider separately the case in which $v' \leadsto w'$ and the one in which this is not true. If $v' \leadsto w'$, then (Rules 5, 6, and 7) $x_v \leq y_{v'}^D \leq \cdots \leq y_{w'}^D < x_w$, which violates the hypothesis that $x_v = x_w$. If $\neg(v' \leadsto w')$ then Rule 8 ensure that, since $x_{v'} = x_{w'}$, we must have $y_{v'}^D \leq y_{w'}^D$, moreover, by Rule 6, we have $y_{w'}^D < x_w$, which leads to $x_v \leq y_{v'}^D \leq y_{w'}^D < x_w$ which violates the hypothesis that $x_v = x_w$.
Minimality is more complicated, and is based on three observations: (1) the ILP solution builds the correct value (as per Definition 5) of the y variables of interest, (2) N_D is the number of partition elements for section D, and (3) the ILP is not over-constrained (or if v and v' could stay in the same partition element, then there is no $<$ among their x).
For point 1, let's assume that there are n vertices $w_1, \ldots, w_n \in D$ such that $w_i \to v$ and $v \notin D$ (the generalisation to $\leadsto$ is trivial due to Rule 7 that propagates the $\leq$ constraints among y variables in presence of a direct arc). Rule

5 sets a constraint $x_{w_i} \leq y_v^D$ for each vertex w_i and at least one strict constraint $y_v^D < x_{v'}$, if there is an edge from v to $v' \in D$. Then the minimization of $x_{v'}$ assigns to y_v^D the minimum possible value, which is the minimum value that satisfies the $x_{w_i} \leq y_v^D$ constraints, which is precisely the maximum over the x_{w_i} values. The proof indicates that the y_v^D value computed by the ILP is exactly equal to the maximum only in presence of a path from v back to D, in all other cases the ILP can assign any value $y \geq max(\ldots)$. But if there is no path from v back to D, then the value of y_v^D is inessential for the definition of the x variables of D. In case instead of $w \overset{*}{\leadsto} v$, the path between w and v is made either of pairs $\langle a_k, a_h \rangle$ such that either $a_k \rightarrow a_h$ (in this case the $y_{a_k}^D$ value, set initially to x_{w_i} propagates according to Rule 7) or, by definition of $\overset{*}{\rightarrow}$, $sect(a_k) = sect(a_h)$ and $x_{a_k} \leq x_{a_h}$. This implies (by Rule 8) that also $y_{a_k}^D \leq y_{a_h}^D$ and again the value of x_{w_i} propagates as if there were an edge between a_k and a_h. As in the previous case, if there is a path between v and a vertex in D, then the y_v^D is set precisely to the maximum among the x_{w_i}.

For point 2, we need to prove that $\forall i \in \{1..N_D\}, \exists v \in D : x_v = i$. This is true since, in the rules of the ILP, the $<$ order between x variables only involves x variables of the same section D, either directly (through Rule 4) or indirectly through y^D variables (Rule 6) which, by definition, carry the value of one of the x variables of D, as proved in point 1.

For point 3, we need to prove that, if w and v are in the same partition element in the optimal partitioning $\mathcal{P}^*$ of $\mathcal{G}$, then they are assigned the same x value by the ILP. For simplicity, let's assume that $\mathcal{P}^*$ is unique in $Parts(\mathcal{G})$. We prove that if $x_v \neq x_w$ then the ILP solution violates a constraint. The only way by which the ILP, given the goal of minimizing N_D, can assign a different value to x_v and x_w is the presence of $<$ among the two variables, either directly, as in Rule 4, or indirectly, through Rule 6. In the case of Rule 4, the constraint is inserted only if there is a non-joinable edge between w and v, which clearly violate the hypothesis that w and v are in the same partition element in $\mathcal{P}^*$. In the latter case, if it is a constraint $y_{v'}^D < x_v$ (of Rule 6) that causes x_w to be different from x_v, it means that $y_{v'}^D \geq x_w$. Definition 5 implies that there is a path between w and v that passes through vertex v'. In that case, w and v could not stay in the same partition element, otherwise acyclicity would be violated. Clearly the path between w and v could be either through $\leadsto$ or through $\overset{*}{\leadsto}$ since we have already shown that both can create a loop among partition elements. □

We now show two small examples of DAG-LJ, whose optimal partitioning have been constructed with the ILP method. The ILPs have been solved using the lp_solve tool.

Example 2. Consider the DAG-LJ $\mathcal{G}$ shown in Fig. 4, left, that has 14 vertices and 3 sections $D_{1\ldots3}$. Each box reports in the first row the vertex and the section. The second line of each box reports the x_v number as computed by the ILP.

The minimal solution of the ILP is found with $N_{D_1} = 3, N_{D_2} = 2$ and $N_{D_3} = 2$, which leads to a partition of the vertices in 7 subsets (partition elements). Observe that v_8 is not a direct successor of v_4, but they cannot form a single component because it would form a loop. Since there is $v_4 \xrightarrow{N} v_6$ an $v_7 \xrightarrow{N} v_8$ with $x_{v_6} \leq x_{v_7}$, Rule 8 adds a constraint $y_4^{D_3} \leq y_8^{D_3}$, to ensure the acyclicity. Figure 4, right, shows the optimal valid partitioning $\mathcal{P}^*$. □

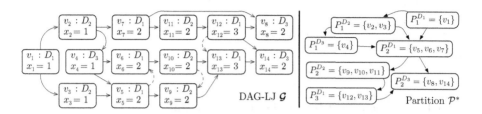

Fig. 4. Example of a DAG-LJ with the x_v values and $\mathcal{P}^*$.

Example 3. Figure 5 reports a rather different DAG-LJ, as there is no connection among the vertices of the same section, but if all the vertices of equal section are put in the same partition element, then acyclicity is violated.

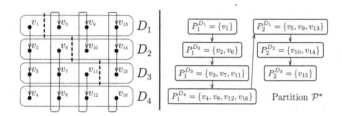

Fig. 5. An example of a DAG-LJ and $\mathcal{P}^*$ with a complex structure.

This is a prototypical example for the need of Rule 8 in the ILP. Without that rule, all the vertices of the same sections would form a single partition element, resulting in a cyclic partitioning. The problem of determining where partition elements are separated, however, is not trivial, since there are many possible combinations. In this case, the optimization problem is crucial in finding the partition boundaries that minimize the total number of components. □

5 Assessment and Conclusions

Since the ILP solution finds the optimal partition, the assessment of the proposed method does not address the quality of the solution, but aims at comparing the ILP solution with the greedy one of [4] (obviously on relatively small examples since ILP solution is known to be NP-hard), to identify the cases in which the greedy approach fails. Table 1 shows such a comparison.

The models used in the comparison are non-ergodic MRP created from Deterministic Stochastic Petri Nets with GreatSPN [6], and could be solved in Great-SPN using any of the implemented techniques for non-ergodic MRP (classical,

Table 1. Result of the ILP method against the greedy method.

Model	$|\mathcal{D}|$	SCC	EJ	EN	Greedy	ILP vars	ILP	Constr.P.
PhMissionA, K=1, NP=2	3	47	36	36	6	2767	6	20
PhMissionB, K=3, M=2	3	52	47	36	6	2492	5	7
PhMissionC, K=6, M=3	3	45	25	27	7	504	7	7
Cross	6	10	0	20	7	108	7	10
MRP of Fig. 5	6	18	0	44	12	372	9	13

matrix-free, or component-based). The partition computed by the ILP (or by the greedy method) is the base for the component method, that usually is the best solver of the three available in GreatSPN. Models can be found at www.di.unito. it/~amparore/QEST16.zip: for each model the zip file includes the pdf of the net (drawn with the GreatSPN editor [1]), and a representation of their DAG-LJ and of their ILP-computed partitioning. The whole process is automatized: from the DSPN description the MRP state space, their SCCs, and the corresponding DAG-LJ is constructed, the ILP is produced and solved with lp_solve, the components are then computed and provided as input to the component method. A similar chain is available for the greedy method. We consider 5 models: the last two have been artificially created to investigate cases in which acyclicity is non-trivial (cases in which Rule 8 plays a significant role in constraining the solution), while the first three are variations of Phased Mission Systems (PMS). In particular they are cases of a *Scheduled Maintenance System* (SMS), inspired by [9], in which a system alternates between two phases: Work and Maintenance, and behaves differently depending on the phase (as typical in PMS). The model is studied for its transient behaviour, the stopping condition for model A is determined by the number NP of phases, while models B and C cycle over the two phases, and the stopping condition is triggered when the system reaches a failure state. K and M are the number of pieces and machines.

The table reports the model name and the number of sections, SCCs, joinable and non-joinable edges. The column 'Greedy' indicates the number of components found by the greedy method, while the two subsequent columns reports the number of variables of the ILP and the number of components found by solving the ILP. Finally, the last column reports the number of components found by applying a constraint propagation method, i.e. by applying the ILP constraint in order to maximize the x and y variables until a fix-point is found. Constraint propagation can be seen as an approximate solution of the ILP, where the found partitioning is always valid but not necessarily optimal.

As the table shows, the greedy method performs reasonably, but it does not always found the optimal solution, although it goes very close to it (a behaviour that has been observed on other cases of "real" systems). It instead performs badly in cases created ad-hoc to experiment with Rule 8 (models Cross and MRP of Fig. 5). The constraint propagation method is consistently the worst one. The MRP size we could solve with standard computer are below a hundred of SCCs (of course the state spaces could be much larger), which is not surprising considering that the ILP size grows rapidly with the number of SCCs (the vertices of the DAG-LJ) and that the problem is NP-hard.

Conclusions. This paper introduces a technique to find the optimal partition of a non-ergodic MRP which is the basis of the Component Method solver for MRPs. The method is both general (can be applied to any non-ergodic MRP) and optimal, as it finds the minimum number of partition elements, and therefore of components. Optimality is important not only for the solution time, but also because it provides a baseline against which to assess the greedy solution. An optimal solution is a prerequisite to compare the component method against specific ad-hoc MRP solver. A typical example are the MRPs generated from Phased Petri nets [8] for which an efficient ad-hoc solution technique was devised in [15]: this technique can be interpreted as a special case of the component method (with roughly one component per phase), moreover with the component method the class of PMS that can be efficiently solved can be enlarged to include, for example, different type of dependencies of the system behaviour from the phase description that we believe are relevant for reliability analysis and that will be investigated in our future research work. Optimality is also a prerequisite when comparing the efficiency of a CSL^{TA} model checker, as the one in [3], on verifying CSL Until formulas. The component method with optimal partitioning reduces the time complexity of the CSL^{TA} model checker to that of a CSL one (CSL model-checking algorithm as described in [7]), as already envisioned in [4].

A question that might arise is whether it is worth to define DAG-LJ, instead of deriving the ILP directly from the SCCs of the MRP. The answer is that the DAG-LJ abstraction may be used for other purposes and we are indeed currently using it in the context of model-checking of CSL^{TA} based on zone graph. The idea here is that the MRP that describes the set of accepting paths of the formula is obtained by a cross product of the Markov chain model and the zone graph (a rather trivial construction since there is a single clock) of the timed automata that describes the CSL^{TA} formula. The MRP is then solved with the component method. But this is an a-posteriori work: the MRP is first built completely, and then solved by component. The solution we are working on translates the zone graph in a DAG-LJ, computes the components of the DAG-LJ and then does the cross-product between a zone graph component and the Markov chain.

Another more than legitimate question is whether it makes sense to rely on ILP solution, in particular as it is not uncommon to have MRPs with thousands of SCCs. But luckily this is not always the case, for example the DAG-LJ of the zoned graph of the timed automata that describes a CSL^{TA} formula typically has a very limited number of SCCs, and the ILP solution can be easily found, while a similar situation arises in PMS, as typically the number of SCCs is related to the number of phases, which is usually significantly less than 10. As future work, we plan nevertheless to experiment with classical approximate ILP solvers, and to compare it with the greedy approach.

References

1. Amparore, E.G.: A new GreatSPN GUI for GSPN editing and CSL^{TA} model checking. In: Norman, G., Sanders, W. (eds.) QEST 2014. LNCS, vol. 8657, pp. 170–173. Springer, Heidelberg (2014)

2. Amparore, E.G., Donatelli, S.: A component-based solution method for non-ergodic Markov regenerative processes. In: Aldini, A., Bernardo, M., Bononi, L., Cortellessa, V. (eds.) EPEW 2010. LNCS, vol. 6342, pp. 236–251. Springer, Heidelberg (2010)

3. Amparore, E.G., Donatelli, S.: MC4CSL^{TA}: an efficient model checking tool for CSL^{TA}. In: International Conference on Quantitative Evaluation of Systems, pp. 153–154. IEEE Computer Society, Los Alamitos (2010)

4. Amparore, E.G., Donatelli, S.: A component-based solution for reducible Markov regenerative processes. Perform. Eval. **70**(6), 400–422 (2013)

5. Amparore, E.G., Donatelli, S.: Improving and assessing the efficiency of the MC4CSL^{TA} model checker. In: Balsamo, M.S., Knottenbelt, W.J., Marin, A. (eds.) EPEW 2013. LNCS, vol. 8168, pp. 206–220. Springer, Heidelberg (2013)

6. Baarir, S., Beccuti, M., Cerotti, D., Pierro, M.D., Donatelli, S., Franceschinis, G.: The GreatSPN tool: recent enhancements. SIGMETRICS Perform. Eval. Rev. **36**(4), 4–9 (2009)

7. Baier, C., Haverkort, B., Hermanns, H., Katoen, J.P.: Model-checking algorithms for continuous-time Markov chains. IEEE Trans. Softw. Eng. **29**(6), 524–541 (2003)

8. Bondavalli, A., Mura, I.: High-level Petri net modelling of phased mission systems. In: 10th European Workshop on Dependable Computing, pp. 91–95, Vienna (1999)

9. Bondavalli, A., Filippini, R.: Modeling and analysis of a scheduled maintenance system: a DSPN approach. Comput. J. **47**(6), 634–650 (2004)

10. Brown, G.G., Dell, R.F.: Formulating integer linear programs: a rogues' gallery. INFORMS Trans. Educ. **7**(2), 153–159 (2007)

11. Cassandras, C.G., Lafortune, S.: Introduction to Discrete Event Systems. Springer, Secaucus (2006)

12. Donatelli, S., Haddad, S., Sproston, J.: Model checking timed and stochastic properties with CSL^{TA}. IEEE Trans. Softw. Eng. **35**(2), 224–240 (2009)

13. German, R.: Performance Analysis of Communication Systems with Non-Markovian Stochastic Petri Nets. Wiley, New York (2000)

14. German, R.: Iterative analysis of Markov regenerative models. Perform. Eval. **44**, 51–72 (2001)

15. Mura, I., Bondavalli, A.: Markov regenerative stochastic petri nets to model and evaluate phased mission systems dependability. IEEE Trans. Comput. **50**(12), 1337–1351 (2001)

16. Stewart, W.J.: Probability, Markov Chains, Queues, and Simulation: The Mathematical Basis of Performance Modeling. Princeton University Press, Princeton (2009)

Data-Efficient Bayesian Verification of Parametric Markov Chains

E. Polgreen[1]([✉]), V.B. Wijesuriya[1], S. Haesaert[2], and A. Abate[1]

[1] Department of Computer Science, University of Oxford, Oxford, UK
elizabeth.polgreen@cs.ox.ac.uk
[2] Department of Electrical Engineering, Eindhoven University of Technology,
Eindhoven, The Netherlands

Abstract. Obtaining complete and accurate models for the formal verification of systems is often hard or impossible. We present a data-based verification approach, for properties expressed in a probabilistic logic, that addresses incomplete model knowledge. We obtain experimental data from a system that can be modelled as a parametric Markov chain. We propose a novel verification algorithm to quantify the confidence the underlying system satisfies a given property of interest by using this data. Given a parameterised model of the system, the procedure first generates a feasible set of parameters corresponding to model instances satisfying a given probabilistic property. Simultaneously, we use Bayesian inference to obtain a probability distribution over the model parameter set from data sampled from the underlying system. The results of both steps are combined to compute a confidence the underlying system satisfies the property. The amount of data required is minimised by exploiting partial knowledge of the system. Our approach offers a framework to integrate Bayesian inference and formal verification, and in our experiments our new approach requires one order of magnitude less data than standard statistical model checking to achieve the same confidence.

1 Introduction

Complex engineering systems, such as autonomous vehicles, are often safety-critical and demand high guarantees of correctness. Given a complete model of the system of interest, these guarantees can be obtained through formal methods, such as model checking [1], though the outcomes of these formal proofs are bound to the model of the system of interest. Obtaining a complete model is not possible for systems with uncertain stochastic dynamics, but we can capture these dynamics with parameterised Markov chains. Model checking now produces a result dependent on knowledge of the value of parameters within the model.

In this work we integrate the use of model checking techniques (for parameter synthesis over the model) with data-based approaches (for parametric Bayesian inference) in order to compute a confidence, based on observed data collected from the system, that the system satisfies a given specification.

The proposed approach is distinctively different from statistical model checking (SMC) [14], a known data-based technique for model verification, and has

© Springer International Publishing Switzerland 2016
G. Agha and B. Van Houdt (Eds.): QEST 2016, LNCS 9826, pp. 35–51, 2016.
DOI: 10.1007/978-3-319-43425-4_3

a distinct set-up and addresses a different objective: The original SMC tools such as *YMer* and *Vesper* target systems with fully known models too large for conventional model checking, and use the known models to generate simulated data; SMC has also been applied in a model-free setting where system-generated data is directly employed towards statistical validation of properties of interest [19]. Our technique instead targets partially known systems, captured as a parameterised model class, and still uses data collected from the original system.

In general SMC requires a large amount of sample data covering the entire system behaviour to obtain good confidence results, our method requires much less sample data, and can accommodate data with only partial coverage.

Our method is elucidated in three phases. In the first phase, having a parameterised model of our partially known system, we use parameter synthesis to determine a set of feasible parameters over the given model class, namely those parameters corresponding to models of the system satisfying the given specification. Among a number of alternatives, we use an existing parameter synthesis method implemented in PRISM [11]. The second phase, executed in parallel with the first, uses Bayesian statistics to infer a distribution over the likely values of the parameters of the model class, based on data collected from the underlying system. Finally, we combine the outputs from the previous two phases to compute the confidence attached to the system satisfying the given specification.

Alongside the new methodology introduced in this work (first presented over different model class and properties in [9]), the key contribution resides in phase two: our algorithm introduces expansions of states and transitions of the parameterised Markov chain, which guarantees the posterior probability distributions over the parameters can be obtained analytically, and integrated easily. The work discusses a case study, demonstrating the implementation of the algorithm, and a comparison with a standard SMC procedure.

Related Work. Statistical Model Checking (SMC) [14] replaces numerical model-based procedures with empirical testing of formalised properties. The original SMC algorithms target fully observable stochastic systems with little non-determinism and may require the generation of large numbers of sample trajectories from a complete system model. SMC techniques have been utilised to tackle verification of black box probabilistic systems [19], with no model of the system available, but this approach requires large amounts of data. Extensions towards the inclusion of non-determinism have been studied in [12], with preliminary steps towards Markov decision processes. Related to SMC techniques, [6,15] assume the system is encompassed by a finite-state Markov chain and efficiently use data to learn and verify the corresponding model. Similarly, [2,4] employ machine learning techniques to infer finite-state Markov models from data over given logical formulae.

Bayesian inference uses Bayes theorem to update the probability distribution of a set of hypotheses based on observed data [3]. Bayesian inference for learning transition probabilities in Markov Processes is presented in [16].

2   Background

2.1   Parametrised Markov Chains – Syntax and Semantics

Let S be a finite, non-empty set of states representing all possible configurations of the system being modelled. A discrete-time Markov chain (DTMC) is a stochastic time-homogeneous process over this set of states [1], as follows.

Definition 1. *A discrete-time Markov chain* $\mathbf{M}$ *is a tuple* $(S, \mathbb{T}, \iota_{init}, \mathrm{AP}, L)$, *where S is a finite, non-empty set of states,* $\mathbb{T} : S \times S \to [0,1]$ *is the transition probability function such that for* $\forall s \in S : \sum_{s' \in S} \mathbb{T}(s, s') = 1$. *The function* $\iota_{init} : S \to [0,1]$ *denotes an initial probability distribution over the states S, such that* $\sum_{s \in S} \iota_{init}(s) = 1$. *The states in S are labelled with atomic propositions* $a \in \mathrm{AP}$ *via the labelling function* $L : S \to 2^{\mathrm{AP}}$.

Consider the evolution of a Markov chain over a time horizon $t = 0, 1, \ldots, N_t$, with $N_t \in \mathbb{N}$. Then an execution of the process is characterised by a state trajectory given as $\{s_t | t = 0, 1, \ldots, N_t\}$. The transition function $\mathbb{T}(s, s')$ specifies for each state s the probability of moving to s' in one step, and hinges on the *Markov Property*, which states that the conditional probability distribution of the future possible states depends only on the current state, namely $\mathbb{P}(s' = s_{t+1} \mid s_t, \ldots s_0) = \mathbb{P}(s' = s_{t+1} \mid s_t)$. Furthermore, the definition of $\mathbf{M}$ requires $\mathbb{T}$ is time homogeneous, that is $\mathbb{P}(s' = s_{t+1} \mid s_t = s) = \mathbb{P}(s' = s_t \mid s_{t-1} = s), \forall t \in \mathbb{N}$. The model is extended with (internal) non-determinism in order to express lack of complete knowledge of the underlying system.

Definition 2. *A discrete-time Parametric Markov chain is defined as a tuple* $\mathbf{M}_\Theta = (S, \mathbb{T}_\theta, \iota_{init}, \mathrm{AP}, L, \Theta)$ *where $S, \iota_{init}, \mathrm{AP}, L$ are as in Definition 1. The entries in $\mathbb{T}_\theta$ are specified in terms of parameters, collected in a parameter vector* $\theta \in \Theta$, *where Θ is the set of all possible evaluations of θ. Each evaluation gives rise to an induced Markov chain* $\mathbf{M}(\theta)$.

Note we require a certain type of well-posedness of the parameterisation, we demand $\forall s \in S, \forall \theta \in \Theta : \sum_{s' \in S} \mathbb{T}_\theta(s, s') = 1$. More precisely, any $\theta \in \Theta$, induces a Markov chain $\mathbf{M}(\theta)$ where the transition function $\mathbb{T}_\theta$ can be represented by a stochastic matrix. Note also, we assume a distribution on the parameters of the model.

We considered two types of parameterised Markov chain. We use the first, simpler type, as a base case to build the method for the more complex linearly parameterised Markov chains.

1. *basic parameterised Markov chains* with independently parameterised transition probabilities. Consider $\mathbf{M}_\Theta = (S, \mathbb{T}_\theta, \iota_{init}, \mathrm{AP}, L, \Theta)$ with $\Theta \subseteq [0,1]^n$ and parameter vector $\theta := (\theta_1, \ldots, \theta_n) \in \Theta$ build up based on individual parameters $\theta_i \in [0,1]$. Then the parameterised MC is considered *basic* if transition probabilities between states are either known and considered constant with a value in $[0,1]$, or have a single parameter θ_i (or $1 - \theta_i$) associated to them and $\forall s \in S, \forall \theta \in \Theta : \sum_{s' \in S} \mathbb{T}_\theta(s, s') = 1$ (cf. Fig. 1, left).

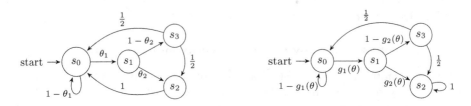

Fig. 1. Two parameterised Markov chains. The nodes of the graph represent states. The labels over the edges provide the probability of taking a transition. The left graph gives parameterised MC with a *basic* parameterisation, where the parameters θ_1, θ_2 are encompassed in the vector $\theta = (\theta_1, \theta_2) \in \Theta = [0,1]^2$. The right graph has a *linear* parameterisation, characterised by affine functions $g_{1,2} : \theta \mapsto [0,1]$.

2. *linearly parameterised Markov chains*, where unknown transition probabilities can be linearly related. Given $\Theta \subseteq [0,1]^n$ and parameter vector $\theta := (\theta_1, \ldots, \theta_n) \in \Theta$ with $\theta_i \in [0,1]$, the parameterised MC is considered *linearly parameterised* if there exists a set of affine functions $g_l(\theta) := k_0 + k_1\theta_1 + \ldots + k_n\theta_n$ with $k_i \in [0,1]$ and $\sum k_i \leq 1$, denoted $g_l(\theta)_{l \in \mathcal{L}}$. All outgoing transition probabilities of states (or, graphically labels of outgoing edges of a node, cf. Fig. 1) have probability $g_l(\theta)$ or $1 - g_l(\theta)$ and $\forall s \in S, \forall \theta \in \Theta : \sum_{s' \in S} \mathbb{T}_\theta(s,s') = 1$.

The basic case leads to simple procedures, and in Sect. 5 we develop the linear structure for Bayesian verification. Parameterisations beyond these two categories, such as *non-linear* ones, are out of the scope of this paper.

2.2 Properties – Probabilistic Computation Tree Logic

We consider system requirements specified in probabilistic logics. As we leverage PRISM's parametric model checking tool [10] for synthesis, we can consider the set of properties supported by the synthesis tool: non-nested Probabilistic Computational Tree Logic (PCTL) [1] formulae. For instance, $\mathbb{P}_{\geq 0.5}(\text{stay } \mathcal{U} \text{ get})$ expresses the property "the probability of remaining in a state labelled with atomic proposition 'stay' until we reach a state labelled as 'get', is bigger or equal to 0.5". PRISM also supports nested PCTL with some restrictions, and a planned extension to this work is to use PROPHESY [8] for parameter synthesis, which supports conditional probabilities and unbounded-time properties. We next define PCTL in nexus to finite discrete-time Markov chains:

Definition 3. *Let a discrete-time Markov chain be given. Let ϕ be a formula interpreted over states $s \in S$, and φ be a formula interpreted on paths of the DTMC. Also, let $\bowtie \in \{<, \leq, \geq, >\}$, $n \in \mathbb{N}$, $p \in [0,1]$, $c \in AP$. The syntax of PCTL is given by:*

$$\phi := \text{True} \mid c \mid \phi \wedge \phi \mid \neg\phi \mid \mathbb{P}_{\bowtie p}(\varphi), \qquad \varphi := \bigcirc \phi \mid \phi \, \mathcal{U} \, \phi.$$

We define the satisfaction function quantifying satisfaction of these properties over the parameter space as follows. We assume it is a measurable function.

Definition 4. *Let* $\mathbf{M}(\theta)$ *be an induced Markov chain of the parametric Markov chain* $\mathbf{M}_\Theta$ *indexed by parameter* $\theta \in \Theta$, *and let* ϕ *be a formula in PCTL. The satisfaction function* $f_\phi : \Theta \to \{0, 1\}$, *defined as* $f_\phi(\theta) = 1$ *if* $\mathbf{M}(\theta) \models \phi$, *and 0 otherwise.*

2.3 Bayesian Inference

Our method uses Bayesian inference to learn the probability distribution of parameters in our model class as more evidence or data becomes available. Bayesian inference derives the posterior probability distribution from a prior probability and a likelihood function derived from a statistical model for the observed data. Bayes' law states that, given observed data D, the posterior probability of a hypothesis $p(H \mid D)$, is proportional to the likelihood $p(D \mid H)$, multiplied by the prior $p(H)$, as

$$p(H \mid D) = \frac{p(D \mid H)p(H)}{p(D)}. \tag{1}$$

D comprises batches of traces of specific length generated by Markov chains instantiated over Θ. The denominator in (1) is an integral over the parameter set Θ, which in general requires numerical approximation. Hence it is of interest to seek a *conjugate* prior $p(H)$ resulting in a closed-form expression for the posterior $p(H \mid D)$: in this work we make use of the Dirichlet distribution, which is conjugate to the multinomial [3]. When insufficient initial knowledge is available, we choose a non-informative prior, which has minimal influence on the posterior, such as a uniform prior.

3 Problem Statement and Overview of the Approach

Consider a partly unknown dynamical system $\mathbf{S}$, and suppose we can gather a limited amount of sample trajectories from this system as data. Assume the knowledge about the system is encompassed within a parametric model class, describing the behaviour of $\mathbf{S}$ up to the unknown parameterisation of some of its transitions. We plan to investigate the following goal: *can we efficiently use the gathered data and the model knowledge of* $\mathbf{S}$ *to formally verify given PCTL properties over* $\mathbf{S}$, *quantifying a confidence in our assertions?*

The three phases of our work are as follows. In the first phase, Sect. 4, we use parameter synthesis to determine a set of feasible parameters for which the system satisfies the given property. The second phase, Sect. 5, uses Bayesian Inference to infer a distribution over the likely value of the parameters given sample data from the system. In the final phase, Sect. 6, we combine the outputs of parametric inference and parameter synthesis to quantify the confidence that the system verifies a PCTL property of interest.

Bayesian probability calculus [3] leads to expressing the confidence in a property as a measure of the uncertainty distribution over the synthesised parameter sets. Uncertainty distributions are handled as probability distributions of random variables. Given a specification ϕ and a data set D, the confidence $\mathbf{S} \models \phi$ can be quantified via inference as $\mathbb{P}(\mathbf{S} \models \phi \mid D) = \int_{\Theta} f_{\phi}(\theta)p(\theta \mid D)\, d\theta$, where $\mathbb{P}(\cdot)$ is a probability measure obtained integrating the distribution $p(\cdot)$ of the uncertainty parameter over $\mathbf{M}_{\Theta}$, expressed as the *a-posteriori* $p(\theta \mid D)$ given the data set D and the uncertainty distribution $p(\theta)$ over the parameter set Θ.

The computation in the third phase is a key challenge for Markov chains with non trivial parameterisation due to the required complex manipulation of Dirichlet posterior distributions. This motivates the introduction of a Markov chain expansion algorithm in Sect. 5.2, which enables us to analytically obtain samples of complex posterior distributions.

4 Parameter Synthesis

The first phase of our method uses parameter synthesis and, given a property and a parameterised Markov chain, synthesises the feasible set of parameters corresponding to models satisfying the given PCTL property. This corresponds to the set of parameters for which the binary satisfaction function, $f_{\phi}(\theta) = \mathbb{P}(\mathbf{M}(\theta) \models \phi)$, is equal to 1. We denote this set Θ_{ϕ}, namely

$$\Theta_{\phi} = \{\theta \in \Theta : \mathbf{M}(\theta) \models \phi\}.$$

We leverage PRISM's parametric model checking functionality based on [11] to perform this synthesis. [11] expresses quantitative specifications as rational functions that are later manipulated. PRISM's parameteric model checking approach can be applied to unbounded until, steady-state probabilities, reachability reward and steady-state reward properties for parametric DTMCs. The result is a mapping from hyper-rectangles (subsets of parameter valuations) to functions over the parameters.

Alternatives to these techniques have not shown to be scalable or sufficiently general. [5] explores the parameter space with the objective of model verification. [13] employs an analytical approach to parameter synthesis for probabilistic transition systems and is bound to at most two parameters. [7] employs a language-theoretical approach based on regular expressions, which however does not scale as the number of transitions of the Markov model increases. [18] synthesise single-parameter Markov models via accurate interval propagation.

5 Bayesian Inference in Parameterised Markov Chains

In this section we consider the application of Bayesian inference to parameterised Markov chains, in order to infer unknown parameter probabilities based on observed data. We will first present the technique for basic parameterised

Markov chains, and then extend the method to linearly related parameterisations in Sect. 5.2, where we show data obtained from a linearly parameterised Markov chain can equally be represented by data complemented with a set of hidden (or unobserved) data of a basic Markov chain. We use $\mathbb{P}(\cdot)$ to denote a probability measure, and $p(\cdot)$ to denote a probability density function.

5.1 Basic Parameterised Markov Chains

Let us consider a basic parameterised Markov chain $\mathbf{M}_\Theta = (S, \mathbb{T}_\theta, \iota_{init}, \mathrm{AP}, L, \Theta)$ (cf. Definition 2). In this basic parameterised Markov chain, every individual parameter θ_i of vector $\theta = (\theta_1, \theta_2, \ldots, \theta_n) \in \Theta$ is exclusively used to assign the outgoing transition probabilities of a single state. We can decompose our parameter vector θ into sub-vectors θ_{s_i}, giving the parameters for the outgoing transitions of the corresponding state s_i.

Consider the parameter vector composed of one parameter, $\theta_{s_k} = \theta_j$, and the corresponding state $s_k \in S$, with outgoing transitions θ_j and $1 - \theta_j$ to states s_1 and s_2, respectively. We denote by $p(\theta_j)$ the prior over θ_j, which fully defines the transition probabilities $\mathbb{T}_\theta(s_k, \cdot)$ at state s_k. Denote a data set D giving transition counts for trajectories generated from the real system $\mathbf{S}$. For any pair $(s_k, s_l) \in S \times S$ the number of transitions $s_k \to s_l$ in D is denoted as $D^{s_l}_{s_k}$. The posterior density $p(\theta_j \mid D)$ over θ_j based on D is

$$p(\theta_j \mid D) = \frac{\mathbb{P}(D \mid \theta_j)p(\theta_j)}{\mathbb{P}(D)} = \frac{p(\theta_j)\prod_{s' \in S} \mathbb{T}_\theta(s_k, s')^{D^{s'}_{s_k}}}{\mathbb{P}(D_{s_k})} \tag{2}$$

and depends only on $D_{s_k} = \{D^{s'}_{s_k}\}_{s' \in S}$, i.e., the counts of transitions leaving state s_k. Note the likelihood function $\prod_{s' \in S}(\mathbb{T}_\theta(s_k, s'))^{D^{s'}_{s_k}}$ takes the form of a multinomial distribution,[1] which reduces to a binomial in the case of two outgoing transitions. A closed-form expression for the posterior is obtained by taking a conjugate prior, which, for the class of multinomial distributions, is a Dirichlet distribution. For the pair $(\theta_j, 1 - \theta_j)$ the Dirichlet distribution with hyperparameters $\alpha = (\alpha_1, \alpha_2)$ has a probability density function given by

$$\mathrm{Dir}(\theta_j \mid \alpha) = \tfrac{1}{B(\alpha)}\theta_j^{\alpha_1 - 1}(1 - \theta_j)^{\alpha_2 - 1}$$

on the open simplex defined by $0 < \theta_j < 1$. The normalising constant, $B(\alpha)$, is a multinomial beta function, and can be written in terms of gamma functions as $B(\alpha) = \Gamma(\alpha_1)\Gamma(\alpha_2)/\Gamma(\alpha_1 + \alpha_2)$. Hence, for a prior $p(\theta_j) = \mathrm{Dir}(\theta_j \mid \alpha)$ we obtain the posterior distribution for $\theta_j \sim p(\theta_j \mid D) = \mathrm{Dir}(\theta_j \mid D_{s_k} + \alpha)$, namely

$$p(\theta_j \mid D) \propto p(\theta_j)\prod_{s' \in S} \mathbb{T}_\theta(s_k, s')^{D^{s'}_{s_k}} \propto \theta_j^{\alpha_1 - 1}(1 - \theta_j)^{\alpha_2 - 1}\theta_j^{D^{s_1}_{s_k}}(1 - \theta_j)^{D^{s_2}_{s_k}} \tag{3}$$

---

[1] A multinomial is defined by its density function $f(\cdot \mid p, N) \propto \prod_{i=1}^{k} p_i^{n_i}$, for $n_i \in \{0, 1, \ldots, N\}$ and such that $\sum_{i=1}^{k} n_i = N$, where $N \in \mathbb{N}$ is a parameter and p is a discrete distribution over k outcomes.

where the normalisation constant of the obtained Dirichlet distribution is $B(\alpha + D_{s_k}) = \Gamma(\alpha_1 + D_{s_k}^{s_1})\Gamma(\alpha_2 + D_{s_k}^{s_2})/\Gamma(\alpha_1 + D_{s_k}^{s_1} + \alpha_2 + D_{s_k}^{s_2})$. In other words, as data is gathered, we analytically update the posterior probability distribution $p(\theta_j \mid D)$ by updating the parameters of a Dirichlet distribution.

This result can be extended to the case of a state s_l with $m > 2$ outgoing transitions. We parameterise the outgoing transitions with the sub-vector $\theta_{s_l} = (\theta_1, \ldots, \theta_{m-1})$ and $1 - \theta_1 - \ldots - \theta_{m-1}$, and obtain the posterior for the sub-vector, $p(\theta_{s_l} \mid D)$. The likelihood function takes the form of an m-dimensional multinomial distribution, and we express the prior as an m-dimensional Dirichlet.

This yields a posterior distribution as an m-dimensional Dirichlet distribution, $p(\theta_{s_l}|D) = \text{Dir}(\theta_{s_l} \mid D_{s_l} + \alpha)$.

The posterior distribution for the entire parameter vector $p(\theta \mid D)$ is equal to the product of the posterior distributions for the sub-vectors of θ. This holds due to the stated independence of the parameters in a basic parameterised Markov chain, which results in independent priors and independent likelihood functions. Hence $p(\theta \mid D) = \prod_{s_i} \text{Dir}(\theta_{s_i} \mid D_{s_i} + \alpha)$.

Transition Grouping. For simplicity, given a state with multiple outgoing transitions we may obtain the distribution for each parameter using marginal distributions. Consider state s_l with $m > 2$ outgoing transitions, parameterised with the sub-vector $\theta_{s_l} = (\theta_1, \ldots, \theta_{m-1})$ and $1 - \theta_1 - \ldots - \theta_{m-1}$ We have shown earlier that, if the parameters are independent, the joint posterior distribution over the transition probabilities for this state is an m-dimensional Dirichlet: $p(\theta_{s_l}|D) = \text{Dir}(\theta_{s_l} \mid D_{s_l} + \alpha)$. The marginal distribution of θ_i is a 2-dimensional Dirichlet, or a beta distribution, $\theta_i \sim \text{Dir}(\alpha_i, (\sum_{i=1}^{m} \alpha_i) - 1)$. We can hence obtain a posterior distribution for each parameter, by effectively grouping the training data together for all transitions except the one we obtain the posterior distribution for.

5.2   Linearly Parameterised Markov Chains

In this section we build on the Bayesian inference for basic parameterisations and tackle linearly parameterised Markov chains. As defined before, in a linear parameterised Markov chain, the transition probabilities will be expressed in the form $g(\theta) = k_0 + k_1\theta_1 + \ldots + k_n\theta_n$. For a given data set D and a linearly parameterised Markov chain we want to use Bayesian inference to get the posterior distribution $p(\theta|D)$ over the parameter set Θ. In order to work with linear parameters we introduce two types of transformations of the Markov chain. In the first, we consider a compression of the data. When two states of the DTMC have "similar" transitions, what can be learned is equivalent. These states are referred to as being *parameter similar* and will be introduced more precisely in the following. Next we show that, by introducing additional, non-observed states, into the Markov chain and the data, the linear parameterised Markov chain can be transformed to a basic Markov chain with unobserved states (and hidden data). After these transformations we can apply the Bayes rule over the expanded Markov chain and hidden data.

Parameter Similar States. If we have the same parameter appearing multiple times in our Markov chain, we must combine the data obtained from all these transitions to obtain a sole posterior distribution for the parameter in our confidence computation. This technique, referred to as "parameter tying", is used in [17]. We can perform this step analytically for Dirichlet distributions over *parameter similar states*, by which we denote states with outgoing transitions having identical parameterisations.

Manipulating posterior Dirichlet distributions is mathematically complex because of the dependence between the variables. However, if states are parameter similar, we can use the result in (3). Consider two parameter similar states, s_1 and s_2, with outgoing transition probabilities θ_j and $1 - \theta_j$, and observed data over the transitions. We combine the data to give one posterior Dirichlet distribution for the parameter, $p(\theta_j) = \text{Dir}(D_{s_1} + D_{s_2} + \alpha_{s_1})$.

Parameterised Markov Chain State Expansions. Consider a parameterised DTMC $\mathbf{M}_\Theta = (S, \mathbb{T}, \iota_{init}, \text{AP}, L, \Theta)$. We wish to define a new parameterised DTMC $\mathbf{M}_\Theta^*$ that produces the same output for our method, but which has a simpler parameterisation. Our method hinges on obtaining a distribution for θ based on collected training data D, and so if $\mathbf{M}_\Theta^*$ is equivalent to $\mathbf{M}_\Theta$, the probabilities of reaching a set of states in $\mathbf{M}_\Theta$ must be the same as reaching the equivalent states in $\mathbf{M}_\Theta^*$, but we may disregard the length of associated paths.

Before introducing the definition of state expansion, we first need to define hidden data. Suppose the two Markov chains have states S and S^*, such that $S \subset S^*$: all states of S^* not in S are defined as *hidden*. Ω denotes the set of finite paths ω in $\mathbf{M}_\Theta$, and Ω^* denotes the set of finite paths ω^* in $\mathbf{M}_\Theta^*$. Then any observed state sequence consists only of states in S, and the states in $S^* \setminus S$ remain hidden from the observations. The data set D over the states S consists of transition counts $D_{s_k}^{s_l}$ for pairs $s_k, s_l \in S$. Observe that for the set of states S^* the data is incomplete, namely it does not represent the actual state transitions but only the observed ones. For an observed transition count $D_{s_k}^{s_l}$, we introduce the extended set $D_{s_k}^{s_l}{}^*$ as the collection of counts over all hidden paths from s_k to s_l. Consider states s_0 and s_2, and hidden state s_0^* in Fig. 3a: hidden paths from s_0 to s_2 can be of the form $\{s_0, s_2\}, \{s_0, s_0^*, s_2\} \in \Omega^*$, with the associated extended data count $D_{s_0}^{s_2}{}^* := \{D_{s_0}^{s_2}, D_{s_0}^{s_0^*}, D_{s_0^*}^{s_2}\}$. The set of possible extended transition counts is denoted as $\mathcal{D}_{s_k}^{s_l}{}^*$ for the pair (s_k, s_l), and $\mathcal{D}^*$ for all transitions – note they are set-valued mappings of $D_{s_k}^{s_l}$ and D, respectively.

Definition 5. *Consider parameterised Markov chains* $\mathbf{M}_\Theta = (S, \mathbb{T}, \iota_{init}, \text{AP}, L, \Theta)$ *and* $\mathbf{M}_\Theta^* = (S^*, \mathbb{T}^*, \iota_{init}^*, \text{AP}, L^*, \Theta)$, *both over set* Θ. *We say* $\mathbf{M}_\Theta^*$ *is an expansion of* $\mathbf{M}_\Theta$ *if, for all D and for all $\theta \in \Theta$,*

$$\mathbb{P}_{\mathbf{M}(\theta)}(D) = \mathbb{P}_{\mathbf{M}^*(\theta)}(\mathcal{D}^*),$$

and if $\iota_{init} = \iota_{init}^*$. *The extended labelling map L^* is a trivial extension of L, assigning labels $L(s)$ for $s \in S$ and an empty label to $S^* \setminus S$.*

Theorem 1. *The expansion relation is transitive; if* $\mathbf{M}_{\Theta,1}, \mathbf{M}_{\Theta,2}, \mathbf{M}_{\Theta,3}$ *are all parameterised with* Θ, $\mathbf{M}_{\Theta,3}$ *is an expansion of* $\mathbf{M}_{\Theta,2}$ *and* $\mathbf{M}_{\Theta,2}$ *is an expansion of* $\mathbf{M}_{\Theta,1}$, *then* $\mathbf{M}_{\Theta,3}$ *is an expansion of* $\mathbf{M}_{\Theta,1}$.

Case I: Transition splitting. We split a transition probability parameterised with $k_0 + \sum_i k_i\theta_i$ into transitions to hidden states with probabilities $k_i\theta_i$, and refer to this operation as *transition splitting*. As a basic example, consider Fig. 2 where state s_0 in M has two outgoing transition probabilities expressed as functions of the parameter vector, $g(\theta)$ and $1 - g(\theta)$, where $g(\theta) = k_0 + k_1\alpha + k_2\beta$. We expand $\mathbf{M}_{\Theta}$ into $\mathbf{M}_{\Theta}^*$ by splitting state s_1 into a set of states, and splitting the transition from $s_0 \to s_1$ into the monomials concerning each parameter in θ, as shown in Fig. 2. $\mathbf{M}_{\Theta}^*$ is an expansion of $\mathbf{M}_{\Theta}$ as per Definition 5.

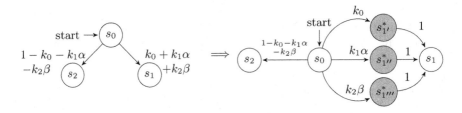

Fig. 2. Case I: transitions splitting

Lemma 1. *Transition splitting of* $\mathbf{M}_{\Theta}$ *(Case I) generates an expansion of* $\mathbf{M}_{\Theta}$.

Case II: State splitting. We present a second case, state splitting, for a parameter θ_i multiplied by a constant, $k_i\theta_i$. Consider the simple DTMC in Fig. 3a, and the state s_0 in $\mathbf{M}_{\Theta}$ with two outgoing transition probabilities expressed as a constant multiplied by one parameter, $k_1\theta_1$ and $1 - k_1\theta_1$, where $0 \le k_1 \le 1$. We expand $\mathbf{M}_{\Theta}$ to give $\mathbf{M}_{\Theta}^*$ by splitting state s_0 into two states, and compute the transition probabilities the imposed expansion demands. As an additional example, notice the transitions studied in Case I are all of the form $k_i\theta_i$. Applying the state splitting to this expanded DTMC we obtain Fig. 3b. The subsequent application of both state splitting cases (cf. Fig. 3b) induces again an expanded parameterised Markov chain as per Definition 5.

Lemma 2. *State splitting of* $\mathbf{M}_{\Theta}$ *(Case II) generates an expansion of* $\mathbf{M}_{\Theta}$.

We are led to the following result.

Theorem 2. *Any linearly parameterised Markov chain can be expanded into a basic parameterised Markov chain by application of Lemmas 1 and 2.*

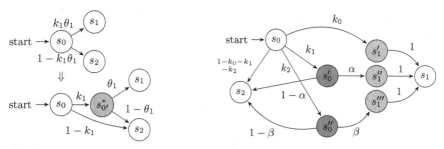

(a) Simple example of state splitting   (b) State splitting of Fig.2 (cf. Case I).

Fig. 3. Case II: state splitting (two examples)

Bayesian Inference with Missing Data. We now consider Bayesian inference on the newly expanded Markov chain $\mathbf{M}_\Theta^*$. The data set D, which is sampled from our system, corresponds to a state trajectory or set of trajectories over the model $\mathbf{M}_\Theta$. This set further comprises only part of the corresponding trajectories in the expanded model $\mathbf{M}_\Theta^*$. For a given trajectory in D, we refer to D^* as the completed trajectory, and to $\mathcal{D}^*$ as the set of all possible completions D^*. Note the expanded parametric Markov chain has a *basic parameterisation*, hence for a given completed data set D^* the Bayes rule as elaborated in (1) can be applied to obtain $p(\theta|D^*)$. For $\mathbf{M}_\Theta^*$ Bayes rule can be applied over the hidden data as follows:

$$p(\theta|D) = \frac{\sum_{D^*\in\mathcal{D}^*} p(\theta, D^*, D)}{\mathbb{P}(D)} = \frac{\sum_{D^*\in\mathcal{D}^*} p(\theta|D^*, D)\,\mathbb{P}(D^*|D)\mathbb{P}(D)}{\mathbb{P}(D)}$$
$$= \sum_{D^*\in\mathcal{D}^*} p(\theta|D^*)\,\mathbb{P}(D^*|D).$$

Completed data sets have a multinomial distribution dependent on the parameterisation, hence the distribution of D^* is given as $\mathbb{P}(D^*) = \int_\Theta \mathbb{P}(D^*|\theta)p(\theta)\,d\theta$. For a given D the conditional distribution $\mathbb{P}(D^*|D)$ is $\mathbb{P}(D^*|D) = \mathbb{P}(D^*)/\mathbb{P}(D)$, with $D^* \in \mathcal{D}^*$ and $\mathbb{P}(D) = \sum_{D^*} \int_\Theta \mathbb{P}(D^*|\theta)p(\theta)\,d\theta$.

Remark 1. Realisations of the posterior can be obtained without computing the entire integral as follows. A set of realisations θ_i for $i \in \{1,\dots,\mathcal{N}\}$ with probability density function $p(\theta|D)$ can be obtained by generating samples D_i^* with distribution $\mathbb{P}(D^*|D)$ and subsequently generating samples θ_i with distribution $p(\theta|D_i^*)$ for all $i \in \{1,\dots,\mathcal{N}\}$. These samples can then directly be used to perform the confidence calculation as in Sect. 6.   □

Algorithm 1 presents the state expansion procedure, and Algorithm 2 in the next section summarises how to obtain a realisation of the posterior $p(\theta \mid D^*)$, and to integrate it with the confidence computation.

Algorithm 1. Markov chain expansion ($\mathbf{M}_\Theta$)

$\mathbf{M}_\Theta^* \leftarrow \mathbf{M}_\Theta$

for all $s_i \in S^*$ **do**   ▷ Case I: transition splitting

    for all $\mathbb{T}_\theta^*(s_i, s_j) = k_0 + \sum_{l \in \mathcal{L}} k_l \theta_l$ **do**

        $S^* \leftarrow \{s_{ij,l}^*\}_{l \in L} \cup s_{ij,0}$

        $\mathbb{T}_\theta^*(s_i, s_j) := 0$

        $\mathbb{T}_\theta^*(s_i, s_{ij,0}^*) := k_0$ and $\mathbb{T}_\theta^*(s_{ij,0}^*, s_j) := 1$

        for all $l \in L$ **do**

            $\mathbb{T}_\theta^*(s_i, s_{ij,l}^*) := k_l \theta_l$ and $\mathbb{T}_\theta^*(s_{ij,l}^*, s_j) := 1$

for all $s_i \in S^*$ **do**   ▷ Case II: state splitting

    if $\exists s_k \in S^* : \mathbb{T}_\theta^*(s_i, s_k) = 1 - k_0 - \sum_{l \in \mathcal{L}} k_l \theta_l$ **then**

        $\mathbb{T}_\theta^*(s_i, s_k) := 1 - k_0 - \sum_{l \in \mathcal{L}} k_l$

        for all $\mathbb{T}_\theta^*(s_i, s_m) = k_l \theta_l$ **do**

            $S^* \leftarrow s_{m'}^*$

            $\mathbb{T}_\theta^*(s_i, s_m) := 0$, $\mathbb{T}_\theta^*(s_i, s_{m'}^*) := k_l$ and $\mathbb{T}_\theta^*(s_{m'}^*, s_k) := 1 - \theta_l$

            $\mathbb{T}_\theta^*(s_{m'}^*, s_m) := \theta_l$

return $\mathbf{M}_\Theta^*$ ▷ return expanded DTMC

---

6 Bayesian Verification: Computation of Confidence

In this section we detail the final phase of our method: a quick procedure computes a confidence estimate for the satisfaction of a PCTL specification formula ϕ by a system $\mathbf{S}$ of interest, namely $\mathbf{S} \models \phi$. Our method takes as input a posterior distribution over Θ, obtained using Bayesian inference in Sect. 5.2, and the feasible set for the parameters, obtained by parameter synthesis in Sect. 4.

Definition 6. *Given a PCTL specification ϕ, a complete trace (sample trajectory) D of the system $\mathbf{S}$ up to time t, and a transition function $\mathbb{T}$, the confidence $\mathbf{S} \models \phi$ can be quantified by Bayesian Inference as*

$$\mathbb{P}(\mathbf{S} \models \phi \mid D) = \int_\Theta f_\phi(\theta) p(\theta \mid D) d\theta. \tag{4}$$

As we only consider the satisfaction of a property $\mathbf{S} \models \phi$ as a binary-valued mapping from the space of parameters, the satisfaction function in (4), $f_\phi : \Theta \to \{0, 1\}$, (4) can be reformulated as:

$$\mathbb{P}(\mathbf{S} \models \phi \mid D) = \int_{\Theta_\phi} p(\theta \mid D) d\theta, \tag{5}$$

where Θ_ϕ denotes the set of parameters corresponding to models verifying the property ϕ (as generated by PRISM). Further, given the independent posterior distributions for each parameter in θ resulting from Sect. 5.2, the confidence can be computed as $\mathbb{P}(\mathbf{S} \models \phi \mid D) = \int_{\Theta_\phi} \prod_{\theta_i \in \theta} p(\theta_i \mid D) d\Theta$. The integral of a Dirichlet distribution can be obtained by iterative or numerical methods: here we use a simple Monte-Carlo approach, which depends on samples of the posterior distribution as clarified in Algorithm 2.

Algorithm 2. Monte-Carlo Integration for linearly parameterised DTMC

$\mathcal{N} :=$ number of Monte-Carlo samples
$\{D_i^*\}_{i \in \{1,...,\mathcal{N}\}} \sim p(D^*|D)$ ▷ hidden data samples
for all $i \in \{1,...,\mathcal{N}\}$ **do**
     Compute $p(\theta|D_i^*)$ ▷ Bayesian inference
     $\theta_i \sim p(\theta|D_i^*)$ ▷ posterior samples
     $j_\# \leftarrow j_\# + \text{Boolean}[\theta_i \in \Theta_\phi]$
$\hat{\mathbb{P}}(\mathbf{S} \models \phi) := \frac{j_\#}{\mathcal{N}}$
return $\hat{\mathbb{P}}(\mathbf{S} \models \phi)$ ▷ estimate of $\mathbb{P}(\mathbf{S} \models \phi)$

7 Experiment Results

We show our approach requires smaller amounts of data than statistical model checking (SMC) to verify the system satisfies a given quantitative specification up to a prescribed confidence level. We further claim our approach is more robust than standard SMC in situations where only data of limited trace length is available.

Experiment Setup. We focus our experimental discussion on the basic parameterised Markov chain $\mathbf{M}_\Theta$ in Fig. 1 and the PCTL property $\phi = \mathbb{P}_{>0.5}[\neg s_3 \mathcal{U} s_2]$.

The ground truth for $\mathbf{S} = \mathbf{M}(\theta)$, namely Y_{true}, is a step function over the parameter θ, namely

$$Y_{true} = \begin{cases} 0 & \text{if } \theta \leq 0.5, \\ 1 & \text{if } \theta > 0.5, \end{cases} \tag{6}$$

so the feasible set is $\Theta_\phi = [0.5, 1]$. We choose a uniform prior for both methods: for our approach $p(\theta \mid D) = \text{Dir}(1,1)$, which, for property ϕ, means $p(\mathbf{M}(\theta) \models \phi) = \text{Dir}(1,1)$; for SMC we set $p(\mathbf{M}(\theta) \models \phi) = \text{Dir}(1,1)$. We run both methods over empirical data obtained from $\mathbf{M}(\theta)$, our "underlying system", for values of $0 < \theta < 1$, i.e., different "underlying systems", and compare the outcomes with the ground truth. We collect data, denoted D, from our underlying system in the form of a set of state trajectories of a set length. We vary trajectory length to test robustness to data with incomplete coverage. We disregard the numerical error in the Monte Carlo approximate integration, which is the same for both techniques.

We compute the mean squared error (MSE) between the confidence outcome and the ground truth from Eq. (6), namely $MSE = \frac{1}{n} \sum_{i=1}^{n} (Y_{true} - Y_i)^2$, where n is the number of experiments run and Y_i is the result $\mathbb{P}(\mathbf{M}_\theta \models \phi)$ for the i-th run.

The SMC we compare our work to is "black box" and collects sample trajectories from the system, then determines whether the trajectories satisfy a given property, and applies statistical techniques (such as hypothesis testing) to decide whether the system satisfies the property or not, with some degree of confidence. Our "grey-box" approach collects data from the system, uses the data to determine a distribution over parameter values in the parameterised model class and applies statistical techniques (in this case, a Bayesian confidence calculation) to decide whether the system satisfies the property or not, with some degree of confidence. We could then additionally apply hypothesis testing to our approach. However, as we do not do this, for a meaningful comparison with our approach we implement the framework of the SMC procedure outlined in [14] and omit the hypothesis testing. Instead, we compute a Bayesian confidence by integrating the posterior distribution given over the [0,1] interval, representing the probability of a trace satisfying the property. The trace generation and trace verification stages of SMC are implemented in the same way in the four statistical model checking methods in PRISM.

Results and Discussion. The first point to note is the confidence is low, and MSE high for parameter values close to $\theta = 0.5$ for both approaches. This is due to $\theta = 0.5$ being on the edge of the feasible set and is consistent with the information we wish to obtain from the confidence calculation: if the parameter value is near the edge of the feasible set, we need to know its value precisely to be sure it falls in the feasible set. Consider that in order to compute the confidence $\mathbf{S} \models \phi$, we integrate the posterior distribution over the feasible set $\Theta_\phi = \{\theta > 0.5\}$. The posterior distribution for $\theta = 0.5$ should have a peak centred at 0.5 with half of the area under the peak in the feasible set, leading to $\mathbb{P}(\mathbf{M}(\theta) \models \phi) = 0.5$. The height and width of the distribution $p(\theta \mid D)$ are characterised by the amount of data available, as well as the consistency of the data, and so we expect the MSE to be higher for parameter values close to the threshold.

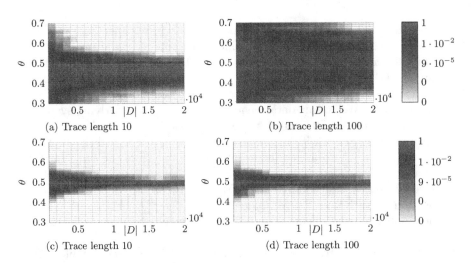

Fig. 4. Outcomes of SMC are given in (a) and (b), outcomes of our approach are given in (c) and (d). The comparison is done over a data set D composed of traces of 10 and 100 transitions. On the x-axis, $1000 \leq |Dt| \leq 20000$. On the y-axis, $0.3 \leq \theta \leq 0.7$. The darker (purple) colour indicates a higher mean squared error. (Color figure online)

The key result is, for both approaches, the mean squared error reduces as $|D|$ increases and the variance decreases, but our approach consistently produces a smaller error and variance than SMC for any parameter values excluding $\theta = 0.5$ (where both approaches perform comparably). Our approach requires an order of magnitude less data than SMC and above $|D| = 2000$, the error for our approach is smaller than the error in the Monte Carlo integration, whereas SMC does not reach this precision threshold in our experiments, which we perform up to $|D| = 200000$.

We ascribe both the reduced error and reduced variance to the data efficiency of our approach: SMC receives the training data in the form of short traces, and discerns whether a trace is a counter example or witness for the property. A trace can, however, be neither, in which case it is discarded, even if that trace contains parameterised transitions. Our approach counts each parameterised transition in the data, and so uses more of the data available than SMC. It is unsurprising accuracy and variance improve when more data is used.

We investigate robustness in a situation where it is only possible to collect short trajectories from the system, whilst verifying an unbounded property. Figure 4a and b show the performance of SMC with $|D|$ made up of trace lengths of 10 and 100 transitions respectively. We show a part of our data set, discarding data above $|D| = 20,000$ where our approach produces no measurable error. The mean squared error in Fig. 4b is 50 % lower than in Fig. 4a over the entire parameter range, but the run with trace lengths of 10 performs better for values of $\theta > 0.55$.

We explain this because, computed using PRISM, the expected length of a witness for our property and Markov Chain ranges between 4.33, for $\theta = 0.3$ and 2.42 for $\theta = 0.7$ (due to the symmetrical structure of our Markov Chain, the lengths of counter-examples are also expected to be the same). Thus a large proportion of the traces of length 10 are discarded, and so SMC has less data to use, explaining the increased error across the parameter range. However, when $\theta > 0.55$, the expected counter-example length is higher, and so the number of traces of length 10 that are useful begins to exceed the total number of traces of length 100 received.

In contrast, the performance of our approach, shown in Fig. 4c and d, yields approximately the same outcomes for both trace lengths, as we consider each transition in the training data individually and only discard non-parameterised transitions. Admittedly it is not always the case that the performance of our method is independent of the length of the traces: consider for example the case of a large Markov chain where a parameterised transition is only reachable after a large number of steps. In this case the performance of our approach would be comparable to SMC.

We run experiments on linearly parameterised Markov chains of a similar scale and obtain comparable results.

8 Conclusions and Future Work

We have presented a data-based verification approach addressing incomplete model knowledge. The method offers a framework to integrate Bayesian inference and formal verification, and in comparison to standard statistical model checking promises to be more parsimonious with the required data.

We plan to investigate extensions in the following directions: performing parameter synthesis with alternative available techniques, such as [8], which builds on the work of [10] using graph topological properties and fixed points); working with non-linearly parameterised Markov chains; inspired by [9], integrating external non-determinism in the form of actions, thus leading to parameterised Markov decision processes. Finally, we are interested in the use of Bayesian hypothesis testing, which will further solidify this method as a provable verification technique even when the prior probability distribution is not reliably known.

References

1. Baier, C., Katoen, J.: Principles of Model Checking. MIT Press, Cambridge (2008)
2. Bartocci, E., Bortolussi, L., Sanguinetti, G.: Learning temporal logical properties discriminating ECG models of cardiac arrhytmias. CoRR abs/1312.7523 (2013)
3. Bernardo, J., Smith, A.: Bayesian Theory. Wiley, Chichester (1994)
4. Bortolussi, L., Sanguinetti, G.: Learning and designing stochastic processes from logical constraints. In: Joshi, K., Siegle, M., Stoelinga, M., D'Argenio, P.R. (eds.) QEST 2013. LNCS, vol. 8054, pp. 89–105. Springer, Heidelberg (2013)
5. Brim, L., Češka, M., Dražan, S., Šafránek, D.: Exploring parameter space of stochastic biochemical systems using quantitative model checking. In: Sharygina, N., Veith, H. (eds.) CAV 2013. LNCS, vol. 8044, pp. 107–123. Springer, Heidelberg (2013)
6. Chen, Y., Nielsen, T.D.: Active learning of Markov decision processes for system verification. In: ICMLA, pp. 289–294. IEEE (2012)
7. Daws, C.: Symbolic and parametric model checking of discrete-time Markov chains. In: Liu, Z., Araki, K. (eds.) ICTAC 2004. LNCS, vol. 3407, pp. 280–294. Springer, Heidelberg (2005)
8. Dehnert, C., Junges, S., Jansen, N., Corzilius, F., Volk, M., Bruintjes, H., Katoen, J.-P., Ábrahám, E.: PROPhESY: a PRObabilistic ParamEter SYnthesis tool. In: Kroening, D., Păsăreanu, C.S. (eds.) CAV 2015. LNCS, vol. 9206, pp. 214–231. Springer, Heidelberg (2015)
9. Haesaert, S., Van den Hof, P.M.J., Abate, A.: Data-driven property verification of grey-box systems by Bayesian experiment design. In: ACC, pp. 1800–1805. IEEE (2015)
10. Hahn, E.M., Hermanns, H., Wachter, B., Zhang, L.: PARAM: a model checker for parametric markov models. In: Touili, T., Cook, B., Jackson, P. (eds.) CAV 2010. LNCS, vol. 6174, pp. 660–664. Springer, Heidelberg (2010)
11. Hahn, E.M., Hermanns, H., Zhang, L.: Probabilistic reachability for parametric Markov models. In: Păsăreanu, C.S. (ed.) Model Checking Software. LNCS, vol. 5578, pp. 88–106. Springer, Heidelberg (2009)

12. Henriques, D., Martins, J., Zuliani, P., Platzer, A., Clarke, E.M.: Statistical model checking for Markov decision processes. In: QEST, pp. 84–93. IEEE (2012)
13. Lanotte, R., Maggiolo-Schettini, A., Troina, A.: Parametric probabilistic transition systems for system design and analysis. Formal Asp. Comput. **19**(1), 93–109 (2007)
14. Legay, A., Delahaye, B., Bensalem, S.: Statistical model checking: an overview. In: Barringer, H., Falcone, Y., Finkbeiner, B., Havelund, K., Lee, I., Pace, G., Roşu, G., Sokolsky, O., Tillmann, N. (eds.) RV 2010. LNCS, vol. 6418, pp. 122–135. Springer, Heidelberg (2010)
15. Mao, H., Jaeger, M.: Learning and model-checking networks of I/O automata. In: ACML. JMLR, vol. 25, pp. 285–300. JMLR.org (2012)
16. Eichelsbacher, P., Ganesh, A.: Bayesian inference for Markov chains. J. Appl. Probab. **39**(1), 91–99 (2002)
17. Poupart, P., Vlassis, N.A., Hoey, J., Regan, K.: An analytic solution to discrete Bayesian reinforcement learning. In: ICML. ACM, vol. 148, pp. 697–704. ACM (2006)
18. Su, G., Rosenblum, D.S.: Nested reachability approximation for discrete-time Markov chains with univariate parameters. In: Cassez, F., Raskin, J.-F. (eds.) ATVA 2014. LNCS, vol. 8837, pp. 364–379. Springer, Heidelberg (2014)
19. Younes, H.L.S.: Probabilistic verification for "black-box" systems. In: Etessami, K., Rajamani, S.K. (eds.) CAV 2005. LNCS, vol. 3576, pp. 253–265. Springer, Heidelberg (2005)

Probabilistic Reasoning Algorithms

Exploiting Robust Optimization for Interval Probabilistic Bisimulation

Ernst Moritz Hahn[1], Vahid Hashemi[2,3]([⊠]), Holger Hermanns[3],
and Andrea Turrini[1]

[1] State Key Laboratory of Computer Science, Institute of Software,
Chinese Academy of Sciences, Beijing, China
[2] Max Planck Institute for Informatics, Saarbrücken, Germany
hashemi@mpi-inf.mpg.de
[3] Department of Computer Science, Saarland University, Saarbrücken, Germany

Abstract. Verification of PCTL properties of MDPs with convex uncertainties has been investigated recently by Puggelli et al. However, model checking algorithms typically suffer from the state space explosion problem. In this paper, we discuss the use of probabilistic bisimulation to reduce the size of such an MDP while preserving the PCTL properties it satisfies. As a core part, we show that deciding bisimilarity of a pair of states can be encoded as adjustable robust counterpart of an uncertain LP. We show that using *affine decision rules*, probabilistic bisimulation relation can be approximated in polynomial time. We have implemented our approach and demonstrate its effectiveness on several case studies.

1   Introduction

Real world systems are usually too complex to be analyzed in full detail. To reduce the complexity of such an analysis, a simplified but accurate enough model of the system has to be constructed and then verified with respect to a number of properties the system is expected to satisfy. Among others, probability, nondeterminism, and uncertainty are core aspects of a real world system that are worth considering in the model. *Probability* represents the fact that the behaviour of the system is not uniquely determined by its status and the action it performs, but depends on random choices as well; these choices may be present by design (as the toss of a coin in a distributed algorithm so as to break symmetry) or to represent general properties such as transmission errors during a communication. *Nondeterminism* can be used whenever a specific behavior is unknown or it is left undetermined by purpose: an example of the former is the unknown relative speed of several distributed systems interacting with each other while an example of the latter is the possibility of leaving some behavior undetermined so an implementation can fix it. *Uncertainty* appears when some information is available but it is not precise enough to be represented as a probability.

A problem that may occur during the formal verification of a system, for instance by model checking it, is the notorious *state-explosion* problem. Such a

© Springer International Publishing Switzerland 2016
G. Agha and B. Van Houdt (Eds.): QEST 2016, LNCS 9826, pp. 55–71, 2016.
DOI: 10.1007/978-3-319-43425-4_4

problem can be mitigated by reducing the size of the model to be verified while preserving its properties. This goal can by achieved by finding another model that is smaller than the original one while behaving the same. Bisimulation allows us to construct such a model; this strategy has been proven very effective [16, 29] in related settings.

Several models have been proposed in literature as frameworks for modelling real world systems, frameworks equipped with bisimulation. Among others, there are Labelled Transition Systems, Probabilistic Automata [42], and Markov Decision Processes (MDPs). In this work we focus on the Interval Markov Decision Processes (IMDPs) model [27,28,38,41,45,46], an extension of classical MDPs where uncertainty is represented by intervals of probability values. It is known that bisimilar IMDPs satisfy the same PCTL properties [27]. As established in [27,28], computing the coarsest bisimulation on a given IMDP is a difficult problem; our aim is to provide a polynomial algorithm that returns a non-trivial bisimulation for the given IMDP. We achieve this goal by taking advantage of the results from the Operations Research community about robust optimization and uncertain Linear Programming (LP) problems.

Summarizing, the main contributions of this paper are as follows.

- We build a bridge between *Probabilistic Verification* and *Robust Optimization* and establish a novel modelling of the probabilistic bisimulation problem for interval MDPs as an instance of an uncertain LP problem.
- We show that, by using *affine decision rules*, the probabilistic bisimulation problem for IMDPs can be approximately decided in polynomial time.
- We show promising results on a number of case studies, obtained by a prototypical implementation of our algorithm.

Related Work. We classify related works in four areas. Firstly, various probabilistic modelling formalisms with uncertain transitions are studied in the literature. Interval Markov chains [31,35] or Abstract Markov chains [20] extend standard discrete-time Markov chains (MCs) with interval uncertainties and thus do not feature the non-deterministic choice of transitions. Uncertain MDPs [38,40,45] allow for more general sets of distributions to be associated with each transition, not only those described by intervals. Usually, this is restricted to *rectangular uncertainty sets* requiring that the uncertainty is linear and independent for any two transitions of any two states. Our general algorithm working with polytopes can be easily adapted to this setting. Parametric MDPs [26] to the contrary allow for such dependencies as every probability is described as a rational function of a finite set of global parameters.

Secondly, computational complexity of the probabilistic bisimulation for uncertain probabilistic models has been studied quite recently in [27,28]. Among similar concepts studied in the literature are simulation [22,47] and refinement [18,19,31] relations for previously mentioned models.

Thirdly, from model checking viewpoint, many new verification algorithms for interval models appeared in last few years. Reachability and expected total reward is addressed for Interval MCs [15] as well as Interval MDPs [46]. PCTL model checking and LTL model checking are studied for Interval MCs [9,14,15] and also for Interval MDPs [41,45]. Among other technical tools, all these approaches make use of (robust) dynamic programming relying on the fact that transition probability distributions are resolved dynamically: a probability distribution is chosen from interval restrictions each time the system enters a state. For the static resolution of distributions, an adaptive discretization technique for PCTL parameter synthesis is given in [26]. Uncertain models are also widely studied in the control community [23,38,46], mainly interested in maximal expected finite-horizon reward or maximal expected discounted reward.

Finally, as regards the application of *Robust Optimization* in *Probabilistic Verification* community, to the best of our knowledge, we are not aware of any work in the literature. Therefore, the current contribution is novel in this matter. On the other hand, the aforementioned theory has been adapted and applied successfully in control theory realm. For instance, Abate and El Ghaoui [5] developed a robust modal predictive control using two-stage robust optimization.

2   Preliminaries

In this paper, the sets of all positive integers, rational numbers, real numbers and non-negative real numbers are denoted by $\mathbb{N}$, $\mathbb{Q}$, $\mathbb{R}$, and $\mathbb{R}_{\geq 0}$, respectively. We denote by $\mathbb{I}$ the set of closed sub-intervals of $[0,1]$ and, for a given $[a,b] \in \mathbb{I}$, we denote by $\inf[a,b]$ the lower bound a and by $\sup[a,b]$ the upper bound b. We denote by b^k the k-th element of a vector $b \in \mathbb{R}^n$. For a set X, we denote by $\Delta(X)$ the set of discrete probability distributions over X; given $\rho \in \Delta(X)$, we denote by $\mathrm{Supp}(\rho) = \{ x \in X \mid \rho(x) > 0 \}$ the support of ρ and we say that ρ is Dirac, denoted δ_x, if $\mathrm{Supp}(\rho) = \{x\}$ with $x \in X$. For an equivalence relation $\mathcal{R}$ on X and $\rho_1, \rho_2 \in \Delta(X)$, we write $\rho_1 \mathcal{L}(\mathcal{R}) \rho_2$ if for each $\mathcal{C} \in X/\mathcal{R}$, it holds that $\rho_1(\mathcal{C}) = \rho_2(\mathcal{C})$. By abuse of notation, we extend $\mathcal{L}(\mathcal{R})$ to distributions over $X/\mathcal{R}$, i.e., for $\rho_1, \rho_2 \in \Delta(X/\mathcal{R})$, we write $\rho_1 \mathcal{L}(\mathcal{R}) \rho_2$ if for each $\mathcal{C} \in X/\mathcal{R}$, it holds that $\rho_1(\mathcal{C}) = \rho_2(\mathcal{C})$.

2.1   Interval Markov Decision Processes

Let us formally define Interval Markov Decision Processes.

Definition 1 (IMDPs). *An* Interval Markov Decision Process *(IMDP)* $\mathcal{M}$ *is a tuple* $\mathcal{M} = (S, \bar{s}, \mathcal{A}, \mathrm{AP}, L, I)$, *where* S *is a finite set of* states, $\bar{s} \in S$ *is the* initial state, $\mathcal{A}$ *is a finite set of* actions, AP *is a finite set of* atomic propositions, $L \colon S \to 2^{\mathrm{AP}}$ *is a* labeling function, *and* $I \colon S \times \mathcal{A} \times S \to \mathbb{I}$ *is an* interval transition probability function.

Given $s \in S$ and $a \in \mathcal{A}$, we write $s \xrightarrow{a} \mu_s$ whenever $\mu_s \in \Delta(S)$ is a *feasible distribution*, i.e., for each $s' \in S$ we have $\mu_s(s') \in I(s, a, s')$. Let $\mathcal{P}^{s,a} = \{ \mu_s \in \Delta(S) \mid s \xrightarrow{a} \mu_s \}$; we denote by $\mathcal{A}(s) = \{ a \in \mathcal{A} \mid \mathcal{P}^{s,a} \neq \emptyset \}$ the set of actions that are enabled from s and we require that $\mathcal{A}(s) \neq \emptyset$ for each $s \in S$.

We extend I to sets of states as follows: given $S' \subseteq S$, we let

$$I(s,a,S') = \left[\min \left\{ 1, \sum_{s' \in S'} \inf I(s,a,s') \right\}, \min \left\{ 1, \sum_{s' \in S'} \sup I(s,a,s') \right\} \right].$$

An interval MDP is initiated in some state s_1 and then moves in discrete steps from state to state forming an infinite path $s_1 s_2 s_3 \dots$. One step, say from state s_i, is performed as follows. First, an action $a \in \mathcal{A}(s_i)$ is chosen probabilistically by *scheduler*. Then, *nature* resolves the uncertainty and chooses nondeterministically one corresponding feasible distribution $\mu_{s_i} \in \mathcal{P}^{s_i,a}$. Finally, the next state s_{i+1} is chosen probabilistically according to the distribution μ_{s_i}.

Let us define the semantics of an *IMDP* formally. A *path* is a finite or infinite sequence of states $\pi = s_1 s_2 \dots$. For a finite path π, we denote by $last(\pi)$ the last state of π. The set of all finite and infinite paths are denoted by $Paths^*$ and $Paths^\omega$, respectively. Furthermore, let $Cyl_\pi = \{ \pi' \in Paths^\omega \mid \pi \leqslant \pi' \}$ denote the set of paths having $\pi \in Paths^*$ as prefix.

Definition 2 (Scheduler and Nature). *Given an IMDP $\mathcal{M}$, a scheduler is a function $\sigma \colon Paths^* \to \Delta(\mathcal{A})$ such that for each $\pi \in Paths^*$, $\mathrm{Supp}(\sigma(\pi)) \subseteq \mathcal{A}(last(\pi))$. Further, a nature is a function $\nu \colon Paths^* \times \mathcal{A} \to \Delta(S)$ such that for each $\pi \in Paths^*$ and $a \in \mathcal{A}(last(\pi))$, $\nu(\pi,a) \in \mathcal{P}^{last(\pi),a}$. We denote by $\mathfrak{S}$ and $\mathfrak{N}$ the set of all schedulers and natures of $\mathcal{M}$, respectively.*

For an initial state s, a scheduler σ, and a nature ν, let $Pr_s^{\sigma,\nu}$ denote the unique probability measure over $(Paths^\omega, \mathcal{B})$[1] such that the probability $Pr_s^{\sigma,\nu}[Cyl_{s'}]$ of starting in s' equals 1 if $s' = s$ and 0 otherwise and the probability $Pr_s^{\sigma,\nu}[Cyl_{\pi s'}]$ of traversing a finite path $\pi s'$ equals $Pr_s^{\sigma,\nu}[Cyl_\pi] \cdot \sum_{a \in \mathcal{A}} \sigma(\pi)(a) \cdot \nu(\pi,a)(s')$.

Observe that the scheduler does not choose an action but a *distribution* over actions. It is well-known [42] that such a randomization is useful in the context of bisimulations as it allows to define coarser equivalence relations. To the contrary, nature is not allowed to randomize over the set of feasible distributions $\mathcal{P}^{s,a}$. This is in fact not necessary, since the set $\mathcal{P}^{s,a}$ is closed under convex combinations. Finally, we call a scheduler σ *deterministic*, or *Dirac* if, for each finite path $\pi \in Paths^*$, $\sigma(\pi)$ is a Dirac distribution.

We determine the size of an *IMDP* $\mathcal{M}$ as follows. Let $|S|$ denote the number of states in $\mathcal{M}$; each state has at most $|\mathcal{A}|$ actions and at most $|\mathcal{A}| \cdot |S|$ transitions, each of which is associated with a probability interval. Therefore, the overall size of $\mathcal{M}$ is $|\mathcal{M}| \in \mathcal{O}(|S|^2 \cdot |\mathcal{A}|)$.

2.2   Robust Optimization

Robust optimization is a new approach in mathematical optimization that is concerned about optimization problems in which a certain level of robustness

---

[1]   Here, $\mathcal{B}$ is the standard σ-algebra over $Paths^\omega$ generated from the set of all cylinder sets $\{ Cyl_\pi \mid \pi \in Paths^* \}$. The unique probability measure is obtained by the application of the extension theorem (see, e.g., [11]).

is desirable against *uncertainty* [6,7]. This modelling methodology is integrated with computational tools to treat optimization problems with uncertain data that is only known to be included in some uncertainty set [3,24,37]. This approach has been shown to be very useful in real-world applications that are entirely or to a certain extent affected by uncertainty [8,10]. In this section, we introduce the concept of Uncertain Linear Programming problems (ULPs) and afterwards, we provide an overview of the essential background required for the rest of the paper. We refer the interested reader to [6,10] for a comprehensive reference on robust optimization.

Uncertain Linear Programming (ULPs). Linear Programming (LP) problems are problems that can be described in canonical form as:

$$\text{Min}_{x \in \mathbb{R}^n} \left\{ c^T x : Ax \leq b \right\}$$

where $x \in \mathbb{R}^n$ is the vector of *decision variables*, $c \in \mathbb{R}^n$ is the vector of *coefficients*, $A \in \mathbb{R}^{m \times n}$ is the constant *coefficient matrix* and $b \in \mathbb{R}^m$ is the *right hand side* vector.

The data of an LP problem, i.e., the collection of tuples $[c, A, b]$, in general are not known precisely when the LP encodes a real-world problem. This issue reveals the need for an approach to produce LP solutions which are immune against uncertainty.

Definition 3 (cf. [6,7]). *An Uncertain Linear Program (ULP) is a family*

$$\left\{ \text{Min}_{x \in \mathbb{R}^n} \{ c^T x : Ax \leq b \} \right\}_{[c,A,b] \in \mathcal{Z}} \tag{1}$$

of LP problems $\text{Min}_{x \in \mathbb{R}^n} \{ c^T x : Ax \leq b \}$ *with the same structure (i.e., same number of constraints and variables) in which the data range over a given nonempty compact uncertainty set* $\mathcal{Z} \subset \mathbb{R}^n \times \mathbb{R}^{m \times n} \times \mathbb{R}^m$.

To simplify the notation, we may write $\left\{ \text{Min}\{ c^T x : Ax \leq b \} \right\}_{\mathcal{Z}}$.

In contrast to an usual single LP problem, it is not possible to associate the notions of feasibility/optimal solutions and optimal objective value with a collection of optimization problems like ULPs. In the setting of ULPs, the feasible solutions are solutions which are *robust feasible*. Roughly speaking, feasible solutions are those which satisfy the set of constraints whatever the realization of uncertain data is. More precisely:

Definition 4 (cf. [6,8]). *A vector* $x \in \mathbb{R}^n$ *is* robust feasible *to an ULP with uncertainty set* $\mathcal{Z}$ *if for each* $[c, A, b] \in \mathcal{Z}$, $Ax \leq b$. *Given a robust feasible solution* x, *the* robust value $\hat{z}(x)$ *of the objective function is* $\hat{z}(x) := \sup_{[c,A,b] \in \mathcal{Z}} c^T x$.

After carefully defining the robust feasible/optimal solutions as well as their robust objective value, we can describe the central concept in robust optimization setting that is the *robust counterpart* (RC) of an uncertain LP problem.

Definition 5 (cf. [8]**).** *Given an ULP problem* $\{\text{Min}\{c^T x : Ax \leq b\}\}_\mathscr{Z}$, *the* Robust Counterpart *of ULP is the optimization problem*

$$\text{Min}_{x \in \mathbb{R}^n} \left\{ \hat{z}(x) = \sup_{[c,A,b] \in \mathscr{Z}} \{c^T x : Ax \leq b\} \right\}$$

that seeks for the best possible value of the objective function among all possible robust feasible solutions to the ULP. Furthermore, the optimal solution/value to the robust counterpart is called the robust optimal solution/value to the ULP.

In the robust counterpart (RC) approach, all the variables are *"here and now decisions"*: they must be decided before the realization of unknown data. However, in some cases, some part of the variables are *"wait and see decisions"*, i.e., they might tune themselves to the varying parameters. In the rest of the paper, we call the variables that may depend on the realizations of the uncertain data as *adjustable*, while other variables are called *non-adjustable*. Therefore, we can split the vector x of Eq. (1) from Defnition 3 as $x = (u, v)^T$ where the sub-vectors u and v indicate the non-adjustable and the adjustable variables, respectively.

Adjustable Robust Counterpart. Splitting the decision variable x to the adjustable and non-adjustable variables allows us to rewrite the uncertain LP (1) as the following equivalent form:

$$\{\text{Min}_{u,v}\{c^T u : Uu + Vv \leq b\}\}_{[c,U,V,b] \in \mathscr{Z}} \tag{2}$$

In the above presentation, without loss of generality, we assume that the objective function is normalized with respect to the non-adjustable variables. Moreover, the matrix V is called *recourse matrix* [17] and when it is not uncertain, we call the uncertain LP (2) a *fixed recourse* one. We can now define the RC and the *Adjustable robust counterpart (ARC)* as follows:

$$\text{RC: } \text{Min}_u \{c^T u : \exists v : \forall [U, V, b] \in \mathscr{Z} : Uu + Vv \leq b\}; \tag{3}$$

$$\text{ARC: } \text{Min}_u \{c^T u : \forall [U, V, b] \in \mathscr{Z} : \exists v : Uu + Vv \leq b\}. \tag{4}$$

It is not difficult to see that ARC is less conservative than RC allowing for better optimal values while still having all realizations of the constraints satisfied. The distinction between RC and ARC can be very significant (see, e.g., [6,7]).

The RC of an uncertain LP is a computationally tractable problem in general [8]. On the contrary, this is not the case with ARC. This fact stimulates a very good reason to introduce the notion of *Affinely Adjustable Robust Counterpart (AARC)* of an uncertain LP in which we make a simplification on how the adjustable variables can tune themselves upon the uncertain data. By posing $v = w + W\xi$, we consider an affine dependency between adjustable variables and uncertain parameter. Therefore, the AARC of the uncertain LP (2) reads as:

$$\begin{aligned} &\text{Min}_{u,w,W} \left\{ c^T u : Uu + V(w + W\xi) \leq b, \forall (\xi \equiv [U, V, b] \in \mathscr{Z}) \right\} \\ &\equiv \text{Min}_u \left\{ c^T u : \forall (\xi \equiv [U, V, b] \in \mathscr{Z}) : \exists (w, W) : Uu + V(w + W\xi) \leq b \right\}. \end{aligned} \tag{5}$$

3   Probabilistic Bisimulation for Interval MDPs

This section revisits required main results on probabilistic bisimulation for interval MDPs, as developed in [27]. In the setting of this paper, we consider the notion of probabilistic bisimulation for the cooperative interpretation of interval MDPs. This semantics is very natural in the context of verification of parallel systems with uncertain transition probabilities in which we assume that scheduler and nature are resolved *cooperatively* in the most *adversarial* way: in the game view of the bisimulation, challenging scheduler and nature work together in order to defeat the defender with a transition that can not be matched.

Besides the cooperative behaviour, the choice of a probability distribution respecting the interval constraints can be done either *statically* [31], i.e., at the beginning once for all, or *dynamically* [30,43], i.e., independently at each computation step. In this paper, we focus on the dynamic approach in resolving the stochastic nondeterminism: it is easier to work with algorithmically and can be seen as a relaxation of the static approach that is often intractable [9,14,19,23].

Let $s \longrightarrow \mu_s$ denote a transition from s to μ_s taken cooperatively, i.e., there is a scheduler $\sigma \in \mathfrak{S}$ and a nature $\nu \in \mathfrak{N}$ such that $\mu_s = \sum_{a \in \mathcal{A}} \sigma(s)(a) \cdot \nu(s,a)$. In other words, $s \longrightarrow \mu_s$ if $\mu_s \in \mathrm{CH}(\bigcup_{a \in \mathcal{A}(s)} \mathcal{P}^{s,a})$ where $\mathrm{CH}(X)$ denotes the convex hull of X.

Definition 6 (cf. [27]). *Given an IMDP $\mathcal{M}$, let $\mathcal{R} \subseteq S \times S$ be an equivalence relation. We say that $\mathcal{R}$ is a* probabilistic bisimulation *if for each $(s,t) \in \mathcal{R}$ we have that $L(s) = L(t)$ and for each $s \longrightarrow \mu_s$ there exists $t \longrightarrow \mu_t$ such that $\mu_s \, \mathcal{L}(\mathcal{R}) \, \mu_t$. Furthermore, we write $s \sim_c t$ if there is a probabilistic bisimulation $\mathcal{R}$ such that $(s,t) \in \mathcal{R}$.*

Intuitively, each (cooperative) step of scheduler and nature from state s needs to be matched by a (cooperative) step of scheduler and nature from state t; symmetrically, s also needs to match t. It is shown in [27] that the bisimulation $\sim_c$ preserves the (cooperative) universally quantified PCTL satisfaction $\models_c$.

Theorem 7 (cf. [27]). *For states $s \sim_c t$ and any PCTL formula φ, we have $s \models_c \varphi$ if and only if $t \models_c \varphi$.*

Computation of probabilistic bisimulation for IMDPs follows the standard partition refinement approach [13,32,39]. However, the core part of the algorithm is to find out whether two states "violate the definition of bisimulation". Verification of this violation amounts to checking inclusion of polytopes defined as follows. For $s \in S$ and $a \in \mathcal{A}(s)$, recall that $\mathcal{P}^{s,a}$ denotes the polytope of feasible successor distributions over *states* with respect to taking the action a in the state s. By $\mathcal{P}_{\mathcal{R}}^{s,a}$, we denote the polytope of feasible successor distributions over *equivalence classes* of $\mathcal{R}$ with respect to taking the action a in the state s. Formally, for $\mu \in \Delta(S/\mathcal{R})$ we set $\mu \in \mathcal{P}_{\mathcal{R}}^{s,a}$ if, for each $\mathcal{C} \in S/\mathcal{R}$, we have $\mu(\mathcal{C}) \in I(s,a,\mathcal{C})$. Furthermore, we define $\mathcal{P}_{\mathcal{R}}^s = \mathrm{CH}(\bigcup_{a \in \mathcal{A}(s)} \mathcal{P}_{\mathcal{R}}^{s,a})$. It is the set of feasible successor distributions over $S/\mathcal{R}$ with respect to taking an *arbitrary* distribution over actions in state s. As specified in [27], checking violation of a given pair of states, amounts to check equality of the corresponding constructed poly-topes for the states. As regards the computational complexity of the proposed algorithm, the following theorem indicates that it is fixed parameter tractable.

Theorem 8 (cf. [27]). *Given an IMDP $\mathcal{M}$, let f be the maximal fanout, i.e., $f = \max_{s \in S, a \in \mathcal{A}(s)} |\{ s' \in S \mid I(s,a,s') \neq [0,0] \}|$. Computing $\sim_c$ can be done in time $|\mathcal{M}|^{\mathcal{O}(1)} \cdot 2^{\mathcal{O}(f)}$.*

The exact time complexity of deciding probabilistic bisimulation for *IMDPs* has recently been explored in [28], leading to the following result.

Theorem 9 *Given an IMDP $\mathcal{M}$, computing $\sim_c$ is coNP-complete.*

4   Computational Tractability

Definition 6 is the central definition around which the paper revolves. Given an *IMDP*, the complexity of computing $\sim_c$ strictly depends on finding $t \longrightarrow \mu_t$: we show how a finer (sub-optimal) equivalence relation can be computed in poly-nomial time. The bisimulation in Definition 6 can be reformulated equivalently as follows:

Definition 10. *Let $\mathcal{R} \subseteq S \times S$ be an equivalence relation. We say that $\mathcal{R}$ is a probabilistic bisimulation if $(s,t) \in \mathcal{R}$ implies that $L(s) = L(t)$ and for each $a \in \mathcal{A}(s)$ and each $\mu_s \in \mathcal{P}_{\mathcal{R}}^{s,a}$, there exists $\mu_t \in \mathcal{P}_{\mathcal{R}}^t$ such that $\mu_s \, \mathcal{L}(\mathcal{R}) \, \mu_t$.*

Recall that a probabilistic bisimulation can be seen as a game between two play-ers: in each round, the challenger, or attacker, s proposes a transition, or step, that has to be matched by the defender t. The two states s and t are bisimilar if the defender is always able to match the challenging transitions proposed by the attacker, that is, the game can be played forever. Correspondingly, in our setting, probabilistic bisimulations require that each transition proposed by the challenger s which is selected from the set $\mathcal{P}_{\mathcal{R}}^{s,a}$, is matched by the defender t via a single (combined) transition. The above definition essentially disallows the state s to randomize over the set of its available actions. Therefore, instead of allow-ing the challenger to pick a probability distribution from $\mathrm{CH}(\bigcup_{a \in \mathcal{A}(s)} \mathcal{P}_{\mathcal{R}}^{s,a})$, we restrict his choice to select a distribution for an action from the polytope $\mathcal{P}_{\mathcal{R}}^{s,a}$. This restriction does not lead to any loss of generality, since it is routine to check that the bisimulation $\mathcal{R}$ from Definition 10 satisfies the condition of Definition 6.

4.1   Robust Methodologies for Probabilistic Bisimulation

We now discuss the key elements of a decision algorithm for probabilistic bisim-ulation on *IMDPs*. As we will see in Sect. 5, the core part—and the main source of the exponential complexity of the decision algorithm in [27]—is the need to repeatedly verify the step condition, that is, given a challenging transition $\mu \in \mathcal{P}_{\mathcal{R}}^s$ and $(s,t) \in \mathcal{R}$, to check if there exists $t \longrightarrow \mu_t$ such that $\mu \, \mathcal{L}(\mathcal{R}) \, \mu_t$. We show that, using some inspiration from network flow problems, it is possible to treat a transition $t \longrightarrow \mu_t$ of the *IMDP* $\mathcal{M}$ as a flow where the initial probability mass δ_t flows and splits along transitions appropriately to the transition target distributions and the resolution of the nondeterminism fulfilled by the scheduler and nature. This intuition essentially enables us to model the probabilistic bisimulation problem as an *adjustable robust counterpart* of an uncertain LP problem that is intractable in general [6,7].

4.2   Adjustable Robust Counterpart for Probabilistic Bisimulation

From now on, we assume that the *IMDP* $\mathcal{M}$, the state t, the probability distribution μ, and the equivalence relation $\mathcal{R}$ on S are given. We intend to verify or refute the existence of a transition $t \longrightarrow \mu_t$ of $\mathcal{M}$ satisfying $\mu \, \mathcal{L}(\mathcal{R}) \, \mu_t$ via the construction of a flow through the network graph $G(t, \mathcal{R}) = (V, E)$ defined as follows: the set of vertices is $V = \{\triangle, \blacktriangledown, t\} \cup S_\mathcal{A} \cup S_\mathcal{R} \cup (S/\mathcal{R})$ where $S_\mathcal{A} = \{ t_a \mid a \in \mathcal{A}(t) \}$ and $S_\mathcal{R} = \{ s_\mathcal{R} \mid s \in S \}$, and the set of arcs is $E = \{(\triangle, t)\} \cup \{ (v_\mathcal{R}, \mathcal{C}), (\mathcal{C}, \blacktriangledown) \mid \mathcal{C} \in S/\mathcal{R}, v \in \mathcal{C} \} \cup \{ (t, t_a), (t_a, v_\mathcal{R}) \mid a \in \mathcal{A}(t), v \in S \}$. In the flow network definition, $\triangle$ and $\blacktriangledown$ are the source node and the sink node of the network, respectively. The set of *transition nodes* $S_\mathcal{A}$ includes vertices that represent the interval transitions of the *IMDP* $\mathcal{M}$. More precisely, each transition labelled by a enabled at state t is represented by a transition node $t_a \in S_\mathcal{A}$. The set $S_\mathcal{R}$ is a copy of the state set S that is used to represent the states reached after having performed the transition; for such states, we connect them to the equivalence class they belong to so to verify the condition of the lifting. The network construction can be seen as an adaptation to the strong case of flow networks used in [21, 44].

We take advantage of the above transformation of the "*IMDP* into a network graph" to generate an optimization problem. To this aim, we adopt the same notation of the network optimization setting so we use $f_{u,v}$ to show the "flow" through the arc from u to v. In formulating the optimization problem, we use in addition the so-called *balancing constraints* [44] in order to reflect the probabilistic choices in the given *IMDP* $\mathcal{M}$ and to ensure the correct splitting of outgoing flows from the transition nodes in the set $S_\mathcal{A}$.

Definition 11. *The optimization problem associated to the network* $G(t, \mathcal{R}) = (V, E)$ *is defined as follows:*

$$\text{Min}_f \quad 0$$
subject to
$$f_{u,v} \geq 0 \qquad\qquad\qquad\qquad \textit{for each } (u, v) \in E$$
$$f_{\triangle,t} = 1$$
$$f_{\mathcal{C},\blacktriangledown} = \mu(\mathcal{C}) \qquad\qquad\qquad\qquad \textit{for each } \mathcal{C} \in S/\mathcal{R}$$
$$\sum_{\{ u \in V \mid (u,v) \in E \}} f_{u,v} - \sum_{\{ w \in V \mid (v,w) \in E \}} f_{v,w} = 0 \qquad \textit{for each } v \in V \setminus \{\triangle, \blacktriangledown\}$$
$$f_{t_a, v_\mathcal{R}} - p_{a,v} \cdot f_{t, t_a} = 0 \qquad\qquad \textit{for each } a \in \mathcal{A}(t) \textit{ and } v \in S$$
$$p_{a,v} \in I(t, a, v) \qquad\qquad\qquad \textit{for each } a \in \mathcal{A}(t) \textit{ and } v \in S$$

It is not difficult to see that the optimization problem just defined is not an LP problem, as there are quadratic constraints where the flow variable f_{t,t_a} is multiplied with the "probability" variable $p_{a,v}$. As a matter of fact, for a given $a \in \mathcal{A}(t)$, the variables $p_{a,v}$ have to lie in the interval defined by the interval transition $I(t,a,v)$ and they have to induce a probability distribution, i.e., $p_{a,v} \geq 0$ for each $v \in S$ and $\sum_{v \in S} p_{a,v} = 1$. The non-negativity of the variables comes for free from the constraints $p_{a,v} \in I(t,a,v)$ since $I(t,a,v) \subseteq [0,1]$; $\sum_{v \in S} p_{a,v} = 1$ follows by the flow conservation constrain $\sum_{\{u \in V | (u,v) \in E\}} f_{u,v} - \sum_{\{w \in V | (v,w) \in E\}} f_{v,w} = 0$ for $v = t_a$. Therefore, the optimization problem can be easily cast as an LP problem by replacing the pair of constraints $f_{t_a, v_{\mathcal{R}}} - p_{a,v} \cdot f_{t,t_a} = 0$ and $p_{a,v} \in I(t,a,v)$ with the pair of constraints $f_{t_a, v_{\mathcal{R}}} - \inf I(t,a,v) \cdot f_{t,t_a} \geq 0$ and $f_{t_a, v_{\mathcal{R}}} - \sup I(t,a,v) \cdot f_{t,t_a} \leq 0$, i.e., the state v is reached from t with probability $p_{a,v} = \frac{f_{t_a, v_{\mathcal{R}}}}{f_{t,t_a}}$ at least $\inf I(t,a,v)$ and at most $\sup I(t,a,v)$, as required. Taking this modification into account, we can reformulate the optimization problem in Definition 11 as the following LP problem.

Definition 12 (The $LP(t,\mu,\mathcal{R})$ LP problem). *The $LP(t,\mu,\mathcal{R})$ LP problem associated to the network graph $G(t,\mathcal{R}) = (V,E)$ is defined as follows:*

$$\text{Min}_f \quad 0$$
subject to

$f_{u,v} \geq 0$	*for each $(u,v) \in E$*
$f_{\triangle,t} = 1$	
$f_{\mathcal{C},\blacktriangledown} = \mu(\mathcal{C})$	*for each $\mathcal{C} \in S/\mathcal{R}$*
$\sum_{\{u \in V | (u,v) \in E\}} f_{u,v} - \sum_{\{w \in V | (v,w) \in E\}} f_{v,w} = 0$	*for each $v \in V \setminus \{\triangle, \blacktriangledown\}$*
$f_{t_a, v_{\mathcal{R}}} - \inf I(t,a,v) \cdot f_{t,t_a} \geq 0$	*for each $a \in \mathcal{A}(t)$ and $v \in S$*
$f_{t_a, v_{\mathcal{R}}} - \sup I(t,a,v) \cdot f_{t,t_a} \leq 0$	*for each $a \in \mathcal{A}(t)$ and $v \in S$*

The feasibility of the resulting LP problem can be seen as an oracle to verify or refute the existence of a probabilistic transition $t \longrightarrow \mu_t$. Formally,

Lemma 13. *Given an IMDP $\mathcal{M}$, $t \in S$, $\mu \in \Delta(S)$, and an equivalence relation $\mathcal{R}$ on S, the $LP(t,\mu,\mathcal{R})$ LP problem has a feasible solution if and only if there exist $\sigma \in \mathfrak{S}$ and $\nu \in \mathfrak{N}$ inducing $t \longrightarrow \mu_t$ such that $\mu \, \mathcal{L}(\mathcal{R}) \, \mu_t$.*

It is worthwhile to be noted that the resulting scheduler and nature are history-independent, i.e., they base their choice only on the current state (and action, for nature). Moreover, solving the generated LP problem from Definition 11 can be done in polynomial time [33,34]. The polynomial time complexity, however, is not preserved when uncertainty affects transition probabilities in the model. In fact, in presence of uncertainty, the step condition needs to be checked for any realization of the probability distribution $\mu_s \in \mathcal{P}_{\mathcal{R}}^{s,a}$. This fact is essentially the main barrier in designing efficient algorithms for probabilistic bisimulation on such uncertain systems which particularly leads the problem to be intractable. To this end, we first model the probabilistic bisimulation problem as the ARC of the uncertain $LP(t,\mu,\mathcal{R})$ LP problem in which the uncertain data is the probability distribution μ. More precisely, by Lemma 13, we can replace in Definition 10 the matching transition $\mu_t \in \mathcal{P}_{\mathcal{R}}^t$ for $\mu_s \in \mathcal{P}_{\mathcal{R}}^{s,a}$ such that $\mu_s \, \mathcal{L}(\mathcal{R}) \, \mu_t$ with the check for feasibility of $LP(t,\mu_s,\mathcal{R})$.

Modelling this probabilistic bisimulation game as ARC of an uncertain LP allows the adjustable flow variables $f_{i,j}$ in the $LP(t,\mu,\mathcal{R})$ LP problem to tune

$\text{Min}_{l,w} \quad 0$

subject to

$$l_{u,v} + \sum_{k=1}^{n} w^k \cdot \mu(\mathcal{C}_k) \geq 0 \qquad \text{for each } (u,v) \in E$$

$$l_{\Delta,t} + \sum_{k=1}^{n} w^k \cdot \mu(\mathcal{C}_k) = 1$$

$$l_{\mathcal{C},\blacktriangledown} + \sum_{k=1}^{n} w^k \cdot \mu(\mathcal{C}_k) = \mu(\mathcal{C}_i) \qquad \text{for each } \mathcal{C}_i \in S/\mathcal{R}, i = 1, \ldots, n$$

$$\sum_{\{u | (u,v) \in E\}} (l_{u,v} + \sum_{k=1}^{n} w^k \cdot \mu(\mathcal{C}_k)) - \sum_{\{u | (v,u) \in E\}} (l_{v,u} + \sum_{k=1}^{n} w^k \cdot \mu(\mathcal{C}_k)) = 0$$

$$\text{for each } v \in V \setminus \{\Delta, \blacktriangledown\}$$

$$l_{t_a,v_{\mathcal{R}}} + \sum_{k=1}^{n} w^k \cdot \mu(\mathcal{C}_k) - \inf I(t,a,v) \cdot (l_{t,t_a} + \sum_{k=1}^{n} w^k \cdot \mu(\mathcal{C}_k)) \geq 0$$

$$\text{for each } a \in \mathcal{A}(t) \text{ and } v \in S$$

$$l_{t_a,v_{\mathcal{R}}} + \sum_{k=1}^{n} w^k \cdot \mu(\mathcal{C}_k) - \sup I(t,a,v) \cdot (l_{t,t_a} + \sum_{k=1}^{n} w^k \cdot \mu(\mathcal{C}_k)) \leq 0$$

$$\text{for each } a \in \mathcal{A}(t) \text{ and } v \in S$$

$$\forall \mu = (\mu(\mathcal{C}_1), \ldots, \mu(\mathcal{C}_n)) \in \mathcal{P}_{\mathcal{R}}^{s,a}$$

Fig. 1. Affinely adjustable robust counterpart of the ULP $\{LP(t, \mu, \mathcal{R})\}_{\mu \in \mathcal{P}_{\mathcal{R}}^{s,a}}$.

themselves to the uncertain probability distribution μ. However, the ARC is in general computationally hard. On the other hand, restricting the adjustable flow variables $f_{i,j}$ to be affinely dependent on the uncertain probability distributions μ allows us to model the bisimulation problem as *affinely adjustable robust counterpart* of an uncertain LP problem and thus to arrive at a polynomial time algorithm to compute the equivalence relation $\mathcal{R}$. From the game semantics viewpoint, such affine dependency restriction reduces the power of the defender to match the challenger's choices and therefore, it leads to a finer (sub-optimal) equivalence relation.

4.3 Affinely Adjustable Robust Counterpart for Probabilistic Bisimulation

In this section, we adapt the ARC theory presented in Sect. 2.2 to the setting of probabilistic bisimulation by imposing a restriction on adjustable flow variables $f_{i,j}$ to tune themselves *affinely* upon the uncertain probability distribution μ in the challenger's uncertainty set $\mathcal{P}_{\mathcal{R}}^{s,a}$. Without loss of generality, we let $\mathcal{C}_1, \ldots, \mathcal{C}_n$ be the equivalence classes induced by $\mathcal{R}$. We encode the *affine dependence* in the network graph $G(t, \mathcal{R}) = (V, E)$ by restricting, for each arch $(i, j) \in E$, the flow variable $f_{i,j}$ to be

$$f_{i,j} = l_{i,j} + \sum_{k=1}^{n} w^k \cdot \mu(\mathcal{C}_k),$$

where the new optimization variables are considered in the vector l and the matrix W. Plugging affine equivalences of flow variables, we end up with the affinely adjustable robust counterpart (AARC) of the ULP problem $\{LP(t, \mu, \mathcal{R})\}_{\mu \in \mathcal{P}_{\mathcal{R}}^{s,a}}$ shown in Fig. 1.

In order to show the computational tractability of the AARC, we need to ensure that the uncertainty set $\mathcal{P}_{\mathcal{R}}^{s,a}$ is itself computationally tractable. Formally, a set $\mathcal{P}_{\mathcal{R}}^{s,a}$ is *computationally tractable* [25] if for any vector μ, there is a tractable "separation oracle" that either decides correctly $\mu \in \mathcal{P}_{\mathcal{R}}^{s,a}$ or otherwise, generates a *separator*, i.e., a non-zero vector r such that $r^T \mu \geq \max_{\gamma \in \mathcal{P}_{\mathcal{R}}^{s,a}} r^T \gamma$.

Proposition 14. *For every state $s \in S$, action $a \in \mathcal{A}(s)$ and equivalence relation $\mathcal{R}$, the polytopic uncertainty set $\mathcal{P}_{\mathcal{R}}^{s,a}$ is computationally tractable.*

Computational tractability of the polytopic uncertainty sets concludes immediately tractability of the AARC. Formally,

Theorem 15. *Given the fixed recourse ULP problem $\{LP(t, \mu, \mathcal{R})\}_{\mu \in \mathcal{P}_{\mathcal{R}}^{s,a}}$, the AARC of $\{LP(t, \mu, \mathcal{R})\}_{\mu \in \mathcal{P}_{\mathcal{R}}^{s,a}}$ is computationally tractable.*

It is not difficult to see that in the setting of probabilistic bisimulation, the polytopic uncertainty sets $\mathcal{P}_{\mathcal{R}}^s$ are closed, convex, and *well structured*, i.e., they can be described by a list of linear inequalities. Thus in our setting, the resulting AARC is also well structured and thus can be solved using highly efficient LP solvers (for instance, CPLEX [2] and Gurobi [1]) even for large-scale cases.

Theorem 16. *Given the fixed recourse ULP problem $\{LP(t, \mu, \mathcal{R})\}_{\mu \in \mathcal{P}_{\mathcal{R}}^{s,a}}$, the AARC of $\{LP(t, \mu, \mathcal{R})\}_{\mu \in \mathcal{P}_{\mathcal{R}}^{s,a}}$ is equivalent to an explicit LP program.*

The "affine decision rules" used to derive the AARC counterpart of the probabilistic bisimulation problem allow us to compute a sub-optimal (finer) probabilistic bisimulation defined as follows.

Definition 17. *Let $\mathcal{R} \subseteq S \times S$ be an equivalence relation. We say that $\mathcal{R}$ is an AARC probabilistic bisimulation if $(s, t) \in \mathcal{R}$ implies that $L(s) = L(t)$ and for each $a \in \mathcal{A}(s)$, the AARC of the ULP problem $\{LP(t, \mu, \mathcal{R})\}_{\mu \in \mathcal{P}_{\mathcal{R}}^{s,a}}$ is feasible.*

Furthermore, we write $s \sim_{AARC} t$ if there exists an AARC probabilistic bisimulation $\mathcal{R}$ such that $(s, t) \in \mathcal{R}$.

An immediate result relating $\sim_{AARC}$ and $\sim_c$ is that the former is a refinement of the latter, as formalized by the following proposition.

Proposition 18. *Given $\mathcal{M}$, if $s \sim_{AARC} t$, then $s \sim_c t$, i.e., $\sim_{AARC} \subseteq \sim_c$.*

5 Decision Algorithm

In this section, we give a polynomial algorithm computing the probabilistic bisimulation $\sim_{AARC}$. The general idea of the algorithm follows the one of the algorithm in [27] and involves the construction of the polytopes of the challenger's probability distributions. In order to compute $\sim_{AARC}$ an *IMDP* $\mathcal{M} = (S, \bar{s}, \mathcal{A}, \text{AP}, L, I)$, we follow the usual partition refinement approach [12,21,32,39,44], formalized by the BISIMULATION procedure in Algorithm 1. Namely, we start with $\mathcal{R}$ being the equivalence relation containing the pairs of states with the same labels; then we iteratively refine $\mathcal{R}$ by splitting the states that violate the definition of bisimulation with respect to $\mathcal{R}$. The core part is to check whether two states "violate the definition of bisimulation". This is where our algorithm differs from the one proposed in [27].

```
┌─────────────────────────────────────────┐
│          BISIMULATION(M)                 │
└─────────────────────────────────────────┘
```

1: $\mathcal{R} \leftarrow \{ (s,t) \in S \times S \mid L(s) = L(t) \}$;
2: **repeat**
3:     $\mathcal{R}' \leftarrow \mathcal{R}$;
4:     **for all** $s \in S$ **do**
5:         $D \leftarrow \emptyset$;
6:         **for all** $t \in [s]_\mathcal{R}$ **do**
7:             **for all** $a \in \mathcal{A}(s)$ **do**
8:                 **if** VIOLATE$(t, \mathcal{R}, \mathcal{P}_\mathcal{R}^{s,a})$
9:                     $D \leftarrow D \cup \{t\}$;
10:         split $[s]_\mathcal{R}$ in $\mathcal{R}$ into D and $[s]_\mathcal{R} \setminus D$;
11: **until** $\mathcal{R} = \mathcal{R}'$;
12: **return** $\mathcal{R}$;

```
┌─────────────────────────────────────────┐
│       VIOLATE(t, R, P_R^{s,a})           │
└─────────────────────────────────────────┘
```

1: Construct the AARC of the ULP $\{LP(t, \mu, \mathcal{R})\}_{\mu \in \mathcal{P}_\mathcal{R}^{s,a}}$ defined in Fig. 1
2: **return** is AARC not feasible?

Algorithm 1: Probabilistic bisimulation algorithm for *IMDP*s

The violation is checked by the procedure VIOLATE. We show that this amounts in solving the AARC of the uncertain LP problem $\{LP(t, \mu, \mathcal{R})\}_{\mu \in \mathcal{P}_\mathcal{R}^{s,a}}$ as follows. Recall that for $s \in S$ and an action $a \in \mathcal{A}(s)$, we denote by $\mathcal{P}_\mathcal{R}^{s,a}$ the polytope of feasible successor distributions over *equivalence classes* of $\mathcal{R}$ with respect to taking the action a in the state s, as discussed in Sect. 3. Note that we require that the probability of each class $\mathcal{C}$ must be in the interval of the sum of probabilities that can be assigned to states of $\mathcal{C}$. As specified in the procedure VIOLATE, we show that it suffices to check the feasibility of the resulting AARC of the constructed uncertain LP problem.

Given an *IMDP* $\mathcal{M}$, let $N = \max\{|S|, |A|\}$. It is not difficult to see that the procedure VIOLATE is called at most N^4 times. In every call to this procedure, we need to generate and solve the explicit form of the AARC which is an LP according to Theorem 16, solvable in polynomial time $\mathcal{O}(poly(N))$ (see, e.g., [25,33]). This means that computing $\sim_{AARC}$ can be done in time $|\mathcal{M}|^{\mathcal{O}(1)} \cdot \mathcal{O}(poly(N))$.

Theorem 19. *Algorithm 1 computes $\sim_{AARC}$ in time polynomial in $|\mathcal{M}|$.*

6   Case Studies

We have written a prototypical implementation for computing the bisimulation presented in this paper. Our tool reads a model specification in the input language of the probabilistic model checker PRISM [36] (extended to support also intervals in the transitions), and constructs an explicit-state representation of the state space. Afterwards, it computes the quotient using Algorithm 1.

Table 1 shows the performance of our prototype on a number of case studies taken from the PRISM website [4], where we have relaxed some of the probabilistic choices to intervals. The machine we used for the experiments is a 3.6 GHz Intel Core i7-4790 with 16 GB 1600 MHz DDR3 RAM of which 12 GB were assigned to the tool. Despite using an explicit representation for the model, the prototype is able to manage cases studies in the order of millions of states and

Table 1. Experimental evaluation of the bisimulation computation

Model	$|S_i|$	$|I_i|$	S/L	$t_\sim$ (s)	$|S_\sim|$	$|I_\sim|$
Consensus-Shared-Coin-3	5 216	13 380	2	0	787	1 770
Consensus-Shared-Coin-4	43 136	144 352	2	2	2 189	5 621
Consensus-Shared-Coin-5	327 936	1 363 120	2	23	5 025	14 192
Consensus-Shared-Coin-6	2 376 448	11 835 456	2	219	10 173	30 861
Crowds-5-10	111 294	261 444	2	1	107	153
Crowds-5-20	2 061 951	7 374 951	2	17	107	153
Crowds-5-30	12 816 233	61 511 033	2	116	107	153
Crowds-5-40	44 045 030	266 812 421	2	464	125	198
Mutual-Exclusion-PZ-3	3 008	10 868	2	0	1 123	3 939
Mutual-Exclusion-PZ-4	48 128	231 040	2	0	7 319	32 630
Mutual-Exclusion-PZ-5	770 048	4 611 072	2	7	32 053	168 151
Mutual-Exclusion-PZ-6	3 377 344	25 470 144	2	98	109 986	649 360
Dining-Phils-LR-nofair-4	9 440	40 120	4	0	1 232	5 037
Dining-Phils-LR-nofair-5	93 068	494 420	4	1	9 408	49 467
Dining-Phils-LR-nofair-6	917 424	5 848 524	4	14	76 925	487 620
Dining-Phils-LR-nofair-7	9 043 420	67 259 808	4	173	646 928	4 804 695

transitions (columns "Model", "$|S_i|$", and "$|I_i|$"). The time in seconds required to compute the bisimulation relation and the corresponding quotient $IMDP$, shown in columns "$t_\sim$", "$|S_\sim|$", and "$|I_\sim|$", is much less than the time expected from the theoretical analysis of the algorithm: this is motivated by the fact that we have implemented optimizations, such as caching equivalent LP problems, which improve the runtime of our algorithm in practice. Because of this, we never had to solve more than 30 LP problems in a single tool run, thereby avoiding the potentially costly solution of LP problems from becoming a bottleneck.

Acknowledgments. We would like to thank Arkadi Nemirovski (Georgia Institute of Technology) and Daniel Kuhn (EPFL) for many invaluable and insightful discussions. This work is supported by the EU 7th Framework Programme under grant agreements 295261 (MEALS) and 318490 (SENSATION), by the DFG as part of SFB/TR 14 AVACS, by the ERC Advanced Investigators Grant 695614 (POWVER), by the CAS/SAFEA International Partnership Program for Creative Research Teams, by the National Natural Science Foundation of China (Grants No. 61472473, 61532019, 61550110249, 61550110506), by the Chinese Academy of Sciences Fellowship for International Young Scientists, and by the CDZ project CAP (GZ 1023).

References

1. Gurobi 4.0.2. http://www.gurobi.com/
2. IBM ILOG CPLEX Optimizer. http://www.ibm.com/software/commerce/optimization/cplex-optimizer/
3. PICOS: A Python interface for conic optimization solvers. http://picos.zib.de/
4. PRISM model checker. http://www.prismmodelchecker.org/
5. Abate, A., El Ghaoui, L.: Robust model predictive control through adjustable variables: an application to path planning. In: CDC, pp. 2485–2490 (2004)
6. Ben-Tal, A., El Ghaoui, L., Nemirovski, A.: Robust Optimization. Princeton University Press, Princeton (2009)
7. Ben-Tal, A., Goryashko, A., Guslitzer, E., Nemirovski, A.: Adjustable robust solutions of uncertain linear programs. Math. Program. **99**(2), 351–376 (2004)
8. Ben-Tal, A., Nemirovski, A.: Robust solutions of uncertain linear programs. Oper. Res. Lett. **25**, 1–13 (1999)
9. Benedikt, M., Lenhardt, R., Worrell, J.: LTL model checking of interval Markov chains. In: Piterman, N., Smolka, S.A. (eds.) TACAS 2013 (ETAPS 2013). LNCS, vol. 7795, pp. 32–46. Springer, Heidelberg (2013)
10. Bertsimas, D., Brown, D.B., Caramanis, C.: Theory and applications of robust optimization. SIAM Rev. **53**(3), 464–501 (2011)
11. Billingsley, P.: Probability and Measure. Wiley, New York (1979)
12. Cattani, S., Segala, R.: Decision algorithms for probabilistic bisimulation. In: Brim, L., Jančar, P., Křetínský, M., Kučera, A. (eds.) CONCUR 2002. LNCS, vol. 2421, pp. 371–385. Springer, Heidelberg (2002)
13. Cattani, S., Segala, R., Kwiatkowska, M., Norman, G.: Stochastic transition systems for continuous state spaces and non-determinism. In: Sassone, V. (ed.) FOSSACS 2005. LNCS, vol. 3441, pp. 125–139. Springer, Heidelberg (2005)
14. Chatterjee, K., Sen, K., Henzinger, T.A.: Model-checking ω-regular properties of interval Markov chains. In: Amadio, R.M. (ed.) FOSSACS 2008. LNCS, vol. 4962, pp. 302–317. Springer, Heidelberg (2008)
15. Chen, T., Han, T., Kwiatkowska, M.: On the complexity of model checking interval-valued discrete time Markov chains. Inf. Process. Lett. **113**(7), 210–216 (2013)
16. Coste, N., Hermanns, H., Lantreibecq, E., Serwe, W.: Towards performance prediction of compositional models in industrial GALS designs. In: Bouajjani, A., Maler, O. (eds.) CAV 2009. LNCS, vol. 5643, pp. 204–218. Springer, Heidelberg (2009)
17. Dantzig, G.B., Madansky, A.: On the solution of two-stage linear programs under uncertainty. In: Proceedings of the Fourth Berkeley Symposium on Mathematical Statistics and Probability, vol. 1. pp. 165–176 (1961)
18. Delahaye, B., Katoen, J.P., Larsen, K.G., Legay, A., Pedersen, M.L., Sher, F., Wasowski, A.: New results on abstract probabilistic automata. In: ACSD, pp. 118–127 (2011)
19. Delahaye, B., Larsen, K.G., Legay, A., Pedersen, M.L., Wąsowski, A.: Decision problems for interval Markov chains. In: Dediu, A.-H., Inenaga, S., Martín-Vide, C. (eds.) LATA 2011. LNCS, vol. 6638, pp. 274–285. Springer, Heidelberg (2011)
20. Fecher, H., Leucker, M., Wolf, V.: *Don't know* in probabilistic systems. In: Valmari, A. (ed.) SPIN 2006. LNCS, vol. 3925, pp. 71–88. Springer, Heidelberg (2006)
21. Ferrer Fioriti, L.M., Hashemi, V., Hermanns, H., Turrini, A.: Deciding probabilistic automata weak bisimulation: theory and practice. Form. Asp. Comput. **28**(1), 109–143 (2016)

22. Gebler, D., Hashemi, V., Turrini, A.: Computing behavioral relations for probabilistic concurrent systems. In: Remke, A., Stoelinga, M. (eds.) Stochastic Model Checking. LNCS, vol. 8453, pp. 117–155. Springer, Heidelberg (2014)

23. Givan, R., Leach, S.M., Dean, T.L.: Bounded-parameter Markov decision processes. Artif. Intell. **122**(1–2), 71–109 (2000)

24. Goerigk, M.: ROPI–a robust optimization programming interface for C++. Optim. Methods Softw. **29**(6), 1261–1280 (2014)

25. Grötschel, M., Lovász, L., Schrijver, A.: The ellipsoid method and its consequences in combinatorial optimization. Combinatorica **1**(2), 169–197 (1981)

26. Hahn, E.M., Han, T., Zhang, L.: Synthesis for PCTL in parametric Markov decision processes. In: Bobaru, M., Havelund, K., Holzmann, G.J., Joshi, R. (eds.) NFM 2011. LNCS, vol. 6617, pp. 146–161. Springer, Heidelberg (2011)

27. Hashemi, V., Hatefi, H., Krčál, J.: Probabilistic bisimulations for PCTL model checking of interval MDPs (extended version). In: SynCoP. EPTCS, vol. 145, pp. 19–33 (2014)

28. Hashemi, V., Hermanns, H., Song, L., Subramani, K., Turrini, A., Wojciechowski, P.: Compositional bisimulation minimization for interval Markov decision processes. In: Dediu, A.-H., Janoušek, J., Martín-Vide, C., Truthe, B. (eds.) LATA 2016. LNCS, vol. 9618, pp. 114–126. Springer, Heidelberg (2016). doi:10.1007/978-3-319-30000-9_9

29. Hermanns, H., Katoen, J.P.: Automated compositional Markov chain generation for a plain-old telephone system. Sci. Comput. Program. **36**(1), 97–127 (2000)

30. Iyengar, G.N.: Robust dynamic programming. Math. Oper. Res. **30**(2), 257–280 (2005)

31. Jonsson, B., Larsen, K.G.: Specification and refinement of probabilistic processes. In: LICS, pp. 266–277 (1991)

32. Kanellakis, P.C., Smolka, S.A.: CCS expressions, finite state processes, and three problems of equivalence. Inf. Comput. **86**(1), 43–68 (1990)

33. Karmarkar, N.: A new polynomial-time algorithm for linear programming. Combinatorica **4**(4), 373–395 (1984)

34. Khachyan, L.G.: A polynomial algorithm in linear programming. Sov. Math. Doklady **20**(1), 191–194 (1979)

35. Kozine, I., Utkin, L.V.: Interval-valued finite Markov chains. Reliable Comput. **8**(2), 97–113 (2002)

36. Kwiatkowska, M., Norman, G., Parker, D.: PRISM 4.0: verification of probabilistic real-time systems. In: Gopalakrishnan, G., Qadeer, S. (eds.) CAV 2011. LNCS, vol. 6806, pp. 585–591. Springer, Heidelberg (2011)

37. Löfberg, J.: Automatic robust convex programming. Optim. Methods Softw. **17**(1), 115–129 (2012)

38. Nilim, A., El Ghaoui, L.: Robust control of Markov decision processes with uncertain transition matrices. Oper. Res. **53**(5), 780–798 (2005)

39. Paige, R., Tarjan, R.E.: Three partition refinement algorithms. SIAM J. Comput. **16**(6), 973–989 (1987)

40. Puggelli, A.: Formal techniques for the verification and optimal control of probabilistic systems in the presence of modeling uncertainties. Ph.D. thesis, University of California, Berkeley (2014)

41. Puggelli, A., Li, W., Sangiovanni-Vincentelli, A.L., Seshia, S.A.: Polynomial-time verification of PCTL properties of MDPs with convex uncertainties. In: Sharygina, N., Veith, H. (eds.) CAV 2013. LNCS, vol. 8044, pp. 527–542. Springer, Heidelberg (2013)

42. Segala, R.: Modeling and verification of randomized distributed real-time systems. Ph.D. thesis, MIT (1995)
43. Sen, K., Viswanathan, M., Agha, G.: Model-checking Markov chains in the presence of uncertainties. In: Hermanns, H., Palsberg, J. (eds.) TACAS 2006. LNCS, vol. 3920, pp. 394–410. Springer, Heidelberg (2006)
44. Turrini, A., Hermanns, H.: Polynomial time decision algorithms for probabilistic automata. Inf. Comput. **244**, 134–171 (2015)
45. Wolff, E.M., Topcu, U., Murray, R.M.: Robust control of uncertain Markov decision processes with temporal logic specifications. In: CDC, pp. 3372–3379 (2012)
46. Wu, D., Koutsoukos, X.D.: Reachability analysis of uncertain systems using bounded-parameter Markov decision processes. Artif. Intell. **172**(8–9), 945–954 (2008)
47. Yi, W.: Algebraic reasoning for real-time probabilistic processes with uncertain information. Formal Techniques in Real-Time and Fault-Tolerant Systems. LNCS, vol. 863, pp. 680–693. Springer, Heidelberg (1994)

Approximation of Probabilistic Reachability for Chemical Reaction Networks Using the Linear Noise Approximation

Luca Bortolussi[3], Luca Cardelli[1,2], Marta Kwiatkowska[2], and Luca Laurenti[2(✉)]

[1] Microsoft Research, Cambridge, UK
[2] Department of Computer Science, University of Oxford, Oxford, UK
luca.laurenti@cs.ox.ac.uk
[3] Department of Mathematics and Geosciences, University of Trieste, Trieste, Italy

Abstract. We study time-bounded probabilistic reachability for Chemical Reaction Networks (CRNs) using the Linear Noise Approximation (LNA). The LNA approximates the discrete stochastic semantics of a CRN in terms of a continuous space Gaussian process. We consider reachability regions expressed as intersections of finitely many linear inequalities over the species of a CRN. This restriction allows us to derive an abstraction of the original Gaussian process as a time-inhomogeneous discrete-time Markov chain (DTMC), such that the dimensionality of its state space is independent of the number of species of the CRN, ameliorating the state space explosion problem. We formulate an algorithm for approximate computation of time-bounded reachability probabilities on the resulting DTMC and show how to extend it to more complex temporal properties. We implement the algorithm and demonstrate on two case studies that it permits fast and scalable computation of reachability properties with controlled accuracy.

1 Introduction

It is well known that a biochemical system evolving in a spatially homogeneous environment, at constant volume and temperature, can be modelled as a continuous-time Markov chain (CTMC) [18]. Stochastic modelling is necessary to describe stochastic fluctuations for low molecular counts [14,16], when deterministic models are not accurate [15]. Computing the probability distributions of the species over time is achieved by solving the Chemical Master Equation (CME) [25]. Unfortunately, numerical solution methods based on uniformisation [4] are often infeasible because of the state space explosion problem. A more scalable transient analysis can be achieved by employing statistical model checking based on the Stochastic Simulation Algorithm (SSA) [17], but to obtain good

This research is supported by a Royal Society Research Professorship and ERC AdG VERIWARE. LB is supported by EU-FET project QUANTICOL (nr 600708).

© Springer International Publishing Switzerland 2016
G. Agha and B. Van Houdt (Eds.): QEST 2016, LNCS 9826, pp. 72–88, 2016.
DOI: 10.1007/978-3-319-43425-4_5

accuracy large numbers of simulations are needed, which for some systems can be very time consuming.

A promising approach is to instead approximate the CTMC induced by a biochemical system as a *continuous state space* stochastic process by means of the *Linear Noise Approximation (LNA)*, a Gaussian process derived as an approximation of the CME [25]. Its solution requires solving a number of differential equations that is quadratic in the number of species and independent of the molecular population. As a consequence, the LNA is generally much more scalable than a discrete state stochastic representation and has been successfully used for model checking of large biochemical systems [7,12]. However, none of these approaches enables the computation of global *probabilistic reachability* properties, that is, the probability of reaching a particular region of the state space in a particular time interval. This property is important not only to analyse biochemical systems, for example to quantify the probability that a particular protein or gene is ever expressed in Gene Regulatory Networks, but is also fundamental for the verification of more complex temporal logic properties, since model checking for CSL [2] or LTL [24] is reduced to the computation of reachability probabilities.

Contributions. We derive an algorithm to compute a fast and scalable approximation of probabilistic reachability using the LNA, where the target region of the state space is given by a polytope, i.e. an intersection of a set of linear inequalities over the species of a CRN. More specifically, we compute the probability that the system falls in the target region during a specified time interval. Given a set of k linear inequalities, and relying on the fact that a linear combination of the components of a Gaussian distribution is still Gaussian, we discretize time and space for the k-dimensional stochastic process defined by the particular linear combinations. This permits the derivation of an abstraction in terms of a time-inhomogeneous *discrete-time Markov chain* (DTMC), whose dimension is independent of the number of species, since a linear combination is always uni-dimensional, and ensures scalability, as in general we are interested in one or at most two linear inequalities. This abstraction can then be used for model checking of complex temporal properties [2,4,21]. In order to compute such an abstraction, the most delicate aspect is to derive equations for the transition kernel of the resulting DTMC. This is given by the conditional probability at the next discrete time step given the system in a particular state. Reachability probabilities are then computed by making the target set absorbing. We use our algorithm to extend the Stochastic Evolution Logic (SEL) introduced in [12] to enable model checking of probabilistic reachability of linear combinations of the species of a CRN. We show the effectiveness of our approach on two case studies, also in cases where existing numerical model checking techniques are infeasible.

Related Work. Algorithms to compute the reachability probabilities over discrete state space Markov processes are well understood [4]. They require computation of transient probabilities in a modified Markov chain, where states in the target region are made absorbing. Unfortunately, their practical use is severely hindered by state space explosion, which in a CRN grows exponentially with the number of molecules when finite, and may be infinite, in which case finite projection methods have to be used [23]. As a consequence, approximate but faster algorithms are appealing, in particular for CRNs, where it is not necessary to provide certified guarantees on reachability probabilities. The mainstream solution is to rely on simulations combined with statistical inference to obtain estimates [9]. These methods, however, are still computationally expensive. A recent trend of works explored as an alternative whether estimates could be obtained by relying on approximations of the stochastic process based on mean-field [6] or linear noise [7,8,12]. However, reachability properties, like those considered here, are very challenging. In fact, most approaches consider either local properties of individual molecules [6], or properties obtained by observing the behaviour of individual molecules and restricting the target region to an absorbing subspace of the (modified) model [7]. The only approach dealing with more general subsets, [8], imposes restrictions on the behaviour of the mean-field approximation, whose trajectory has to enter the reachability region in a finite time.

Our approach differs in that it is based on the LNA and considers regions defined by polytopes, which encompasses most properties of practical interest. The simplest idea would be to consider the LNA and compute reachability probabilities for this stochastic process, invoking convergence theorems for the LNA to prove the asymptotic correctness. Unfortunately, there is no straightforward way to do this, since dealing with a continuous space and continuous time diffusion process, e.g., Gaussian, is computationally hard, and computing reachability is challenging (see [10]). As a consequence, discrete abstractions are appealing.

2   Background

Chemical Reaction Networks. A *chemical reaction network (CRN)* $C = (\Lambda, R)$ is a pair of finite sets, where Λ is a set of *chemical species*, $|\Lambda|$ denotes its size, and R is a set of reactions. Species in Λ interact according to the reactions in R. A *reaction* $\tau \in R$ is a triple $\tau = (r_\tau, p_\tau, k_\tau)$, where $r_\tau \in \mathbb{N}^{|\Lambda|}$ is the *reactant complex*, $p_\tau \in \mathbb{N}^{|\Lambda|}$ is the *product complex* and $k_\tau \in \mathbb{R}_{>0}$ is the coefficient associated with the rate of the reaction. r_τ and p_τ represent the stoichiometry of reactants and products. Given a reaction $\tau_1 = ([1,1,0]^T, [0,0,2]^T, k_1)$, where $\cdot^T$ is the transpose of a vector, we often refer to it as $\tau_1 : \lambda_1 + \lambda_2 \rightarrow^{k_1} 2\lambda_3$. The *state change* associated to a reaction τ is defined by $v_\tau = p_\tau - r_\tau$. For example, for τ_1 as above, we have $v_{\tau_1} = [-1, -1, 2]^T$. Assuming well mixed environment, constant volume V and temperature, a *configuration* or *state* $x \in \mathbb{N}^{|\Lambda|}$ of the system is given by a vector of the number of molecules of each species. Given a configuration x then $x(\lambda_i)$ represents the number of molecules of λ_i in the configuration and $\frac{x(\lambda_i)}{N}$ is the concentration of λ_i in the same configuration, where $N = V \cdot N_A$ is the volumetric factor or system size, V is the volume and N_A Avogadro's number. The *deterministic* semantics approximates the concentrations of species over time as the solution $\Phi(t)$ of the rate equations [11], a set of differential equations of the form:

$$\frac{d\Phi(t)}{dt} = F(\Phi(t)) = \sum_{\tau \in R} \upsilon_\tau \cdot (k_\tau \prod_{i=1}^{|A|} \Phi_i^{r_{i,\tau}}(t)) \tag{1}$$

where $\Phi_i^{r_{i,\tau}}(t)$ is the ith component of vector $\Phi(t)$ raised to the power of $r_{i,\tau}$, ith component of vector r_τ. The initial condition is $\Phi(0) = \frac{x_0}{N}$. It is known that Eq. (1) is accurate in the limit of high population [15].

Stochastic Semantics. The propensity rate α_τ of a reaction τ is a function of the current configuration x of the system such that $\alpha_\tau(x)dt$ is the probability that a reaction event occurs in the next infinitesimal interval dt. We assume mass action kinetics, therefore $\alpha_\tau(x) = k_\tau \frac{\prod_{i=1}^{|A|} r_{i,\tau}!}{N^{|r_\tau|-1}} \prod_{i=1}^{|A|} \binom{x(\lambda_i)}{r_{i,\tau}}$, where $r_{i,\tau}!$ is the factorial of $r_{i,\tau}$, and $|r_\tau| = \sum_{i=1}^{|A|} r_{i,\tau}$ [1]. To simplify the notation, N is considered embedded inside the coefficient k for any reaction. The stochastic semantics of the CRN $C = (A, R)$ is represented by a *time-homogeneous continuous-time Markov chain* (CTMC) [15] $(X^N(t), t \in \mathbb{R}_{\geq 0})$ with state space S, where in X^N we made explicit the dependence by N. $X^N(t)$ is a random vector describing the molecular population of each species at time t. Let $x_0 \in \mathbb{N}^{|A|}$ be the initial condition of X^N then $P(X^N(0) = x_0) = 1$. For $x \in S$, we define $P(x,t) = P(X^N(t) = x \mid X^N(0) = x_0)$. The transient evolution of X^N is described by the Chemical Master Equation (CME), a set of differential equations

$$\frac{d}{dt}(P(x,t)) = \sum_{\tau \in R} \{\alpha_\tau(x - \upsilon_\tau)P(x - \upsilon_\tau, t) - \alpha_\tau(x)P(x,t)\}. \tag{2}$$

Solving Eq. (2) requires computing the solution of a differential equation for each reachable state. The size of the reachable state space depends on the number of species and molecular populations and can be huge or even infinite. As a consequence, solving the CME is generally feasible only for CRNs with very few species and small molecular populations.

Linear Noise Approximation. The *Linear Noise Approximation* (LNA) is a continuous state space approximation of the CME, which approximates the CTMC induced by a CRN as a Gaussian process [25]. In [26], the LNA has been derived as a linearized solution of the Chemical Langevin Equation (CLE) [19]. This derivation shows that the LNA is accurate if the two *leap conditions* on the reactions are satisfied. The leap conditions are satisfied at time t if (i) there exists an infinitesimal time interval dt such that the propensity rate of each reaction is approximately constant during dt and if (ii) each reaction fires many times during dt. It is possible to show that, assuming mass action kinetics, in the limit of high volume these conditions are always satisfied. The LNA at time t approximates the distribution of $X^N(t)$ with the distribution of the random vector $Y^N(t)$ such that

$$X^N(t) \approx Y^N(t) = N\Phi(t) + N^{\frac{1}{2}}G(t) \tag{3}$$

where $G(t) = (G_1(t), G_2(t), \ldots, G_{|A|})$ is a random vector, independent of N, representing the stochastic fluctuations at time t and $\Phi(t)$ is the solution of Eq. (1). The probability distribution of $G(t)$ is then given by the solution of a linear Fokker-Planck equation [26]. As a consequence, for every time instant t, $G(t)$ has a multivariate normal distribution whose expected value $E[G(t)]$ and covariance matrix $C[G(t)]$ are the solution of the following differential equations:

$$\frac{dE[G(t)]}{dt} = J_F(\Phi(t))E[G(t)] \tag{4}$$

$$\frac{dC[G(t)]}{dt} = J_F(\Phi(t))C[G(t)] + C[G(t)]J_F^T(\Phi(t)) + W(\Phi(t)) \tag{5}$$

where $J_F(\Phi(t))$ is the Jacobian of $F(\Phi(t))$, $J_F^T(\Phi(t))$ its transpose, $W(\Phi(t)) = \sum_{\tau \in R} \upsilon_\tau \upsilon_\tau^T \alpha_{c,\tau}(\Phi(t))$ and $F_j(\Phi(t))$ the jth component of $F(\Phi(t))$. We assume $X^N(0) = x_0$ with probability 1; as a consequence $E[G(0)] = 0$ and $C[G(0)] = 0$, which implies $E[G(t)] = 0$ for every t. The following theorem illustrates the nature of the approximation using the LNA.

Theorem 1. *[15] Let $C = (A, R)$ be a CRN and X^N the discrete state space Markov process induced by C. Let $\Phi(t)$ be the solution of rate equations with initial condition $\Phi(0) = \frac{x_0}{N}$ and G be the Gaussian process with expected value and variance given by Eqs. (4) and (5). Then, for any $t < \infty$ and $N \to \infty$,*

$$N^{\frac{1}{2}} \left| \frac{X^N(t)}{N} - \Phi(t) \right| \Rightarrow_N G(t). \tag{6}$$

In the above, $\Rightarrow$ indicates convergence in distribution [5]. The LNA is exact in the limit of high populations, but can also be used in different scenarios if the leap conditions are satisfied [20,26]. To compute the LNA it is necessary to solve $O(|A|^2)$ first order differential equations, and the complexity is independent of the initial number of molecules of each species. Therefore, one can avoid the exploration of the state space that methods based on uniformization rely upon.

3   Linear Noise Approximation of Reachability Probabilities

We are interested in computing the probability that the CTMC induced by a biochemical network enters a region of the state space at some time instant between t_1 and t_2. In order to exploit the LNA, we will first discretize time for the Gaussian process given by the LNA, with a fixed (or adaptive) step size h, which we can do effectively owing to the Markov property and the knowledge of its mean and covariance. As a result, we obtain a *discrete-time, continuous space*, Markov process with a Gaussian transition kernel. Then, by resorting to state space discretization, we compute the reachability probability on this new process, obtaining an approximation converging to the LNA approximation as h tends to zero.

At first sight, there seems to be little gain, as we now have to deal with a $|\Lambda|$-dimensional continuous state space. Indeed, for general regions this can be the case. However, if we restrict to regions defined by linear inequalities, we can exploit properties of Gaussian distributions (i.e. their closure wrt linear combinations), reducing the dimension of the continuous space to the number of different linear combinations used in the definition of the linear inequalities (in fact, the same hyperplane can be used to fix both an upper and a lower bound). As typically we are interested in regions defined by one or two inequalities, the complexity will then be dramatically reduced.

3.1   Reachability Problem: Formal Definition

Recall that, given a CRN $C = (\Lambda, R)$ with initial configuration x_0, its stochastic behaviour is described by the CTMC X^N. A path of X^N is a sequence $\omega = x_0 t_1 x_1 t_1 x_2 \ldots$ where $x_i \in \mathbb{N}^{|\Lambda|}$ is a state and $t_i \in \mathbb{R}_{>0}$ is the time spent in the state x_i. A path is finite if there is a state x_k that is absorbing. $\omega(t)$ is the state of the path at time t. $Path(X^N, x_0)$ is the set of all (finite and infinite) paths of the CTMC starting in x_0. We work with the standard probability measure $Prob$ over paths $Path(X^N, x_0)$ defined using cylinder sets [21].

We now formalize the reachability problem we want to solve. For a simpler presentation, we restrict to a single linear inequality over the species. This still covers many practical scenarios, in particular in systems biology. Next, we show how to generalise the method to regions specified by the intersection of more than one hyperplane, though the complexity of our method will grow exponentially with the number of different hyperplanes, unless additional approximations are introduced.

Definition 1. *Let $C = (\Lambda, R)$ be a CRN with initial state x_0, fix vector of weights $B \in \mathbb{Z}^{|\Lambda|}$, finite set of disjoint intervals $I = [l_1, u_1] \cup \ldots \cup [l_k, u_k], k \geq 1$, such that, for $i \in [1, k]$, $[l_i, u_i] \subseteq \mathbb{R} \cup [-\infty, +\infty]$, and an interval $[t_1, t_2] \subset \mathbb{R}_{\geq 0}$. The reachability probability of B-weighted linear combination of species falling in the target set I in time interval $[t_1, t_2]$, for initial condition x_0, is*

$$P_{reach}(B, x_0, I, [t_1, t_2]) = Prob\{\omega \in Path(X^N, x_0) | B \cdot \omega(t) \in I, t \in [t_1, t_2]\}. \quad (7)$$

3.2   LNA and Dimensionality Reduction

In order to approximate the reachability probability in Eq. (7), we rely on the LNA $Y^N(t)$ of $X^N(t)$ Eqs. (4) and (5). By Eq. (3), we have that the distribution of $Y^N(t)$ is Gaussian with expected value and covariance matrix given by

$$E[Y^N(t)] = N\Phi(t)$$
$$C[Y^N(t)] = N^{\frac{1}{2}} C[G(t)] N^{\frac{1}{2}} = NC[G(t)].$$

Let $B \in \mathbb{Z}^{|\Lambda|}$, then $Z^N = B \cdot Y^N$ is a uni-dimensional process and for any t it represents the time evolution of the linear combination of the species defined by

B over time. Furthermore, $Z^N(t)$ is also Gaussian distributed, being the linear combination of Gaussian variables. In particular, $Z^N(t)$ is characterised by the following mean and covariance:

$$E[Z^N(t)] = BE[Y^N(t)] \tag{8}$$

$$C[Z^N(t)] = BC[Y^N(t)]B^T \tag{9}$$

Note also that the distribution of Z^N depends on Y^N *only via its mean and covariance*, which are obtained by solving ODEs (4) and (5). This is the key feature that enables an effective dimensionality reduction.

3.3 Time Discretization Scheme

We now introduce an exact time discretization scheme for Z^N. Fix a small time step $h > 0$. By sampling Y^N at step h and invoking the Markov property,[1] we obtain a *discrete-time Markov process* (DTMP) $\bar{Y}^N(k) = Y^N(kh)$ on continuous space. Applying the linear projection mapping Z^N to $\bar{Y}^N(k)$, and leveraging its Gaussian nature, we obtain a process $\bar{Z}^N(k) = Z^N(kh)$ which is also a DTMP, though with a kernel depending on time through the mean and variance of Y^N.

Definition 2. *A (time-inhomogeneous) discrete-time Markov process (DTMP) $(\bar{Z}^N(k), k \in \mathbb{N})$ is uniquely defined by a triple (S, σ, T), where (S, σ) is a measurable space and $T : \sigma \times S \times \mathbb{N} \to [0, 1]$ is a transition kernel such that, for any $z \in S$, $A \in \sigma$ and $k \in \mathbb{N}$, $T(A, z, k)$ is the probability that $\bar{Z}^N(k+1) \in A$ conditioned on $\bar{Z}^N(k) = z$. S is the state space of $\bar{Z}^N$.*

In order to characterise $\bar{Z}^N$, we need to compute its transition kernel. This can be done by computing $f_{Z^N(t+h)|Z^N(t)=\bar{z}}(z)$, i.e. the density function of $Z^N(t+h)$ given the event $Z^N(t) = \bar{z}$.

Consider the joint distribution $Y^N(t), Y^N(t + h)$, which is Gaussian. Its projected counterpart $Z^N(t), Z^N(t + h)$ is thus also Gaussian, with covariance function $cov(Z^N(t), Z^N(t + h)) = Bcov(Y^N(t), Y^N(t + h))B^T$, where $cov(Y^N(t), Y^N(t+h))$ is the covariance function of Y^N at times t and $t+h$. It follows by the linearity of B that $f_{Z^N(t+h)|Z^N(t)=\bar{z}}$ is Gaussian too, and to fully characterize it we need to compute $E[Z^N(t+h)|Z^N(t) = \bar{z}]$ and $C[Z^N(t+h)|Z^N(t) = \bar{z}]$. To this end, we need to derive $cov(Y^N(t), Y^N(t+h))$. From now on, we denote $cov(Y^N(t+h), Y^N(t)) = C_Y(t+h, t)$ and $cov(Z^N(t+h), Z^N(t)) = C_Z(t+h, t)$. Following [15], we introduce the following matrix differential equation

$$\frac{d\Omega(t, s)}{dt} = J_F(\Phi(t))\Omega(t, s) \tag{10}$$

with $t \geq s$ and initial condition $\Omega(s, s) = Id$, where Id is the identity matrix of dimension $|\Lambda|$. Then, as illustrated in [15], we have

$$C_Y(t, t + h) = \int_0^t \Omega(t, s)J_F(\Phi(s))[\Omega(t + h, s)]^T ds. \tag{11}$$

---

[1] The Gaussian process obtained by linear noise approximation is Markov, as it is the solution of a linear Fokker-Planck equation (stochastic differential equation) [25].

This is an integral equation, which has to be computed numerically. To simplify this task, we derive an equivalent representation in terms of differential equations. This is given by the following lemma.

Lemma 1. *Solution of Eq. (11) is given by the solution of the following differential equations*

$$\frac{dC_Y(t, t+h)}{dt} = W(\Phi(t))\Omega^T(t+h, t) + J_F(\Phi(t))C_Y(t, t+h) + C_Y(t, t+h)J_F^T(\Phi(t+h)) \quad (12)$$

with initial condition $C_Y(0, h)$ *computed as the solution of*

$$\frac{dC_Y(0, s)}{ds} = C_Y(0, 0 + s)J_F^T(\Phi(s)).$$

$\Omega(t+h, t)$ can be computed by solving Eq. (10). Knowledge of $C_Y(t, t+h)$ allows us to directly compute $C_Z(t, t+h) = BC_Y(t, t+h)B^T$. Then, by using the law for conditional expectation of a Gaussian distribution, we finally have

$$E[Z^N(t+h)|Z^N(t) = \bar{z}] = $$
$$E[Z^N(t+h)] + C_Z(Z(t+h), Z(t))C[Z(t)]^{-1}(\bar{z} - E[Z^N(t)]) \quad (13)$$

$$C[Z^N(t+h)|Z^N(t) = \bar{z}] = C[Z^N(t+h)] - C_Z(t, t+h)C[Z^N(t)]^{-1}C_Z(t, t+h). \quad (14)$$

Note that the resulting kernel is time-inhomogeneous. The dependence on t is via the mean and covariance of Y^N, which are functions of time and define completely the distribution of Y^N.

3.4   Computation of Reachability Probabilities

In order to compute the reachability probability for the DTMP $\bar{Z}^N(k)$, we discretize its continuous state space, obtaining an abstraction in terms of a discrete-time Markov chain (DTMC) $Z^{N,D}(k)$ with state space S. That is, the states of the original Markov process are partitioned into a countable set of non-overlapping sets. We assume an order relation between elements of each set and, for each set, we consider a representative point, given by the median of the set. S is given by the set of representative points. In particular, we partition the state space of $\bar{Z}^N$ in intervals of length $2\Delta z$, where Δz defines how fine our space discretization is. A possible choice is $\Delta z = 0.5$, which basically means $S \subseteq \mathbb{Z}$. For $\Delta z \to 0$ the error introduced by the space discretization goes to zero. However, when the molecular population is of the order of hundreds or thousands, it can be beneficial to consider $\Delta z > 0.5$, since a coarser state space aggregation is reasonable.

Then, we solve the reachability problem on the resulting DTMC. For $z', z \in S$, the transition kernel of $Z^{N,D}(k)$ is defined as

$$T(z', z, k) = \int_{z'-\Delta z}^{z'+\Delta z} f_{Z^N(hk+h)|Z^N(hk)=z}(x)dx, \quad (15)$$

where h is the discrete time step, assumed to be fixed for a simpler notation. Finally, in order to compute the reachability of the target set I we make all the states $z \in I$ absorbing. That is, for $z \in I$

$$T(z', z, k) = \begin{cases} 1 & \text{if } z' = z \\ 0 & \text{otherwise} \end{cases}$$

Algorithm 1 illustrates our approach for computing reachability probabilities. In Line 1, we initialize the system at time 0. In the context of the algorithm, S is a set containing the reachable states at a particular time with probability mass greater than the threshold $\mathcal{TH}$. $\mathcal{TH}$ equals 10^{-14} in all our experiments. This guarantees that the algorithm always terminates in finite time even if the state space is not finite. Initially, we have that S contains only one state $B \cdot x_0$. Then, in Lines $3 - 10$, we propagate the probability for any discrete step until $t < t_1$, as illustrated in [21]. For generality, we assume that the time step h is chosen adaptively, according to the system dynamics. Propagating probability is possible, as for any $z' \in S$, $T(z', z, k)$, which has a Gaussian nature, defines the probability of being in z' in the next discrete time step by $Z^{N,D}(k) = z$. From Line 12 to 20, we compute probabilistic reachability $P_{reach}(B, x_0, I, [t_1, t_2])$ by propagating the probability only for states that are not in I. When we reach $t \geq t_2$, we have that $P_{reach}(B, x_0, I, [t_1, t_2]) \approx \sum_{z \in I} P(Z^{N,D}(t) = z | Z^{N,D}(0) = B \cdot x_0)$.

Algorithm 1. Compute Time-Bounded Probabilistic Reachability

Require: A CRN $C = (\Lambda, R)$ with initial condition x_0, $B \in \mathbb{Z}^{|\Lambda|}$, a finite time interval $[t_1, t_2]$, a target set I and a threshold $\mathcal{TH}$.

1: **function** COMPUTEREACH(C, B, x_0, I, $[t_1, t_2]$, $\mathcal{TH}$)
2:      Set $t = 0$, $S = \{B \cdot x_0\}$ and $P(Z^{N,D}(0) = B \cdot x_0) = 1$
3:      **while** $t < t_1$ **do**
4:          Compute time step h
5:          **for each** $z \in S$ **do**
6:              Propagate probability at time $t + h$ and update S
7:          **for each** $z \in S$ **do**
8:              **if** $P(Z^{N,D}(t + h) = z) < \mathcal{TH}$ **then**
9:                  $S \leftarrow S - \{z\}$
10:         $t \leftarrow t + h$
11:      **while** $t < t_2$ **do**
12:          Compute time step h
13:          **for each** $z \in S/I$ **do**
14:              Propagate probability at time $t + h$ and update S
15:          **for each** $z \in S/I$ **do**
16:              **if** $P(Z^{N,D}(t + h) = z) < \mathcal{TH}$ **then**
17:                  $S \leftarrow S - \{z\}$
18:         $t \leftarrow t + h$
19:      **return** $P_{reach}(B, x_0, I, [t_1, t_2]) = \sum_{z \in I} P(Z^{N,D}(t) = z)$

3.5 Correctness

The method we present is approximate. In particular, errors are introduced in two ways: by resorting to the LNA and by discretisation of time and space of the LNA. The quality of these two approximations is controlled by three parameters: (a) N, the system size, which influences the accuracy of LNA, (b) h, the step size, and (c) Δz, the discretization step, which influences the quality of the approximation of the reachability probability of the LNA.

Recall that X^N and $Z^{N,D}$ are, respectively, the CTMC induced by a CRN and the DTMC obtained by discretization of the LNA of X^N for a particular N. Fix $B \in \mathbb{Z}^{|A|}$ and I, a set of disjoint closed intervals of reals, and denote by $P_{X^N}(B, t_1, t_2)$ and $P_{Z^{N,D}}(B, t_1, t_2)$, $t_1 < t_2$, the reachability probabilities for $Z^{N,D}$ and X^N. Then, we have the following result

Theorem 2. *With the notation above, for $t_1 \leq t_2 < \infty$:*

$$\lim_{N \to \infty} \lim_{h \to 0} \lim_{\Delta z \to 0} \{|P_{X^N}(B, t_1, t_2) - P_{Z^{N,D}}(B, t_1, t_2)|\} = 0.$$

The convergence stated in Theorem 2 means that, since N is fixed for a given CRN, that even if we have control over h and Δz, the quality of the approximation depends on how well the LNA approximated the CRN. Error bounds would be a viable companion to estimate the committed error, but we are not aware of any explicit formulation of them for the convergence of the LNA. However, experimental results in Sect. 5 show that the error committed is generally limited also for moderately small N and quite large h.

3.6 Complexity

Complexity of the method depends on the following: (a) the equations we need to solve, (b) the step size h, and (c) the space discretization step Δz. Algorithm 1 requires solving Eqs. (5) and (12), that is, a set of differential equations quadratic in the number of species. In fact, solving these equations requires computing J_F, Jacobian of F. However, the number of equations we need to solve is independent of the number of molecules in the system. This guarantees the scalability of our approach. An important point is that Eq. (12) requires solving Eq. (11) once for each sampling point of the numerical solution of Eq. (12). A possible way to avoid this is to consider approximate solutions of Eq. (11), which are accurate in the limit of $h \to 0$. However, to keep this approximation under control, h has to be chosen really small, slowing down the computation. Moreover, for any sample point, Eq. (11) is solved only for a small time interval (between t and $t+h$). As a consequence, in practice, the computational cost introduced in solving Eq. (11) is under control.

A smaller value of h implies that, for a given time interval, we have a greater number of discrete time steps, which can slow down the computation in some cases. The value of Δz determines the number of states of the resulting DTMC. However, we stress that we discretize $Z^N(t)$, a uni-dimensional distribution (or m-dimensional in the case we have $m > 1$ linear inequalities). As a consequence, the number of reachable states with probability mass is generally limited and under control. Obviously, if the number of molecules is large and Δz extremely small, then this is detrimental on performance.

3.7   Extensions

Remark 1. Our approach can be easily extended to target regions defined by intersections of finitely many linear inequalities over species. That is, we consider a set of linear predicates $Z_j^N = B_j \cdot X^N(t) \in I_j$, $j = 1 \dots, m$ with $m > 1$, and ask what is the probability that during a finite time interval we are in a state where each predicate is verified. In order to do that, we can define $B = (B_1, ..., B_m)^T \in \mathbb{Z}^{m \times |\Lambda|}$, a matrix where each row is a vector specifying a different linear combination. As a consequence, $Z^N = B \cdot Y^N$ is an m dimensional Gaussian process and all the properties we used for the unidimensional case remain valid in this extended scenario. The resulting DTMC $Z^{N,D}$ is m−dimensional. However, note that m is generally equal to 1 or 2 in practical applications (see Remark 2).

Remark 2. The method presented here can be extended to compute the probability of a non-nested until formula of CSL [3], that is, a formula of the type

$$P_{\sim p}[\Psi_1 U^{[t_1, t_2]} \Psi_2].$$

This formula is satisfied if the probability of a path such that there exists $t \in [t_1, t_2]$ for which Ψ_2 is satisfied and, for all $t' \in [0, t]$, Ψ_1 is satisfied meets the bound p. We restrict Ψ_1, Ψ_2 to linear inequalities over species. Computing this probability, as explained in [21], requires computing two terms: (a) the probability of reaching a state between $[0, t_1)$ such that $\neg \Psi_1$ is satisfied, and (b) the probability of reaching a state during $[t_1, t_2]$ where $\neg \Psi_1 \wedge \Psi_2$ is satisfied. The former is simply reachability on $\neg \Psi_1$. The latter can be computed by considering reachability over the bi-dimensional system given by the joint distribution of the linear combinations associated to $\neg \Psi_1$ and Ψ_2.

4   Stochastic Evolution Logic (SEL)

The method presented here permits an extension of the Stochastic Evolution Logic (SEL) introduced in [12] for approximate model checking of CRNs based on the LNA. Here, we extend the original formulation of the logic with an operator for computing (time-bounded) probabilistic reachability. However, as explained in Remark 2, more complex temporal behaviours could be introduced as well.

Let $C = (\Lambda, R)$ be a CRN with initial state x_0, then SEL enables evaluation of the probability, reachability, variance and expectation of linear combinations of populations of the species of C. The syntax of SEL is given by

$$\eta := P_{\sim p}[B, I]_{[t_1, t_2]} \quad | \quad F_{\sim p}[B, I]_{[t_1, t_2]} \quad | \quad Q_{\sim v}[B]_{[t_1, t_2]} \quad | \quad \eta_1 \wedge \eta_2 \quad | \quad \eta_1 \vee \eta_2$$

where $Q = \{supV, infV, supE, infE\}$, $\sim = \{<, >\}$, $p \in [0,1]$, $v \in \mathbb{R}$, $B \in \mathbb{Z}^{|A|}$, $I = [l_1, u_1] \cup \ldots \cup [l_k, u_k]$, $k \geq 1$, such that, for $i \in [1, k]$, $[l_i, u_i] \subseteq \mathbb{R} \cup [-\infty, +\infty]$ is a finite set of disjoint intervals and $[t_1, t_2]$ is a closed time interval, with the constraint that $t_1 \leq t_2$ and $t_1, t_2 \in \mathbb{R}_{\geq 0}$. If $t_1 = t_2$ the interval reduces to a singleton.

Formulae η describe global properties of the stochastic evolution of the system. (B, I) specifies a linear combination of the species, where $B \in \mathbb{Z}^{|A|}$ is a vector of weights defining the linear combination and I is a set of disjoint closed real intervals. $P_{\sim p}[B, I]_{[t_1, t_2]}$ is the probabilistic operator, which specifies the average value of the probability that the linear combination defined by B falls within the range I over the time interval $[t_1, t_2]$. Given $Pr_{B,I}^{X^N}(t) = Prob\{\omega \in Path(X^N, x_0) \mid B \cdot \omega(t) \in I\}$, then, for $t_1 = t_2$, its semantics is defined as

$$X^N, x_0 \models P_{\sim p}[B, I]_{[t_1, t_1]} \quad \leftrightarrow \quad Pr_{B,I}^{X^N}(t_1) \sim p.$$

Instead, for $t_1 < t_2$ we have

$$X^N, x_0 \models P_{\sim p}[B, I]_{[t_1, t_2]} \quad \leftrightarrow \quad \frac{1}{t_2 - t_1} \int_{t_1}^{t_2} Pr_{B,I}^{X^N}(t)\, dt \sim p.$$

$F_{\sim p}[B, I]_{[t_1, t_2]}$ is the new probabilistic reachability operator, which specifies the probability that the linear combination of species defined by B reaches I during $[t_1, t_2]$. Its semantics can be defined as

$$X^N, x_0 \models F_{\sim p}[B, I]_{[t_1, t_2]} \leftrightarrow Prob(\omega \in Path(X^N, x_0) \mid B \cdot \omega(t) \in I, t \in [t_1, t_2]) \sim p$$

The operators $supE, infE, infV, supV$, see [12], respectively, yield the supremum and infimum of expected value and variance of the random variables associated to B within the specified time interval. The quantitative value associated to a formula can be computed by writing $=?$ instead of $\sim p$ or $\sim v$. For instance, $F_{=?}[B, I]_{[t_1, t_2]}$ gives the probability value associated to the reachability property. The following example illustrates that the P and F operators differ.

Example 1. Consider the following CRN, taken from [13], modelling a phosphorelay network

$$\tau_1 : L1 + ATP \to^{0.01} L1p + ATP; \quad \tau_2 : L1p + L2 \to^{0.01} L2p + L2;$$
$$\tau_3 : L2p + L3 \to^{0.01} L3p + L2; \quad \tau_4 : L3p \to^{0.1} L3;$$

with initial conditions $x_0(L1) = x_0(L2) = x_0(L3) = 50$, $x_0(ATP) = 150$ and all other species equal 0. Then, if we consider $P_{>0.3}[L3p, [40, \infty]]_{[0,10]}$, which is true if the average probability that $L3p > 40$ is greater that 0.3. Then, this is evaluated to false. Instead, $F_{>0.3}[L3p, [40, \infty]]_{[0,10]}$, which models the probability of being in a state where $L3p > 40$ during the first 10 seconds, is evaluated as true.

5 Experimental Results

We implemented Algorithm 1 in Matlab and evaluated it on two case studies. All the experiments were run on an Intel Dual Core i7 machine with 8 GB of RAM. The first case study is a Phospohorelay Network with 7 species. We use this example to show the trade-off between the different parameters and the molecular population. More precisely, we show that the accuracy of our approach increases as the number of molecules grows, but can still give fast and accurate results when the molecular population is not large. The second example is a Gene Regulatory network. We use this example to show how our approach is more powerful than existing approximate techniques, and is able to accurately handle properties where existing techniques fail. We validate our results by comparing our method with statistical model checking (SMC) as implemented in PRISM [22]. In fact, for both examples, exact numerical computation of the reachability probabilities on the CTMC is infeasible because of state space explosion.

5.1 Phosphorelay Network

The first case study is a three-layer phosphorelay network as shown in Example 1. There are 3 layers, $(L1, L2, L3)$, which can be found in phosphorylate form $(L1p, L2p, L3p)$, and the ligand B. We consider the initial condition $x_0 \in \mathbb{N}^7$ such that $x_0(L1) = x_0(L2) = x_0(L3) = L \in \mathbb{N}$, $x_0(L1p) = x_0(L2p) = x_0(L3p) = 0$ and $x_0(B) = 150$. In Fig. 1, we compare the estimates obtained by our approach for two different initial conditions ($L = 100$ and $L = 200$) with statistical model checking as implemented in PRISM [22], with 30000 simulations and confidence interval equal to 0.01. In both experiments we consider $\Delta z = 0.5$.

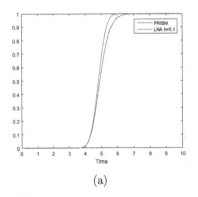
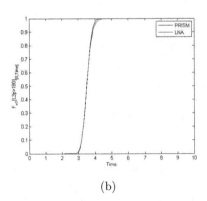

(a)   (b)

Fig. 1. Comparison of the evaluation of $F_{[0,Time]}[L3p > 80]$ (Fig. 1a) and $F_{[0,Time]}[L3p > 180]$ (Fig. 1b) using statistical model checking as implemented in PRISM and our approach. In Fig. 1a, we used $h = 0.1$, $\Delta z = 0.5$, and $L = 100$. In Fig. 1b, we considered $h = 0.1$, $\Delta z = 0.5$ and $L = 200$. (Color figure online)

In Fig. 1a we can see that, if we increase the time interval of interest, the error tends to increase. This is because, for $L = 100$, the LNA and CME do not have perfect convergence. As a consequence, at every step of the discretized DTMC, a small error is introduced. This source of error is present until we enter the target region with probability 1. If we increase L this error disappears, and the inaccuracies are due to the finiteness of h and Δz. However, already for $h = 0.1$ and $L = 100$, the LNA produces a fast and reasonably accurate approximation. In the following table we compare our approach and PRISM evaluations for different values of L and h and $\Delta z = 0.5$. In order to compare the accuracy we consider the absolute average error, $||\epsilon||_1$, and the maximum absolute error, $||\epsilon||_\infty$. $||\epsilon||_1 = \frac{1}{|\Sigma|} \sum_{n \in \Sigma} |F^Y_{[0,n]} - F^X_{[0,n]}|$ and $||\epsilon||_\infty = max_{n \in \Sigma}\{|F^Y_{[0,n]} - F^X_{[0,n]}|\}$, where Σ is the set of discrete times between 0 and 10, and $F^Y_{[0,n]}$ and $F^X_{[0,n]}$ are the evaluation of the particular reachability formula in the interval $[0, n]$ according to the LNA and PRISM.

Property	Ex. time	h	L	$||\epsilon||_1$	$||\epsilon||_\infty$
$F_{=?}[L3p > 80]_{[0,Time]}$, $Time \in [0, 10]$	97 s	0.1	100	0.0088	0.11
$F_{=?}[L3p > 180]_{[0,Time]}$, $Time \in [0, 10]$	130 s	0.1	200	0.0015	0.0217
$F_{=?}[L3p > 80]_{[0,Time]}$, $Time \in [0, 10]$	28 s	0.5	100	0.0381	0.24
$F_{=?}[L3p > 180]_{[0,Time]}$, $Time \in [0, 10]$	39 s	0.5	200	0.0289	0.14

The results show that the best accuracy is obtained for $h = 0.1$ and $L = 200$, where $h = 0.1$ induces a finer time discretization, whereas the worst are for $h = 0.5$ and $L = 100$. We comment that the numerical solution of the CME using PRISM is not feasible for this model, and our method is several orders of magnitude faster than statistical model checking with PRISM (30000 simulations for each time point).

5.2 Gene Expression

We consider the following gene expression model, as introduced in [27]:

$$\tau_1 :\to^{0.5} mRNA; \quad \tau_2 : mRNA \to^{0.0058} mRNA + P;$$
$$\tau_3 : mRNA \to^{0.0029} W; \quad \tau_4 : P \to^{0.0001} W;$$

with initial condition x_0 such that all the species have initial concentrations equal to 0. We consider the property $F_{=?}[\geq 175]_{[0,Time]}$, which quantifies the probability that the $mRNA$ is produced for at least 175 molecules during the first $Time$ seconds, for $Time \in [0, 1000]$. This is a particularly difficult property because the trajectory of the mean-field of the model, and so the expected value of the LNA, does not enter the target region. As a consequence, approximate approaches introduced in [15] and [8], which are based on the hitting times of the mean-field model, fail and evaluate the probability as always equal to 0.

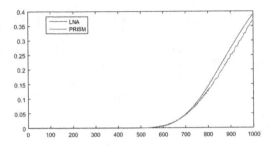

Fig. 2. The figure plots $F_{=?}[mRNA \geq 174]_{[0,Time]}$ for $h = 1.85$ and $\Delta z = 0.5$. The x-axis represents the value of $Time$ and the y-axis the quantitative value of the formula for that value of $Time$. (Color figure online)

Conversely, our approach is able to evaluate correctly such a property. Figure 2 compares the value computed by our approach with statistical model checking of the same property as implemented in PRISM over 30000 simulations for each time point and confidence interval 0.01. In Fig. 2 we consider $h = 1.8$ and $\Delta z = 0.5$ and demonstrate that our approach is able to correctly estimate such a difficult property. Note that, as the mean-field does not enter the target region, for each time point the probability to enter the target region depends on a portion of the tail of the Gaussian given by the LNA. As a consequence, the accuracy of our results strictly depends on how well the LNA approximates the original CTMC, much more than for properties where the mean-field enters the target region. In the following table, we evaluate our results for two different values of h and $\Delta z = 0.5$.

Property	Ex. time	h	$\|\|\epsilon\|\|_1$	$\|\|\epsilon\|\|_\infty$
$F_{=?}[mRNA \geq 174]_{[0,Time]}$, $Time \in [0, 100]$	298 s	1.85	0.0075	0.022
$F_{=?}[mRNA \geq 174]_{[0,Time]}$, $Time \in [0, 100]$	152 s	5	0.0147	0.13

6   Conclusion

We presented a method for computing (time-bounded) probabilistic reachability for CRNs based on the LNA, which is challenging because the LNA yields a continuous time and uncountable state space stochastic process. As a consequence, existing methods that rely on finite state spaces cannot be used directly and discretizing the uncountable state space defined by the LNA will lead to state space explosion. In order to overcome these issues, we considered reachability regions defined as polytopes. Using the fact that the LNA is a solution of a linear Fokker-Planck equation, and so a Gaussian Markov process, for a given linear combination of the species of a CRN, we are able to project the original, multi-dimensional Gaussian process onto a uni-dimensional stochastic process. We then derived an abstraction in terms of a time-inhomogeneous DTMC, whose state space is independent of the number of the species of a CRN, as it is derived by discretizing linear combinations of the species. This ensures scalability. Finally, we used our approach to extend the Stochastic Evolution Logic in order to verify complex temporal properties. On two case studies, we showed that our approach permits fast and scalable probabilistic analysis of CRNs. The accuracy depends on parameters controlling space and time discretization, as well as the accuracy of the LNA.

References

1. Anderson, D.F., Kurtz, T.G.: Continuous time Markov chain models for chemical reaction networks. In: Koeppl, H., Setti, G., di Bernardo, M., Densmore, D. (eds.) Design and Analysis of Biomolecular Circuits, pp. 3–42. Springer, New York (2011)
2. Aziz, A., Sanwal, K., Singhal, V., Brayton, R.: Verifying continuous time Markov chains. In: Alur, R., Henzinger, T.A. (eds.) CAV 1996. LNCS, vol. 1102, pp. 269–276. Springer, Heidelberg (1996)
3. Aziz, A., Sanwal, K., Singhal, V., Brayton, R.: Model-checking continuous-time Markov chains. ACM Trans. Comput. Logic (TOCL) 1(1), 162–170 (2000)
4. Baier, C., Haverkort, B., Hermanns, H., Katoen, J.-P.: Model-checking algorithms for continuous-time markov chains. IEEE Trans. Software Eng. 29(6), 524–541 (2003)
5. Billingsley, P.: Convergence of probability measures. Wiley, Hoboken (1999)
6. Bortolussi, L., Hillston, J.: Fluid model checking. In: Koutny, M., Ulidowski, I. (eds.) CONCUR 2012. LNCS, vol. 7454, pp. 333–347. Springer, Heidelberg (2012)
7. Bortolussi, L., Lanciani, R.: Model checking Markov population models by central limit approximation. In: Joshi, K., Siegle, M., Stoelinga, M., D'Argenio, P.R. (eds.) QEST 2013. LNCS, vol. 8054, pp. 123–138. Springer, Heidelberg (2013)
8. Bortolussi, L., Lanciani, R.: Stochastic approximation of global reachability probabilities of Markov population models. In: Horváth, A., Wolter, K. (eds.) EPEW 2014. LNCS, vol. 8721, pp. 224–239. Springer, Heidelberg (2014)
9. Bortolussi, L., Milios, D., Sanguinetti, G.: Smoothed model checking for uncertain continuous-time Markov chains. Information and Computation (2016)
10. Bujorianu, L.M.: Stochastic Reachability Analysis of Hybrid Systems. Springer, London (2012)
11. Cardelli, L.: On process rate semantics. Theoret. Comput. Sci. 391(3), 190–215 (2008)
12. Cardelli, L., Kwiatkowska, M., Laurenti, L.: Stochastic analysis of chemical reaction networks using linear noise approximation. In: Roux, O., Bourdon, J. (eds.) CMSB 2015. LNCS, vol. 9308, pp. 64–76. Springer, Heidelberg (2015)
13. Csikász-Nagy, A., Cardelli, L., Soyer, O.S.: Response dynamics of phosphorelays suggest their potential utility in cell signalling. J. R. Soc. Interface 8(57), 480–488 (2011)
14. Eldar, A., Elowitz, M.B.: Functional roles for noise in genetic circuits. Nature 467(7312), 167–173 (2010)
15. Ethier, S.N., Kurtz, T.G.: Markov Processes: Characterization and Convergence, vol. 282. Wiley, New York (2009)

16. Fedoroff, N., Fontana, W.: Small numbers of big molecules. Science **297**(5584), 1129–1131 (2002)
17. Gillespie, D.T.: Exact stochastic simulation of coupled chemical reactions. J. Phys. Chem. **81**(25), 2340–2361 (1977)
18. Gillespie, D.T.: A rigorous derivation of the chemical master equation. Phys. A Stat. Mech. Appl. **188**(1), 404–425 (1992)
19. Gillespie, D.T.: The chemical Langevin equation. J. Chem. Phys. **113**(1), 297–306 (2000)
20. Grima, R.: Linear-noise approximation and the chemical master equation agree up to second-order moments for a class of chemical systems. Phys. Rev. E **92**(4), 042–124 (2015)
21. Kwiatkowska, M., Norman, G., Parker, D.: Stochastic model checking. In: Bernardo, M., Hillston, J. (eds.) SFM 2007. LNCS, vol. 4486, pp. 220–270. Springer, Heidelberg (2007)
22. Kwiatkowska, M., Norman, G., Parker, D.: PRISM 4.0: verification of probabilistic real-time systems. In: Gopalakrishnan, G., Qadeer, S. (eds.) CAV 2011. LNCS, vol. 6806, pp. 585–591. Springer, Heidelberg (2011)
23. Munsky, B., Khammash, M.: The finite state projection algorithm for the solution of the chemical master equation. J. Chem. Phys. **124**(4), 044–104 (2006)
24. Pnueli, A.: The temporal logic of programs. In: 18th Annual Symposium on Foundations of Computer Science 1977, pp. 46–57. IEEE (1977)
25. Van Kampen, N.G.: Stochastic Processes in Physics and Chemistry, vol. 1. Elsevier, Amsterdam (1992)
26. Wallace, E., Gillespie, D., Sanft, K., Petzold, L.: Linear noise approximation is valid over limited times for any chemical system that is sufficiently large. IET Syst. Biol. **6**(4), 102–115 (2012)
27. Wolf, V., Goel, R., Mateescu, M., Henzinger, T.A.: Solving the chemical master equation using sliding windows. BMC Syst. Biol. **4**(1), 1 (2010)

Polynomial Analysis Algorithms for Free Choice Probabilistic Workflow Nets

Javier Esparza[1], Philipp Hoffmann[1(✉)], and Ratul Saha[2]

[1] Technische Universität München, Munich, Germany
esparza@in.tum.de, ph.hoffmann@tum.de
[2] National University of Singapore, Singapore, Singapore
ratul@comp.nus.edu.sg

Abstract. We study Probabilistic Workflow Nets (PWNs), a model extending van der Aalst's workflow nets with probabilities. We give a semantics for PWNs in terms of Markov Decision Processes and introduce a reward model. Using a result by Varacca and Nielsen, we show that the expected reward of a complete execution of the PWN is independent of the scheduler. Extending previous work on reduction of non-probabilistic workflow nets, we present reduction rules that preserve the expected reward. The rules lead to a polynomial-time algorithm in the size of the PWN (not of the Markov decision process) for the computation of the expected reward. In contrast, since the Markov decision process of PWN can be exponentially larger than the PWN itself, all algorithms based on constructing the Markov decision process require exponential time. We report on a sample implementation and its performance on a collection of benchmarks.

1 Introduction

Workflow Petri Nets are a class of Petri nets for the representation and analysis of business processes [1,2,5]. They are a popular formal back-end for different notations like BPMN (Business Process Modeling Notation), EPC (Event-driven Process Chain), or UML Activity Diagrams.

There is recent interest in extending these notations, in particular BPMN, with the concept of cost (see e.g. [16,18,19]). The final goal is the development of tool support for computing the worst-case or the average cost of a business process. A sound foundation for the latter requires to extend Petri nets with probabilities and rewards. Since Petri nets can express complex interplay between nondeterminism and concurrency, the extension is a nontrivial semantic problem which has been studied in detail (see e.g. [3,4,7,21]).

Fortunately, giving a semantics to probabilistic Petri nets is much simpler for *confusion-free* Petri nets [3,21], a class that already captures many control-flow constructs of BPMN. In particular, confusion-free Petri nets strictly contain Workflow Graphs, also called free-choice Workflow Nets [1,9,12,13]. In this

This work was funded by the DFG Project "Negotiations: A Model for Tractable Concurrency".

© Springer International Publishing Switzerland 2016
G. Agha and B. Van Houdt (Eds.): QEST 2016, LNCS 9826, pp. 89–104, 2016.
DOI: 10.1007/978-3-319-43425-4_6

paper we study free choice Workflow Nets extended with rewards and probabilities. Rewards are modeled as real numbers attached to the transitions of the workflow, while, intuitively, probabilities are attached to transitions modeling nondeterministic choices. Our main result is the first polynomial algorithm for computing the expected reward of a workflow.

In order to define expected rewards, we give untimed, probabilistic confusion-free nets a semantics in terms of Markov Decision Processes (MDP), with rewards captured by a reward function. In a nutshell, at each reachable marking the enabled transitions are partitioned into *clusters*. All transitions of a cluster are in conflict, while transitions of different clusters are concurrent. In the MDP semantics, a scheduler selects one of the clusters, while the transition inside this cluster is chosen probabilistically.

In our first contribution, we prove that the expected reward of a confusion-free workflow net is independent of the scheduler resolving the nondeterministic choices, and so we can properly speak of *the* expected reward of a free-choice workflow. The proof relies on a result by Varacca and Nielsen [20] on Mazurkiewicz equivalent schedulers.

Since MDP semantics of concurrent systems captures all possible interleavings of transitions, the MDP of a free-choice workflow can grow exponentially in the size of the net, and so MDP-based algorithms for the expected reward have exponential runtime. In our second contribution we provide a polynomial-time *reduction algorithm* consisting of the repeated application of a set of *reduction rules* that simplify the workflow while preserving its expected reward. Our rules are an extension to the probabilistic case of a set of rules for free-choice Colored Workflow Nets recently presented in [9]. The rules allow one to merge two alternative tasks, summarize or shortcut two consecutive tasks by one, and replace a loop with a probabilistic guard and an exit by a single task. We prove that the rules preserve the expected reward. The proof makes crucial use of the fact that the expected reward is independent of the scheduler.

Finally, as a third contribution we report on a prototype implementation, and on experimental results on a benchmark suite of nearly 1500 free-choice workflows derived from industrial business processes. We compare our algorithm with the different algorithms based on the construction of the MDP implemented in PRISM [15].

Due to space limitations, the proofs have been deferred to the extended version [10].

2   Workflow Nets

We recall the definition of a workflow net, and the properties of soundness and 1-safeness.

Definition 1 (Workflow Net [1]). *A workflow net is a tuple* $\mathcal{W} = (P, T, F, i, o)$ *where P is a finite set of places, T is a finite set of transitions ($P \cap T = \emptyset$), $F \subseteq (P \times T) \cup (T \times P)$ is a set of arcs, $i, o \in P$ are distinguished initial and* final *places such that i has no incoming arcs and o has no outgoing arcs and the graph $(P \cup T, F \cup (o, i))$ is strongly connected.*

We write $\bullet p$ and $p \bullet$ to denote the input and output transitions of a place p, respectively, and similarly $\bullet t$ and $t \bullet$ for the input and output places of a transition t. A *marking* M is a function from P to the natural numbers that assigns a number of tokens to each place. A transition t is *enabled* at M if all places of $\bullet t$ contain at least one token in M. An enabled transition may *fire*, removing a token from each place of $\bullet t$ and adding one token to each place of $t \bullet$. We write $M \xrightarrow{t} M'$ to denote that t is enabled at M and its firing leads to M'. The *initial marking* (*final marking*) of a workflow net, denoted by i (o), puts one token on place i (on place o), and no tokens elsewhere. A sequence of transitions $\sigma = t_1 t_2 \cdots t_n$ is an *occurrence sequence* or *firing sequence* if there are markings $M_1, M_2, \ldots, M_n$ such that $i \xrightarrow{t_1} M_1 \cdots M_{n-1} \xrightarrow{t_n} M_n$. $Fin_{\mathcal{W}}$ is the set of all firing sequences of $\mathcal{W}$ that end in the final marking. A marking is *reachable* if some occurrence sequence ends in that marking.

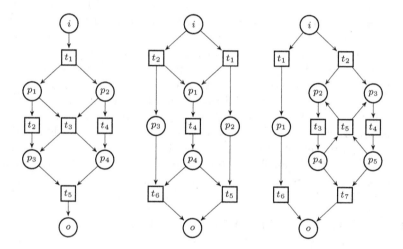

Fig. 1. Three workflow nets

Definition 2 (Soundness and 1-safeness [1]). *A workflow net is* sound *if the final marking is reachable from any reachable marking, and for every transition t there is a reachable marking that enables t. A workflow net is* 1-safe *if $M(p) \leq 1$ for every reachable marking M and for every place p.*

Figure 1 shows three sound and 1-safe workflow nets. In this paper we only consider 1-safe workflow nets, and identify a marking with the set of places that are marked. Markings which only mark a single place are written without brackets and in bold, like the initial marking i. In general, deciding if a workflow net is sound and 1-safe is a PSPACE-complete problem. However, for the class of *free-choice* workflow nets, introduced below, and for which we obtain our main result, there exists a polynomial algorithm [6].

2.1   Confusion-Free and Free-Choice Workflow Nets

We recall the notions of independent transitions and transitions in conflict.

Definition 3 (Independent Transitions, Conflict). *Two transitions t_1, t_2 of a workflow net are* independent *if $\bullet t_1 \cap \bullet t_2 = \emptyset$. Two transitions are* in conflict *at a marking M if M enables both of them and they are not independent. The set of transitions in conflict with a transition t at a marking M is called the* conflict set *of t at M.*

In Fig. 1 transitions t_2 and t_4 of the left workflow are independent, while t_2 and t_3 are in conflict. The conflict set of t_2 at the marking $\{p_1, p_2\}$ is $\{t_2, t_3\}$, but at the marking $\{p_1, p_4\}$ it is $\{t_2\}$.

It is easy to see that in a 1-safe workflow net two transitions enabled at a marking are either independent or in conflict. Assume that a 1-safe workflow net satisfies that for every reachable marking M, the conflict relation at M is an equivalence relation. Then, at every reachable marking M we can partition the set of enabled transitions into equivalence classes, where transitions in the same class are in conflict and transitions of different classes are independent. Such nets have a simple stochastic semantics: at each reachable marking an equivalence class is selected nondeterministically, and then a transition of the class is selected stochastically with probability proportional to a *weight* attached to the transition. However, not every workflow satisfies this property. For example, the workflow on the left of Figure 1 does not: at the reachable marking marking $\{p_1, p_2\}$ transition t_3 is in conflict with both t_2 and t_4, but t_2 and t_4 are independent. Confusion-free nets, whose probabilistic semantics is studied in [20], are a class of nets in which this kind of situation cannot occur.

Definition 4 (Confusion-Free Workflow Nets). *A marking M of a workflow net is* confused *if there are two independent transitions t_1, t_2 enabled at M such that $M \xrightarrow{t_1} M'$ and the conflict sets of t_2 at M and at M' are different. A 1-safe workflow net is* confusion-free *if no reachable marking is confused.*

The workflows in the middle and on the right of Fig. 1 are confusion-free.

Lemma 1 [20]. *Let W be a 1-safe, confusion-free workflow net. For every reachable marking of W the conflict relation on the transitions enabled at M is an equivalence relation.*

Unfortunately, deciding if a 1-safe workflow net is confusion-free is a PSPACE-complete problem (this can be proved by an easy reduction from the reachability problem for 1-safe Petri nets, see [8] for similar proofs). Free-choice workflow nets are a syntactically defined class of confusion-free workflow nets.

Definition 5 (Free-Choice Workflow Nets [1,6]**).** *A workflow net is* free-choice *if for every two places* p_1, p_2 *either* $p_1^\bullet \cap p_2^\bullet = \emptyset$ *or* $p_1^\bullet = p_2^\bullet$.

The workflow in the middle of Fig. 1 is not free-choice, e.g. because of the places p_3 and p_4, but the one on the right is.

It is easy to see that free-choice workflow nets are confusion-free, but even more: in free-choice workflow nets, the conflict set of a transition t is the same at all reachable markings that enable t. To formulate this, we use the notion of a cluster.

Definition 6 (Transition Clusters). *Let* $W = (P, T, F, i, o)$ *be a free-choice workflow net. The* cluster *of* $t \in T$ *is the set of transitions* $[t] = \{t' \in T \mid {}^\bullet t \cap {}^\bullet t' \neq \emptyset\}$.[1]

By the free-choice property, if a marking enables a transition of a cluster, then it enables all of them. We say that the marking enables the cluster; we also say that a cluster fires if one of its transitions fires.

Proposition 1.

- *Let* t *be a transition of a free-choice workflow net. For every marking that enables* t, *the conflict set of* t *at* M *is the cluster* $[t]$.
- *Free-choice workflow nets are confusion-free.*

3 Probabilistic Workflow Nets

We introduce Probabilistic Workflow Nets, and give them a semantics in terms of Markov Decision Processes. We first recall some basic definitions.

3.1 Markov Decision Processes

For a finite set Q, let $dist(Q)$ denote the set of probability distributions over Q.

Definition 7 (Markov Decision Process). *A* Markov Decision Process *(MDP) is a tuple* $\mathcal{M} = (Q, q_0, Steps)$ *where* Q *is a finite set of states,* $q_0 \in Q$ *is the initial state, and* $Steps \colon Q \to 2^{dist(Q)}$ *is the probability transition function.*

For a state q, a probabilistic transition corresponds to first nondeterministically choosing a probability distribution $\mu \in Steps(q)$ and then choosing the successor state q' probabilistically according to μ.

A path is a finite or infinite non-empty sequence $\pi = q_0 \xrightarrow{\mu_0} q_1 \xrightarrow{\mu_1} q_2 \dots$ where $\mu_i \in Steps(q_i)$ for every $i \geq 0$. We denote by $\pi(i)$ the i-th state along π (i.e., the state q_i), and by π^i the prefix of π ending at $\pi(i)$ (if it exists). For a finite path π, we denote by $last(\pi)$ the last state of π. A *scheduler* is a function that maps every finite path π of $\mathcal{M}$ to a distribution of $Steps(last(\pi))$.

For a given scheduler S, let $Paths^S$ denote all infinite paths $\pi = q_0 \xrightarrow{\mu_0} q_1 \xrightarrow{\mu_1} q_2 \dots$ starting in s_0 and satisfying $\mu_i = S(\pi^i)$ for every $i \geq 0$. We define a probability measure $Prob^S$ on $Paths^S$ in the usual way using cylinder sets [14].

We introduce the notion of rewards for an MDP.

---

[1] In [6] clusters are defined in a slightly different way.

Definition 8 (Reward). *A* reward function *for an MDP is a function* $rew \colon S \to \mathbb{R}_{\geq 0}$. *For a path π and a set of states F, the reward $R(F, \pi)$ until F is reached and the expected reward $E^S(F)$ to reach F are defined as*

$$R(F, \pi) := \sum_{i=0}^{\min\{j \mid \pi(j) \in F\}} rew(\pi(i)) \qquad E^S(F) := \int_{\pi \in Paths^S} R(F, \pi) dProb^S$$

where $R(F, \pi)$ is ∞ if the minimum does not exist.

3.2 Syntax and Semantics of Probabilistic Workflow Nets

We introduce Probabilistic Workflow Nets with Rewards, just called Probabilistic Workflow Nets or PWNs in the rest of the paper.

Definition 9 (Probabilistic Workflow Net with Rewards). *A Probabilistic Workflow Net with Rewards(PWN) is a tuple (P, T, F, i, o, w, r) where (P, T, F, i, o) is a 1-safe confusion-free workflow net, and $w, r \colon T \to \mathbb{R}^+$ are a weight function and a reward function, respectively.*

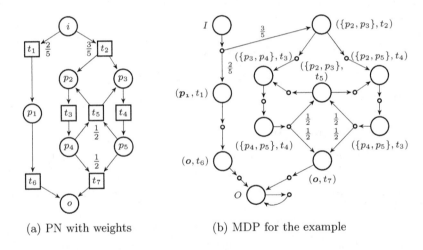

(a) PN with weights   (b) MDP for the example

Fig. 2. Running example

Figure 2a shows a free-choice PWN. All transitions have reward 1, and so only the weights are represented. Unlabeled transitions have weight 1.

The semantics of a PWN is an MDP with a reward function. Intuitively, the states of the MDP are pairs (M, t), where M is a marking, and t is the transition that was fired to reach M (since the same marking can be reached by firing different transitions, the MDP can have states (M, t_1), (M, t_2) for $t_1 \neq t_2$). Additionally there is a distinguished initial and final states I, O. The transition relation *Steps* is independent of the transition t, i.e., $Steps((M, t_1)) = Steps((M, t_2))$ for any two transitions t_1, t_2, and the reward of a state (M, t) is the reward of the transition t. Figure 2b shows the MDP of the PWN of Fig. 2a, representing only the states reachable from the initial state.

Definition 10 (Probability Distribution). *Let $W = (P, T, F, i, o, w, r)$ be a PWN, let M be a 1-safe marking of W enabling at least one transition, and let C be a conflict set enabled at M. The probability distribution $P_{M,C}$ over T is obtained by normalizing the weights of the transitions in C, and assigning probability 0 to all other transitions.*

Definition 11 (MDP and Reward Function of a PWN). *Let $W = (P, T, F, i, o, w, r)$ be a PWN. The MDP $M_W = (Q, q_0, Steps)$ of W is defined as follows:*

- $Q = (\mathcal{M} \times T) \cup \{I, O\}$ *where $\mathcal{M}$ are the 1-safe markings of W, and $q_0 = I$.*
- *For every transition t:*
  - *$Steps((o, t))$ contains exactly one distribution, which assigns probability 1 to state o, and probability 0 to all other states.*
  - *For every marking $M \neq o$ enabling no transitions, $Steps((M, t))$ contains exactly one distribution, which assigns probability 1 to (M, t), and probability 0 to all other states.*
  - *For every marking M enabling at least one transition, $Steps((M, t))$ contains a distribution μ_C for each conflict set C of transitions enabled at M. The distribution μ_C is defined as follows. For the states I, O: $\mu_C(I) = 0 = \mu_C(O)$. For each state (M', t') such that $t' \in C$ and $M \xrightarrow{t'} M'$: $\mu_C((M', t')) = P_{M,C}(t')$. For all other states (M', t'): $\mu_C((M', t')) = 0$.*
  - *$Steps(I) = Steps((i, t))$ for any transition t.*
  - *$Steps(O) = Steps((o, t))$ for any transition t.*

The reward function rew_W of W is defined by: $rew_W(I) = 0 = rew_W(O)$, and $rew_W((M, t)) = r(t)$.

In Fig. 2a, $Steps(i)$ is a singleton set that contains the probability distribution which assigns probability $\frac{2}{5}$ to the state (p_1, t_1) and probability $\frac{3}{5}$ to the state $(\{p_2, p_3\}, t_2)$. $Steps((\{p_2, p_3\}, t_2))$ contains two probability distributions, that assign probability 1 to $(\{p_5, p_3\}, t_4)$ and $(\{p_2, p_6\}, t_4)$, respectively.

We define a correspondence between firing sequences and MDP paths.

Definition 12. *Let W be a PWN, and let M_W be its associated MDP. Let $\sigma = t_1 t_2 \ldots t_n$ be a firing sequence of W. The path $\Pi(\sigma)$ of M_W corresponding to σ is $\pi_\sigma = I \xrightarrow{\mu_0} (M_1, t_1) \xrightarrow{\mu_1} (M_2, t_2) \xrightarrow{\mu_2} \ldots$, where $M_0 = i$ and for every $1 \leq k$:*

- *M_k is the marking reached by firing $t_1 \ldots t_k$ from i, and*
- *μ_k is the unique distribution of $Steps(M_{k-1}, t_{k-1})$ such that $\mu(t_k) > 0$.*

Let $\pi = I \xrightarrow{\mu_0} (M_1, t_1) \cdots (M_n, t_n)$ be a path of M_W. The sequence $\Sigma(\pi)$ corresponding to π is $\sigma_\pi = t_1 \ldots t_n$.

It follows immediately from the definition of M_W that the functions Π and Σ are inverses of each other. For a path π of the MDP that ends in state $last(\pi)$, the distributions in $Steps(last(\pi))$ are obtained from the conflict sets enabled after $\Sigma(\pi)$ has fired, if any. If no conflict set is enabled the choice is always trivial by construction. Therefore, a scheduler of the MDP $\mathcal{M}_W$ can be equivalently defined as a function that assigns to each firing sequence $\sigma \in T^*$ one of the conflict sets enabled after σ has fired. In our example, after t_2 fires, the conflict sets $\{t_3\}$ and $\{t_4\}$ are concurrently enabled. A scheduler chooses either $\{t_3\}$ or $\{t_4\}$. A possible scheduler always chooses $\{t_3\}$ every time the marking $\{p_2, p_3\}$ is reached, and produces sequences in which t_3 always occurs before t_4, while others may behave differently.

Convention: In the rest of the paper we define schedulers as functions from firing sequences to conflict sets.

In particular, this definition allows us to define the *probabilistic language* of a scheduler as the function that assigns to each finite firing sequence σ the probability of the cylinder of all paths that "follow" σ. Formally:

Definition 13 (Probabilistic Language of a Scheduler [20]). *The probabilistic language ν_S of a scheduler S is the function $\nu_S \colon T^* \to \mathbb{R}^+$ defined by $\nu_S(\sigma) = Prob^S(cyl^S(\Pi(\sigma)))$. A transition sequence σ is produced by S if $\nu_S(\sigma) > 0$.*

The reward function extends to transition sequences in the natural way by taking the sum of all rewards. In pictures, we label transitions with pairs (w, c), where w is a weight and c a reward. See for example Fig. 3a.

We now introduce the expected reward of a PWN under a scheduler.

Definition 14 (Expected Reward of a PWN Under a Scheduler). *Let W be a PWN, and let S be a scheduler of its MDP $\mathcal{M}_W$. The expected reward $V^S(W)$ of W under S is the expected reward $E^S(O)$ to reach the final state O of $\mathcal{M}_W$.*

Given a firing sequence σ, we have $r(\sigma) = R(O, \Pi(\sigma))$ by the definition of the reward function and the fact that O can only occur at the very end of π_σ.

Lemma 2. *Let W be a sound PWN, and let S be a scheduler. Then $V^S(W)$ is finite and $V^S(W) = \sum_{\pi \in \Pi} R(O, \pi) \cdot Prob^S(cyl^S(\pi)) = \sum_{\sigma \in Fin_W} r(\sigma) \cdot \nu_S(\sigma)$, where Π_O are the paths of the MDP $\mathcal{M}_W$ leading from the initial state I to the state O (without looping in O).*

3.3 Expected Reward of a PWN

Using a result by Varacca and Nielsen [20], we prove that the expected reward of a PWN is the same for all schedulers, which allows us to speak of "the" expected reward of a PWN. We first define partial schedulers.

Definition 15 (Partial Schedulers). *A partial scheduler of length n is the restriction of a scheduler to firing sequences of length less than n. Given two partial schedulers S_1, S_2 of lengths n_{S_1}, n_{S_2}, we say that S_1 extends S_2 if $n_{S_1} \geq$*

n_{S_2} and S_2 is the restriction of S_1 to firing sequences of length less than n_{S_2}. The probabilistic language ν_S of a partial scheduler S of length n is the function $\nu_S \colon T^{\leq n} \to \mathbb{R}^+$ defined by $\nu_S(\sigma) = Prob^S(cyl^S(\Pi(\sigma)))$. A transition sequence σ is produced by S if $\nu_S(\sigma) > 0$.

Observe that if σ is not a firing sequence, then $\nu_S(\sigma) = 0$ for every scheduler S. In our running example there are exactly two partial schedulers S_1, S_2 of length 2; after t_2 they choose t_3 or t_4, respectively: $S_1 \colon \epsilon \mapsto \{t_1, t_2\}$ $t_1 \mapsto \{t_6\}$ $t_2 \mapsto \{t_3\}$ and $S_2 \colon \epsilon \mapsto \{t_1, t_2\}$ $t_1 \mapsto \{t_6\}$ $t_2 \mapsto \{t_4\}$. We have $\nu_{S_1}(t_2 t_3) = 3/5$, and $\nu_{S_2}(t_2 t_3) = 0$.

For finite transition sequences, Mazurkiewicz equivalence, denoted by $\equiv$, is the smallest congruence such that $\sigma t_1 t_2 \sigma' \equiv \sigma t_2 t_1 \sigma'$ for every $\sigma, \sigma' \in T^*$ and for any two *independent* transitions t_1, t_2 [17]. We extend Mazurkiewicz equivalence to partial schedulers.

Definition 16 (Mazurkiewicz Equivalence of Partial Schedulers). *Given a partial scheduler S of length n, we denote by F_S the set of firing sequences σ of $\mathcal{W}$ produced by S such that either $|\sigma| = n$ or σ leads to a marking that enables no transitions.*

Two partial schedulers S_1, S_2 with probabilistic languages ν_{S_1} and ν_{S_2} are Mazurkiewicz equivalent, denoted $S_1 \equiv S_2$, if they have the same length and there is a bijection $\phi \colon F_{S_1} \to F_{S_2}$ such that $\sigma \equiv \phi(\sigma)$ and $\nu_{S_1}(\sigma) = \nu_{S_2}(\phi(\sigma))$ for every $\sigma \in F_n$.

The two partial schedulers of our running example are not Mazurkiewicz equivalent. Indeed, we have $F_{S_1} = \{t_1 t_6, t_2 t_3\}$ and $F_{S_2} = \{t_1 t_6, t_2 t_4\}$, and no bijection satisfies $\sigma \equiv \phi(\sigma)$ for every $\sigma \in F_{S_1}$. We can now present the main result of [20], in our terminology and for PWNs.[2]

Theorem 1 (Equivalent Extension of Schedulers [20][3]). *Let S_1, S_2 be two partial schedulers. There exist two partial schedulers S_1', S_2' such that S_1' extends S_1, S_2' extends S_2 and $S_1' \equiv S_2'$.*

In our example, S_1 can be extended to S_1' by adding $t_1 t_6 \mapsto \emptyset$ and $t_2 t_3 \mapsto t_4$, and S_2 to S_2' by adding $t_1 t_6 \mapsto \emptyset$ and $t_2 t_4 \mapsto t_3$. Now we have $F_{S_1'} = \{t_1 t_6, t_2 t_3 t_4\}$ and $F_{S_2'} = \{t_1 t_6, t_2 t_4 t_3\}$. The obvious bijection shows $S_1' \equiv S_2'$, because we have $t_2 t_3 t_4 \equiv t_2 t_4 t_3$ and $\nu_{S_1'}(t_2 t_3 t_4) = 3/5 = \nu_{S_2'}(t_2 t_4 t_3)$.

Using Theorem 1, we are able to prove one of our central theorems.

Theorem 2. *Let $\mathcal{W}$ be a PWN. There exists a value v such that for every scheduler S of $M_{\mathcal{W}}$, the expected reward $V^S(\mathcal{W})$ is equal to v.*

---

[2] In [20], enabled conflict sets are called actions, and markings are called cases.

[3] Stated as Theorem 2, the original paper gives this theorem with S_1' and S_2' being (non-partial) schedulers. However, in the paper equivalence is only defined for partial schedulers and the schedulers constructed in the proof are also partial.

Proof Sketch. Given two schedulers, we construct a bijection between the transition sequences they produce that end in the final marking. This bijection maps each transition sequence to a Mazurkiewicz equivalent one. To do this, for each $k \geq 0$ we reduce the schedulers to partial schedulers of length k, extend them to equivalent schedulers using Theorem 1, and map every sequence of length k one of them produces to a Mazurkiewicz equivalent one of the other. Since equivalent transition sequences have the same reward, applying Lemma 2 yields that the values of the two schedulers are equal.

Free-choice PWNs. By Proposition 1, in free-choice PWNs the conflict set of a given transition is its cluster, and so its probability is the same at any reachable marking enabling it. We label a transition directly with this probability.

Convention: We assume that the weights of the transitions of a cluste are normalized, i.e. the weights are already a probability distribution.

In the next section we present a reduction algorithm that decides if a given free-choice PWN is sound or not, and if sound computes its expected reward. If the PWN is unsound, then we just apply the following lemma:

Lemma 3. *The expected reward of an unsound free-choice PWN is infinite.*

4   Reduction Rules

We transform the reduction rules of [9] for non-probabilistic (colored) workflow nets into rules for probabilistic workflow nets.

Definition 17 (Rules, Correctness, and Completeness). *A rule R is a binary relation on the set of PWNs. We write $\mathcal{W}_1 \xrightarrow{R} \mathcal{W}_2$ for $(\mathcal{W}_1, \mathcal{W}_2) \in R$.*

A rule R is correct *if $\mathcal{W}_1 \xrightarrow{R} \mathcal{W}_2$ implies that $\mathcal{W}_1$ and $\mathcal{W}_2$ are either both sound or both unsound, and have the same expected reward.*

A set $\mathcal{R}$ of rules is complete *for a class of PWNs if for every sound PWN $\mathcal{W}$ in that class there exists a sequence $\mathcal{W} \xrightarrow{R_1} \mathcal{W}_1 \cdots \xrightarrow{R_n} \mathcal{W}'$ such that $\mathcal{W}'$ is a PWN consisting of a single transition t between the two only places i and o.*

As in [9], we describe rules as pairs of a *guard* and an *action*. $\mathcal{W}_1 \xrightarrow{R} \mathcal{W}_2$ holds if $\mathcal{W}_1$ satisfies the guard, and $\mathcal{W}_2$ is a possible result of applying the action to $\mathcal{W}_1$.

Merge Rule. The *merge rule* merges two transitions with the same input and output places into one single transition. The weight of the new transition is the sum of the old weights, and the reward is the weighted average of the reward of the two merged transitions.

Guard: $\mathcal{W}$ contains two transitions $t_1 \neq t_2$ such that ${}^\bullet t_1 = {}^\bullet t_2$ and $t_1^\bullet = t_2^\bullet$.

Action: (1) $T := (T \setminus \{t_1, t_2\}) \cup \{t_m\}$, where t_m is a fresh name.
(2) $t_m^\bullet := t_1^\bullet$ and ${}^\bullet t_m := {}^\bullet t_1$.
(3) $r(t_m) := w(t_1) \cdot r(t_1) + w(t_2) \cdot r(t_2)$.
(4) $w(t_m) = w(t_1) + w(t_2)$.

Iteration Rule. Loosely speaking, the iteration rule removes arbitrary iterations of a transition by adjusting the weights of the possible successor transitions. The probabilities are normalized again and the reward of each successor transition increases by a geometric series dependent on the reward and weight of the removed transition.

Guard: $\mathcal{W}$ contains a cluster c with a transition $t \in c$ such that $t^\bullet = {}^\bullet t$.

Action: (1) $T := (T \setminus \{t\})$.

    (2) For all $t' \in c \setminus \{t\}$: $r(t') := \frac{w(t)}{1-w(t)} \cdot r(t) + r(t')$

    (3) For all $t' \in c \setminus \{t\}$: $w(t') := \frac{w(t')}{1-w(t)}$

Observe that $\frac{w(t)}{1-w(t)} \cdot r(t) = (1 - w(t)) \cdot \sum_{i=0}^{\infty} w(t)^i \cdot i \cdot r(t)$ captures the fact that t can be executed arbitrarily often, each execution yields the reward $r(t)$, and eventually some other transition occurs. For an example of an application of the iteration rule, consult Fig. 3b and c. Transition t_9 has been removed and as a result the label of transition t_7 changed.

Shortcut rule. The shortcut rule merges transitions of two clusters into one single transition with the same effect. The reward of the new transition is the sum of the rewards of the old transitions, and its weight the product of the old weights.

A transition t *unconditionally enables* a cluster c if ${}^\bullet t' \subseteq t^\bullet$ for some transition $t' \in c$. Observe that if t unconditionally enables c then any marking reached by firing t enables every transition in c.

Guard: $\mathcal{W}$ contains a transition t and a cluster $c \neq [t]$ such that t unconditionally enables c.

Action: (1) $T := (T \setminus \{t\}) \cup \{t'_s \mid t' \in c\}$, where t'_s are fresh names.

    (2) For all $t' \in c$: ${}^\bullet t'_s := {}^\bullet t$ and $t'^\bullet_s := (t^\bullet \setminus {}^\bullet t') \cup t'^\bullet$.

    (3) For all $t' \in c$: $r(t'_s) := r(t) + r(t')$.

    (4) For all $t' \in c$: $w(t'_s) = w(t) \cdot w(t')$.

    (5) If ${}^\bullet p = \emptyset$ for all $p \in c$, then remove c from $\mathcal{W}$.

For an example shortcut rule application, compare the example of Fig. 2a with the net in Fig. 3a. The transition t_1 which unconditionally enabled the cluster $[t_6]$ has been shortcut, a new transition t_8 has been created, and t_1, p_1 and t_6 have been removed.

Theorem 3. *The merge, shortcut and iteration rules are correct for PWNs.*

Proof. That the rules preserve soundness is shown in [9]. To show that the rules preserve the expected reward we use Theorem 2: For each rule, we carefully choose schedulers for the PWNs before and after the application of the rule, and show that their expected rewards are equal. We sketch the idea for the shortcut rule. Let $\mathcal{W}_1$, $\mathcal{W}_2$ be such that $\mathcal{W}_1 \xrightarrow{\text{shortcut}} \mathcal{W}_2$. Let c, t be as in Definition 4. Let S_1 be a scheduler for $\mathcal{W}_1$ such that $S_1(\sigma_1) = c$ if σ_1 ends with t. We define a bijection ϕ that maps firing sequences in $\mathcal{W}_2$ to firing sequences in $\mathcal{W}_1$ by replacing every occurrence of t'_s by $t t'$. We define a scheduler S_2 for $\mathcal{W}_2$ by $S_2(\sigma_2) = S_1(\phi(\sigma_2))$. Let now σ_2 be a firing sequence in $\mathcal{W}_2$ and let $\sigma_1 = \phi(\sigma_2)$.

We prove that σ_1 and σ_2 have the same reward and also $\nu_{S_1}(\sigma_1) = \nu_{S_2}(\sigma_2)$. Indeed, since the only difference is that every occurrence of t'_s is replaced by $t\,t'$ and $r(t'_s) = r(t) + r(t')$ and $w(t'_s) = w(t)w(t')$ by the definition of the shortcut rule, the reward must be equal and $\nu_{S_1}(\sigma_1) = \nu_{S_2}(\sigma_2)$. We now use these equalities, the fact that there is a bijection between firing sequences that end with the final marking, and Lemma 2:

$$V(\mathcal{W}_2) = \sum_{\sigma_2 \in Fin_{\mathcal{W}_2}} r(\sigma_2) \cdot \nu_{S_2}(\sigma) = \sum_{\sigma_2 \in Fin_{\mathcal{W}_2}} r(\phi(\sigma_2)) \cdot \nu_{S_1}(\phi(\sigma_2))$$

$$= \sum_{\sigma_1 \in Fin_{\mathcal{W}_1}} r(\sigma_1) \cdot \nu_{S_1}(\sigma_1) = V(\mathcal{W}_1). \qquad \square$$

In [9] we provide a reduction algorithm for non-probabilistic free-choice workflow, and prove the following result.

Theorem 4 (Completeness [9]). *The reduction algorithm summarizes every sound free choice workflow net in at most $\mathcal{O}(|C|^4 \cdot |T|)$ applications of the shortcut rule and $\mathcal{O}(|C|^4 + |C|^2 \cdot |T|)$ applications of the merge and iteration rules, where C is the set of clusters of the net. Any unsound free-choice workflow nets can be recognized as unsound in the same number of rule applications.*

We illustrate a complete reduction by reducing the example of Fig. 2a. We set the reward for each transition to 1, so the expected reward of the net is the expected number of transition firings until the final marking is reached. Initially, t_1 unconditionally enables $[t_6]$ and we apply the shortcut rule. Since $[t_6] = \{t_6\}$, exactly one new transition t_8 is created. Furthermore t_1, p_1 and t_6 are removed (Fig. 3a). Now, t_5 unconditionally enables $[t_3]$ and $[t_4]$. We apply the shortcut rule twice and call the result t_9 (Fig. 3b). Transition t_9 now satisfies the guard of the iteration rule and can be removed, changing the label of t_7 (Fig. 3c). Since t_2 unconditionally enables $[t_3]$ and $[t_4]$, we apply the shortcut rule twice and call the result t_{10} (Fig. 3d). After short-cutting t_{10}, we apply the merge rule to the two remaining transitions, which yields a net with one single transition labeled by $(1, 5)$ (Fig. 3e). So the net terminates with probability 1 after firing 5 transitions in average.

Fixing a Scheduler. Since the expected reward of a PWN $\mathcal{W}$ is independent of the scheduler, we can fix a scheduler S and compute the expected reward $V^S(\mathcal{W})$. This requires to compute only the Markov chain induced by S, which can be much smaller than the MDP. However, it is easy to see that this idea does not lead to a polynomial algorithm. Consider the free-choice PWN of Fig. 4, and the scheduler that always chooses the largest enabled cluster according to the order $\{t_{11}, t_{12}\} > \cdots > \{t_{n1}, t_{n2}\} > \{u_{11}\} > \{u_{12}\} > \cdots > \{u_{n1}\} > \{u_{n2}\}$. Then for every subset $K \subset \{1, \ldots, n\}$ the Markov chain contains a state enabling $\{u_{i1} \mid i \in K\} \cup \{u_{i2} \mid i \notin K\}$, and has therefore exponential size. There might be a procedure to find a suitable scheduler for a given PWN such that the Markov chain has polynomial size, but we do not know of such a procedure.

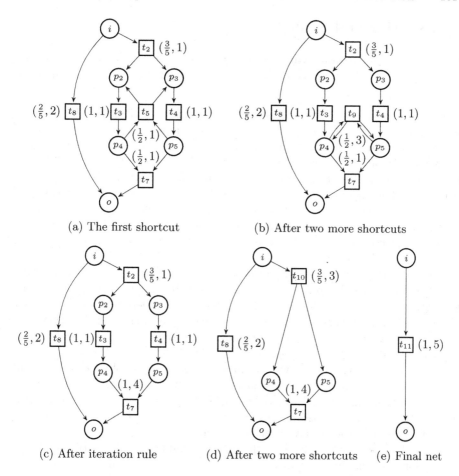

Fig. 3. Example of reduction

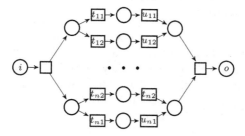

Fig. 4. Example

5 Experimental Evaluation

We have implemented our reduction algorithm as an extension of the algorithm described in [9]. In this section we report on its performance and on a comparison with PRISM [15].

Industrial benchmarks. The benchmark suite consists of 1385 free-choice workflow nets, previously studied in [11], of which 470 nets are sound. The workflows correspond to business models designed at IBM. Since they do not contain probabilistic information, we assigned to each transition t the probability $1 / |[t]|$ (i.e., the probability is distributed uniformly among the transitions of a cluster). We study the following questions, which can be answered by both our algorithm and PRISM: Is the probability to reach the final marking equal to one (equivalent to "is the net sound?"). And if so, how many transitions must be fired in average to reach the final marking? (This corresponds to a reward function assigning reward 1 to each transition.)

All experiments were carried out on an i7-3820 CPU using 1 GB of memory.

PRISM has three different analysis engines able to compute expected rewards: explicit, sparse and symbolic (bdd). In a preliminary experiment with a timeout of 30 s, we observed that the explicit engine clearly outperforms the other two: It solved 1309 cases, while the bdd and sparse engines only solved 636 and 638 cases, respectively. Moreover, 418 and 423 of the unsolved cases were due to memory overflow, so even with a larger timeout the explicit engine is still leading. For this reason, in the comparison we only used the explicit engine.

After increasing the timeout to 10 min, the explicit engine did not solve any further case, leaving 76 cases unsolved. This was due to the large state space of the nets: 69 out of the 76 have over 10^6 reachable states.

The 1309 cases were solved by the explicit engine in 353 s, with about 10 s for the larger nets. Our implementation solved all 1385 cases in 5 s combined. It never needs more than 20 ms for a single net, even for those with more than 10^7 states.

In the unsound case, our implementation still reduces the reachable state space, which makes it easier to apply state exploration tools. After reduction, the 69 nets with at least 10^6 states had an average of 5950 states, with the largest at 313443 reachable states.

An Academic Benchmark. Many workflows in our suite have a large state space because of fragments modeling the following situation. Multiple processes do a computation step in parallel, after which they synchronize. Process i may execute its step normally with probability p_i, or a failure may occur with probability $1 - p_i$, which requires to take a recovery action and therefore has a higher cost. Such a scenario is modeled by the free-choice PWNs net of Fig. 5a, where the probabilities and costs are chosen at random. The scenario can also be easily modeled in PRISM. Figure 5b shows the time needed by the three PRISM engines and by our implementation for computing the expected reward using a time limit of 10 min. The number of reachable states grows exponentially in the number processes, and the explicit engine runs out of memory for 15 processes, the symbolic engine times

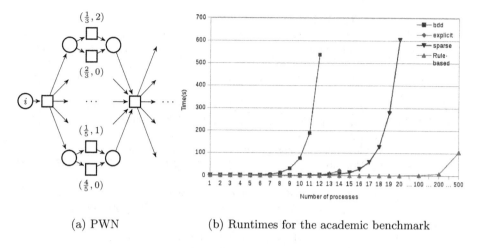

(a) PWN   (b) Runtimes for the academic benchmark

Fig. 5. Academic benchmark

out for 13 processes, and the sparse engine reaches the time limit at 20 processes. However, since the rule-based approach does not need to construct the state space, we can easily solve the problem with up to 500 processes.

6   Conclusion

We have presented a set of reduction rules for PWNs with rewards that preserve soundness and the expected reward of the net, and are complete for free-choice PWNs. While the semantics and the expected reward are defined via an associated MDP, our rules work directly on the workflow net. The rules lead to the first polynomial-time algorithm to compute the expected reward.

In future work we want to generalize our algorithm to compute the probability of non-termination and the conditional expected reward under termination, which is of interest in the unsound case, and also to compute the expected time to termination for timed workflow nets.

Acknowledgments. We thank the anonymous referees for their comments, and especially the one who helped us correct a mistake in Lemmas 2 and 3.

References

1. Van der Aalst, W.: The application of Petri nets to workflow management. J. Circuits Syst. Comput. **8**(1), 21–66 (1998)
2. Van der Aalst, W., Van Hee, K.M.: Workflow Management: Models, Methods, and Systems. MIT press, Cambridge (2004)
3. Abbes, S., Benveniste, A.: True-concurrency probabilistic models: Markov nets and a law of large numbers. Theoret. Comput. Sci. **390**(2–3), 129–170 (2008)
4. Abbes, S., Benveniste, A.: Concurrency, σ-algebras, and probabilistic fairness. In: de Alfaro, L. (ed.) FOSSACS 2009. LNCS, vol. 5504, pp. 380–394. Springer, Heidelberg (2009)

5. Desel, J., Erwin, T.: Modeling, simulation and analysis of business processes. In: Aalst, W.M.P., Desel, J., Oberweis, A. (eds.) Business Process Management: Models, Techniques, and Empirical Studies. LNCS, vol. 1806, pp. 129–141. Springer, Heidelberg (2000)
6. Desel, J., Esparza, J.: Free Choice Petri Nets, vol. 40. Cambridge University Press, Cambridge (2005)
7. Eisentraut, C., Hermanns, H., Katoen, J.-P., Zhang, L.: A semantics for every GSPN. In: Colom, J.-M., Desel, J. (eds.) PETRI NETS 2013. LNCS, vol. 7927, pp. 90–109. Springer, Heidelberg (2013)
8. Esparza, J.: Decidability and complexity of Petri net problems – an introduction. In: Reisig, W., Rozenberg, G. (eds.) APN 1998. LNCS, vol. 1491, pp. 374–428. Springer, Heidelberg (1998)
9. Esparza, J., Hoffmann, P.: Reduction rules for colored workflow nets. In: Stevens, P., et al. (eds.) FASE 2016. LNCS, vol. 9633, pp. 342–358. Springer, Heidelberg (2016). doi:10.1007/978-3-662-49665-7_20
10. Eszarza, J., Hoffmann, P., Saha, R.: Polynomial analysis algorithms for free choice probabilistic workflow nets (2016). arXiv:1606.00175 [cs.LO]
11. Fahland, D., Favre, C., Jobstmann, B., Koehler, J., Lohmann, N., Völzer, H., Wolf, K.: Instantaneous soundness checking of industrial business process models. In: Dayal, U., Eder, J., Koehler, J., Reijers, H.A. (eds.) BPM 2009. LNCS, vol. 5701, pp. 278–293. Springer, Heidelberg (2009)
12. Favre, C., Fahland, D., Völzer, H.: The relationship between workflow graphs and free-choice workflow nets. Inf. Syst. **47**, 197–219 (2015)
13. Favre, C., Völzer, H., Müller, P.: Diagnostic information for control-flow analysis of workflow graphs (a.k.a. Free-Choice workflow nets). In: Chechik, M., Raskin, J.-F. (eds.) TACAS 2016. LNCS, vol. 9636, pp. 463–479. Springer, Heidelberg (2016). doi:10.1007/978-3-662-49674-9_27
14. Kemeny, J.G., Snell, J.L., Knapp, A.W.: Denumerable Markov Chains: With a Chapter of Markov Random Fields by David Griffeath. GTM, vol. 40. Springer Science & Business Media, New York (2012)
15. Kwiatkowska, M., Norman, G., Parker, D.: PRISM 4.0: verification of probabilistic real-time systems. In: Gopalakrishnan, G., Qadeer, S. (eds.) CAV 2011. LNCS, vol. 6806, pp. 585–591. Springer, Heidelberg (2011)
16. Magnani, M., Cucci, F.: BPMN: how much does it cost? An incremental approach. In: Alonso, G., Dadam, P., Rosemann, M. (eds.) BPM 2007. LNCS, vol. 4714, pp. 80–87. Springer, Heidelberg (2007)
17. Mazurkiewicz, A.: Trace theory. In: Brauer, W., Reisig, W., Rozenberg, G. (eds.) APN 1986. LNCS, vol. 255, pp. 278–324. Springer, Heidelberg (1987)
18. Saeedi, K., Zhao, L., Sampaio, P.R.F.: Extending BPMN for supporting customer-facing service quality requirements. In: ICWS 2010, pp. 616–623. IEEE Computer Society (2010)
19. Sampath, P., Wirsing, M.: Evaluation of cost based best practices in business processes. In: Halpin, T., Nurcan, S., Krogstie, J., Soffer, P., Proper, E., Schmidt, R., Bider, I. (eds.) BPMDS 2011 and EMMSAD 2011. LNBIP, vol. 81, pp. 61–74. Springer, Heidelberg (2011)
20. Varacca, D., Nielsen, M.: Probabilistic Petri nets and Mazurkiewicz equivalence. Unpublished Manuscript (2003). http://www.lacl.fr/~dvaracca/works.html. Accessed 27 May 2016
21. Varacca, D., Völzer, H., Winskel, G.: Probabilistic event structures and domains. Theoret. Comput. Sci. **358**(2–3), 173–199 (2006)

Queueing Models

Energy-Aware Server with SRPT Scheduling: Analysis and Optimization

Misikir Eyob Gebrehiwot[(✉)], Samuli Aalto, and Pasi Lassila

School of Electrical Engineering, Aalto University, Espoo, Finland
{misikir.gebrehiwot,samuli.aalto,pasi.lassila}@aalto.fi

Abstract. We consider the optimal energy-aware control of a single server in a server farm. The server is modeled as an M/G/1 queue with a particular control policy that allows to put the server to a sleep mode to save energy with an additional delay cost, the setup delay, after the server is turned on again. Our main result is the derivation of mean response time for such a system under SRPT scheduling. In particular, we show that the mean response time can be decomposed into two parts: the mean response time of an ordinary M/G/1-SRPT, and an additional penalty term for switching the server to a sleep state. Furthermore, we study the energy-performance optimization of the system and prove that, for the Energy Response time Weighted Sum (ERWS) and Energy Response time Product (ERP) cost metrics, the optimal control either puts the server into a sleep state immediately when it becomes idle or keeps it idling until the next job arrives.

Keywords: Performance-energy trade-off $\cdot$ $M/G/1$-SRPT $\cdot$ Setup delay

1 Introduction

Server farms in data centres are known to spend a substantial proportion of time in an idle state, having a utilization factor in the range of 10–20 % [1]. While in this state, a server wastes about 60–70 % of the peak power it draws to process requests [2]. This has inspired the study of low energy sleep states to be used whenever the server becomes idle [7,8].

However, the energy saving attained by using a sleep state comes at a performance cost. This is due to the fact that such sleep states are characterized by long setup delays required to turn the server back on to a functional state. Many stochastic models that study this trade-off between performance and energy consumption have been developed and studied by the research community [3,5,6,9,13]. The most common cost metrics utilized to capture the trade-off are the Energy Response time Weighted Sum (ERWS),

$$w_1\mathrm{E}[T] + w_2\mathrm{E}[P],\tag{1}$$

and the Energy Response time Product (ERP),

$$\mathrm{E}[T]\mathrm{E}[P],\tag{2}$$

© Springer International Publishing Switzerland 2016
G. Agha and B. Van Houdt (Eds.): QEST 2016, LNCS 9826, pp. 107–122, 2016.
DOI: 10.1007/978-3-319-43425-4_7

where $E[T]$ and $E[P]$ are the mean response time and mean power consumption of the server. The weights w_1 and w_2 are constants that can be chosen based on the need to emphasize either performance or energy saving.

In the case of an ordinary M/G/1 system, Shortest Remaining Processing Time (SRPT) scheduling is known to minimize mean response time [10]. However, its behaviour in such an energy-aware setup is largely unknown. In this paper, we study the energy-performance optimization of an energy-aware M/G/1 system under SRPT scheduling. In addition to the common *busy* and *idle* states, we assume the system is capable of switching to a *sleep* state, which should be followed by a transient *setup* state.

We apply a similar approach as in [11] to study the energy-aware system, although in this case, the derivation of mean response time is rather involved, and requires dividing the process into parts. We show that the final form of the mean response time has a simple structure consisting of the mean response time of the ordinary M/G/1 SRPT queue and an additional penalty term related to the setup delay. Similar decomposition results for the mean response time of systems with setup delay have been identified, e.g., in [3–5] for other scheduling policies. Moreover, the form turns out to be such that finally the optimal control remains the same as for the FIFO and PS systems studied in [5,6]. That is, depending on the parameters, either the server is immediately switched off to sleep state after becoming idle, or the server never goes to sleep state. Additionally, our comparison shows that SRPT still outperforms FIFO and PS.

The paper is organized as follows. In the following section a formal definition of the model is given. In Sect. 3 we analyse the mean response time of the system, and Sect. 4 briefly discusses optimization. Numerical examples are given in Sect. 5 and in Sect. 6 we conclude the paper.

2   Model

We consider an energy-aware $M/G/1$-SRPT queue to which jobs arrive according to a Poisson process with rate λ. Let S denote a generic service time having a continuous valued cumulative distribution function $F(s)$ and density $f(s)$. The tail probability is denoted by $\bar{F}(s)$. The load on the system, ρ, is given by $\rho = \lambda E[S]$. We assume a stable system, i.e., $\rho < 1$.

The energy-aware system is controlled as follows. The server is kept busy until all the jobs in the system are served according to the SRPT service discipline. When the last job leaves the system and the server becomes idle, a timer I, which is a generally distributed random variable, is set. If a job arrives before the timer expires, the timer is reset and the server starts serving the job. On the other occasion where the timer expires before a job arrives, the server is switched into a sleep state where it cannot serve any more jobs until it is turned back on. The length of the sleep period is controlled by counting the number of arrivals. The server is switched on as soon as k jobs have been accumulated. Even then, the server has to transit through a setup period, during which it cannot serve any job. We represent this period by a generally distributed random variable D.

In addition, let I^{tot} denote the total idle period accumulated between successive expirations of the idling timer. Finally, we denote by Π the whole family of these control policies parameterized by I and k.

Note that in this model we have four states characterized by their own power consumption values of P_{busy}, P_{idle}, P_{sleep}, and P_{setup} with a natural ordering $P_{\text{busy}} \geq P_{\text{setup}} > P_{\text{idle}} > P_{\text{sleep}} = 0$. We denote the mean power consumption of the system by $E[P]$, and the mean response time of a job by $E[T]$.

3   Analysis

In this section we derive the mean response time of the energy-aware SRPT queue described in the preceding section.

Schrage and Miller [11] derived the well known closed form expression for the mean response time of an ordinary M/G/1-SRPT queue, with no sleep and setup states,

$$E[T_{\text{M/G/1-SRPT}}] = \frac{\lambda}{2} \int_0^\infty \frac{\int_0^s t^2 f(t)\, dt + s^2 \bar{F}(s)}{(1 - \rho(s))^2} f(s)\, ds + \int_0^\infty \frac{\bar{F}(s)}{1 - \rho(s)} ds, \quad (3)$$

where $\rho(s)$ denotes the proportion of time the server is busy processing jobs that are originally shorter than s given by

$$\rho(s) = \lambda \int_0^s t f(t)\, dt. \quad (4)$$

Here the response time of a job is composed of waiting time and residence time. The *waiting time* of a job is defined as the time elapsed between its arrival time and the time at which it gets service for the first time. The *residence time* covers the remainder of the response time, which includes its processing time plus any additional time spent waiting due to pre-emption imposed by SRPT.

Let us now consider the mean response time in the energy-aware system using the same decomposition. By denoting the waiting and residence times by W and R, respectively, we have $E[T] = E[W] + E[R]$.

The energy-aware system under study is work conserving in the sense that it goes to sleep only when there is no job to serve. Hence, it is easy to see that the mean residence time is not affected by the introduction of sleep and setup states so that

$$E[R] = \int_0^\infty \frac{\bar{F}(s)}{1 - \rho(s)} ds. \quad (5)$$

So we can focus on the derivation of the mean waiting time.

3.1   Mean Waiting Time

To determine the mean waiting time $E[W]$, we apply the theory of regenerative processes by setting the regeneration point at the time epoch where the idling timer expires. Following the same approach as in [11], we first derive the mean conditional waiting time, $E[W(s)]$, for a *test job* of size s. Then the mean waiting time can be calculated as

$$E[W] = \int_0^\infty E[W(s)]f(s)\,ds.$$

We refer to the test job as a type-s job.

During one full regenerative cycle, the server goes to *sleep*, waits for k jobs to arrive, then transits through *setup*, and starts serving the accumulated jobs in the *busy* state. After alternating between busy and *idle* states, it eventually goes back to the sleep state, marking the end of one (regenerative) cycle. The test job (of size s) may arrive during any of these periods. As illustrated in Fig. 1, the entire busy period is further divided into several parts depending on when the test job arrives to the system to systematically address the derivation of mean waiting time of the test job.

We refer to the duration spent in the sleep and setup states as *period 1* and *period 2*, respectively. The busy period, i.e., time during which the server is in the busy state, is composed of several parts. *Period 3*, *period 4*, and *period 5* denote the durations of the associated type-s busy periods to be discussed below during which jobs smaller than s are served. Note that, as shown in the figure, period 3 consists of a continuous interval of time, but periods 4 and 5 in fact may consist of several parts and the period length is the sum of the respective parts. During a cycle, *period 6* has a similar structure and it corresponds to the time during which jobs larger than s are being served. Finally, *period 7* represents the time the system is in the idle state, which may also consist of several parts as the system may visit the idle state several times before the eventual timer I expiration happens. Recall that the total length of period 7 is denoted by I^{tot}.

Let B denote an entire busy period, and $B(s)$ be a *type-s busy period*, in which all the jobs getting service have the remaining size less than s. In particular, all jobs with size less than s that arrive during a type-s busy period are served in this busy period. A type-s busy period can be started in three different ways during the busy period.

Out of the jobs that accumulate during the sleep and setup states, each job with original size shorter than s starts its own type-s sub-busy period. The complete length of the type-s busy period induced by all jobs with sizes less than s that arrived during periods 1 and 2 is denoted by $B_3(s)$. Note that during a cycle there is at most one such type-s busy period, see period 3 in Fig. 1.

On the other hand, all the remaining jobs accumulated in the sleep and setup states, with sizes larger than s, would eventually start their own type-s busy periods when the remaining size is reduced to s. We denote this kind of a type-s busy period as $B_4(s)$. Here the type-s busy period is started by exactly one job, and the whole period 4 may consist of multiple $B_4(s)$ busy periods, see period 4 in Fig. 1. However, they all will be served before the system enters the idle state for the first time after setup.

The third kind of type-s busy period is started due to an arrival in the busy state after the sleep and setup or during the idle period. It is similar to the type-s busy period in an ordinary M/G/1-SRPT queue (denoted by $Y(p)$ in [11]).

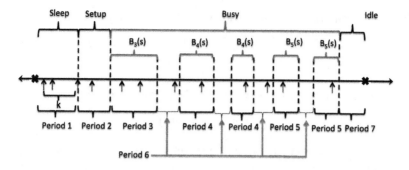

Fig. 1. A complete cycle between two expirations of idling timer (represented by the cross marks at either ends of the horizontal line).

We denote it by $B_5(s)$ in this paper. Here again the type-s busy period is started by exactly one job, and period 5 may consist of multiple $B_5(s)$ busy periods, see period 5 in Fig. 1. Also note that when the system becomes idle for the first time during a cycle, all type-s busy periods are $B_5(s)$ busy periods after that.

Consider now a test job of size s that arrives at a random time instant. Let p_i represent the probability that the test job arrives during period i. Thus, the mean conditional waiting time of the type-s job can be derived by conditioning over its arrival time as

$$E[W(s)] = \sum_{i=1}^{7} p_i E[W_i(s)], \tag{6}$$

where $E[W_i(s)]$ is the mean conditional waiting time of the job assuming it arrives in period i. We provide the derivation of these arrival probabilities and mean conditional waiting times below in 1° and 2°, respectively.

1° *Arrival Probabilities.* Let N denote the number of jobs served during an entire regeneration cycle. The mean number of arrivals $E[N]$ has been derived in [5] for FIFO, and since it is the same for all work-conserving policies, we have

$$E[N] = \frac{k + \lambda E[D] + \lambda E[I^{tot}]}{1 - \rho}, \tag{7}$$

where $E[I^{tot}]$ is the mean cumulative idling time in a cycle. Note that I^{tot} consists of i.i.d. idling times distributed as $\min\{I, A\}$, where A denotes an inter-arrival time. The number of such idling times has a geometric distribution with success probability $P\{I < A\}$. Thus, $E[I^{tot}]$ is given by

$$E[I^{tot}] = \frac{E[\min\{I, A\}]}{P\{I < A\}}. \tag{8}$$

The probability p_i that the test job arrives during period i is given by the ratio of the mean number of arrivals during period i to $E[N]$.

Now, the proportion of time the server spends on jobs whose remaining processing time is less than s is given by

$$a(s) = \rho(s) + \lambda s \bar{F}(s). \tag{9}$$

On the other hand, $p_3 + p_4 + p_5$ represents the same proportion of time. Thus,

$$p_3 + p_4 + p_5 = a(s).$$

Finally, the proportion of time that the server is busy, $p_3 + p_4 + p_5 + p_6$, must be equal to ρ,

$$p_3 + p_4 + p_5 + p_6 = \rho.$$

When considering p_4, we observe that the random number of jobs accumulated at the start of the busy period (i.e., after sleep and setup) consists of the k arrivals during period 1 (sleep) and the Poisson distributed number of arrivals during period 2 (setup). The mean value of the number of arrivals is then simply $k + \lambda E[D]$. Out of these arrivals, each one has its size greater than s with probability $\bar{F}(s)$. Thus, the mean number of $B_4(s)$ busy periods is given by $(k + \lambda E[D])\bar{F}(s)$.

With all the required variables defined, the probability that an arriving job would find the system in each period is given by

$$
\begin{aligned}
p_1 &= \frac{k}{E[N]}, & p_2 &= \frac{\lambda E[D]}{E[N]}, \\
p_3 &= \frac{\lambda E[B_3(s)]}{E[N]} & p_4 &= \frac{\lambda(k + \lambda E[D])\bar{F}(s)E[B_4(s)]}{E[N]} \\
p_5 &= a(s) - p_3 - p_4, & p_6 &= \rho - a(s), \\
p_7 &= \frac{\lambda E[I^{\text{tot}}]}{E[N]}.
\end{aligned} \tag{10}
$$

Note that above $E[B_3(s)]$ and $E[B_4(s)]$ are still unknown. We will derive these below in Sect. 3.2.

2° *Mean Conditional Waiting Times.* A test job of size s would have to wait until the current type-s busy period is completed if it arrives during periods 3, 4, or 5. Thus, the mean waiting time is the mean remaining type-s busy period,

$$E[W_i(s)] = \frac{E[B_i(s)^2]}{2E[B_i(s)]}, \tag{11}$$

for $i \in \{3, 4, 5\}$. Derivation of the moments of these busy periods is not quite straightforward and will be done below in Sect. 3.2.

On the other hand, if the test job arrives during periods 1 or 2 (sleep/setup), the test job will experience the remaining time until the end of setup and after that also the corresponding busy period related to $B_3(s)$. We will return to the derivation of $E[W_1(s)]$ and $E[W_2(s)]$, when we analyze the moments of $B_3(s)$.

Finally, as discussed earlier, during period 6 jobs larger than s are being served. Hence, a type-s job that arrives during period 6 preempts the job being processed and starts service immediately without any wait. Similarly, period 7 being the idle period, the test job would start service immediately upon arrival. Hence, we obtain

$$W_6(s) = W_7(s) = 0. \tag{12}$$

3.2   Moments of $B_3(s)$, $B_4(s)$, and $B_5(s)$ Busy Periods

In this section we derive the first two moments of $B_3(s)$, $B_4(s)$, and $B_5(s)$ together with the mean values $\mathrm{E}[W_1(s)]$ and $\mathrm{E}[W_2(s)]$. In the analysis, we need the random variable $\tilde{S}$ defined as the conditional service time S given that $S \leq s$. Specifically, we need its mean value given by

$$\mathrm{E}[\tilde{S}] = \mathrm{E}[S|S \leq s] = \frac{1}{F(s)} \int_0^s t f(t)\, dt,$$

and the second moment, which equals

$$\mathrm{E}[\tilde{S}^2] = \mathrm{E}[S^2|S \leq s] = \frac{1}{F(s)} \int_0^s t^2 f(t)\, dt.$$

Note also that

$$\rho(s) = \lambda F(s)\mathrm{E}[\tilde{S}].$$

The result for $B_3(s)$ is given first in Proposition 1 below.

Proposition 1. *The first and second moments of $B_3(s)$ are given by*

$$\mathrm{E}[B_3(s)] = (k + \lambda\mathrm{E}[D]) \frac{F(s)\mathrm{E}[\tilde{S}]}{1 - \rho(s)} = \left(\frac{k}{\lambda} + \mathrm{E}[D]\right) \frac{\rho(s)}{1 - \rho(s)}, \tag{13}$$

$$\mathrm{E}[B_3(s)^2] = (k + \lambda\mathrm{E}[D]) \frac{F(s)\mathrm{E}[\tilde{S}^2]}{(1 - \rho(s))^3}$$

$$+ \left(k(k-1) + 2k\lambda\mathrm{E}[D] + \lambda^2\mathrm{E}[D^2]\right) \left(\frac{F(s)\mathrm{E}[\tilde{S}]}{1 - \rho(s)}\right)^2. \tag{14}$$

Proof. Recall that $B_3(s)$ is a type-s busy period that is started by those jobs that arrived during period 1 (sleep) and 2 (setup) and are in size smaller than s. The number of such jobs is a random variable and we denote it by N_3. Conditioned on the value of the set up delay D, we have

$$N_3 \,|\, D \sim \mathrm{Bin}(k, F(s)) + \mathrm{Poi}(\lambda F(s)D),$$

i.e., given D, the conditional value of N_3 is the sum of a binomially distributed number of jobs smaller than s from period 1 with parameters k and $F(s)$ and a Poisson distributed number of jobs smaller than s with parameter $\lambda F(s)D$. These two numbers are also independent from each other. By conditioning on D, the first and second moments of N_3 are given by

$$\mathrm{E}[N_3] = (k + \lambda\mathrm{E}[D])F(s), \tag{15}$$

$$\mathrm{E}[N_3^2] = (k + \lambda\mathrm{E}[D])F(s) + \left(k(k-1) + 2k\lambda\mathrm{E}[D] + \lambda^2\mathrm{E}[D^2]\right)F(s)^2. \tag{16}$$

By the end of $B_3(s)$ there is no type s job in the system. The busy period $B_3(s)$ consists of exactly N_3 sub-busy periods $B_{3,n}$ initiated by the N_3 jobs,

$$B_3(s) = \sum_{n=1}^{N_3} B_{3,n}. \tag{17}$$

In SRPT scheduling, the shortest of the N_3 jobs starts $B_{3,1}$, the second shortest job starts $B_{3,2}$ and so on. But since the complete length of the $B_3(s)$ busy period is the same for any work-conserving policy, we may assume that the N_3 jobs are randomly selected to start the sub-busy periods $B_{3,n}$. Now the key observation is that each sub-busy period $B_{3,n}$ is i.i.d. It is initiated by a job distributed as $\tilde{S}$, i.e., from the conditional distribution that the size is less than s. In addition, all the subsequent arriving jobs during the sub-busy period are also from the same distribution. Thus, such a sub-busy period behaves the same as a busy period in a standard M/G/1 queue with arrival rate $\lambda F(s)$ and service times $\tilde{S}$ so that the first and second moments are given by

$$E[B_{3,n}] = \frac{E[\tilde{S}]}{1-\rho(s)}, \quad E[B_{3,n}^2] = \frac{E[\tilde{S}^2]}{(1-\rho(s))^3}. \tag{18}$$

In (17), the random variable N_3 is independent from the distribution of $B_{3,n}$ and thus applying Wald's equation to (17) gives the first moment of $B_3(s)$ as follows,

$$E[B_3(s)] = E[N_3]\,E[B_{3,n}] = (k + \lambda E[D])F(s)\frac{E[\tilde{S}]}{1-\rho(s)},$$

which completes the proof for the first moment.

To find the second moment of $B_3(s)$ one can condition on the value of N_3 and apply the conditional variance formula on the random sum (17), from which we can determine the second moment as

$$E[B_3(s)^2] = E[N_3]E[B_{3,n}^2] + (E[N_3^2] - E[N_3])E[B_{3,n}]^2.$$

Using (15), (16) and (18) in the above we arrive at

$$E[N_3]E[B_{3,n}^2] = (k + \lambda E[D])F(s)\frac{E[\tilde{S}^2]}{(1-\rho(s))^3},$$

$$(E[N_3^2] - E[N_3])E[B_{3,n}]^2 = (k(k-1)+2k\lambda E[D]+\lambda^2 E[D^2])F(s)^2\left(\frac{E[\tilde{S}]}{1-\rho(s)}\right)^2,$$

which completes the proof.   □

With the moments of $B_3(s)$ now available we can analyze the conditional waiting time $W_1(s)$ and $W_2(s)$ that the test job experiences if it arrives during period 1 (sleep) or 2 (setup), respectively. The results are stated below in Corollaries 1 and 2.

Corollary 1. *For a test job with size s that arrives during period 1 (sleep), the mean conditional waiting time* $E[W_1(s)]$ *is given by*

$$E[W_1(s)] = \frac{k-1}{2\lambda} \cdot \frac{1+\rho(s)}{1-\rho(s)} + \frac{E[D]}{1-\rho(s)}. \tag{19}$$

Proof. A test job that arrives during period 1 would be one of the first k jobs. So, assuming it is the ith arrival in the sleep state, it would have to wait for an aggregate time of $k - i$ arrivals and the entire setup time D. On average these correspond to $(k-1)/(2\lambda)$ and $E[D]$ amounts of additional waiting time for the test job. Additionally, the test job needs to wait until the end of a slightly modified $B_3(s)$ busy period, which we denote by $B_3^{(1)}(s)$. Thus,

$$E[W_1(s)] = \frac{k-1}{2\lambda} + E[D] + E[B_3^{(1)}(s)].$$

The busy period that the test job experiences is otherwise exactly as in $B_3(s)$ except that the test job is one of the k jobs that arrive during period 1. Thus, the mean busy period $E[B_3^{(1)}(s)]$ in this case is given by (13) with k replaced by $(k-1)$, from which we arrive at the final result after some simplifications.   □

Corollary 2. *For a test job with size s arriving in period 2 (setup), the mean conditional waiting time* $E[W_2(s)]$ *is given by*

$$E[W_2(s)] = \frac{k}{\lambda} \cdot \frac{\rho(s)}{1-\rho(s)} + \frac{E[D^2]}{E[D]} \cdot \frac{1+\rho(s)}{1-\rho(s)}. \tag{20}$$

Proof. A test job arriving during period 2 needs to wait for the remainder of the setup time, which equals on average $E[D^2]/(2E[D])$. In addition, the test job again needs to wait until the end of a slightly differently (compared to the previous proof) modified $B_3(s)$ busy period, which we denote by $B_3^{(2)}(s)$. Thus,

$$E[W_2(s)] = \frac{E[D^2]}{2E[D]} + E[B_3^{(2)}(s)].$$

In this case, the test job arrives somewhere between the start and end of setup. By a standard renewal argument, we know that the number of other jobs that arrived during this setup is on average $\lambda E[D^2]/E[D]$. Thus, the mean busy period $E[B_3^{(2)}(s)]$ in this case is given by (13) with $E[D]$ replaced by $E[D^2]/E[D]$, which, after some simplifications, completes the proof.   □

In our analysis, we next consider the first and second moments of the $B_4(s)$ busy period. The results are stated in Proposition 2 below.

Proposition 2. *The first and second moments of $B_4(s)$ are given by*

$$E[B_4(s)] = \frac{s}{1-\rho(s)}, \tag{21}$$

$$E[B_4(s)^2] = \frac{\lambda s F(s) E[\tilde{S}^2]}{(1-\rho(s))^3} + \frac{s^2}{(1-\rho(s))^2}. \tag{22}$$

Proof. Recall that $B_4(s)$ busy period is initiated by exactly one job that arrived during periods 1 or 2, had an original size greater than s, and has by the beginning of the $B_4(s)$ busy period received service so that its remaining size has reduced to s. Thus, such a busy period behaves the same as an *initial busy period* in a standard M/G/1 queue with arrival rate $\lambda F(s)$, service times $\tilde{S}$, and an initial workload of size s, which gives (21) and (22), see, e.g., [12].   □

Finally, we analyze the $B_5(s)$ busy period, and give its moments in Proposition 3.

Proposition 3. *The first and second moments of $B_5(s)$ are given by*

$$\mathrm{E}[B_5(s)] = \frac{\mathrm{E}[R(s)]}{1 - \rho(s)}, \tag{23}$$

$$\mathrm{E}[B_5(s)^2] = \frac{\lambda \mathrm{E}[R(s)]F(s)\mathrm{E}[\tilde{S}^2]}{(1 - \rho(s))^3} + \frac{\mathrm{E}[R(s)^2]}{(1 - \rho(s))^2}, \tag{24}$$

where

$$\mathrm{E}[R(s)] = \frac{(\rho + \frac{\lambda \mathrm{E}[I^{\mathrm{tot}}]}{\mathrm{E}[N]} - a(s))F(s)\mathrm{E}[\tilde{S}] + (\rho + \frac{\lambda \mathrm{E}[I^{\mathrm{tot}}]}{\mathrm{E}[N]})\bar{F}(s)s}{\rho + \frac{\lambda \mathrm{E}[I^{\mathrm{tot}}]}{\mathrm{E}[N]} - a(s)F(s)}, \tag{25}$$

$$\mathrm{E}[R(s)^2] = \frac{(\rho + \lambda\frac{\lambda \mathrm{E}[I^{\mathrm{tot}}]}{\mathrm{E}[N]} - a(s))F(s)\mathrm{E}[\tilde{S}^2] + (\rho + \frac{\lambda \mathrm{E}[I^{\mathrm{tot}}]}{\mathrm{E}[N]})\bar{F}(s)s^2}{\rho + \frac{\lambda \mathrm{E}[I^{\mathrm{tot}}]}{\mathrm{E}[N]} - a(s)F(s)}. \tag{26}$$

Proof. The $B_5(s)$ busy period is initiated by a job that has arrived during the time that the server is busy (i.e., after periods 1 and 2), or during the time system is idle (i.e., during period 7). Therefore, it is similar to the analysis of the type-s busy period as done for the ordinary M/G/1-SRPT queue in [11]. The difference is that while in the ordinary SRPT analysis a type-s busy period can be initiated by an arriving job at any time, in our system we must exclude arrivals that occurred during periods 1 and 2.

An arriving job with size smaller than s will initiate a $B_5(s)$ busy period whenever it arrives during period 6, i.e., the time during which jobs with size greater than s are served. This happens with probability $p_6 F(s)$. Also, when the system is busy, if the size of an arriving job is originally greater than s, it will eventually have remaining size s and initiate a $B_5(s)$ busy period. This happens with probability $\rho \bar{F}(s)$. Finally, with probability p_7, a job arrives during period 7 and will initiate immediately or eventually a $B_5(s)$ busy period. The total probability that an arriving job begins a $B_5(s)$ busy period is thus

$$p_6 F(s) + \rho \bar{F}(s) + p_7 = \rho + \lambda \mathrm{E}[I^{\mathrm{tot}}]/\mathrm{E}[N] - a(s)F(s).$$

Let $R(s)$ denote the remaining service time of the job that initiates a $B_5(s)$ busy period. For a job with size $t < s$ arriving during period 6 or period 7, which happens with probability

$$(p_6 + p_7)F(s) = (\rho + \lambda \mathrm{E}[I^{\mathrm{tot}}]/\mathrm{E}[N] - a(s))F(s),$$

we have $R(s) = t$. On the other hand, a job originally greater than s arriving when the system is busy or idle, which happens with probability

$$(\rho + p_7)\bar{F}(s) = (\rho + \lambda E[I^{\text{tot}}]/E[N])\bar{F}(s),$$

eventually has remaining size s and starts a $B_5(s)$ busy period with $R(s) = s$. Thus, the first and second moments of $R(s)$ are given by (25) and (26), respectively.

The jobs served in a $B_5(s)$ busy period following the initiating job all have sizes smaller than s, i.e., they are samples of $\tilde{S}$. Thus, such a busy period behaves the same as an *initial busy period* in a standard M/G/1 queue with arrival rate $\lambda F(s)$, service times $\tilde{S}$, and an initial workload of size $R(s)$, which gives (23) and (24), see, e.g., [12].   □

Having completed the analysis, we are now ready to state the main contribution of the paper, i.e., Theorem 1 which gives the complete expression for the mean waiting time $E[W(s)]$.

Theorem 1. *For the energy-aware SRPT queue under study, the mean conditional waiting time of a type-s job is given by*

$$E[W(s)] = E[W(s)_{\text{M/G/1-SRPT}}]$$
$$+ \frac{1}{E[N]}\left(\frac{k(k-1)}{2\lambda} + kE[D] + \frac{\lambda}{2}E[D^2]\right)\frac{1}{(1-\rho(s))^2}, \quad (27)$$

where $E[W(s)_{\text{M/G/1-SRPT}}]$ *is the mean conditional waiting time in the ordinary M/G/1-SRPT queue given by*

$$E[W(s)_{\text{M/G/1-SRPT}}] = \frac{\lambda}{2}\frac{\int_0^s t^2 f(t)\,dt + s^2\bar{F}(s)}{(1-\rho(s))^2}. \quad (28)$$

Proof. The proof follows directly by applying the derived results on (6). The probabilities of arriving in a given period are expressed in equations (10) together with (4), (7) and (9). The mean conditional waiting times for periods 1 and 2 are from Corollaries 1 and 2, while Eq. (11) combined with Propositions 1, 2, and 3 can be used to determine the mean conditional waiting times for periods 3, 4, and 5, respectively. With some algebraic manipulations we finally arrive at the surprisingly simple form given in (27).   □

The resulting form has a strikingly compact form, consisting of the waiting time as in the ordinary M/G/1-SRPT queue plus an additional waiting time due to idling timer and the setup delay. Note also that, when $E[I^{\text{tot}}] \to \infty$ the formula reduces to the one for the ordinary M/G/1-SRPT queue, and also when $k = 1$ and $E[D] = 0$. Recall that the overall mean delay satisfies $E[T] = E[W] + E[R]$, where $E[R]$ is the mean residence time given by (5). Since $E[R]$ is the same for the ordinary M/G/1-SRPT queue as for our system, we can conclude that the expression for the overall mean delay $E[T]$ in our system is as expressed in Theorem 2 below.

Theorem 2. *For the energy-aware SRPT queue under study, the mean delay of jobs* $E[T]$ *is given by*

$$E[T] = E[T_{M/G/1-SRPT}] +$$
$$+ \frac{1}{E[N]} \left(\frac{k(k-1)}{2\lambda} + kE[D] + \frac{\lambda}{2}E[D^2] \right) \int_0^\infty \frac{f(s)}{(1-\rho(s))^2} \, ds, \quad (29)$$

where $E[T_{M/G/1-SRPT}]$ *is the mean delay in the ordinary M/G/1-SRPT queue given in (3).*

Thus, the mean delay $E[T]$ in our energy-aware system given by (29) consists of the mean delay in the ordinary M/G/1-SRPT queue plus an additional penalty term for the idling and the setup delay. A similar structure holds for the mean delay in the corresponding energy-aware M/G/1-FIFO and M/G/1-PS queues, as well, see [5,6].

4   Optimization

Here we address energy-performance optimization of the energy-aware SRPT queue by applying the popular ERWS (1) and ERP (2) cost metrics. We apply the same method as in [5,6].

Theorem 3. *The optimal control policy in* Π *sets either* $I = 0$ *or* $I = \infty$ *for both ERWS and ERP cost metrics.*

Proof. By [5, Propositions 1 and 2] it suffices to show that both $E[P]$ and $E[T]$ can be expressed in the form

$$E[T] = A_1 + \frac{B_1}{C_0+E[I^{tot}]}, \quad E[P] = A_2 + \frac{B_2}{C_0+E[I^{tot}]}, \quad (30)$$

where constants $A_1, A_2, B_1, C_0 > 0$, but B_2 can be negative.

First, we consider $E[T]$.

Looking at (29), $E[I^{tot}]$ appears only in the denominator of the second term. Thus, $E[T]$ is already in the form of (30).

The mean power consumption of this system can be given as

$$E[P] = \rho P_{busy} + \frac{\frac{k}{\lambda}P_{sleep} + E[D]P_{setup} + E[I^{tot}]P_{idle}}{E[C]},$$

where $E[C]$ denotes the mean length of one regeneration cycle, which is given by $E[C] = E[N]/\lambda$. Substituting this in the above equation, we have

$$E[P] = E[P_{M/G/1}] + \frac{1}{E[N]} \left(k(P_{sleep} - P_{idle}) + \lambda E[D](P_{setup} - P_{idle}) \right), \quad (31)$$

where $E[P_{M/G/1}]$ is the power consumption of an ordinary M/G/1 system given by

$$E[P_{M/G/1}] = \rho P_{busy} + (1 - \rho)P_{idle}.$$

Clearly, the $E[P]$ expression in (31) is also in the form of (30). $\qquad \square$

For the ERP and ERWS cost metrics, the optimal control remains the same as for energy-aware FIFO and PS systems, studied in [5,6]. We conjecture that this optimality result holds for any work conserving scheduling policy. The optimality result might seem to be an obvious consequence of the memoryless property of Poisson arrivals. However, it is possible to construct counter-examples, like in [5], where the idle timer has finite optimal value, even under a Poisson arrival process, when a more general cost metric is considered. Thus, one cannot determine the optimal control solely based on the arrival process.

5   Numerical Results

Next we give a numerical illustration of the results obtained in the analysis and optimization sections. For this, we use among the S-states implemented in modern servers, the suspend sleep state, with power consumption and setup delay values of $P_{\text{sus}} = 15$ W and $D_{\text{sus}} = 10$ s. In [8], it has been shown to give a nice balance between sleeping power consumption and a reasonably low setup delay. Moreover, we use $P_{\text{busy}} = P_{\text{setup}} = 200$ W and $P_{\text{idle}} = 120$ W. Throughout the illustration, we also assume the turn on threshold $k = 1$ and $E[S] = 1$ s.

We confirmed the validity of the mean response time expression in (29) by developing a simulator for the energy-aware M/G/1-SRPT system. Figure 2 shows the analysis and simulation results as a function of load. Pareto, with shape parameter $\alpha = 2.5$, and exponentially distributed service times are considered. The confidence intervals on the simulation results are omitted because they are too narrow to show in the same figure. Here it is interesting to see that the mean response time decreases as a function of load, except at high load values. However, this is intuitive since the response time is already dominated by the setup delay of the system, so that the waiting time increment due to increasing load has little effect. The impact of setup delay is more visible at lower load since the system becomes idle more often, causing frequent idle-sleep-setup transitions.

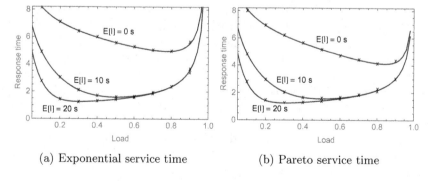

(a) Exponential service time   (b) Pareto service time

Fig. 2. Mean response time comparison between analytic and simulation results. The blue curves represent results obtained analytically. The red cross points are mean response time values obtained using a simulator at the respective load. We assume a deterministic setup delay of $D = 10$ s. (Color figure online)

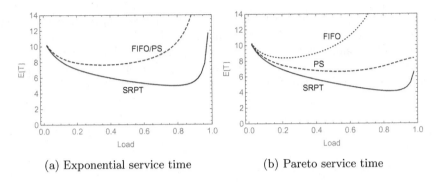

Fig. 3. Mean response time comparison among energy-aware FIFO, PS, and SRPT queues, with $I = 0$ and $D = 10$ s.

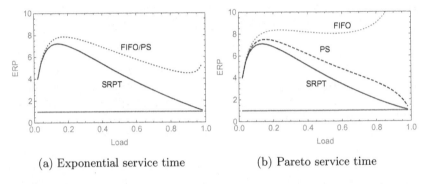

Fig. 4. ERP comparison of energy-aware FIFO, PS, and SRPT queues. All values are normalized with respect to the ERP of an ordinary M/G/1-SRPT queue.

For an ordinary M/G/1 queue, SRPT scheduling is known to give the minimum response time [10]. Figure 3 depicts how the energy-aware M/G/1-SRPT system performs compared to energy-aware M/G/1-FIFO and M/G/1-PS systems. Mean response time of the latter two is studied in [5] and [6] respectively. Figure 3 shows SRPT still provides the lowest mean response time for for such a system for both exponentially and Pareto distributed service times.

We showed in Theorem 3 that either $I = 0$ or $I = \infty$ are the optimal values under the ERP and ERWS cost metrics. Figure 4 illustrates the ERP of energy aware systems with SRPT, FIFO and PS scheduling ($I = 0$), normalized with respect to the ERP of an ordinary M/G/1-SRPT ($I = \infty$). For both exponentially (Fig. 4a) and Pareto (Fig. 4b) distributed service times, the ERP of the energy-aware systems is much higher than that of the ordinary M/G/1-SRPT system. This is due to the high setup delay of the *suspend* state.

Figure 5 gives a similar comparison for the ERWS cost metric, for $w_1 = 1$ and $w_2 = 1$. Except at very low load, the energy-aware system still under performs compared to the ordinary M/G/1-SRPT system. We experimented with wider range of w_2 values and the ordinary M/G/1 SRPT still resulted in the lowest

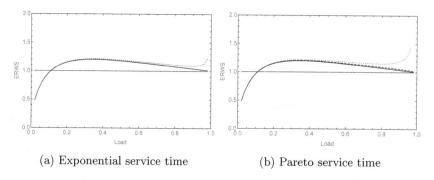

(a) Exponential service time   (b) Pareto service time

Fig. 5. ERWS comparison of energy-aware FIFO, PS, and SRPT queues, with $w_1 = 1$ and $w_2 = 1$. All values are normalized with respect to the ERWS of an ordinary M/G/1-SRPT queue. The dotted, dashed, and solid lines represent energy-aware FIFO, PS, and SRPT systems respectively.

ERWS for most load values. However, at around 0.1 load, the ERWS of the energy-aware system is about 10 % lower than that of the ordinary SRPT.

6   Conclusion

We studied the analysis and optimal control of an energy-aware server, modelled as an M/G/1 SRPT system. Energy can be saved by putting the server into a sleep state when it remains idle for a certain period of time. Turning the server back on incurs a setup cost, in the form of delay and power consumption.

Our analysis, by applying the theory of regenerative processes, shows that the mean response time of such an energy-aware system can be decomposed into two terms: the mean response time of an ordinary M/G/1 SRPT queue, and an additional delay penalty introduced due to the use of a sleep state.

We also considered the Energy-performance optimization of this system under the Energy Response time Weighted Sum (ERWS) and Energy Response time Procuct (ERP) cost metrics. The optimal control policy lies in a set of two distinct policies: one that switches the server to a sleep mode immediately when it becomes idle and the other which leaves it idling. This might seem intuitive, given the memoryless property of a Poisson arrival process. However, these results do not necessarily hold for a more general cost metric, highlighting the fact that the optimality results cannot be readily deduced from the arrival process. Numerical study of these policies with real setup delay and power consumption values show that the idling policy has a consistently lower ERWS and ERP costs compared to the sleeping policy, except at very low load values. This is mainly due to the high setup delay needed by a typical present day server.

The optimality result in this paper for SRPT, in line with our earlier results for FIFO and PS disciplines, has relied on the explicit form of the mean response time. The results anyway suggest that the optimality holds more generally, even for any work-conserving discipline, but proving it remains a challenging open problem.

Acknowledgement. This research was partially supported by the TOP-Energy project funded by Academy of Finland (grant no. 268992).

References

1. Armbrust, M., Fox, A., Griffith, R., Joseph, A.D., Katz, R., Konwinski, A., Lee, G., Patterson, D., Rabkin, A., Stoica, I., Zaharia, M.: A view of cloud computing. Commun. ACM **53**, 50–58 (2010)
2. Barroso, L.A., Hölzle, U.: The case for energy-proportional computing. Computer **40**(12), 33–37 (2007)
3. Gandhi, A., Gupta, V., Harchol-Balter, M., Kozuch, M.A.: Optimality analysis of energy-performance trade-off for server farm management. Perform. Eval. **67**(11), 1155–1171 (2010)
4. Gandhi, A., Harchol-Balter, M., Adan, I.: Server farms with setup costs. Perform. Eval. **67**(11), 1123–1138 (2010)
5. Gebrehiwot, M.E., Aalto, S., Lassila, P.: Optimal sleep-state control of energy-aware M/G/1 queues. In: Proceedings of the 8th International Conference on Performance Evaluation Methodologies and Tools, VALUETOOLS 2014, pp. 82–89 (2014)
6. Gebrehiwot, M.E., Aalto, S., Lassila, P.: Energy-performance trade-off for processor sharing queues with setup delay. Oper. Res. Lett. **44**(1), 101–106 (2016)
7. Gough, C., Steiner, I., Saunders, W.: Energy Efficient Servers: Blueprints for Data Center Optimization. Apress, New York (2015)
8. Isci, C., McIntosh, S., Kephart, J., Das, R., Hanson, J., Piper, S., Wolford, R., Brey, T., Kantner, R., Ng, A., Norris, J., Traore, A., Frissora, M.: Agile, efficient virtualization power management with low-latency server power states. In: Proceedings of the 40th Annual International Symposium on Computer Architecture (ISCA 2013), pp. 96–107, June 2013
9. Maccio, V.J., Down, D.G.: On optimal policies for energy-aware servers. In: Proceedings of IEEE 21st International Symposium on Modeling, Analysis and Simulation of Computer and Telecommunication Systems (MASCOTS 2013), pp. 31–39, August 2013
10. Schrage, L.E.: A proof of the optimality of the shortest remaining processing time discipline. Oper. Res. **16**, 687–690 (1968)
11. Schrage, L.E., Miller, L.W.: The queue M/G/1 with the shortest remaining processing time discipline. Oper. Res. **14**(4), 670–684 (1966)
12. Takács, L.: Introduction to the Theory of Queues. Oxford University Press, Oxford (1962)
13. Wierman, A., Andrew, L.L.H., Tang, A.: Power-aware speed scaling in processor sharing systems: optimality and robustness. Perform. Eval. **69**(12), 601–622 (2012)

Dynamic Control of the Join-Queue Lengths in Saturated Fork-Join Stations

Andrea Marin$^{(\boxtimes)}$ and Sabina Rossi

DAIS - Università Ca' Foscari, Venezia, Italy
{marin,srossi}@dais.unive.it

Abstract. The analysis of fork-join queueing systems has played an important role for the performance evaluation of distributed systems where parallel computations associated with the same job are carried out and a job is considered served only when all the parallel tasks it consists of are served and then joined. The fork-join nodes that we consider consist of $K \geq 2$ parallel servers each of which is equipped with two FCFS queues, namely the service-queue and the join-queue. The former stores the tasks waiting for being served while the latter stores the served tasks waiting for being joined. When the queueing station is saturated, i.e., the service-queues are never empty, we observe that the join-queue sizes tend to grow infinitely even if the expected service times at the servers are the same. In fact, this is due to the variance of the service time distribution. To tackle this problem, we propose a simple service-rate control mechanism, and show that under the exponential assumption on the service times, we can analytically study a set of relevant performance indices. We show that by selectively reducing the speed of some servers, significant energy saving can be achieved.

1 Introduction

Fork-join queueing stations have been extensively studied in the literature because of their wide applications in the context of distributed and parallel systems. Such queueing stations behave as follows: jobs arrive according to a certain arrival process and are forked into K tasks that are enqueued in the *service-queues* and then served by independent servers. Once a task is served, it is enqueued in the *join-queue* waiting for the service completions of all the other tasks of the job it belongs to. Once all the tasks of a job are served, the *join* operation is performed and the job leaves the system. In this work we assume that all the queues implement a First Come First Served (FCFS) discipline.

Fork-join queues have found applications in a wide variety of domains in computer science and telecommunication networks. For instance, in [21] the authors study the response times of multiprocessor systems by means of fork-join networks, in [10] the authors consider parallel communication systems and in [12] a RAID system is studied by simulating a fork-join station.

Unfortunately, despite their importance, few analytical results are known for fork-join stations. One of the reasons is the complexity of the model consisting

© Springer International Publishing Switzerland 2016
G. Agha and B. Van Houdt (Eds.): QEST 2016, LNCS 9826, pp. 123–138, 2016.
DOI: 10.1007/978-3-319-43425-4_8

of two sets of queues, the service-queues and the join-queues, and no general decomposition result is available at the state of the art [1]. Many works have considered the fork-join station under heavy traffic (see, e.g., [13]) and provided approximations of the expected response time based on the analysis of the associated reflecting Brownian motion [18]. In this scenario we observe that when $K \gg 2$ the join-queues tend to be very long because each served task has to wait for the completion of the slowest of its siblings (which may also be enqueued at their servers). In [20] the authors observe that such a system can be highly inefficient both because it handles long join-queues and because the servers work at maximum speed even if their join-queue length is very long. Significant energy saving can be obtained by slowing down the servers that have already served more tasks than others.

1.1 Contribution

In this work we introduce a rate control mechanism for the station's servers that allows us to control the join-queue lengths and to reduce the system's power consumption. The importance of containing the size of the output buffer and reducing the energy consumption is well-known in the literature, e.g., [20,22,23]. In contrast with [20], we do not require the estimation of the amount of work needed by a task, but we base our algorithm on a single state variable associated with each server. We assume that each server has a neighbour defined to form a circular dependency. For instance, the neighbour of server i can be server $(i \bmod K)+1$. If a server has completed less or equal tasks than its neighbour then it works at maximum speed, otherwise it reduces its speed by a certain factor. Therefore, each server has to maintain a single variable that is incremented by 1 at each local task completion, while it is decremented by 1 when a task completion occurs at its neighbour. Our contribution includes an analytically tractable model of such a rate control mechanism. We start by considering the Flatto-Hahn-Wright (FHW) model [8,25] in saturation, i.e., the service times are modelled by independent and identically distributed (i.i.d.) exponential random variables, the join operation is instantaneous, and the service-queues are never empty. We show that even in the case of two servers ($K = 2$), the stochastic process modelling the join-queue lengths is unstable because of the variance in the service times. Conversely, by the introduction of our rate-control mechanism we show that, for any $K \geq 2$, the process underlying the join-queue lengths becomes stable and their expectation is finite. Moreover, we are able to derive an analytical expression for the system's throughput. The stationary probabilities, the marginal stationary probabilities and the throughput are expressed in terms of Kummer's confluent hypergeometric functions. In general, the evaluation of such functions can be done by numerical approximations, but in our case the evaluation points are such that a closed form expression is always known.

Finally, we study by simulation the behaviour of our algorithm when the service times are not exponentially distributed and show the impact of the service times' coefficients of variation (CV) on the performance indices.

1.2 Related Work

In [9] the authors extend their previous work on fork-join queueing networks in order to include join nodes and apply an approximate analysis to study their stationary performance indices based on a decomposition technique or an iterative solution of tractable models. In [8,25] the authors introduce the so called Flatto-Han-Wright model [18] consisting of only two exponential servers. They derive the stability conditions and propose an approximate analysis as well as some exact results on the conditional join-queue lengths. In [17] the authors provide the exact expression of the mean response time for the FHW model, when $K = 2$ and the service times are i.i.d. exponential random variables. They also give an approximation technique to study the models with $K > 2$. In [2,3] the authors study the stability conditions for a set of fork-join queueing networks. In [18] the author applies the method based on the heavy traffic assumption that lead to important results in queueing network analysis for studying the fork-join queueing nodes. Order statistics has been used to solve a class of fork-join queues with block-regular structure in [7].

The work that is probably closer to the one proposed here is [20] where the authors propose to reduce the energy consumption of a fork-join station by slowing down the servers that work on tasks with lower needs. They devise a scheduling algorithm and prove an optimality property. However, in contrast to what we propose here, the method requires the estimation of the tasks' service demands which is not always possible. In [22,23] the authors propose an approach based on the order statistics that introduces deterministic delays at the servers aiming at reducing the task dispersion. The delays are determined so that the 100αth percentile of variability of the distributions obtained once the delays are inserted is minimised.

1.3 Structure of the Paper

The paper is structured as follows. In Sect. 2 we introduce the problem that we aim to address and describe the algorithm that we propose. In Sect. 3 we provide an analytical model for the performance evaluation of the algorithm under the assumptions of saturated station and exponential service time distributions. Section 4 studies the performance of the rate-control algorithm by using the results of the previous section and the stochastic simulation. Finally, Sect. 5 gives some concluding remarks.

2 Rate-Control Algorithm

In this section we formally introduce the problem we are studying and the rate-control algorithm that we propose. In the following sections we study the performance of such an algorithm in terms of throughput and energy saving.

2.1   Problem Statement

Let us consider a fork-join queueing system with K servers as depicted in Fig. 1. We consider a saturated model, i.e., there is always a job waiting to be processed. As a consequence the service-queues always contain at least one task. The service times are modelled by i.i.d. continuous time random variables and we initially assume that the join operation occur immediately after all the tasks belonging to the same job are served. All the queues follow a FCFS discipline. Clearly, if the expected service time at the servers is not the same, and if a rate-control mechanism is not applied, then the join-queue length of the fastest server tend to grow infinitely as time $t \to \infty$. Less obvious is the case in which all the service times are independent and identically distributed, i.e., with the same mean. In these cases, the variance of the service time causes an unbounded growth of the join-queue population, i.e., the expected join-queue lengths at the servers tend to infinity as $t \to \infty$. In Fig. 2 we show a transient simulation of the saturated model with three service time distributions: Erlang-2, hyperexponential and exponential. The confidence intervals have been build on 15 independent executions of the simulation with a confidence of 95 %. The plot supports the intuition that higher coefficient of variations in the service times make the expected queue lengths grow faster. We formally prove the model instability if the service times are exponentially distributed.

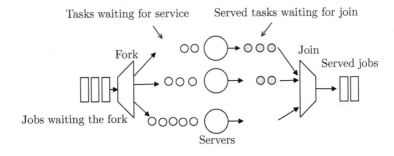

Fig. 1. Fork-join queueing station with $K = 3$ servers

Proposition 1. *In the long run, the saturated fork-join model with $K \geq 2$, i.i.d. exponential service times, immediate join, has an infinite expectation of the join-queue length.*

Proof. For brevity, we give the proof for $K = 2$. The state space of the model is

$$\mathcal{S} = \{(n_1, n_2) : n_1 = 0 \lor n_2 = 0, n_i \in \mathbb{N}\},$$

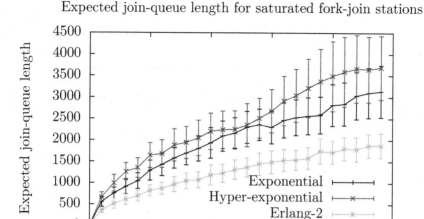

Fig. 2. Growth of the expected join-queue length for $K = 20$ servers, exponential ($CV = 1$), Erlang-2 ($CV = \sqrt{2}/2$), Hyper-exponential ($CV = 1.31$)

where n_i denotes the join-queue length of server i. The transitions are from state $(0, n_2)$ to $(0, n_2 + 1)$ or to $(0, n_2 - 1)$ and from state $(n_1, 0)$ to $(n_1 + 1, 0)$ or $(n_1 - 1, 0)$. Since the service times are exponentially distributed, then the stochastic process is a continuous time Markov chain, and specifically it is a random walk on the line. In this CTMC all the rates are equal and hence the states are not positive recurrent. Therefore, let Q be the random variable associated with the join-queue length for one of the two servers at a time t_0, with $t_0 \to \infty$, then $E[Q] = \infty$. If $K > 2$ the proof is similar but the CTMC is multidimensional. □

We devise an algorithm that dynamically controls the service rates (e.g., by scaling the operating frequency of the processors) with the following aims:

- Having a finite expectation of the join-queue lengths;
- Maintaining the throughput at reasonable high levels;
- Reducing the overall energy consumption by controlling the servers' rates.

Moreover, we will see that if the service rates are exponentially distributed, then a Markovian model with analytically tractable solution exists, therefore one can tackle problems of optimisation or capacity planning that would be expensive to address by stochastic simulation.

2.2   The Rate-Control Algorithm

The main idea of the algorithm is to slow down the servers that have already completed their work on many tasks whereas the servers that have served less tasks will work at maximum speed. Since it would be unrealistic to assume that each server can take a decision about its own speed by knowing the global state of the system, we introduce a policy that implements a rate-control strategy by just maintaining a single integer state variable. Let us label each of the K servers with integer numbers in $\{1, \ldots, K\}$ and define the following neighbourhood relation: for each server k we define its neighbour $ne(k)$ as:

$$ne(k) = \begin{cases} k+1 & \text{if } k < K \\ 1 & \text{if } k = K \end{cases}.$$

Let n_k denote the state variable of each server. When server k completes a task, then n_k is increased by 1, while when its neighbour completes a task n_k is decreased by 1. In other words, n_k maintains the difference between the join-queue length of server k and $ne(k)$. Let $\mu(n_k)$ be the local state dependent service rate at a server (recall that they are all stochastically identical), then:

$$\mu(n_k) = \begin{cases} \frac{\mu}{n_k+1} & \text{if } n_k \geq 0 \\ \mu & \text{otherwise} \end{cases}. \tag{1}$$

Intuitively, when a server k has completed less or the same number of tasks than $ne(k)$ then it works at its full service speed, otherwise it slows down in a proportional way with the number of exceeding jobs. Notice that for server k, the key point for regulating the join-queue length is to consider the difference in the queue lengths of the servers rather than the total length of its join queue. Indeed this latter value could be high because of some delay in the join operation, while the mechanism that we propose is based on balancing the number of tasks served by each server.

3   Analytical Model for the Rate-Control Mechanism

In this section we consider the FHW model equipped with our rate control mechanism, i.i.d. exponentially distributed service times, immediate join and in saturation. Let us consider the vector $\mathbf{n} = (n_1, \ldots, n_K)$ of the state variables of each server, and observe that at each time epoch we have $\sum_{k=1}^{K} n_k = 0$. We aim at studying the stochastic process $\mathbf{n}(t)$ on the state space $\mathcal{S} = \{\mathbf{n} = (n_1, \ldots, n_K) : n_k \in \mathbb{Z}, \sum_{k=1}^{K} n_k = 0\}$. Since the service rates are the only events that cause a state change, from the fact that they are exponentially distributed we conclude that $\mathbf{n}(t)$ is a homogeneous CTMC. Although we will derive a product-form expression for the invariant measure of $\mathbf{n}(t)$, it is worth of notice that $\mathbf{n}(t)$ is *not* reversible for $K > 2$. In fact, consider state $(0, 0, 0)$ and assume that server 2 completes a task taking the state of the process to

$(-1, 1, 0)$. It should be clear that there does not exist any transition bringing back the model to $(0, 0, 0)$. One path that brings back the model state to $(0, 0, 0)$ is that consisting of a sequence of transitions associated with one task completion at servers 1 and 3.

Before proceeding with the analysis we have to introduce the regularized Kummer's confluent hypergeometric function $\mathbf{M}(a, b, x)$ defined as follows (the first equality shows an alternative common notation):

$$\mathbf{M}(a, b, x) = {}_1\tilde{F}_1(a; b; x) = \frac{1}{\Gamma(b)} M(a, b, x) \qquad a, b \in \mathbb{N}^+, \tag{2}$$

where $M(a, b, x)$ is the Kummer's confluent hypergeometric function defined by the series

$$M(a, b, x) = {}_1F_1(a; b; x) = \sum_{k=0}^{\infty} \frac{(a)_k}{(b)_k} \frac{x^k}{k!} \qquad a, b \in \mathbb{N}^+, \tag{3}$$

Γ is the Euler's Gamma function and $(y)_k$ is the Pochhammer's symbol, i.e., $(y)_k = y(y + 1) \cdots (y + k - 1)$.

Theorem 1. *Given the CTMC $\mathbf{n}(t)$, we have that:*

1. *$\mathbf{n}(t)$ is ergodic, i.e., it admits a unique stationary distribution $\pi_K(\mathbf{n})$;*
2. *The stationary distribution is given by the following expression:*

$$\pi_K(\mathbf{n}) = \frac{1}{G_K} \frac{1}{\prod_{i=1}^{K} (n_i \delta_{n_i > 0})!} \tag{4}$$

where we assume that empty products are equal to 1 and δ_P is 1 if proposition P is true, 0 otherwise and

$$G_K = 1 + \sum_{j=1}^{K-1} \binom{K}{j} j^{K-j} \mathbf{M}(K - j, K - j + 1, j). \tag{5}$$

We base the proof of the theorem on few lemmas: first we assume the ergodicity and derive the model's product-form expression. Then, we show that the normalising constant G_K is finite (thanks to the properties of the Kummer's confluent hypergeometric function) for finite K and hence the CTMC must be ergodic.

Lemma 1. *Assume that $\mathbf{n}(t)$ is ergodic and hence admits a unique stationary distribution. Then, its expression is that of Eq. (4) where:*

$$G_K = \sum_{\mathbf{n} \in \mathcal{S}} \frac{1}{\prod_{i=1}^{K} (n_i \delta_{n_i > 0})!}. \tag{6}$$

Proof. The proof can be obtained by substitution of Eq. (4) in the system of global balance equations of the CTMC or by noticing that the process is dynamically reversible [11, 14–16]. Let $\mathbf{n} = (n_1, \ldots, n_K)$ and let its renaming be $\rho(\mathbf{n}) = (n_K, \ldots, n_1)$, then by [11, Theorem 1.14] we have to prove that Eq. (4) satisfies:

$$\pi(\mathbf{n})\mu(n_k) = \pi(\rho(\mathbf{n} + \mathbf{1}_k - \mathbf{1}_{k-1}))\mu(n_{k-1} - 1),$$

where $\mathbf{1}_k$ is a K-size vector with a 1 in the k-th position and zeros elsewhere and we assumed $\mathbf{1}_0 = \mathbf{1}_K$ and $n_0 = n_K$.   □

Notice that since $\mathcal{S}$ is an infinite set, at the moment the fact that G_K is finite, i.e., the infinite series (6) converges, depends on the assumption of ergodicity. We now algebraically prove that (6) and (5) are equivalent and converge. As a consequence the CTMC $\mathbf{n}(t)$ is ergodic.

Lemma 2. *The series (6) is equivalent to the expression given by Eq. (5) which is finite for any $K \in \mathbb{N}, K \geq 2$.*

Proof. Let $\mathcal{P}(\mathbf{n})$ be the multiset with all the non-negative components of $\mathbf{n}$, i.e., $\mathcal{P}(\mathbf{n}) = \{n_i : n_i \geq 0\}$ and observe that for all the states $\mathbf{n}'$ such that $\mathcal{P}(\mathbf{n}') = \mathcal{P}(\mathbf{n})$ the expression under the sum symbol of Eq. (6) is the same. Let $1 \leq j \leq K - 1$ and $(x_1, \ldots, x_j)$ be a tuple such that $x_i \geq 0$ for all $i = 1, \ldots, j$ and $\sum_{i=1}^{j} x_j = n$, with $n \geq 0$. Basically, j denotes the number of non-negative components in a state and n their sum. Notice that, given j and n we can count how many states have exactly j non-negative components whose sum is n. This is given by the product of the number of non-negative solutions of the Diophantine's equation $y_1 + \ldots + y_j = n$ multiplied by the number of strictly positive solutions of the Diophantine's equation $y_1 + \ldots + y_{K-j} = n$ (since the sum of all the state components is 0), i.e., we can rewrite the normalising constant as:

$$G_K = 1 + \sum_{j=1}^{K-1} \sum_{n=K-j}^{\infty} \sum_{\mathbf{x}:x_1+\ldots+x_j=n} \frac{1}{\prod_{t=1}^{j} x_t!} \binom{K}{j}$$

$$\cdot \binom{n-1}{K-j-1} = 1 + \sum_{j=1}^{K-1} \binom{K}{j} \sum_{n=K-j}^{\infty} \frac{j^n}{n!} \binom{n-1}{K-j-1},$$

where the last equality follows from the multinomial theorem. Notice that the boundaries of j in the external summatory start from 1 (there cannot be any state with all negative components) and terminate at $K - 1$. Indeed, the only state with all non-negative components is $\mathbf{0}$ that we take into account by summing 1 at the beginning of the right-hand-side.

We can rewrite Eq. (2) as:

$$\mathbf{M}(a, b, x) = \sum_{k=0}^{\infty} \frac{(a)_k}{\Gamma(b+k)} \frac{x^k}{k!} \qquad b \in \mathbb{N}^+. \tag{7}$$

So we have:

$$
\begin{aligned}
G_K &= 1 + \sum_{j=1}^{K-1} \binom{K}{j} \sum_{w=0}^{\infty} \frac{j^{w+K-j}}{(w+K-j)!} \binom{w+K-j-1}{K-j-1} \\
&= 1 + \sum_{j=1}^{K-1} \binom{K}{j} \sum_{w=0}^{\infty} \frac{j^{w+K-j}}{(w+K-j)!} \frac{(K-j)_w}{w!} \\
&= 1 + \sum_{j=1}^{K-1} \binom{K}{j} j^{K-j} \sum_{w=0}^{\infty} \frac{j^w}{\Gamma(w+K-j+1)} \frac{(K-j)_w}{w!} \\
&= 1 + \sum_{j=1}^{K-1} \binom{K}{j} j^{K-j} \mathbf{M}(K-j, K-j+1, j)
\end{aligned}
$$

where the last equality follows from Eq. (7) with $a = K - j$, $b = K - j + 1$ and $x = j$. Finally, we observe that $1 < G_K < \infty$ since its definition does not involve any infinite sum and function $\mathbf{M}$ evaluated at the specified integer parameters is always finite and non-negative.   □

Proof of Theorem 1. The theorem follows straightforwardly by Lemmas 1 and 2.   □

In order to derive the expression for the marginal distribution of the join-queue lengths we have to consider that although the state space of each single queue ranges from $-\infty$ to $+\infty$, the joint state space is not the Cartesian product of the single state spaces. Therefore, the knowledge of G_K is not sufficient to obtain the marginal distribution. A similar situation arises when studying closed queueing networks. However, while for closed product-form queueing networks several algorithms have been proposed, e.g., [4–6], in our case we are able to express the marginal distributions in terms of (regularized) Kummer's hypergeometric functions evaluated in points whose closed-form solution is known.

Let us consider the definition of G_K given by Eq. (6), and let G_k^N be the normalising constant defined as:

$$
G_k^N = \sum_{\mathbf{n} \in \mathcal{S}_k^N} \frac{1}{\prod_{i=1}^{k} (n_i \delta_{n_i > 0})!},
$$

where $\mathcal{S}_k^N = \{(n_1, \ldots, n_k) : \sum_{i=1}^{k} n_i = N\}$. Note that $G_K = G_K^0$. Then, we can write the marginal distribution as:

$$
\pi_K^*(n) = \frac{1}{(n\delta_{n>0})!} \frac{G_{K-1}^{-n}}{G_K^0}. \tag{8}
$$

The following Lemma gives the expression for G_k^N for arbitrary $k \geq 1$ and $N \in \mathbb{Z}$.

Lemma 3. *The expression for G_k^N is:*

- *If $N \geq 0$:*

$$G_k^N = \frac{(k\mu)^N}{N!} + \mu^N \sum_{j=1}^{k-1} \binom{k}{j} j^{N+k-j} \mathbf{M}(k-j, N+k-j+1, j).$$

- *If $N < 0$ and $2 \leq k \leq -N$:*

$$G_k^N = \binom{-N-1}{k-1}\mu^N + \mu^N \sum_{j=1}^{k-1} \binom{k}{j}\binom{-N-1}{k-j-1} M(-N, -N-k+j+1, j).$$

- *If $N < 0$ and $k > -N$:*

$$G_k^N = \mu^N \sum_{j=1}^{k+N-1} \binom{k}{j} j^{N+k-j} \mathbf{M}(k-j, N+k-j+1, j)$$

$$+\mu^N \sum_{j=k+N}^{K-1} \binom{k}{j}\binom{-N-1}{k-j-1} M(-N, -N-k+j+1, j)$$

- *If $k = 1$:*

$$G_1^N = \begin{cases} \mu^N/N! & \text{if } N \geq 0 \\ \mu^N & \text{if } N < 0 \end{cases}$$

Proof. The proof is based on hypergeometric function manipulations.

In Fig. 3b we show the distribution of $\pi_K^*(n)$ for $K = 2, 5, 10, 15$. Notice that while for $K = 2$ the distribution is symmetric with respect to $n = 0$, this is not true for $K > 2$. Moreover, by increasing the value of K, numerical evidences suggest that there may exist a limiting distribution for the marginal probabilities (and hence for the throughput and the power consumption). Another important aspect is the observation that the expression of π_K and π_K^* in terms of (regularized) Kummer's confluent functions allows us to have a symbolic expression for the stationary probabilities as shown in Fig. 3a for $K = 3$.

One of the most important performance indices for a rate-control algorithm is the throughput, i.e., the number of join performed by the station per unit of time. In fact, by slowing down some servers we surely decrease the system's throughput. We are able to provide an analytical expression for the station's throughput that depends on the number of servers K and the service rate μ.

Lemma 4. *The throughput $X_K(\mu)$ of the model in steady-state is:*

$$X_K(\mu) = \frac{\mu}{KG_K}\left(K + \sum_{j=1}^{K-1} \binom{K}{j} j \Big(j^{K-j+1}\mathbf{M}(K-j, K-j+2, j)\right.$$

$$-(j-1)^{K-j+1} \mathbf{M}(K-j, K-j+2, j-1)$$

$$\left.+(K-j)j^{K-j-1}\mathbf{M}(K-j, K-j+1, j)\Big)\right). \quad (9)$$

Proof. The proof is based on hypergeometric function manipulations.

In Table 1 we show the analytical expression of the throughput for some values of K.

Table 1. Analytical expression of the throughput for the FHW model with K severs.

K	$X_K(\mu)$
2	$\dfrac{4\mu(e-1)}{K(2e-1)}$
3	$\dfrac{9\mu(e^2-e+1)}{K(1+3e^2)}$
4	$\dfrac{8\mu(2e^3+3e-2)}{K(4e^3+6e^2+2e-1)}$
5	$\dfrac{25\mu(6e^4+12e^3-11e+6)}{2K(15e^4+60e^35e+30E^2-5e+3)}$
6	$\dfrac{6\mu(24e^5+120e^4+120e^3-40e^2+53e-24)}{K(24e^5+180e^4+200e^3+20e^2+9e-4)}$
7	$\dfrac{147\mu(40e^6+360e^5+600e^4+100e^3+120e^2-103e+40)}{4K(210e^6+2520e^5+5250e^4+2100e^3+210e^2-77e+30)}$

The numerical evaluations of both G_k^N and of $X_K(\mu)$ rely on the computation of the confluent hypergeometric function $\mathbf{M}(a,b,z)$ with parameters $a \in \mathbb{N}^+$, $b \in \mathbb{N}^+$ and $b > a$. Indeed, if a and b are non-negative integers, then the series (3) converges for all finite x. In particular, for $b > a$, $\mathbf{M}(a,b,z)$ converges to [19]:

$$\mathbf{M}(a,b,x) = \left(e^x \sum_{k=0}^{a-1} \frac{(1-a)_k\,(-x)^k}{k!\,(2-b)_k} \right.$$

$$\left. - \sum_{k=0}^{b-a-1} \frac{(1-b+a)_k\,x^k}{k!\,(2-b)_k} \right) \frac{(2-b)_{a-1}\,x^{1-b}}{(a-1)!}. \quad (10)$$

4   Numerical Evaluation

In this section we study the sensitivity of the throughput, the expected join-queue length and the power consumption with respect to the distribution of the service times. Then, we study the performance in terms of throughput and energy consumption of the model implementing the rate-control algorithm under the assumptions introduced in Sect. 3. We consider three important performance indices: the system throughput, the expected join-queue lengths and the power consumption. While for the first index Lemma 4 gives us its analytical expression, for the latter two indices we rely on the stochastic simulation and on the bounded approximation described in Sect. 4.1, respectively.

4.1   The Power Consumption

Since our rate-control mechanism reduces the computation speed of the severs, this can be interpreted as a reduction of the operating frequency leading to a reduction of the overall server power consumption. Clearly, the minimum power consumption with maximum throughput corresponds to a situation in which the servers work at a constant maximum rate, but we have already discussed that the drawback of this approach is the infinite growth of the join-queue length in saturated models.

Under the assumptions of Sect. 3 we know the analytical expression of the marginal stationary distribution for each server (see Eq. (8) and Lemma 3). This allows us to define a lower and upper bound of the energy consumption by truncation of the probabilities. Given an integer $E > 0$, the expected power consumption in steady-state $\overline{P}_K$ is bounded by:

$$\sum_{i=-E}^{-1} \pi_K^*(i) + \sum_{i=0}^{E-1} \pi_K^*(i) \frac{1}{(i+1)^3} < \overline{P}_K < \sum_{i=-E}^{-1} \pi_K^*(i)$$
$$+ \sum_{i=0}^{E-1} \pi_K^*(i) \frac{1}{(i+1)^3} + (1 - \sum_{i=-E}^{E-1} \pi_K^*(i)),$$

where we have assumed that the sever at maximum speed consumes 1 unit of energy for unit of time, and that the power consumption depends on the cube of the operating frequency, i.e.:

$$\overline{P}_K = \sum_{i=-\infty}^{-1} \pi_K^*(i) + \sum_{i=0}^{\infty} \pi_K^*(i) \frac{1}{(i+1)^3}.$$

Clearly, more accurate models of the relation between operating frequency and power consumption can be considered, but this is out of the scope of this paper, especially because this relation depends on the intrinsic characteristics of the processors [20]. It is important to notice that with small values of $E \simeq 10$ we obtain tight bounds for the energy consumption as shown in Fig. 3c.

4.2   Sensitivity Analysis

The analytical model proposed in Sect. 3 requires that the service times are state dependent i.i.d. exponential random variables. Under this assumption, and by considering a saturated model with immediate join, we proved the stability of the process modelling the join-queue lengths. Clearly, we expect to find a sensitivity of the performance indices on the distribution of the service times, because it is its variance the cause of the join-queue length growth in the model without the rate-control mechanism. Figures 3d–f show the three considered performance indices for a saturated model with immediate join. The indices with exact or approximated analytical expression have not been simulated, while the others have been obtained via stochastic simulation. For each scenario we run

15 independent experiments and considered the confidence interval of 95 %. The widths of the confidence intervals are all below 1 % of the measure and are too small to be visible in the plots. The warm up periods have been removed by using the Welch's method [24]. The service time distributions have mean 1 and the Erlang 2 has a coefficient of variations of $\sqrt{2}/2$ while the Hyper-Exponential has a coefficient of variation of 1.31.

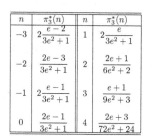

n	$\pi_3^*(n)$	n	$\pi_3^*(n)$
-3	$2\dfrac{e-2}{3e^2+1}$	1	$2\dfrac{e}{3e^2+1}$
-2	$\dfrac{2e-3}{3e^2+1}$	2	$\dfrac{2e+1}{6e^2+2}$
-1	$2\dfrac{e-1}{3e^2+1}$	3	$\dfrac{e+1}{9e^2+3}$
0	$\dfrac{2e-1}{3e^2+1}$	4	$\dfrac{2e+3}{72e^2+24}$

(a) Marginal distribution for $K = 3$

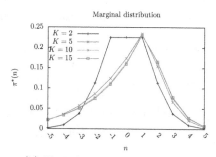

(b) Plot of marginal distributions

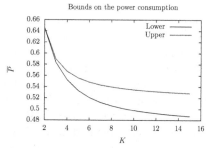

(c) Bounds of the power consumption

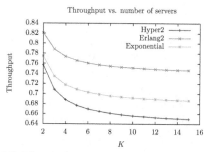

(d) Throughput as function of K with $\mu = 1$

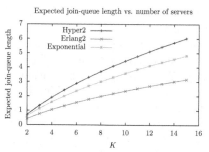

(e) Expected join-queue length as function of K with $\mu = 1$

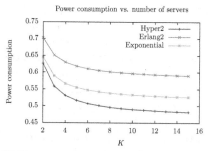

(f) Power consumption as function of K with $\mu = 1$

Fig. 3. Numerical evaluation.

4.3   Performance of the Algorithm as Function of the Number of Servers

In this section we focus on the saturated FHW model with immediate join and study the impact of the number of servers K on the performance indices. Figures 3d, e and f show the system's throughput, the expected queue length for the join-queues and the power consumption for each server when the maximum service rate is $\mu = 1$. Notice that the expected queue length is for each server and is obtained by stochastic simulation. We notice that while the throughput decreases very slowly with the growth of the number of servers (e.g., for $K = 150$ servers we compute a throughput of 0.677), the expected join-queue lengths tend to grow with the number of servers and hence for large models the benefits of the rate-adaptation algorithm are lower. As for the power consumption, the power consumption is significantly lower than the reference value of the model without rate-control, 1. For instance for $K = 6$ the throughput is $X_K(1) \simeq 0.70$ while the power consumption $\overline{P}_K \simeq 0.54$.

5   Conclusion

In this paper we have proposed a rate-control mechanism for fork-join stations designed to maintain the join-queue lengths finite in the long run, even when the station is saturated. We observed that the variance in the service time distribution causes an unbounded increase of the join-queue lengths. Informally, the idea behind our rate-control mechanism is to reduce the operating speed of the servers that have served more customers while maintaining at the maximum level the speed of the other servers. Each server maintains a state variable which is incremented at a local service completion event and is decremented at a service completion event occurring at a neighbour server. The servers maintain their maximum speed if the state variable is not positive, otherwise they reduce their speed. This allows for both a control of the join-queue length and a reduction on the system's power consumption. However, we also observed a reduction in the system's throughput. Despite the few analytical results available for fork-join stations, we have provided the analytical expression for the steady-state distribution of the rate-control model and derived the marginal distributions for each server and the system's throughput under the FHW assumptions. The stationary distributions and the performance indices are expressed in terms of Kummer's confluent hypergeometric functions which are evaluated at special points that require the computations of finite sum. We resorted to the simulation for studying the impact of the rate-control algorithm on stations with different service time distributions and the experiments have supported the intuition that the performance degrades with the increase of the variance in the service time distribution. The main strengths of the proposed mechanism are the easiness of implementation, since the algorithm is basically stateless and does not require nor the estimation of the jobs' service times as in [20], neither the knowledge of the service time distributions as in [22,23], and the effectiveness in drastically reducing the expected join-queue lengths with respect to the models not implementing any rate-control mechanism for the servers.

With respect to a solution which addresses the problem of containing the join-queue length based on a rate adaptation mechanism that considers for each server its associated join-queue length, our approach has the advantage that its implementation is independent of the system's parameters since it aims at balancing the total work performed by each server. Conversely, the join-queues may be long because the join operation's rate is close to the system's throughput and hence considering only its instantaneous state for deciding the service rate can be counter-productive.

Future work includes the derivation of the analytical expression for other performance indices in the case of the saturated FHW model. Moreover, we aim at introducing a parameterisation of the algorithm so that we can control the servers' speed more accurately, e.g., by reducing the service rate for positive states n by $\alpha n + 1$, where $0 < \alpha < 1$ is a parameter that regulates the trade-off between the throughput and the expected join-queue length. However, at the moment, no analytical solution for such a model is known.

References

1. Baccelli, F., Makowski, M.A.: Queueing models for systems with synchronization constraints. Proc. IEEE **77**(1), 138–161 (1989)
2. Baccelli, F., Liu, Z.: On the execution of parallel programs on multiprocessor systems: a queuing theory approach. J. ACM **37**(2), 373–414 (1990)
3. Baccelli, F., Massey, W.A., Towsley, D.: Acyclic fork-join queuing networks. J. ACM **36**(3), 615–642 (1989)
4. Bruell, S.C., Balbo, G., Afshari, P.V.: Mean value analysis of mixed, multiple class BCMP networks with load dependent service stations. Perform. Eval. **4**, 241–260 (1984)
5. Buzen, J.P.: Computational algorithms for closed queueing networks with exponential servers. Commun. ACM **16**(9), 527–531 (1973)
6. Casale, G.: A generalized method of moments for closed queueing networks. Perform. Eval. **68**(2), 180–200 (2011)
7. Fiorini, P.M., Lipsky, L.: Exact analysis of some split-merge queues. SIGMETRICS Perform. Eval. Rev. **43**(2), 51–53 (2015)
8. Flatto, L., Hahn, S.: Two parallel queues created by arrivals with two demands. SIAM J. Appl. Math. **44**(5), 1041–1053 (1984)
9. Heidelberger, P., Trivedi, K.: Analytic queueing models for programs with internal concurrency. IEEE Trans. Comput. **C–32**, 73–82 (1983)
10. Hoekstra, G.J., van der Mei, R.D., Bhulai, S.: Optimal job splitting in parallel processor sharing queues. Stoch. Models **28**, 144–166 (2012)
11. Kelly, F.: Reversibility and Stochastic Networks. Wiley, New York (1979)
12. Lebrecht, A.S., Dingle, N.J., Knottenbelt, W.J.: Modelling zoned RAID systems using fork-join queueing simulation. In: Bradley, J.T. (ed.) EPEW 2009. LNCS, vol. 5652, pp. 16–29. Springer, Heidelberg (2009)
13. Lu, H., Pang, G.: Gaussian limits for a fork-join network with nonexchangeable synchronization in heavy traffic. Math. Oper. Res. **41**(2), 560–595 (2016)

14. Marin, A., Rossi, S.: On discrete time reversibility modulo state renaming and its applications. In: 8th International Conference on Performance Evaluation Methodologies and Tools, VALUETOOLS, pp. 1–8 (2014)
15. Marin, A., Rossi, S.: On the relations between lumpability and reversibility. In: Proceedings of the IEEE 22nd International Symposium on Modeling, Analysis and Simulation of Computer and Telecommunication Systems (MASCOTS 2014), pp. 427–432 (2014)
16. Marin, A., Rossi, S.: On the relations between Markov chain lumpability and reversibility. Acta Inform. 1–39 (2016)
17. Nelson, R., Tantawi, A.N.: Approximate analysis of fork/join synchronization in parallel queues. IEEE Trans. Comput. **37**(6), 739–743 (1986)
18. Nguyen, V.: Processing networks with parallel and sequential tasks: heavy traffic analysis and Browinian limits. Ann. Appl. Probab. **3**(1), 28–55 (1993)
19. Olver, F.W., Lozier, D.W., Boisvert, R.F., Clark, C.W.: NIST Handbook of Mathematical Functions, 1st edn. Cambridge University Press, New York (2010)
20. Rauber, T., Rünger, G.: Energy-aware execution of fork-join-based task parallelism. In: Proceedings of the 20th IEEE International Symposium on Modeling, Analysis and Simulation of Computer and Telecommunication Systems, (MASCOTS), pp. 231–240 (2012)
21. Towsley, D., Romel, G., Astankovic, J.: Analysis of fork-join program response times on multiprocessors. IEEE Trans. Parallel Distrib. Syst. **1**(3), 286–303 (1990)
22. Tsimashenka, I., Knottenbelt, W., Harrison, P.G.: Controlling variability in split-merge systems. In: Al-Begain, K., Fiems, D., Vincent, J.-M. (eds.) ASMTA 2012. LNCS, vol. 7314, pp. 165–177. Springer, Heidelberg (2012)
23. Tsimashenka, I., Knottenbelt, W.J., Harrison, P.G.: Controlling variability in split-merge systems and its impact on performance. Ann. Oper. Res. **239**(2), 569–588 (2016)
24. Welch, P.D.: On the problem of the initial transient in steady-state simulations. Technical report, IBM Watson Research Center, Yorktown Heights, NY (1981)
25. Wright, P.E.: Two parallel processors with coupled inputs. Adv. Appl. Probab. **24**, 986–1007 (1992)

Moment-Based Probabilistic Prediction of Bike Availability for Bike-Sharing Systems

Cheng Feng$^{(\boxtimes)}$, Jane Hillston, and Daniël Reijsbergen

LFCS, School of Informatics, University of Edinburgh, Scotland, UK
s1109873@sms.ed.ac.uk, jane.hillston@ed.ac.uk, dreijsbe@inf.ed.ac.uk

Abstract. We study the problem of future bike availability prediction of a bike station through the moment analysis of a PCTMC model with time-dependent rates. Given a target station for prediction, the moments of the number of available bikes in the station at a future time can be derived by a set of moment equations with an initial set-up given by the snapshot of the current state of all stations in the system. A directed contribution graph with contribution propagation method is proposed to prune the PCTMC to make it only contain stations which have significant contribution to the journey flows to the target station. The underlying probability distribution of the available number of bikes is reconstructed through the maximum entropy approach based on the derived moments. The model is parametrized using historical data from Santander Cycles, the bike-sharing system in London. In the experiments, we show our model outperforms the classic time-inhomogeneous queueing model on several performance metrics for bike availability prediction.

Keywords: Availability prediction · PCTMC models · Moment analysis · Maximum entropy reconstruction

1 Introduction

In recent years, we have seen significant growth of bike-sharing programs all over the world [1]. Public bike-sharing systems have been launched in many major cities such as London, Paris, and Vienna. Indeed, they have become an important part of urban transportation which provides improved connectivity to other modes of public transit. The concept of bike-sharing systems is rather simple: the system consists of a number of bike stations distributed over a geographic area (city). Each station is equipped with a limited number of bike slots in which public bikes can be parked. When users arrive at a station, they pick up a bike, use it for a while, and then return it to another station of their choice.

With the increasing popularity of the smart transport theme, there has been great interest from the research community in the intelligent management of bike-sharing systems. Topics include, but are not limited to, policy design [2,3], intelligent bike redistribution [4–6], and user journey planning [7,8]. The focus of this paper is on the probabilistic prediction of the number of available bikes in

© Springer International Publishing Switzerland 2016
G. Agha and B. Van Houdt (Eds.): QEST 2016, LNCS 9826, pp. 139–155, 2016.
DOI: 10.1007/978-3-319-43425-4_9

stations. Having a predictive model is of vital interest to both the user and the system administrator. The user can use it to identify likely origin/destination stations for which a trip can be successfully made. System administrators can use the model to undertake service level agreement checking, and plan bike redistribution for stations which are likely to break the service level requirement.

In this paper we present a novel moment-based prediction model that can provide probabilistic forecasts for the number of available bikes in a bike station. By representing the bike-sharing system as a Population Continuous Time Markov Chain (PCTMC) with time-dependent rates, our model is explanatory as the dynamics of the system is explicitly given. Gast *et al.* [8] show the benefits of predicting (*forecasting*) the entire probability distributions of possible bike availabilities in a station, compared with previous models that were only able to produce point estimates, often using time-series-based techniques [7,9,10]. However, unlike [8], in which all the considered forecasting methods worked on the level of isolated stations, our model also captures the journey dynamics between stations. Guenther and Bradley [11] also provide a inhomogeneous-time PCTMC model with time-dependent rates for bike availability prediction, however there are several key differences between that model and ours. Firstly, our model provides the full probability distribution of the number of available bikes in a station whereas their model only provides a point estimate. Secondly, we use a model reduction method to prune our PCTMC such that the significant journey dynamics with respect to the target station are guaranteed to be preserved. However, their model aggregates stations which are spatially close, assuming that they have similar journey durations to the target station, which causes the information about the emptiness and fullness of stations to be lost.

We summarize the contribution of our paper as follows. Firstly, a novel PCTMC model with time-dependent rates is presented to successfully capture the journey dynamics between bike stations. Secondly, we propose a novel model reduction technique to prune the PCTMC model based on the directed contribution graph with a contribution propagation method for a given target station for bike availability prediction. Finally, we reconstruct the underlying probability distribution of the number of available bikes in the target station using the maximum entropy principle based on a few moments generated from fluid approximation of the PCTMC, and show that the model has a better performance on a set of metrics for bike availability prediction compared with the classic Markov single-station queueing model.

The rest of this paper is structured as follows. We briefly introduce the concepts of PCTMC with time-dependent rates in the next section. Section 3 gives the introduction of the classic Markov queueing model for bike availability prediction. In Sect. 4, we present our PCTMC model for the bike-sharing scenario. In the next section we show how to reconstruct the probability distribution of number of available bikes using the maximum entropy approach. Section 6 presents the experimental results of our model on the London bike-sharing system compared with the classic Markov queueing model. Finally, Sect. 7 discusses possible extensions of our model and draws final conclusions.

2 PCTMC with Time-Dependent Rates

A PCTMC is a stochastic process which consists of a number of distinct agent populations and a set of transition classes. The state of a PCTMC is captured by an integer vector counting the number of each agent type. The model evolves with the firing of transitions. When a transition fires, one or more agent populations are updated. Each transition is associated with a rate function, which assigns a rate governed by an exponential distribution to the transition based on the current state of the PCTMC. In this paper, we specifically consider time-inhomogeneous PCTMCs, in which transition rates can also be time-dependent. Specifically, a PCTMC with time-dependent rates can be expressed as a tuple $\mathcal{P} = (\mathbf{X}(t), \mathcal{T}, \mathbf{X_0})$:

- $\mathbf{X}(t) = (X_1(t), ..., X_n(t)) \in \mathbb{Z}_{\geq 0}^n$ is an integer vector with the ith ($1 \leq i \leq n$) component representing the current number of an agent type S_i.
- $\mathcal{T} = \{\tau_1, ..., \tau_m\}$ is the set of transition classes, of the form $\tau = (r_\tau(\mathbf{X}, t), \mathbf{d}_\tau)$, where:
  1. $r_\tau(\mathbf{X}, t) \in \mathbb{R} \geq 0$ is a time-dependent rate function, associating with each transition the rate of an exponential distribution, depending on the state of the PCTMC $\mathbf{X}$ as well as the current time t.
  2. $\mathbf{d}_\tau \in \mathbb{Z}^n$ is the update vector which gives the net change for each element of $\mathbf{X}$ caused by transition τ.
- $\mathbf{X_0} \in \mathbb{Z}_{\geq 0}^n$ is the initial state of the model.

Transition rules can be easily expressed in the chemical reaction style, as

$$\ell_1 S_1 + \ldots + \ell_n S_n \longrightarrow_\tau \bar{\ell}_n S_1 + \ldots + \bar{\ell}_n S_n \quad \text{at rate } r_\tau(\mathbf{X}, t)$$

where the net change of agents of type S_i due to transition τ is given by $d_\tau^i = \bar{\ell}_i - \ell_i$ ($1 \leq i \leq n$), and the transition rate is
$$\begin{cases} r_\tau(\mathbf{X}, t) & \text{if } X_i \geq \ell_i \;\; \forall i = 1, 2, \ldots, n \\ 0 & \text{otherwise.} \end{cases}$$

As the state space of PCTMC models is often very large or even infinite, numerical techniques traditionally used for performance analysis, based on a Markovian approach, are entirely infeasible. Stochastic simulation is feasible, but deriving useful metrics such as mean, variance, probability distribution of populations often requires a large number of simulation runs, thus making this approach extremely costly in terms of computational resources, particularly when estimating full probability distributions over large state spaces. In this paper, we will adopt a much more computationally efficient approach to analyse the PCTMC for the bike-sharing model. Specifically, we approximate the evolution of the moments of the underlying population-level stochastic process of a PCTMC model by the following set of ODEs [12]:

$$\frac{\mathrm{d}}{\mathrm{d}t}\mathbb{E}[M(\mathbf{X}(t))] = \sum_{\tau \in \mathcal{T}} \mathbb{E}[(M(\mathbf{X}(t) + \mathbf{d}_\tau) - M(\mathbf{X}(t)))r_\tau(\mathbf{X}, t)] \tag{2.1}$$

where $M(\mathbf{X})$ denotes the moment to be calculated. For instance, by substituting $M(\mathbf{X})$ with X_i, X_i^2 and X_iX_j, we get the set of ODEs to describe the first moment, second moment and second-order joint moment respectively, of population variables in an arbitrary PCTMC model. The set of ODEs can be directly solved by numerical simulation as long as there is no transition rate in the PCTMC with non-linear polynomials. With time-dependent rates, the system becomes hybrid with discrete jumps of rates at some specific points of numerical simulation.

3   Markov Queueing Model

Before introducing our model, we first give the traditional Markov queueing model for bike stations which is going to serve as our comparator.

The most straightforward way to evaluate the behaviour of a station is to analyse it in isolation. In this case, a station can be modelled as a time-inhomogeneous Markov queue $M/M/1/k_i$, illustrated in Fig. 1.

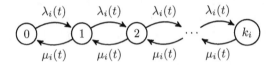

Fig. 1. The time-inhomogeneous Markov queue for station i

Specifically, k_i denotes the capacity of a station i, $\lambda_i(t)$ and $\mu_i(t)$ are the time-dependent bike arrival and pickup rates of station i at time t of a day. Usually, the time of a day is split into n even slots.

Then, using the transition rate matrix for station i: $Q(\lambda_i(t), \mu_i(t))$, where

$$Q(\lambda, \mu) = \begin{pmatrix} -\mu & \mu & & & \\ \lambda & -(\mu+\lambda) & \mu & & \\ & \ddots & \ddots & \ddots & \\ & & \lambda & -(\mu+\lambda) & \mu \\ & & & \lambda & -\lambda \end{pmatrix},$$

one can predict the probability that there are y bikes in station i at time $t + h$ given the station has x bikes at time t, by the following equation:

$$\Pr(y \mid x, t, h) = \exp\left(\int_0^h Q(t + s)ds\right)_{x,y}$$

where $\exp(M)_{x,y}$ is the element at row x and column y of the matrix exponential of M. Such a model has been used to make bike availability or station inventory level predictions in several papers in the literature (e.g. [6,8,13]).

Two assumptions are made in this model. First, the bike arrivals and pickups at stations form Poisson processes. Second, the state of a particular station does not depend on the state of the others. The first assumption is successfully validated for busy stations in [8], using historical data from the Velib bike-sharing system in Paris. However, we conjecture that the second assumption is generally not true in practice. For example, when a station is empty, no bikes can depart from it, therefore the arrival rate at other stations should be reduced. Hence, we seek a more realistic model, which captures the journey dynamics between stations.

4   PCTMC of Bike-Sharing Model

4.1   A Naive PCTMC Model

To faithfully represent the journey dynamics between bike stations in a bike-sharing system with N stations, we first propose a naive PCTMC model which contains the following transitions:

$$Bike_i \longrightarrow Slot_i + Journey_j^i@P_1 \quad \text{at } \mu_i(t)p_j^i(t) \qquad\qquad \forall i,j \in (1,N)$$

$$Journey_j^i@P_l \longrightarrow Journey_j^i@P_{l+1} \quad \text{at } (P_j^i/d_j^i)\,\#(Journey_j^i@P_l)$$

$$l \geq 1 \wedge l < P_j^i, \forall i,j \in (1,N)$$

$$Journey_j^i@P_{P_j^i} + Slot_j \longrightarrow Bike_j \quad \text{at } (P_j^i/d_j^i)\,\#(Journey_j^i@P_{P_j^i}) \quad \forall i,j \in (1,N)$$

where $Bike_i$, $Slot_i$ represent a bike and a slot agent in station i respectively; $Journey_j^i@P_l$ represents a bike agent which is currently on a journey from station i to station j at phase l. Note that since journey durations are generally not exponentially distributed, we fit the journey duration from station i to station j as an Erlang distribution with P_j^i phases each with rate P_j^i/d_j^i, where d_j^i is the mean journey duration. $\mu_i(t)$ is the bike pickup rate in station i at time t, p_j^i is the probability that a journey will end at station j given that it started from station i at time t. $\#(S)$ denotes the population of an agent type S.

Obviously, the above model is not scalable. Since the total number of bike stations N is usually very large (for example there are around 750 bike stations in London), it is computationally infeasible to analyse a model which captures the full set of bike stations. Fortunately, since we are only interested in the prediction of bike availability of a single target station at a time, we only need to model stations which have a significant contribution to the journey flows to the target station (knowing the state of a station which has a very small contribution to the journey flows to the target station will have negligible impact on the accuracy of bike availability prediction for the target station). Thus, a directed contribution graph together with a contribution propagation method is proposed to automatically identify the set of stations which need to be modelled with respect to a given target station for bike availability prediction.

4.2 Directed Contribution Graph with Contribution Propagation

Here, we show how to derive a set of bike stations $\Theta(v)$ in which all stations have a significant contribution to the journey flows to a given target station $v \in (1, 2, \ldots, N)$ for bike availability prediction. Concretely, we first need a way to quantify the contribution of one station to the journey flows to another station. Specifically, we let C_{ij} denote the *contribution coefficient* of station j to station i which quantifies the contribution of station j to the journey flows to station i.

One station can contribute to the journey flows to another station both directly and indirectly. The definition of a *direct contribution coefficient* at time t is given by the following simple formula:

$$c_{ij}(t) = \lambda_i^j(t)/\lambda_i(t)$$

in which $\lambda_i^j(t)$ represents the bike arrival rate from station j to station i at time t and $\lambda_i(t) = \sum_j \lambda_i^j(t)$. Then, it is clear that $c_{ij}(t) \in [0, 1]$, $0 \le \sum_{j \ne i} c_{ij}(t) \le 1$.

With the definition of directed contribution coefficient, we can construct a directed contribution graph for the bike-sharing system at each time slot of a day. The definition of the directed contribution graph is given as follows (for convenience, we abbreviate $c_{ij}(t)$ to c_{ij}):

Definition 1. *For an arbitrary time t, the directed contribution graph for a bike-sharing system at time t is a graph in which nodes represent the stations in the system, and there is a weighted directed edge from node i to node j if $c_{ij} > 0$, and in this case the weight of the edge is c_{ij}. Thus, the direction of edges is the inverse of contribution flows.*

Figure 2 shows a sample directed contribution graph which consists of six bike stations.

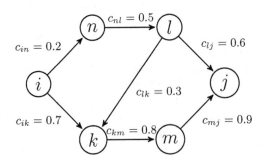

Fig. 2. An example directed contribution graph with six stations

For those stations which are not directly connected in the directed relation graph, by using a contribution propagation method, we can evaluate the *indirect contribution coefficient* of one station on the journey flows to another station. Specifically, the indirect contribution coefficient is quantified by a path dependent coefficient $c_{ij,\gamma}$, which is the product of the direct contribution coefficients along an acyclic path γ from node i to node j. Then, the contribution coefficient of station j to station i is characterized by the maximum of the path dependent coefficients:

$$c_{ij,\gamma} = \prod_{kl \in \gamma} c_{kl}$$

$$C_{ij} = \begin{cases} \max\limits_{all\ paths\ \gamma} c_{ij,\gamma} & \text{if there exists a path from node } i \text{ to node } j \\ 0, & \text{otherwise} \end{cases}$$

For example, according to Fig. 2, the contribution coefficient of station j to station i is $C_{ij} = c_{ik} \times c_{km} \times c_{mj} = 0.504$, since $c_{ik} \times c_{km} \times c_{mj} > c_{in} \times c_{nl} \times c_{lj} > c_{in} \times c_{nl} \times c_{lk} \times c_{km} \times c_{mj}$.

With the contribution coefficient, given a target station v, then for $i \in (1, 2, \ldots, N)$, we can infer:

$$i \in \Theta(v) \qquad\qquad \text{if } C_{vi} > \theta$$
$$i \notin \Theta(v) \qquad\qquad \text{if } C_{vi} \leq \theta$$

where $\theta \in (0, 1)$ is threshold value which can be used to control the extent of model reduction. A point to note is that we choose to characterize contribution coefficients by the maximum instead of the sum of path dependent coefficients because we only want to model stations which have at least a significant (direct or indirect) journey flow to the target station. To model stations which have many small journey flows to the target station is costly but the impact is rather unpredictable. Moreover, the maximum of path dependent coefficients has another nice property that if $i \in \Theta(v)$ and $C_{vi} = c_{vi,\gamma}$, then for a station j which is on the path γ, it is certain that $C_{vj} > \theta$, thus $j \in \Theta(v)$. As a result, for all stations which have a significant journey flow to the target station, that journey flow will certainly be captured in the resulting reduced PCTMC. However, this property will not be preserved if we use the sum of path dependent coefficients. For example in Fig. 2, if we set $\theta = 0.55$, then $\sum_\gamma c_{ij,\gamma} > \theta$, thus station j is included in the reduced PCTMC. However, since $\sum_\gamma c_{il,\gamma} < \theta$, station l will not be included, thus $\sum_\gamma c_{ij,\gamma} < \theta$ will not be satisfied in the reduced PCTMC.

As an illustration of the extent of model reduction, Fig. 3 shows the empirical cumulative distribution function of contribution coefficients between all bike stations during all time slots (which is computed by journey data from the London Santander Bike-sharing system, with 20 min slot duration). It can be seen that more than 96 % stations can be excluded even if θ is set to the small value 0.01.

4.3   The Reduced PCTMC Model

Given a target station v and current time t, suppose we are interested in the number of bikes at the station at time $t + h$, then let $s = (s_1, s_2, \ldots, s_n)$ be the

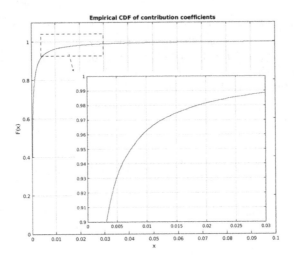

Fig. 3. The empirical cumulative distribution function of contribution coefficients (x is the value of contribution coefficients)

minimal set of time slots which covers $[t, t + h]$, we obtain $\Theta(v) = \Theta(v, s_1) \cup \Theta(v, s_2) \cup \ldots \cup \Theta(v, s_n) \cup v$, where $\Theta(v, s_i)$ is the set of bike stations which have significant contribution to the journey flows to the target station at time slot s_i.

Therefore, the PCTMC for the prediction of bike availability at station v at time $t + h$ can be represented as follows:

$$Bike_i \longrightarrow Slot_i \quad \text{at } \mu_i(t)\left(1 - \sum_{j \notin \Theta(v) \vee c_{ji} \leq \theta} p_j^i(t)\right) \qquad \forall i \in \Theta(v) \qquad (4.1)$$

$$Slot_i \longrightarrow Bike_i \quad \text{at } \sum_{j \notin \Theta(v) \vee c_{ij} \leq \theta} \lambda_i^j(t) \qquad \forall i \in \Theta(v) \qquad (4.2)$$

$$Bike_i \longrightarrow Slot_i + Journey_j^i@P_1 \quad \text{at } \mu_i(t)p_j^i(t) \quad \forall i, j \in \Theta(v) \wedge c_{ji} > \theta \quad (4.3)$$

$$Journey_j^i@P_l \longrightarrow Journey_j^i@P_{l+1} \quad \text{at } (P_j^i/d_j^i) \, \#(Journey_j^i@P_l)$$
$$l \geq 1 \wedge l < P_j^i, \forall i, j \in \Theta(v) \wedge c_{ji} > \theta \quad (4.4)$$

$$Slot_j + Journey_j^i@P_{P_j^i} \longrightarrow Bike_j \quad \text{at } (P_j^i/d_j^i) \, \#(Journey_j^i@P_{P_j^i})$$
$$\forall i, j \in \Theta(v) \wedge c_{ji} > \theta \quad (4.5)$$

$$Journey_j^i@P_{P_j^i} \longrightarrow \varnothing \quad \text{at } \mathbb{1}\big(Slot_j(t) = 0\big)(P_j^i/d_j^i) \, \#(Journey_j^i@P_{P_j^i})$$
$$\forall i, j \in \Theta(v) \wedge c_{ji} > \theta \quad (4.6)$$

where (4.1) represents a bike in station i is picked up for a journey to a station outside $\Theta(v)$ or a station to which the journey flow is negligible (the direct contribution coefficient $c_{ji} \leq \theta$ indicates that journey flow from i to j must not be a significant journey flow); (4.2) represents a bike is returned to station i from a station outside $\Theta(v)$ or a station from which the journey flow is negligible;

(4.3) represents a bike in station i is picked up for a journey to a station j inside $\Theta(v)$ and the journey flow is significant; (4.4), (4.5) represent progress and completion of the journey, respectively; (4.6) assumes a bike in transit from station i to station j will be returned to another station outside $\Theta(v)$ when there is no empty slot in station j, where $\mathbb{1}(Slot_j(t) = 0)$ is an indicator function which returns 1 when the number of empty slots at station j at time t is zero, otherwise returns 0.

Dealing with Indicator Function. Since we are going to numerically solve the PCTMC using moment ODEs as illustrated in Eq. (2.1), we can only access the moments of the number of empty slots at a station i at time t, denoted as u_i^m, during numerical simulation (here we let u_i^m denote $\mathbb{E}[(Slot_i(t))^m]$, where m is the order of the moment), whereas the number of empty slots at station i at time t is a random variable. Thus, we propose a method to approximate the indicator function by a function of the moments u_i^m of the number of empty slots and the capacity of the station: $\mathbb{1}(Slot_i(t) = 0) \sim f(u_i^1, u_i^2, \ldots, u_i^m, k_i)$. Concretely, given the first m moments of the random variable $Slot_i(t)$, and the value domain $Slot_i(t) \in [0, 1, \ldots, k_i]$, we can approximate the probability distribution of $Slot_i(t)$ by a discrete distribution with finite support k_i. For example, if we only know the first moment of $Slot_i(t)$ (which is u_i^1), we can fit a binomial distribution $Slot_i(t) \sim Binomial(k_i, u_i^1/k_i)$ to the probability distribution of $Slot_i(t)$. In this case, we get $Pr(Slot_i(t) = 0) = (1 - u_i^1/k_i)^{k_i}$. Furthermore, if we know the first two moments (u_i^1, u_i^2), then we can fit a beta-binomial distribution $Slot_i(t) \sim BetaBinomial(k_i, \alpha, \beta)$, where

$$\alpha = \frac{u_i^1 u_i^2 - k_i(u_i^1)^2}{k_i(u_i^1)^2 + k_i u_i^1 - k_i u_i^2 - (u_i^1)^2} \qquad \beta = \frac{(k_i - u_i^1)(k_i u_i^1 - u_i^2)}{k_i(u_i^1)^2 + k_i u_i^1 - k_i u_i^2 - (u_i^1)^2}$$

Thus, we get

$$Pr(Slot_i(t) = 0) = \frac{B(\alpha, k_i + \beta)}{B(\alpha, \beta)}$$

where $B(a, b)$ is a beta function. Theoretically, with knowledge of more moments of $Slot_i(t)$, the estimation of $Pr(Slot_i(t) = 0)$ will be more accurate. Finally, we let

$$\mathbb{1}(Slot_i(t) = 0) = \begin{cases} 1 & \text{if } Pr(Slot_i(t) = 0) > p \\ 0 & \text{if } Pr(Slot_i(t) = 0) \le p \end{cases}$$

where $Pr(Slot_i(t) = 0) = f(u_i^1, u_i^2, \ldots, u_i^m, k_i)$, p is a threshold value beyond which we believe the number of empty slots in station i is zero. In general p should be set to a value close to 1. In our later experiments, we explicitly set $p = 0.9$.

Specifying the Initial State. Given a snapshot of the bike-sharing system at a time instant t which contains the following information[1]:

$$Bike_i(t), \ldots, Slot_i(t), \ldots, Journey^i(t, \Delta t), \ldots$$

---

[1] This information is actually recorded for the London bike-sharing system.

where $Bike_i(t)$ and $Slot_i(t)$ are the current number of available bikes and empty slots at a station i; $Journey^i(t, \Delta t)$ represents there is a bike currently en route from station i, and the journey started at time $t - \Delta t$. Then, for each $Journey^i(t, \Delta t)$, we use a random number to determine the destination of the journey, and the time Δt to determine the appropriate phase of the journey time. Thus we generate a random number α uniformly distributed in $(0, 1)$, and let $p_k^i(t - \Delta t), \forall k$ be the probability that the journey will end at station k given that the journey started from station i at time $t - \Delta t$. Then

$$Journey^i(t, \Delta t) = Journey_j^i(t, \Delta t) \text{ if } \alpha \geq \sum_{k=0}^{j-1} p_k^i(t - \Delta t) \text{ and } \alpha < \sum_{k=0}^{j} p_k^i(t - \Delta t).$$

Furthermore, we let

$$Journey_j^i(t, \Delta t) = Journey_j^i @ P_l \text{ if } \Delta t \geq (l-1) d_j^i / P_j^i \text{ and } \Delta t < l \times d_j^i / P_j^i,$$

where $l \leq P_j^i$. Otherwise, if $l > P_j^i$, we let $Journey_j^i(t, \Delta t) = Journey_j^i @ P_{P_j^i}$.

Solving the Moment ODEs. We derive the moment ODEs following Eq. (2.1) for the above PCTMC for the first m order of moments. Furthermore, using the correlation heuristics introduced in [14], we can make a further reduction on the size of the moment ODEs, utilizing the neighbourhood relation between agents in the above PCTMC. Specifically, we let $\mathbb{E}[(X_i)^{m_i}(X_j)^{m_j}] \approx \mathbb{E}[(X_i)^{m_i}]\mathbb{E}[(X_j)^{m_j}]$ if there does not exist a transition in the PCTMC in which both agent S_i and S_j are directly involved. Due to limited space, we refer to [14] for more detail of the reduction algorithm. The moment ODEs can be solved by numerical simulation using standard methods.

5   Reconstructing the Probability Distribution Using the Maximum Entropy Approach

From the moment analysis of the PCTMC for bike-sharing model, we gain the first m moments of the number of available bikes in the target station at the prediction time $t+h$, i.e. $\left(\left(Bike_v(t+h) \right)^1, \left(Bike_v(t+h) \right)^2, \ldots, \left(Bike_v(t+h) \right)^m \right)$, which we denote as $(u^1, u^2, \ldots, u^m)$ in the following. Our goal is to predict the probability that the station has a specific number of bikes at time $t + h$. This means the problem is to reveal $\Pr \left(Bike_v(t + h) = i \mid u^1, u^2, \ldots, u^m, k_v \right)$, where $i \in (1, 2, \ldots, k_v)$. Therefore, we need to reconstruct the entire probability distribution of the random variable $Bike_v(t + h)$ based on its first m moments. The corresponding distribution is generally not uniquely determined. Hence, to select a particular distribution, we apply the maximum entropy principle to minimize the amount of bias in the reconstruction process. In this way, we assume the least amount of prior information about the true distribution. Note that the maximum entropy approach has been successfully applied to reconstruct distributions based on moments in many areas, e.g. physics [15], stochastic chemical kinetics [16], and performance analysis [17].

5.1 Reconstruction Algorithm

Let X_v denote $Bike_v(t+h)$ for convenience, $\mathcal{G}$ be the set of all possible probability distributions for X_v. Then, based on the maximum entropy principle, the goal is to select a distribution g to maximize the entropy $H(g)$ over all distributions in $\mathcal{G}$. The problem can be denoted as follows:

$$\arg\max_{g \in \mathcal{G}} H(g) = \arg\max_{g \in \mathcal{G}} \left(-\sum_{x=0}^{k_v} g(x) \ln g(x) \right)$$

Furthermore, given $(u^1, u^2, \ldots, u^m)$, we know the following constraints should be satisfied:

$$\sum_{x=0}^{k_v} x^n g(x) = u^n, \quad n = 0, 1, \ldots, m$$

where $u^0 = 1$ to ensure that g is a probability distribution. Now, the problem becomes a constrained optimization program. Thus to perform the constrained maximization of the entropy, we introduce one Lagrange multiplier λ_n per moment constraint. We thus seek extrema of the Lagrangian functional:

$$L(g, \lambda) = -\sum_{x=0}^{k_v} g(x) \ln g(x) - \sum_{n=0}^{m} \lambda_n \left(\sum_{x=0}^{k_v} x^n g(x) - u^n \right)$$

Functional variation with respect to the unknown distribution function $g(x)$ yields:

$$\frac{\partial L}{\partial g(x)} = 0 \implies g(x) = \exp\left(-1 - \lambda_0 - \sum_{n=1}^{m} \lambda_n x^n \right)$$

Since $u^0 = 1$, we get

$$\sum_{x=0}^{k_v} \exp\left(-1 - \lambda_0 - \sum_{n=1}^{m} \lambda_n x^n \right) = 1.$$

Thus we can express λ_0 in terms of the remaining Lagrange multipliers

$$e^{1+\lambda_0} = \sum_{x=0}^{k_v} \exp\left(-\sum_{n=1}^{m} \lambda_n x^n \right) \equiv Z$$

Then, the general form of $g(x)$ can be given as follows:

$$g(x) = \frac{1}{Z} \exp\left(-\sum_{n=1}^{m} \lambda_n x^n \right)$$

Insert the preceding equation into the Lagrangian, we can then transform the problem into an unconstrained minimization problem of the following function with respect to variables $\lambda_1, \lambda_2, \ldots, \lambda_n$:

$$\Gamma(\lambda_1, \lambda_2, \ldots, \lambda_n) = \ln Z + \sum_{n=1}^{m} \lambda_n u^n$$

The convexity of the function Γ is proved in [15], which guarantees the existence of a unique solution. Thus, a close approximation $(\lambda_1^*, \lambda_2^*, \ldots, \lambda_n^*)$ of the true solution can be obtained by the classic gradient descent approach [18].

Thus, after finding $(\lambda_1^*, \lambda_2^*, \ldots, \lambda_n^*)$ through gradient descent, we can finally predict

$$\Pr\left(X_v = x\right) = \frac{\exp\left(-\sum_{n=1}^{m} \lambda_n^* x^n\right)}{\sum_{i=0}^{k_v} \exp\left(-\sum_{n=1}^{m} \lambda_n^* i^n\right)}, \quad \forall x \in (1, 2, \ldots, k_v).$$

6   Experiments

In this section, we test the time cost and accuracy of our prediction model in different cases and compare the accuracy of our model with the classic Markov queueing model. We use the historic journey data and bike availability data from January 2015 to March 2015 from the London Santander Cycles Hire scheme to train our PCTMC model as well as the Markov queueing model, and the data in April 2015 to test their prediction accuracy. As in [11], we fit the number of journey phases between stations using the HyperStar tool [19] command line interface. Specifically, we set the maximum value of P_j^i to 20 to make our model compact and also avoid overfitting. Moreover, for parameters estimation, we split a day into slots of 20 min duration. In our experiments, given the bike availability in a station at time t, we predict the probability distribution of the number of available bikes in that station at time $t + h$, where h is set to 10 min for short range prediction and 40 min for long range prediction.

The evaluation of our model is twofold. The first is accuracy, the second is efficiency. These two aspects are both influenced by the value of two important parameters, namely m, the highest order of moments being derived, and θ, the coefficient threshold for the identification of bike stations which have significant contribution to the journey flow to the target station. For higher values of m, the solution cost of our model becomes larger since more moment ODEs are derived, however the model should become more accurate due to more constraints in the probability distribution reconstruction based on the maximum entropy principle. For higher values of θ, more stations are excluded in the reduced PCTMC for a target station whereas the model accuracy can be potentially reduced. Thus, to observe the effects on these two parameters, we do experiments with values $m = 1, 2, 3$, $\theta = 0.01, 0.02, 0.03$.

6.1   Root Mean Square Error

For prediction accuracy, we first consider the classic criterion based on root mean square error (RMSE), a commonly used metric for evaluating point predictions (i.e., predictions that only state the expected number of bikes).

Table 1. The calculated RMSE on the prediction of the number of available bikes

	10 min	40 min	
Markov queueing model	1.52	3.03	
PCTMC with $\theta = 0.03$	1.49	2.81	$m = 1, 2, 3$
PCTMC with $\theta = 0.02$	1.49	2.81	$m = 1, 2, 3$
PCTMC with $\theta = 0.01$	1.48	2.79	$m = 1, 2, 3$

Table 1 compares the RMSE of the prediction results of our PCTMC model with the Markov queueing model. As can be seen, the PCTMC model outperforms the Markov queueing model in both prediction ranges. Especially in the long range, a considerable improvement is observed. For the PCTMC models, smaller values of θ only reduce the RMSE slightly. This means capturing less significant journey flows will have little impact on the prediction accuracy. Moreover, we find that the derived highest moments have almost no impact on the RMSE. This is obvious since the expected number of available bikes is only decided by the first moment.

6.2 Probability of Making a Right Recommendation

Predicting the expected number of available bikes is important for system administrators when they want to decide how to redistribute bikes in the system. However, a user is interested in whether there is a bike in the target station when she wants to pick up a bike from there, or whether there is a free slot in the target station when she wants to return a bike to that station. We are specifically interested in being able to make correct recommendations for the queries "Will there be a bike?" and "Will there be a slot?"[2] to measure the accuracy of our model. Specifically, for the "Will there be a bike?" query, we respond "Yes" if the predicted probability of that station having more than one bike is greater than 0.8, and respond "No" if the predicted probability of that station having more than one bike is less than 0.8. As is argued in [8], the root mean square error is not an appropriate evaluation metric in this setting. After all, we need a prediction of the probability of the recommendation being correct rather than just a point estimate of the number of available bikes/slots. Instead, a suitable evaluation scheme is proposed in [8] that ensures that the best prediction algorithm can always be expected to obtain the highest score. Such a scheme is called a *proper scoring rule*. For the setting described above, the following scoring rule is proper:

$$\text{Score} = \begin{cases} 1 & \text{if } \Pr(X_v > 0) > 0.8 \wedge x_v > 0 \\ -4 & \text{if } \Pr(X_v > 0) > 0.8 \wedge x_v = 0 \\ 1 & \text{if } \Pr(X_v > 0) < 0.8 \wedge x_v = 0 \\ -\frac{1}{4} & \text{if } \Pr(X_v > 0) < 0.8 \wedge x_v > 0 \end{cases}$$

---

[2] These queries can be readily extended to "Will there be n bikes?" and "Will there be n slots?".

Table 2. Average score of making a recommendation to the "Will there be a bike?" query with 95 % confidence interval

	10 min	40 min	
Markov queueing model	0.9 ± 0.05	0.87 ± 0.06	
PCTMC with $\theta = 0.03$	0.91 ± 0.04	0.89 ± 0.05	$m = 2$
	0.92 ± 0.04	0.91 ± 0.04	$m = 3$
PCTMC with $\theta = 0.02$	0.91 ± 0.04	0.89 ± 0.05	$m = 2$
	0.92 ± 0.04	0.91 ± 0.04	$m = 3$
PCTMC with $\theta = 0.01$	0.92 ± 0.04	0.89 ± 0.05	$m = 2$
	0.93 ± 0.04	0.91 ± 0.04	$m = 3$

Table 3. Average score of making a recommendation to the "Will there be a slot?" query with 95 % confidence interval

	10 min	40 min	
Markov queueing model	0.91 ± 0.04	0.88 ± 0.05	
PCTMC with $\theta = 0.03$	0.91 ± 0.04	0.9 ± 0.05	$m = 2$
	0.92 ± 0.04	0.91 ± 0.04	$m = 3$
PCTMC with $\theta = 0.02$	0.91 ± 0.04	0.9 ± 0.05	$m = 2$
	0.92 ± 0.04	0.91 ± 0.04	$m = 3$
PCTMC with $\theta = 0.01$	0.92 ± 0.04	0.91 ± 0.05	$m = 2$
	0.93 ± 0.04	0.92 ± 0.04	$m = 3$

Note that incorrect predictions need to be penalised by a negative score for the rule to be proper. The evaluation of recommendations to the "Will there be a slot?" query follows a similar pattern. Tables 2 and 3 show the experimental results for different models and parameters. Note that the PCTMC model with $m = 1$ is excluded since at least two moments are needed to make a meaningful reconstruction of the probability distribution. As can be seen from the tables, the PCTMC model clearly has a better performance in making such recommendations. Moreover, we also observe that with higher values of m, the average score increases. This is because, with higher values of m, the reconstructed probability distribution is closer to the true distribution.

6.3 Time Cost

The time cost of making a prediction is also important. Table 4 shows the time cost for making a prediction using our PCTMC model with different parameters (we do not show the time costs for the Markov queueing model since they are negligible due to its small state space because of independence assumption). For real time application, we assume that the time cost of making a prediction must be less than one second. Thus, for point prediction, we recommend to set

Table 4. Time cost to make a prediction with 95 % confidence interval

	10 min	40 min	
PCTMC with $\theta = 0.03$	1.76 ± 0.2ms	6.98 ± 0.77ms	$m = 1$
	103 ± 13.7 ms	328 ± 43 ms	$m = 2$
	2.2 ± 0.2 s	8.9 ± 0.83 s	$m = 3$
PCTMC with $\theta = 0.02$	4.25 ± 0.4 ms	15.72 ± 1.42 ms	$m = 1$
	251 ± 25.5 ms	1.1 ± 0.1 s	$m = 2$
	8.9 ± 1.2 s	37 ± 3.5 s	$m = 3$
PCTMC with $\theta = 0.01$	13.5 ± 0.9 ms	49.1 ± 3.92 ms	$m = 1$
	8.8 ± 1.1 s	30.1 ± 0.31 s	$m = 2$
	33.9 ± 5.4 s	157 ± 17.8 s	$m = 3$

$\theta = 0.01, m = 1$ for both prediction ranges. For probability distribution prediction, we recommend to set $\theta = 0.02, m = 2$ for short range prediction, $\theta = 0.03, m = 2$ for long range prediction. Note that we used an Intel CORE i7 laptop with 8 GB RAM to run our experiments, the time cost could be considerably reduced if a more powerful machine, e.g. a server, were used.

7   Conclusion

We have presented a moment-based approach to make predictions of availability in bike-sharing systems. The moments of the number of available bikes are automatically derived via a PCTMC with time-inhomogeneous rates, fitted from historical data. The entire probability distribution is reconstructed using a maximum entropy approach. Our model is easy to understand since it explicitly captures the dynamics of the bike-sharing system. We demonstrated that it outperforms the classic Markov queueing model in several performance metrics for prediction accuracy. Moreover we have also shown that by using the direct contribution graph and the contribution propagation method, the model size can be significantly reduced to such an extent that it is suitable for real time application.

In future work we plan to explore the impact of neighbouring stations, and extend our model to capture their effects. For example, if a station is empty, then the user is likely to pick up a bike from a neighbouring station, thus increasing the pickup rate at the neighbouring station. Conversely, if a station is full, then the user is likely to return a bike to a neighbouring station, increasing the bike arrival rate there. We think another merit of our PCTMC model is that it can be easily extended to capture such impact by using the indicator function to check whether a neighbouring station is empty or full in order to alter the bike arrival and pickup rate of a station. Unfortunately we do not currently have data to capture the impact of neighbouring stations.

Acknowledgement. This work is supported by the EU project QUANTICOL, 600708.

References

1. Fishman, E.: Bikeshare: a review of recent literature. Transp. Rev. **36**(1), 1–22 (2015)
2. Lin, J.R., Yang, T.H.: Strategic design of public bicycle sharing systems with service level constraints. Transp. Res. Part E: Logist. Transp. Rev. **47**(2), 284–294 (2011)
3. Pfrommer, J., Warrington, J., Schildbach, G., Morari, M.: Dynamic vehicle redistribution and online price incentives in shared mobility systems. IEEE Trans. Intell. Transp. Syst. **15**(4), 1567–1578 (2014)
4. Nair, R., Miller-Hooks, E.: Fleet management for vehicle sharing operations. Transp. Sci. **45**(4), 524–540 (2011)
5. Contardo, C., Morency, C., Rousseau, L.M.: Balancing a Dynamic Public Bike-Sharing System, vol. 4. CIRRELT, Montreal (2012)
6. Schuijbroek, J., Hampshire, R., van Hoeve, W.J.: Inventory rebalancing and vehicle routing in bike sharing systems. In: Technical report, Schuijbroek (2013)
7. Yoon, J.W., Pinelli, F., Calabrese, F.: Cityride: a predictive bike sharing journey advisor. In: 2012 IEEE 13th International Conference on Mobile Data Management (MDM), pp. 306–311. IEEE (2012)
8. Gast, N., Massonnet, G., Reijsbergen, D., Tribastone, M.: Probabilistic forecasts of bike-sharing systems for journey planning. In: The 24th ACM International Conference on Information and Knowledge Management (CIKM 2015) (2015)
9. Froehlich, J., Neumann, J., Oliver, N.: Sensing and predicting the pulse of the city through shared bicycling. IJCAI **9**, 1420–1426 (2009)
10. Kaltenbrunner, A., Meza, R., Grivolla, J., Codina, J., Banchs, R.: Urban cycles and mobility patterns: exploring and predicting trends in a bicycle-based public transport system. Pervasive Mob. Comput. **6**(4), 455–466 (2010)
11. Guenther, M.C., Bradley, J.T.: Journey data based arrival forecasting for bicycle hire schemes. In: Dudin, A., De Turck, K. (eds.) ASMTA 2013. LNCS, vol. 7984, pp. 214–231. Springer, Heidelberg (2013)
12. Engblom, S.: Computing the moments of high dimensional solutions of the master equation. Appl. Math. Comput. **180**(2), 498–515 (2006)
13. Raviv, T., Kolka, O.: Optimal inventory management of a bike-sharing station. IIE Trans. **45**(10), 1077–1093 (2013)
14. Feng, C., Hillston, J., Galpin, V.: Automatic moment-closure approximation of spatially distributed collective adaptive systems. ACM Trans. Model. Comput. Simul. (TOMACS) **26**(4), 26 (2016)
15. Mead, L.R., Papanicolaou, N.: Maximum entropy in the problem of moments. J. Math. Phys. **25**(8), 2404–2417 (1984)
16. Andreychenko, A., Mikeev, L., Wolf, V.: Model reconstruction for moment-based stochastic chemical kinetics. ACM Trans. Model. Comput. Simul. (TOMACS) **25**(2), 12 (2015)
17. Tari, Á., Telek, M., Buchholz, P.: A unified approach to the moments based distribution estimation – unbounded support. In: Bravetti, M., Kloul, L., Zavattaro, G. (eds.) EPEW/WS-EM 2005. LNCS, vol. 3670, pp. 79–93. Springer, Heidelberg (2005)

18. Snyman, J.: Practical Mathematical Optimization: An Introduction to Basic Optimization Theory and Classical and New Gradient-Based Algorithms, vol. 97. Springer Science & Business Media, Heidelberg (2005)
19. Reinecke, P., Krauss, T., Wolter, K.: Hyperstar: phase-type fitting made easy. In: 2012 Ninth International Conference on Quantitative Evaluation of Systems, pp. 201–202. IEEE (2012)

Tools

Attack Trees for Practical Security Assessment: Ranking of Attack Scenarios with ADTool 2.0

Olga Gadyatskaya, Ravi Jhawar, Piotr Kordy, Karim Lounis, Sjouke Mauw, and Rolando Trujillo-Rasua[(✉)]

SnT, University of Luxembourg, Luxembourg City, Luxembourg
{olga.gadyatskaya,ravi.jhawar,Piotr.Kordy,karim.lounis, sjouke.mauw,rolando.trujillo}@uni.lu

Abstract. In this tool demonstration paper we present the ADTool2.0: an open-source software tool for design, manipulation and analysis of attack trees. The tool supports ranking of attack scenarios based on quantitative attributes entered by the user; it is scriptable; and it incorporates attack trees with sequential conjunctive refinement.

1 Introduction

Attack trees are a well-known and established methodology for security assessment that facilitates brainstorming, structures available information, and assists human experts in analysis. An attack tree is a graphical model, and as such it is better comprehensible than pure text-based approaches. However, graphical models require usable and efficient tools with suitable Graphical User Interfaces (GUIs) in order to be practical. Moreover, recent advances in automated risk assessment techniques now call for tool support to handle automatically generated attack trees with many thousands of nodes [2,3]. Therefore, the need for more comprehensive analysis tools emerged in the community. In this paper we present the ADTool2.0 that provides advanced capabilities for design, visualization, and analysis of attack trees [9], attack-defense trees [6], and attack trees with sequential conjunctive refinement (SAND attack trees for short) [4].

The ADTool2.0 is not a simple extension of the previous tool [5], but a fully revamped, more advanced system. It has been reimplemented using the advanced cross-platform Docking Frames library[1]. The new version of the tool brings in many new features, including ranking of critical attack scenarios, attack trees with the sequential AND (SAND) operator, and scriptability.

In contrast to many commercial tools, such as SecurITree[2] and AttackTree+[3], the ADTool2.0 is an open source software, freely available to the community[4].

The research leading to the results presented in this work received funding from the European Commission's Seventh Framework Programme (FP7/2007–2013) under grant agreement number 318003 (TREsPASS) and Fonds National de la Recherche Luxembourg under the grant C13/IS/5809105 (ADT2P).

[1] http://www.docking-frames.org/.
[2] http://www.amenaza.com.
[3] http://www.isograph.com/software/.
[4] https://github.com/tahti/ADTool2.

G. Agha and B. Van Houdt (Eds.): QEST 2016, LNCS 9826, pp. 159–162, 2016.
DOI: 10.1007/978-3-319-43425-4_10

Moreover, it continues to be the only software tool providing support for the attack-defense tree modeling language [6]. In that sense, the ADTool2.0 provides unique features in comparison to integration frameworks (e.g., the Möbius framework [1]) and tools based on attack graphs (e.g., ADVISE [8]).

2   Main Features of the ADTool2.0

Sequential Conjunct Refinements in Attack Trees. The ADTool2.0 integrates a crucial modelling aspect: creation of attack trees with SAND refinements (consistent with the graphical language and semantics described in [4]) and their quantitative analysis. Usage of the SAND refinement allows the analyst to model and analyze attack scenarios involving several attack steps that need to be all executed in a specific order, as opposed to the standard AND refinement used to model execution of several attack steps in parallel.

After constructing a SAND attack tree, the user can assign an attribute domain (e.g., minimum time for the attack, probability of success) to the tree. Each leaf node is then initialized with a default value representing the worst case scenario (e.g., ∞ as the minimum time for the attack), and all other nodes are automatically assigned using an n-ary function, depending on the type of attribute and refinement operator, in order to evaluate the security scenario. The ADTool2.0 will automatically compute new attribute values using a bottom-up algorithm.

Ranking Attack Trees. Human ability to visualize and understand attack trees quickly decreases with the increase in size and complexity of the tree. Identifying important portions of an attack tree is therefore of paramount importance for security analysts; it allows to prioritize and focus on those branches that contribute most to the attacker goal. A systematic approach to prioritization is *ranking*, whereby a set of elements is sorted with respect to a total order. In attack graphs, a modelling language similar to attack trees, several ranking approaches have been defined [10]. In attack trees, however, ranking has been mostly neglected by both quantification methods and tools.

The ADTool2.0 implements an efficient and formal approach to rank attack scenarios. In particular, we have extended the bottom-up computation approaches proposed for attack trees [9], attack-defense trees [6], and SAND attack trees [4], in order to efficiently rank attack scenarios, where an attack scenario is either a *bundle* as in the formalisms in [6,9] or an *SP graph* as in [4]. Our approach works intuitively as follows. Given a set of quantitative values V for attack scenarios and a total order $\leq$ on V, we store at every node of the tree n least attacks with respect to the total order $\leq$, where n is a natural number representing a bound on the number of attack scenarios to be ranked.

Ranking results in the ADTool2.0 are shown in the *Ranking View* window, which can be opened from the menu Windows $\rightarrow$ Ranking View. As in the Attribute window, the Ranking window gives the option to open or create an attribute domain. By default, the ADTool2.0 uses as a total order the operator assigned to the OR gate in the attribute domain. A screenshot of the ADTool2.0

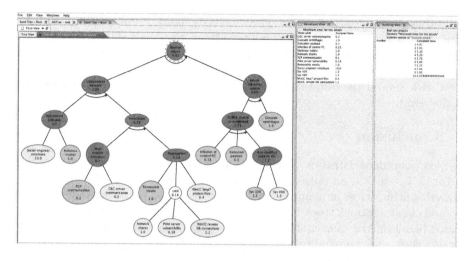

Fig. 1. Screenshot of the ADTool2.0 with the ranking feature. The SAND attack tree used represents the Stuxnet attack, and the ranking is based on the minimal time of attack parameter. The attack scenario (all its attack nodes) with the minimal time of execution is highlighted in green by the tool.

provided in Fig. 1 shows an example of the ranking feature applied to a SAND attack tree modelling the Stuxnet attack (inspired by [7]).

In order to rank attack scenarios up to a given node in the tree, we ought to click that node in the domain for which we want to see the ranking. Doing so, the Ranking view window will automatically update with a table containing optimal attacks with respect to the chosen attribute domain. The ADTool2.0 also offers the option to highlight those nodes that contribute most to the attack, which can be done by clicking on attack scenarios in the ranking table.

Scripting. Scriptability, whereby a tool can be run by scripts and without a GUI, is an important feature of security assessment tools. It allows sensitivity analysis (a standard technique to automatically assess how changes in some attribute values affect the overall security posture) and integration into tool chains. With the current version of the tool, it is now also possible to experiment with countermeasure selection: we can write scripts that will input several attack-defense trees with different defense scenarios applied to a particular attack, and output the best countermeasure set based on the results of the ranking.

In the scripting mode, which is typically executed from the command line[5], the ADTool2.0 supports input files of different formats (e.g., XML files) containing any of the supported attack trees (e.g., SAND trees), and provides various types of outputs such as the most critical attacks or the result of a bottom-up calculation. By using this scriptability feature, the ADTool2.0 has been integrated into the TREsPASS project tool chain[6], where it is used to visualize attack-defense scenarios and automatically or manually produced attack trees.

---

[5] Execute `java -jar ADTool-2.0.jar --help` from the command line for basic help.
[6] http://www.trespass-project.eu/.

Usability Features. The ADTool2.0 includes many usability features, e.g., copy-paste of subtrees, handling of multiple trees, reorder of children nodes, and extended input format (automatically generated attack trees [3] not conforming to the ADTool2.0 XML schema). The ADTool2.0 can handle and analyze *large trees* with several thousand nodes (automatically generated trees are typically of that size).

3 Conclusion

In this tool demonstration paper we presented the main features of the ADTool2.0, which is an open-source software tool for displaying, designing and analyzing attack trees in many flavors (SAND attack trees [4], attack-defense trees [6], and classical attack trees [9]). The ADTool2.0 supports ranking of attack scenarios based on the quantitative values selected by the end-user (e.g., time of attack, cost, and probability). In addition, it can be scripted for performing sensitivity analysis or running in tool chains.

References

1. Deavours, D.D., Clark, G., Courtney, T., Daly, D., Derisavi, S., Doyle, J.M., Sanders, W.H., Webster, P.G.: The möbius framework and its implementation. IEEE Trans. Softw. Eng. **28**(10), 956–969 (2002)
2. Gadyatskaya, O.: How to generate security cameras: towards defence generation for socio-technical systems. In: Mauw, S., et al. (eds.) GraMSec 2015. LNCS, vol. 9390, pp. 50–65. Springer, Heidelberg (2016). doi:10.1007/978-3-319-29968-6_4
3. Ivanova, M.G., Probst, C.W., Hansen, R.R., Kammüller, F.: Transforming graphical system models to graphical attack models. In: Mauw, S., et al. (eds.) GraMSec 2015. LNCS, vol. 9390, pp. 82–96. Springer, Heidelberg (2016). doi:10.1007/978-3-319-29968-6_6
4. Jhawar, R., Kordy, B., Mauw, S., Radomirović, S., Trujillo-Rasua, R.: Attack trees with sequential conjunction. In: Federrath, H., Gollmann, D., Chakravarthy, S.R. (eds.) SEC 2015. IFIP AICT, vol. 455, pp. 339–353. Springer, Heidelberg (2015). doi:10.1007/978-3-319-18467-8_23
5. Kordy, B., Kordy, P., Mauw, S., Schweitzer, P.: ADTool: security analysis with attack–defense trees. In: Joshi, K., Siegle, M., Stoelinga, M., D'Argenio, P.R. (eds.) QEST 2013. LNCS, vol. 8054, pp. 173–176. Springer, Heidelberg (2013)
6. Kordy, B., Mauw, S., Radomirović, S., Schweitzer, P.: Attack-defense trees. J. Log. Comput. **24**(1), 55–87 (2014)
7. Kriaa, S., Bouissou, M., Pietre-Cambacedes, L.: Modeling the Stuxnet attack with BDMP: towards more formal risk assessments. In: Proceedings of the CRiSIS (2012)
8. LeMay, E., Ford, M.D., Keefe, K., Sanders, W.H., Muehrcke, C.: Model-based security metrics using ADversary VIew Security Evaluation (ADVISE). In: Proceedings of QEST 2011, pp. 191–200. IEEE Computer Society, Washington, DC (2011)
9. Mauw, S., Oostdijk, M.: Foundations of attack trees. In: Won, D.H., Kim, S. (eds.) ICISC 2005. LNCS, vol. 3935, pp. 186–198. Springer, Heidelberg (2006)
10. Mehta, V., Bartzis, C., Zhu, H., Clarke, E.: Ranking attack graphs. In: Zamboni, D., Kruegel, C. (eds.) RAID 2006. LNCS, vol. 4219, pp. 127–144. Springer, Heidelberg (2006)

Spnps: A Tool for Perfect Sampling in Stochastic Petri Nets

Simonetta Balsamo$^{(\boxtimes)}$, Andrea Marin, and Ivan Stojic

Università Ca' Foscari Venezia, DAIS, via Torino 155, Venice, Italy
{balsamo,marin,stojic}@dais.unive.it

Abstract. This paper presents a tool spnps for perfect sampling (PS) in stochastic Petri nets (SPN). SPNs are an important formalism for performance evaluation of telecommunication systems and computer hardware and software architectures. Stochastic process underlying an SPN is a continuous time Markov chain, and the tool obtains samples from this chain, distributed according to its stationary probability distribution. The tool is implemented in C++ and is based on an efficient implementation of coupling from the past, an algorithm for PS in Markov chains. It can be obtained at http://www.dais.unive.it/~stojic/soft.html.

Keywords: stochastic Petri net · Perfect sampling · Stationary performance analysis

1 Introduction

Simulation is widely used in stationary performance analysis of stochastic Petri nets (SPN) with large state spaces, when exact numerical solution is infeasible. When simulation is used for stationary analysis, the *warm-up period* of a simulation run—during which state of the simulated system strongly depends on the initial state—needs to be discarded. This requires estimation of the length of the warm-up period, which can in some cases be done prior to simulation and in general can be performed during the simulation [7,9].

An alternative to the above approach is to estimate the stationary performance indices by sampling directly from stationary probability distribution of the model; this is referred to as *perfect sampling* (PS) [11]. The samples obtained by PS are distributed according to the exact stationary distribution of the model, and not according to an approximate stationary distribution which is obtained in simulation runs. If performance of a PS algorithm is not good enough to obtain enough samples needed to achieve required precision of estimates of performance indices, a smaller number of samples can be obtained and used as initial states of the simulation runs, obviating the need for the simulation warm-up period.

The presented tool, called spnps, implements PS for SPN models that have finite state spaces. In contrast to previous approaches [2–5], no special model structure is required and general SPNs with very large state spaces can be handled. The tool is implemented in C++ and is based on an optimised version [1] of

© Springer International Publishing Switzerland 2016
G. Agha and B. Van Houdt (Eds.): QEST 2016, LNCS 9826, pp. 163–166, 2016.
DOI: 10.1007/978-3-319-43425-4_11

a classic perfect sampling algorithm, *coupling from the past* [11]. The optimised algorithm uses multi-valued decision diagrams (MDD) [8] and exploits certain regularities present in Markov chains underlying SPN models to greatly speed up the coupling from the past; in some cases, models with over 10^{100} states can be handled by the tool [1].

2   Objectives

Spnps is targeted at researchers and analysts that perform stationary performance analyses of SPN models. Primary purpose of the tool is obtaining samples from stationary probability distributions of SPNs with finite state spaces. State spaces of some models can be decomposed into subsets between which there is very little communication, and stochastic process defined by such models will, with very high probability, stay constrained to one of the subsets for a large number of steps. This complicates the stationary simulation, especially if this property of the model is unknown to the analyst. Such multimodal behaviour can be heuristically detected by the tool by performing incomplete PS runs [1] (Table 1).

Table 1. Objectives of `spnps`

Application domain	Stationary performance analysis of SPN models
Targeted users	Researchers and analysts
Primary purpose	Sampling from stationary distribution of SPN models
Secondary purpose	Detection of multimodal model behaviour

3   Functionality

Spnps is a command line application that allows the user to load an SPN model from a file and generate a chosen number of perfect samples from the reachability set of the SPN model. Stopping criterion can be changed so that PS runs are stopped before completion, which can be useful in the analysis of models that exhibit multimodal behaviour [1].

3.1   Program Options

Program options are set via command line switches. Switch -x selects an input file containing a description of an SPN model. Format and contents of the input file are described in the next subsection. Switch -r allows selection of quality level of a high quality pseudorandom number generator [10] (PRNG) and -s allows selection of PRNG seed; if the seed is omitted or set to 0, local time is used as the seed. Switch -n specifies the number of sampling runs that are to be performed by the tool. Switch -c sets stopping cardinality; when this is set to 1 samples from the stationary probability distribution are obtained, and larger values can be used to examine multimodal models. Finally, switch -v enables verbose output, and switches -d and -t enable output of diagnostic information.

3.2   Input

Input file is assumed to be in Petri Net Markup Language (PNML) [12] format. PNML is an XML-based syntax for Petri nets which aims at becoming the standard interchange format for Petri net tools. Since PNML is very flexible and extendable, leaving many format details to be defined based on specific needs of a tool using the format, an XML Schema defining the particular structure of PNML files supported by the present tool is also included in the distribution archive. In addition to entries describing structure and initial marking of the SPN model, additional data loaded from the input file are firing rates and semantics for transitions and marking bounds and ordering of places.

3.3   Output

In normal usage, only the results of the sampling runs—perfect samples or sets of markings, depending on the stopping cardinality—are output to the standard output stream of the tool process. Using the verbose output switch -v causes additional output, such as lengths of sampling runs, to be produced during the sampling procedure. Switch -d enables output of program options and loaded model data, and switch -t enables output of timing information. When the tool is invoked without any parameters, a message explaining its usage is output.

4   Installation

Spnps is distributed in the form of C++ source code, and is licensed under GNU General Public Licence, version 3[1]. The distribution archive containing the source code and build system files, along with compilation and usage instructions, can be obtained from the web page[2] of one of the authors.

5   Conclusion

In this paper we have presented spnps, a tool for perfect sampling from the stationary distribution of stochastic Petri nets with finite reachability sets. The obtained samples can be used in stationary performance analysis. Secondary use of the tool is detection and analysis of models with multimodal behaviour, which is a class of models which are especially hard to simulate.

Further development of the tool is expected in several directions. It is known that there are more efficient encodings of the subsets of the reachability set by MDDs, than the encoding used in the present tool [6]. Use of these encodings is expected to lead to better performance of the tool. Additionally, the tool could be adapted to other structured formalisms used in performance evaluation.

Acknowledgment. Work partially supported by MIUR fund Fondo per il sostegno dei giovani "Programma strategico: ICT e componentistica elettronica".

---

[1] The GNU General Public License v3.0 - GNU Project - Free Software Foundation, http://www.gnu.org/licenses/gpl-3.0.en.html.

[2] Ivan Stojic - software, http://www.dais.unive.it/~stojic/soft.html.

References

1. Balsamo, S., Marin, A., Stojic, I.: Perfect sampling in stochastic Petri nets using decision diagrams. In: Modeling, Analysis and Simulation of Computer and Telecommunication Systems (MASCOTS), 2015 IEEE 23rd International Symposium on. pp. 126–135., October 2015
2. Bouillard, A., Bušić, A., Rovetta, C.: Perfect sampling for closed queueing networks. Performance Evaluation 79, 146–159 , special Issue: Performance 2014(2014)
3. Bouillard, A., Gaujal, B.: Backward coupling in Petri nets. In: Proceedings of the 1st International Conference on Performance Evaluation Methodologies and Tools. VALUETOOLS '06, NY, USA. ACM, New York (2006)
4. Bušić, A., Gaujal, B., Perronnin, F.: Perfect Sampling of Networks with Finite and Infinite Capacity Queues. In: Al-Begain, K., Fiems, D., Vincent, J.-M. (eds.) ASMTA 2012. LNCS, vol. 7314, pp. 136–149. Springer, Heidelberg (2012)
5. Bušić, A., Gaujal, B., Vincent, J.M.: Perfect simulation and non-monotone Markovian systems. In: VALUETOOLS '08: Proceedings of the 3rd International Conference on Performance Evaluation Methodologies and Tools. pp. 1–10. ICST (Institute for Computer Sciences, Social-Informatics and Telecommunications Engineering), ICST, Brussels, Belgium, Belgium (2008)
6. Ciardo, G.: Reachability Set Generation for Petri Nets: Can Brute Force Be Smart? In: Cortadella, J., Reisig, W. (eds.) ICATPN 2004. LNCS, vol. 3099, pp. 17–34. Springer, Heidelberg (2004)
7. Haas, P.J.: Stochastic Petri nets: Modelling, stability, simulation. Springer-Verlag, New York (2002)
8. Kam, T., Villa, T., Brayton, R., Sangiovanni-Vincentelli, A.: Multi-valued decision diagrams: theory and applications. Multiple-Valued Logic 4(1), 9–62 (1998)
9. Law, A.M., Kelton, D.M.: Simulation Modeling and Analysis. McGraw-Hill Higher Education, 3rd edn. (1999)
10. Panneton, F., L'Ecuyer, P., Matsumoto, M.: Improved long-period generators based on linear recurrences modulo 2. ACM Trans. Math. Softw. 32(1), 1–16 (2006). http://acm.org/10.1145/1132973.1132974
11. Propp, J.G., Wilson, D.B.: Exact sampling with coupled Markov chains and applications to statistical mechanics. Random Struct. Algorithms 9(1–2), 223–252 (1996)
12. Weber, M., Kindler, E.: The Petri net markup language. In: Ehrig, H., Reisig, W., Rozenberg, G., Weber, H. (eds.) Petri Net Technology for Communication-Based Systems. LNCS, vol. 2472, pp. 124–144. Springer, Berlin Heidelberg (2003)

CARMA Eclipse Plug-in: A Tool Supporting Design and Analysis of Collective Adaptive Systems

Jane Hillston[1] and Michele Loreti[2(✉)]

[1] Laboratory for Foundations of Computer Science,
University of Edinburgh, Edinburgh, UK
[2] Dipartimento di Statistica, Informatica, Applicazioni "G. Parenti",
Università di Firenze, Firenze, Italy
michele.loreti@unifi.it

Abstract. Collective Adaptive Systems (CAS) are heterogeneous populations of autonomous task-oriented agents that cooperate on common goals forming a collective system. This class of systems is typically composed of a huge number of interacting agents that dynamically adjust and combine their behaviour to achieve specific goals. Existing tools and languages are typically not able to describe the complex interactions that underpin such systems, which operate in a highly dynamic environment. For this reason, recently, new formalisms have been proposed to model CAS. One such is CARMA, a process specification language that is equipped with linguistic constructs specifically developed for modelling and programming systems that can operate in open-ended and unpredictable environments. In this paper we present the CARMA Eclipse plug-in, a toolset integrated in Eclipse, developed to support the design and analysis of CAS.

1 Introduction

Collective adaptive systems (CAS) typically consist of very large numbers of components which exhibit autonomic behaviour depending on their properties, objectives and actions. Decision-making in such systems is complicated and interaction between their components may introduce new and sometimes unexpected behaviours. CAS are open, in the sense that components may enter or leave the collective at any time. Components can be highly heterogeneous (machines, humans, networks, etc.) each operating at different temporal and spatial scales, and having different (potentially conflicting) objectives. We are still far from being able to design and engineer real collective adaptive systems, or even specify the principles by which they should operate.

Existing tools and languages are challenged by the complex and evolving interaction patterns that occur within CAS. Nevertheless, the pervasive yet transparent nature of these applications makes it of paramount importance that their behaviour is thoroughly assessed during their design, prior to deployment, and throughout their lifetime.

This work is partially supported by the EU project QUANTICOL, 600708.

G. Agha and B. Van Houdt (Eds.): QEST 2016, LNCS 9826, pp. 167–171, 2016.
DOI: 10.1007/978-3-319-43425-4_12

Within the QUANTICOL project[1], the definition of a formal language to capture CAS has been investigated. Our objective was to develop a coherent, integrated set of linguistic primitives, methods and tools to build systems that can operate in open-ended, unpredictable environments. We named this language CARMA, Collective Adaptive Resource-sharing Markovian Agents. CARMA combines the lessons we learnt from other stochastic process algebras such as PEPA [8], EMPA [2], MTIPP [7] and MoDEST [3], with those learnt from languages specifically designed to model CAS, such as SCEL [5], the AbC calculus [1], PALOMA [6], and the Attributed Pi calculus [9], which feature attribute-based communication and explicit representation of locations.

To support analysis of CARMA models a prototype simulator has been also developed. This software tool, which has been written in Java, can be used to perform stochastic simulation and will also form the basis for implementing further analysis techniques in the future. An Eclipse plug-in, integrating an editor, static analysis tools and various views on a model, has also been developed. Using this plug-in, CARMA systems can be specified by means of an appropriate high-level language, which is mapped to the CARMA process algebra to enable qualitative and quantitive analysis of CAS.

In this paper we first briefly describe the basic ingredients of CARMA. After that an overview of the CARMA Eclipse plug-in and its features is provided.

2   CARMA in a Nutshell

CARMA is a new stochastic process algebra for the representation of systems developed in the CAS paradigm [4]. The language offers a rich set of communication primitives, and exploits *attributes*, captured in a *store* associated with each component, to enable attribute-based communication. For example, for many CAS systems the location is likely to be one of the attributes. Thus it is straightforward to model systems in which, for example, there is limited scope of communication, or interaction is restricted to co-located components, or where there is spatial heterogeneity in the behaviour of agents.

A CARMA system consists of a *collective* operating in an *environment*. The collective is a multiset of components that models the behaviour of a system; it is used to describe a group of interacting *agents*. The environment models all those aspects which are intrinsic to the context where the agents are operating. The environment mediates agent interactions. This is one of the key features of CARMA. It is not a centralised controller but rather something more pervasive and diffusive — the physical context of the real system — which is abstracted within the model to be an entity which exercises influence and imposes constraints on the different agents in the system. The role of the environment is also related to the spatially distributed nature of CAS — we expect that the location *where* an agent is will have an effect on *what* an agent can do.

---

[1] http://www.quanticol.eu.

A CARMA component captures an *agent* operating in the system. It consists of a process, that describes the agent's behaviour, and of a store, that models its *knowledge*. A store is a function which maps *attribute names* to *basic values*.

Processes located within a CARMA component interact with other components via a rich set of communication primitives. Specifically, CARMA supports both unicast and broadcast communication, and permits locally synchronous, but globally asynchronous communication. Distinct predicates (boolean expressions over attributes), associated with senders and potential receivers are used to filter possible interactions. Thus, a component can receive a message only when its store satisfies the target predicate. Similarly, a receiver also uses a *predicate* to identify accepted sources. The execution of communicating actions takes time, which is assumed to be an exponentially distribution random variable whose parameter is determined by the environment.

3 CARMA Eclipse Plug-in

An Eclipse plug-in for supporting the specification and analysis of CAS in CARMA has been developed. A screenshot of the plug-in is presented in Fig. 1.

The CARMA Eclipse plug-in is available at http://quanticol.sourceforge.net/. At the same site detailed installation instructions can be found together with a set of case studies that shows how CAS can be modelled and verified with the provided tool.

The CARMA Eclipse plug-in provides a rich editor for CAS specification using an appropriate high-level language, called the CARMA *Specification Language* (CASL). This high-level language is not intended to add to the expressiveness

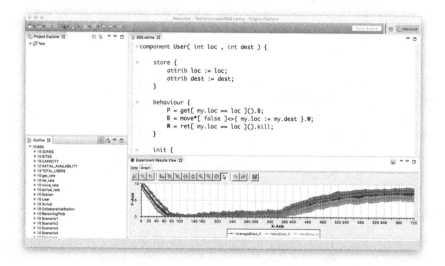

Fig. 1. A screenshot of the CARMA Eclipse plug-in.

of CARMA, which we believe to be well-suited to capturing the behaviour of CAS, but rather to ease the task of modelling for users who are unfamiliar with process algebra and similar formal notations. Each CARMA specification provides definitions for: structured *data types* and the relative *functions*; prototypes of *components* occurring in the system; *systems* composed by collective and environment; and the *measures*, that identify the relevant data to *measure* during simulation runs.

Given a CARMA specification, the CARMA Eclipse Plug-in automatically generates the Java classes needed to simulate the model. This generation procedure can be specialised to different kinds of simulators. Currently, a simple ad-hoc simulator is used. The simulator provides generic classes for representing *models* to be simulated. To perform the simulation each *model* provides a collection of *activities* each of which has its own *execution rate*. The simulation environment applies a standard *kinetic Monte-Carlo* algorithm to select the next activity to be executed and to compute the execution time. The execution of an *activity* triggers an update in the simulation model and the simulation process continues until a given simulation time is reached. From a CARMA specification, these activities correspond to the *actions* that can be executed by processes located in the system components. Indeed, each such activity mimics the execution of a transition of the CARMA operational semantics. Specific *measure functions* can be passed to the simulation environment to collect simulation data at given intervals. To perform statistical analysis of collected data the *Statistics package* of *Apache Commons Math Library* is used[2].

The results are reported within the *Experiment Results View* (see Fig. 2). Two possible representations are available. The former, on the left side of Fig. 2, provides a graphical representation of collected data; the latter, on the right side of Fig. 2, shows average and standard deviation of the collected values, which correspond to the *measures* selected during the simulation set-up, and are reported in a tabular form. These values can then be exported in CSV format.

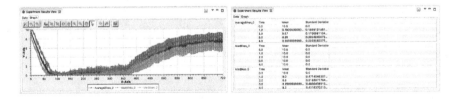

Fig. 2. CARMA Eclipse plug-in: experiment results view.

References

1. Abd Alrahman, Y., De Nicola, R., Loreti, M., Tiezzi, F., Vigo, R.: A calculus for attribute-based communication. In: Proceedings of 30th Annual ACM Symposium on Applied Computing, Salamanca, Spain, pp. 1840–1845, 13–17 April 2015

---

[2] http://commons.apache.org.

2. Bernardo, M., Gorrieri, R.: A tutorial on EMPA: a theory of concurrent processes with nondeterminism, priorities, probabilities and time. Theoret. Comput. Sci. **202**(1–2), 1–54 (1998)
3. Bohnenkamp, H.C., D'Argenio, P.R., Hermanns, H., Katoen, J.-P.: MODEST: a compositional modeling formalism for hard and softly timed systems. IEEE Trans. Softw. Eng. **32**(10), 812–830 (2006)
4. Bortolussi, L., De Nicola, R., Galpin, V., Gilmore, S., Hillston, J., Latella, D., Loreti, M., Massink, M.: CARMA: collective adaptive resource-sharing Markovian agents. In: Proceedings of Workshop on Quantitative Analysis of Programming Languages (2015)
5. De Nicola, R., Loreti, M., Pugliese, R., Tiezzi, F.: A formal approach to autonomic systems programming: the SCEL language. TAAS **9**(2), 7 (2014)
6. Feng, C., Hillston, J.: PALOMA: a process algebra for located Markovian agents. In: Norman, G., Sanders, W. (eds.) QEST 2014. LNCS, vol. 8657, pp. 265–280. Springer, Heidelberg (2014)
7. Hermanns, H., Rettelbach, M.: Syntax, semantics, equivalences and axioms for MTIPP. In: Herzog, U., Rettelbach, M. (eds.) Proceedings of 2nd Process Algebra and Performance Modelling Workshop (1994)
8. Hillston, J.: A Compositional Approach to Performance Modelling. CUP, Cambridge (1995)
9. John, M., Lhoussaine, C., Niehren, J., Uhrmacher, A.M.: The attributed Pi calculus. In: Heiner, M., Uhrmacher, A.M. (eds.) CMSB 2008. LNCS (LNBI), vol. 5307, pp. 83–102. Springer, Heidelberg (2008)

Sampling, Inference, and Optimization Methods

Uniform Sampling for Timed Automata with Application to Language Inclusion Measurement

Benoît Barbot[1], Nicolas Basset[1(✉)], Marc Beunardeau[2,3],
and Marta Kwiatkowska[1]

[1] Department of Computer Science, University of Oxford, Oxford, UK
bassetnicolas@yahoo.fr
[2] Ingenico Labs, Paris, France
[3] École Normale Supérieure, Paris, France

Abstract. Monte Carlo model checking introduced by Smolka and Grosu is an approach to analyse non-probabilistic models using sampling and draw conclusions with a given confidence interval by applying statistical inference. Though not exhaustive, the method enables verification of complex models, even in cases where the underlying problem is undecidable. In this paper we develop Monte Carlo model checking techniques to evaluate quantitative properties of timed languages. Our approach is based on uniform random sampling of behaviours, as opposed to isotropic sampling that chooses the next step uniformly at random. The uniformity is defined with respect to volume measure of timed languages previously studied by Asarin, Basset and Degorre. We improve over their work by employing a zone graph abstraction instead of the region graph abstraction and incorporating uniform sampling within a zone-based Monte Carlo model checking framework. We implement our algorithms using tools PRISM, SageMath and COSMOS, and demonstrate their usefulness on statistical language inclusion measurement in terms of volume.

1 Introduction

Since the seminal work of Alur and Dill [1], timed automata (TAs) have been widely studied in the context of real-time systems verification. Several algorithms from the classical automata-theoretic verification were successfully lifted to the timed case. In spite of this, many problems become undecidable, the most important being the inclusion of timed languages. One way to circumvent undecidability is to employ statistical methods, where results are given with some confidence level. However, timed automata are non-stochastic models and it is not clear a priori with what probability to sample runs when performing statistical experiments. A natural answer is given by the maximal entropy principle: "without

---

This work is supported by ERC AdG VERIWARE.

B. Barbot—Now in LACL, Université Paris Est Créteil, France

M. Beunardeau—Contributed to the work during an internship funded by ERC AdG VERIWARE.

G. Agha and B. Van Houdt (Eds.): QEST 2016, LNCS 9826, pp. 175–190, 2016.
DOI: 10.1007/978-3-319-43425-4_13

knowledge a priori on the distribution of probability to be taken, the one with maximal entropy should be preferred" [15]. A maximal entropy stochastic process for timed automata was recently proposed in [7]. Essentially, this is the stochastic process that yields the most uniform sampling when the length of the timed words tends to infinity. By uniform sampling we mean that all timed words of a given length have the same density of probability to be chosen.

In this paper we propose several algorithms to achieve uniform sampling of timed words in timed languages. The methods are based on the theory of volumetry of timed languages recently developed by Asarin et al. [3], which provides means for quantitative measurement of languages in terms of volume. Here, we employ this theory to achieve statistical estimation of volume and demonstrate its usefulness for language inclusion measurement. The accuracy of statistical estimation depends on the ability to uniformly sample the executions. The method provided in [7], where the transitions of a TA were annotated with probability functions so that the resulting stochastic process enables random simulation in the most uniform way possible, is based on spectral attributes of a functional operator Ψ (an analogue in the TA context of the adjacency matrix of a graph) [3]. Unfortunately, it is not practical, as it relies on the region graph abstraction and the computation of eigenfunctions. In this paper, we overcome this problem by adopting a zone-based approach and approximating the probability functions of [7] with quotients of the volume functions.

Contributions. (i) We provide a zone-based computation of volume functions for TAs, which enables the first practical implementation of volumetry of timed languages. (ii) We develop three methods (Methods 1, 2 and 3) to sample in a (quasi) uniform manner timed words in a language recognised by a deterministic timed automaton (DTA). In particular, we propose a receding horizon framework that allows us to approximate the maximal entropy stochastic process discussed above. (iii) We apply uniform sampling for DTAs to uniform sampling and volume measurement for arbitrary timed languages, provided the membership problem for the language is decidable. (iv) We have implemented the algorithms presented here in PRISM [16] (for the splitting of the DTA into zones), SageMath [20] (for the computation of volume functions) and COSMOS [4] (for the random generation of timed words and property checking) and illustrate them on several examples, with encouraging results. Omitted proofs and further details can be found in [5].

Related Work. The theory of volumetry of timed languages has been studied and applied to robustness analysis [3], timed channel coding [2] and combinatorics of permutations [6], but has not yet been applied in practice.

The *recursive method* for uniform sampling is a well-known method in discrete combinatorics [12] whose generalisation to the timed case (Method 1 here) was already done for very specific timed languages in [6].

Monte Carlo model checking was proposed in [13] for discrete models to randomly explore their behaviour by means of simulating execution paths. Similarly, statistical model checking [21] uses simulation to verify temporal logic properties with statistical guarantees, and has been applied to stochastic timed/hybrid systems [10]. This avoids state-space explosion, thus ensuring the feasibility of verification of complex models, and has also been used to check undecidable properties [10]. Here we implement Monte Carlo techniques for TAs.

Monte Carlo or statistical model checking usually employs an *isotropic* random walk to explore the executions (as explained in [11,19] for discrete models). This involves choosing uniformly at random, at each step of the simulation, the next transition from those available. It has been argued that the isotropic methods are not able to efficiently perform uniform sampling of the behaviours (see e.g. the pathological examples in [19] for sampling of lassos and [11] for sampling paths in a finite-state automaton). Here we implement uniform sampling based on the tool COSMOS, but the techniques are more generally applicable and can be implemented in other tools, for example UPPAAL-SMC [10], which supports user-defined distributions.

Statistical model checkers such as UPPAAL-SMC consider timed automata augmented with probability distribution on transitions that are either user-defined or given "by default". Thus, the model to verify is already probabilistic and specifications are written in temporal logic with probabilistic operators. Our work addresses a different and novel question: how can one use statistical experiments on a non-probabilistic timed language and draw conclusions about that language, without being given probability distributions on it?

2 Preliminaries

2.1 Timed Languages and Volumetry

A *timed word* $\boldsymbol{\alpha} = (t_1, a_1) \ldots (t_n, a_n)$ is a word over the alphabet $\mathbb{R}_{\geq 0} \times \Sigma$, where $\mathbb{R}_{\geq 0}$ denotes the set of non-negative reals and Σ is a finite alphabet of *events*. Times t_i represent *delays* between events a_{i-1} and a_i. Throughout this paper, delays will be bounded[1] by an integer constant M. A *timed language* L is a set of timed words. Given $n \geq 0$, we denote by L_n the timed language L restricted to timed words of length n. For every timed language L and every word $w = a_1 \ldots a_n \in \Sigma^n$, we define $P_w^L = \{(t_1, \ldots, t_n) \mid (t_1, a_1) \ldots (t_n, a_n) \in L\}$, and denote by $\mathtt{Vol}(P_w^L)$ its (hyper-)volume.

Example 1 (Running example). Examples of such hyper-volumes are given in Fig. 1. Anticipating what follows, these sets correspond to the timed language restricted to timed words of length 2 of the TA depicted in Fig. 2 (Left).

For a fixed n, we define the *n-volume* of L as follows:

$$\mathtt{Vol}(L_n) = \sum_{w \in \Sigma^n} \mathtt{Vol}(P_w^L) = \sum_{a_1 \in \Sigma} \int_0^M \cdots \sum_{a_n \in \Sigma} \int_0^M \mathbf{1}_{P_w^L}(t) dt_1 \cdots dt_n.$$

Continuing the example; the hyper-volume for dimension 2 is calculated as

$$\mathtt{Vol}(L_2) = \mathtt{Vol}(P_{ab}^L) + \mathtt{Vol}(P_{aa}^L) + \mathtt{Vol}(P_{ba}^L) + \mathtt{Vol}(P_{bb}^L) = 3.5 + 2 + 4 + 2 = 11.5.$$

---

[1] Our approach to timed languages is based on volume and does not apply, in its present form, to unbounded delays that result in innite volume.

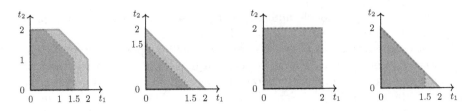

Fig. 1. From left to right, languages P_{ab}^L, P_{aa}^L, P_{ba}^L and P_{bb}^L for the running example (Example 1). The darker areas corresponds to initial clock vector $(x, y) = (0.5, 0)$.

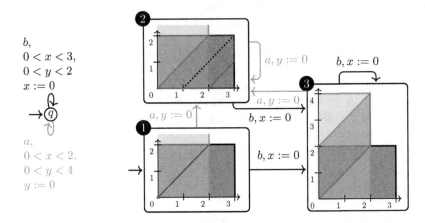

Fig. 2. Left: a DTA. Right: the same DTA obtained after applying the forward reachability algorithm. Entry zones are represented in red. Guards for a and b are the same in the two TAs. The blue part represents clock vectors reachable through entry zones by time elapsing. In location 2, the guard of transition b should be split along the dotted line to obtain the split DTA of Fig. 3. (Color figure online)

We define the *uniform probability distribution* on a timed language L by assigning weight $1/\texttt{Vol}(L_n)$ to every timed word of length n. The main purpose of this article is to show how to sample according to that distribution when the language is recognised by a timed automaton. For instance, the probability of a uniformly sampled timed word to fall in the set $E = \{(t_1, b)(t_2, a) \mid t_1 \in (0, 1), t_2 \in (0, 2)\}$ is $\texttt{Vol}(E)/\texttt{Vol}(L_2) = 2/11.5 \approx 0.17$.

Given two timed languages L, L' over the same alphabet of events Σ, we say that L' is a *tight under-approximation* of L if, for all $w \in \Sigma^*$, $P_w^{L'} \subseteq P_w^L$ and $\texttt{Vol}(P_w^L \setminus P_w^{L'}) = 0$; hence $\texttt{Vol}(P_w^L) = \texttt{Vol}(P_w^{L'})$. In particular, timed words uniformly sampled in L' are uniformly sampled in L.

2.2 Timed Automata

Let X be a finite set of non-negative real-valued variables called *clocks*. Here we assume that clocks remain bounded by a constant $M \in \mathbb{N}$. A *clock constraint* has the form $x \sim c$ or $x - y \sim c$ where $\sim \in \{\leq, <, =, >, \geq\}$, $x, y \in X$, $c \in \mathbb{N}$. A *guard* is a finite conjunction of clock constraints; it is called open if its constraints involve only strict inequalities. A *zone* is a set of clock vectors $x \in [0, M]^X$ satisfying a guard. For a clock vector $x \in [0, M]^X$ and a non-negative real t, we denote by $x + t$ (resp. $x - t$) the vector $x + (t, \ldots, t)$ (resp. $x - (t, \ldots, t)$).

A *timed automaton* (TA) $\mathcal{A}$ is a tuple $(\Sigma, X, Q, i_0, F, \Delta)$ where Σ is a finite set of events; X is a finite set of clocks; Q is the finite set of *locations*; i_0 is the initial location; $F \subseteq Q$ is the set of final locations; and Δ is the finite set of *transitions*. Any transition $\delta \in \Delta$ has an *origin* $\delta^- \in Q$, a *destination* $\delta^+ \in Q$, a label $a_\delta \in \Sigma$, a *guard* $\mathfrak{g}_\delta$ and a *reset*-function $\mathfrak{r}_\delta$ determined by a subset of clocks $B \subseteq X$: it resets to 0 all the clocks in B and does not modify the value of the other clocks.

A *timed transition* is an element (t, δ) of $\mathbb{A} \overset{\text{def}}{=} [0, M] \times \Delta$. The *delay* t represents the time before firing the transition δ. A *state* $s = (q, x) \in Q \times [0, M]^X$ is a pair of a location and a clock vector. Given a state $s = (q, x)$ and a timed transition $\alpha = (t, \delta) \in \mathbb{A}$, the *successor* of s by α is denoted by s_α and defined as follows. If $\delta^- = q$ and $x + t$ satisfies the guard $\mathfrak{g}_\delta$ then $s_\alpha = (\delta^+, \mathfrak{r}_\delta(x + t))$ else $s_\alpha = \bot$. Here and in the rest of the paper $\bot$ represents undefined states. A sequence of timed transitions is called a *timed path*. We extend the successor action to timed paths by induction: $s_\varepsilon = s$ and $s_{(\alpha\alpha')} = (s_\alpha)_{\alpha'}$ for all states s, timed transitions $\alpha \in \mathbb{A}$ and timed paths $\alpha' \in \mathbb{A}^*$. The initial state of the timed automaton is $s = (i_0, \mathbf{0})$. The *labelling* of a timed path $(t_1, \delta_1) \ldots (t_n, \delta_n)$ is the timed word $(t_1, a_{\delta_1}) \ldots (t_n, a_{\delta_n}) \in ([0, M] \times \Sigma)^*$. The *timed language* $L(\mathcal{A})$ of a timed automaton $\mathcal{A}$ is the set of timed words that are labellings of timed paths α such that $s_\alpha \in F \times [0, M]^X$. We also write $L_n(\mathcal{A})$ instead of $(L(\mathcal{A}))_n$.

For a guard g, we denote by $\mathrm{TE}^{-1}(g)$ the set of clock vectors from which g can be reached when time elapses; formally, $\mathrm{TE}^{-1}(g) = \{x \mid \exists t \geq 0, x + t \in g\}$. Given a state $s = (q, x)$ we denote by $\Delta(s)$ the set of transition *available* from s, that is such that $\delta^- = q$ and $x \in \mathrm{TE}^{-1}(\mathfrak{g}_\delta)$. Given a state $s = (q, x)$ and a transition $\delta \in \Delta(s)$, we define $\mathrm{lb}_\delta(s) \overset{\text{def}}{=} \inf\{t \mid x + t \in \mathfrak{g}_\delta\}$ and $\mathrm{ub}_\delta(s) \overset{\text{def}}{=} \sup\{t \mid x + t \in \mathfrak{g}_\delta\}$ so that the condition $x + t \in \mathfrak{g}_\delta$ is equivalent to $t \in (\mathrm{lb}_\delta(s), \mathrm{ub}_\delta(s))$.

A *deterministic timed automaton* (DTA) is a TA such that no clock vector can satisfy guards of pairwise distinct transitions with the same label and origin. This implies that timed words and timed paths of a DTA are in one-to-one correspondence. We are interested in the prefixes of infinite timed words of a DTA. To be sure that $L_n(\mathcal{A})$ contains exactly the prefixes of size n, we consider only DTAs that satisfy the two following conditions: (i) every location is final, (ii) from every reachable state, there is a timed transition that can be taken.

2.3   Equations on Timed Languages and Volumes

Given a DTA $\mathcal{A}$, we denote by $L_n(s)$ the n-th *timed language* recognised from a state s and defined inductively as follows: $L_0(s) = \{\varepsilon\}$, and

$$L_{n+1}(s) = \bigcup_{\delta \in \Delta(s)} \bigcup_{t \in I(s, \delta)} (t, a_\delta) L_n(s_{(t, \delta)}). \tag{1}$$

For the running example and initial state $[q, (0.5, 0)]$ we have:

$$L_2([q, (0.5, 0)]) = \bigcup_{t \in (0, 1.5)} (t, a) L_1([q, (0.5 + t, 0)]) \cup \bigcup_{t \in (0, 2)} (t, b) L_1([q, (0, t)]). \quad (2)$$

The language $L_2([q, (0.5, 0)])$ is depicted in Fig. 1.

We also parametrise the volume by the initial state and define the n-th volume function as $v_n(s) = \mathtt{Vol}(L_n(s))$. These functions can be defined recursively by replacing union over intervals by integrals and union over transitions by finite sums in (1). We obtain $v_0(s) = 1$ and

$$v_{n+1}(s) = \sum_{\delta \in \Delta(s)} \int_{\mathrm{lb}_\delta(s)}^{\mathrm{ub}_\delta(s)} v_n(s_{(t, \delta)}) dt. \quad (3)$$

For the running example, passing to volumes in (2) yields

$$v_2([q, (0.5, 0)]) = \int_0^{1.5} v_1([q, (0.5 + t, 0)]) dt + \int_0^2 v_1([q, (0, t)]) dt. \quad (4)$$

A key idea used in [3, 7] is to rewrite (3) as

$$v_{n+1}(s) = \Psi(v_n)(s) \quad (5)$$

where Ψ is an integral operator defined by

$$\Psi(f)(s) = \sum_{\delta \in \Delta(s)} \Psi_\delta(f)(s) \quad \text{with} \quad (6)$$

$$\Psi_\delta(f)(s) = \int_{\mathrm{lb}_\delta(s)}^{\mathrm{ub}_\delta(s)} f(s_{(t, \delta)}) dt. \quad (7)$$

Thus, volume functions are defined via iteration of the operator Ψ on the constant function 1: $v_n = \Psi^n(1)$. In [3, 7], the state space was decomposed into regions, which guaranteed algebraic properties such as polynomial volume functions at the price of an explosion of the number of locations of the TA. A TA before such a decomposition into regions has volume functions that are complicated (piecewise defined), and hence difficult to handle in practice. Here we want to keep volume functions simple (polynomial) while keeping the set of locations small. For this we adopt a zone-based approach.

Table 1. First volume functions $v_n[l_i, (x, y)]$ associated to the TA of Fig. 3.

	$[l_0, (0, 0)]$	$[l_1, (x, 0)]$	$[l_2, (0, y)]$	$[l_3, (x, 0)]$
v_0	1	1	1	1
v_1	4	$-x + 4$	$-y + 4$	$-2x + 5$
v_2	15	$-4x + 15$	$\frac{1}{2}y^2 - 4y + 15$	$-\frac{1}{2}x^2 - 6x + \frac{35}{2}$
v_3	$\frac{335}{6}$	$-15x + \frac{335}{6}$	$-\frac{1}{6}y^3 + 2y^2 - 15y + \frac{335}{6}$	$-\frac{1}{6}x^3 - \frac{1}{2}x^2 - 25x + \frac{133}{2}$

The idea of the zone-based decomposition described in the next section is to split the state space into several pieces in which the functions $\mathrm{lb}_\delta(s)$ and $\mathrm{ub}_\delta(s)$ have simple form, ensuring that every volume function $v_n = \Psi^n(1)$ restricted to any location is polynomial (see Table 1).

3   Volume Function Computation for DTAs

In this section we explain how to transform a DTA $\mathcal{A}$ into a DTA $\mathcal{A}'$ called *split DTA* that facilitates efficient volume computation.

Decomposition into Zones. We first apply a forward reachability algorithm, implemented for instance in PRISM [16], which returns the so-called *forward-reachability graph*, that is, a finite graph with annotations, which we view as a DTA (the annotations are essentially, for each edge δ, the guard $\mathfrak{g}_\delta$ and label a_δ and, for each location l, the zone Z_l which is entered). Formally, we say that a TA is *decomposed into zones* if, for every $l \in Q$, there is a zone Z_l called the *entry zone* of l, such that the entry zone of the initial state is $\{0\}$ and, for every transition δ, the successors of states in $\{\delta^-\} \times Z_{\delta-}$ through δ with some delay are in $\{\delta^+\} \times Z_{\delta+}$, that is, $\{\mathfrak{r}_\delta(\boldsymbol{x}+t) \mid \boldsymbol{x} \in Z_{\delta-}, \boldsymbol{x}+t \in \mathfrak{g}_\delta\} \subseteq Z_{\delta+}$. We denote by $\mathbb{S} = \cup_{l \in Q}\{l\} \times Z_l$ the set of states corresponding to entry zones. The forward-reachability graph for the running example is given in Fig. 2 (Right).

Guard Split. Let δ be the transition from location 2 to location 3 in the automaton of Fig. 2 (Right), then $g_\delta \stackrel{\mathrm{def}}{=} (0 < x < 3) \wedge (0 < y < 2)$. Then one can see that $\mathrm{ub}_\delta(2, (x,0)) = 2$ if $x \in (0,1)$ (due to guard $y < 2$) and $\mathrm{ub}_\delta(2,(x,0)) = 3 - x$ if $x \in (1,2)$ (due to guard $x < 3$). The guard g_δ thus needs to be split into two (along the dotted line in the figure) to achieve a simpler form for ub_δ. It is well known how to get the tightest constraints of a guard and get rid of redundant constraints using the Floyd-Warshall algorithm (see e.g. [8]). A guard is said to be *upper-split* (*lower-split*) if there is at most one useful constraint (that is, not implied by other constraints) of the form $x_j < a$ $(x_j > a)$. The guard g_δ discussed above is not upper-split as the two constraints $x < 3$ and $y < 2$ are both useful. Analogous definitions hold for lower-bounds and a guard is said to be *split* if it is both lower-split and upper-split.

Pre-stability. A second phenomenon we want to avoid is when the set of available transitions $\Delta(q, \boldsymbol{x})$ is not constant on the entry zone of q. A TA decomposed into zones is called *pre-stable* if, for every location q and clock vector $\boldsymbol{x} \in Z_q$, the set of transitions $\Delta(q, \boldsymbol{x})$ is exactly the set of transitions δ whose origin is q. Equivalently, a TA is pre-stable if $Z_{\delta-} \subseteq \mathrm{TE}^{-1}(\mathfrak{g}_\delta)$ for every δ. In case we detect a transition such that $Z_{\delta-} \not\subseteq \mathrm{TE}^{-1}(\mathfrak{g}_\delta)$ we will split the zone $Z_{\delta-}$ to isolate $\mathrm{TE}^{-1}(\mathfrak{g}_\delta) \cap Z_{\delta-}$ from its complement. Continuing the example above, after splitting g_δ the functions associated to each new guard are null for $x \in (0,1)$ or $x \in (1,2)$. Location 2 is split into two locations of the final TA of Fig. 3: l_1 for $(0,1)$ and l_3 for $(1,2)$. Every incoming transition to location 2 is split accordingly into two transitions (one orange to l_1 and one purple to l_3).

Trimming. Last but not least, we say that a TA is *trimmed* if the set of outgoing transitions of each location is non-empty. A TA is called *split* if it is pre-stable, trimmed and all the guards of its transitions are split and open. It implies, in particular, that, for every entry state $s \in \mathbb{S}$, $\Delta(s)$ is not empty and for all transition $\delta \in \Delta(s)$ it holds that $\mathrm{ub}_\delta(s) - \mathrm{lb}_\delta(s) > 0$. Note that *opening guards*, that is, transforming non-strict inequalities into strict ones is made wlog., as it only removes part of the language that has a null volume measure.

Splitting Algorithm. We propose an algorithm to transform a DTA into a split DTA such that the language of the latter is a tight under-approximation of the language of the former (see Theorem 1). First, we apply a forward reachability algorithm to obtain a DTA decomposed into zones and open its guards. Then we successively split zones that falsify pre-stability and guard split conditions, until the conditions are satisfied in the DTA. The splitting algorithm maintains a stack of transitions that need to be checked, which initially contains all the transitions. As the algorithm proceeds, transitions are popped from the stack and are checked against pre-stability and guard split conditions. If one test fails, the zone (or guard) is split accordingly into several zones (or open guards) and the transitions that are affected are added to the stack (incoming transitions to, and outgoing transitions from the split zone). When no more transition need to be checked (i.e. the stack is empty), the TA is split and the algorithm terminates. This occurs in a finite number of steps since transitions are added to the stack only when a zone is split into strictly smaller sub-zones, and there are finitely many zones (as the clocks are bounded by a constant M).

Theorem 1. *Given a DTA $\mathcal{A}$, one can construct (using the algorithm sketched above) a split DTA $\mathcal{A}'$ that recognises a tight under-approximation of $L(\mathcal{A})$.*

The splitting algorithm and the proof can be found in the technical report [5].

Volume Function of a Split DTA. We have the following result.

Proposition 1. *Given a split DTA $\mathcal{A}$ and $n \in \mathbb{N}$, denote by c the maximal affine dimension of an entry zone of $\mathcal{A}$. One can compute the volume function v_k for $k \leq n$ in time and space complexity $O(n^{c+2}|Q_\mathcal{A}|)$ using dynamic programming based on the recursive equation (3). Each volume function v_k restricted to a location q is a polynomial of degree at most k that is positive on Z_q.*

Example 2. We have implemented the splitting algorithm sketched in Sect. 3 and applied it to the DTA of Fig. 2 (Right) to obtain the DTA of Fig. 3. Our program also returns for each transition δ of the output DTA the interval $(\mathrm{lb}_\delta, \mathrm{ub}_\delta)$, allowing us to compute with SageMath the operator Ψ as well as the volume functions. On the example, for $f : \mathbb{S} \to \mathbb{R}$, $(x, y) \in Z_l$ with $l \in \{l_0, \dots, l_3\}$,

$$\Psi(f)[l_0, (0,0)] = \int_0^1 f(l_1, (t, 0))\mathrm{d}t + \int_0^2 f(l_2, (0, t))\mathrm{d}t + \int_1^2 f(l_3, (t, 0))\mathrm{d}t;$$
$$\Psi(f)[l_1, (x, 0)] = \int_0^{1-x} f(l_1, (x+t, 0))\mathrm{d}t + \int_0^2 f(l_2, (0, t))\mathrm{d}t + \int_{1-x}^{2-x} f(l_3, (x+t, 0))\mathrm{d}t;$$
$$\Psi(f)[l_2, (0, y)] = \int_0^1 f(l_1, (t, 0))\mathrm{d}t + \int_0^{2-y} f(l_2, (0, y+t))\mathrm{d}t + \int_1^2 f(l_3, (t, 0))\mathrm{d}t;$$
$$\Psi(f)[l_3, (x, 0)] = \int_0^{3-x} f(l_2, (0, t))\mathrm{d}t + \int_0^{2-x} f(l_3, (x+t, 0))\mathrm{d}t.$$

First volume functions computed using Eq. (5) are given in Table 1.

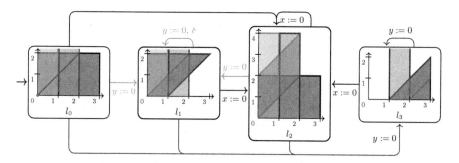

Fig. 3. The split form of the running example (Example 1).

4   Sampling Methods for Timed Languages of DTAs

In this section we consider random sampling of timed words. We first give a method that achieves exact uniform sampling when the length of timed words to be generated is finite; we speak of finite horizon. When the length is infinite or too long to be treated by the previous method, we consider a receding horizon method, where, at the k-th step of the generation, the next timed letter is chosen according to the volume of the timed words for the next m steps; these possible futures constitute a finite receding horizon. At the limit, where the receding horizon becomes infinite ($m \to \infty$), this can be interpreted as a stochastic process over runs of maximal entropy [7].

Parametric Probability Distributions. A *discrete probability distribution* (DPD) on a finite set A is a function $\mathtt{dpd} : A \to [0,1]$ such that $\sum_{a \in A} \mathtt{dpd}(a) = 1$. A *probability density function* (PDF) on an interval (a,b) is a Lebesgue measurable function $\mathtt{pdf} : (a,b) \to \mathbb{R}_{\geq 0}$ such that $\int_a^b \mathtt{pdf}(t)\mathrm{d}t = 1$. Values of DPD and PDF are referred to as weights. The DPD $\mathtt{isoDPD}(A)$ on a set A (resp. the PDF $\mathtt{isoPDF}(a,b)$ on an interval (a,b)) that attributes the same weight to every $a \in A$ (resp. $t \in (a,b)$) is called *isotropic*. In other words, $\mathtt{isoDPD}(A)(a) = 1/|A|$ for every $a \in A$ (resp. $\mathtt{isoPDF}(a,b)(t) = 1/(b-a)$ for every $t \in (a,b)$). PDFs considered in the following are just polynomials on the delay variable t. Their coefficients depend on the current state (location and clock values) and on the transition to fire. Choosing a delay t according to a PDF can be done using the *inverse method*: a random number r is drawn uniformly in $(0,1)$, and the output $t \in (a,b)$ is the unique solution of $\int_a^t \mathtt{pdf}(t')\mathrm{d}t' - r = 0$. In the case of the isotropic PDF on (a,b), the output t is just $a + r(b-a)$.

Random generation of timed words in $L_n(s)$ for a given state $s \in \mathbb{S}$ is done as follows: for $k = 1..n$, pick randomly the next transition δ according to a DPD $\mathtt{dpd}_s^k$ parametrised by the current state s, then chose the delay t in $(\mathrm{lb}_\delta(s), \mathrm{ub}_\delta(s))$ according to a PDF $\mathtt{pdf}_{s,\delta}^k$ parametrised by the current state s and the transition just chosen; take the successor of s by (t, δ) as the new current state s; output (t, a_δ) and repeat the loop.

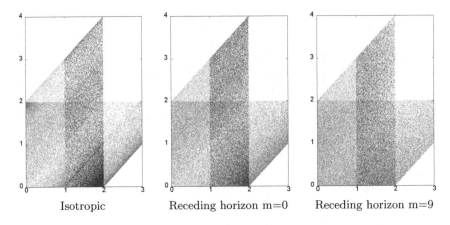

Isotropic   Receding horizon m=0   Receding horizon m=9

Fig. 4. Trajectories of the running example (Fig. 3) sampled using isotropic sampling (Left) and Method 2 with receding horizon $m = 0$ (Middle) and $m = 9$ (Right). Each point of a given colour corresponds to a clock vector where a transition of that colour occurs. Each plot visualises a single trajectory with 200,000 transitions. The receding horizon $m = 9$ visibly yields the most uniform sampling. The receding horizon sampling with $m = 0$ is already more uniform than the isotropic sampling as the former assigns weights to transitions proportional to lengths of intervals $\left(\mathtt{dpd}_s = \frac{\mathrm{ub}_\delta(s) - \mathrm{lb}_\delta(s)}{v_1(s)} \right)$. (Color figure online)

This random generation outputs timed words of $L_n(s)$ with weights given by

$$Weight[(t_1, a_1) \cdots (t_n, a_n)] \stackrel{\text{def}}{=} \prod_{k=1}^{n} \mathtt{dpd}^k_{s_{k-1}}(\delta_k) \mathtt{pdf}^k_{s_{k-1}, \delta_k}(t_k). \qquad (8)$$

where, for every $k = 1..n$, s_{k-1} is the state before the kth sampling loop, (t_k, δ_k) is the kth timed transition randomly picked during the kth sampling loop and a_k is the label of δ_k.

Isotropic and Uniform Sampling. Isotropic sampling[2] relies on using in each step the isotropic DPD $\mathtt{isoDPD}(\Delta(s))$ and the isotropic PDF $\mathtt{isoPDF}(I(s, \delta))$. These distributions are particularly simple to sample, but when the length of samples grows the probability concentrates on small sections of the runs, see Fig. 4 (Left). By contrast, *uniform sampling* for $L_n(s)$ assigns the same weight $1/v_n(s)$ to every timed word. In other words, for any measurable set $B \subseteq L_n(s)$ the probability $\mathtt{Vol}(B)/v_n(s)$ to fall in this set is proportional to its measure.

The Recursive Method for Uniform Sampling. The idea of the recursive method for uniform sampling of n-length timed words from a state s is to choose the first delay t and transition δ according to well chosen DPD and PDF that depend on the volume functions v_n and v_{n-1}, and then recursively apply uniform sampling to generate an $(n-1)$-length timed word from the updated state $s_{(t,\delta)}$.

---

[2] Note that some works, consider instead sampling the delay first and then the transitions available in the state updated by the delay (see [9]).

Define, for every function $f : \mathbb{S} \to \mathbb{R}_{>0}$ and state s, the DPD $\omega(f, s) : \delta \mapsto \frac{\Psi_\delta(f)(s)}{\Psi(f)(s)}$. If moreover δ is given, define the PDF $\varphi(f, s, \delta) : t \mapsto \frac{f(s_{(t,\delta)})}{\Psi_\delta(f)(s)}$ from $(\mathrm{lb}_\delta(s), \mathrm{ub}_\delta(s))$ to $\mathbb{R}_{>0}$.

Method 1 (Exact uniform sampling). *Given a split DTA and $n \in \mathbb{N}$, precompute the volume functions $v_0 = 1, \ldots, v_n = \Psi^n(1)$ (see Proposition 1), then the uniform sampling of n-length timed words can be achieved in linear time using the following sequences of DPDs and PDFs: $(\omega(v_{n-k}, s), \varphi(v_{n-k}, s, \delta))_{k=1..n}$*

Proof. Using the same notation as in (8), it holds that

$$Weight[(t_1, a_1) \cdots (t_n, a_n)] = \prod_{k=1}^{n} \omega(v_{n-k}, s_{k-1})(\delta_k)\varphi(v_{n-k}, s_{k-1}, \delta_k)(t_k)$$

$$= \prod_{k=1}^{n} \frac{\Psi_\delta(v_{n-k})(s_{k-1})}{v_{n-k+1}(s_{k-1})} \frac{v_{n-k}(s_k)}{\Psi_\delta(v_{n-k})(s_{k-1})} = \frac{v_0(s_{n-1})}{v_n(s_0)} = \frac{1}{v_n(s_0)}.$$

Example 3. We illustrate the DPDs and PDFs used in the last but one step of the uniform random sampling for the running example, obtained from volume functions of Table 1. Consider the state $s = (l_1, (x, 0))$ with $x \in (0, 1)$ and δ the self-loop on l_1 (see Fig. 3). Then $(\mathrm{lb}_\delta(s), \mathrm{ub}_\delta(s)) = (0, 1-x)$ and $s_{(t,\delta)} = (l_1, (x+t, 0))$. The DPD used to choose δ is

$$\mathbf{dpd}_s^{n-1}(\delta) = \frac{1}{v_2(s)} \int_{\mathrm{lb}_\delta(s)}^{\mathrm{ub}_\delta(s)} v_1(s_{(t,\delta)})dt = \int_0^{1-x} \frac{4-x-t}{15-4x}dt = \frac{7-8x+x^2}{30-8x}$$

The PDF used to choose t is

$$\mathbf{pdf}_{s,\delta}^{n-1}(t) = \frac{1_{t \in (\mathrm{lb}_\delta(s), \mathrm{ub}_\delta(s))}}{\mathbf{dpd}_s^{n-1}(\delta)} \frac{v_1(s_{(t,\delta)})}{v_2(s)} = 1_{t \in (0,x)} \frac{8-2x-2t}{7-6x+x^2}$$

Random Sampling with Finite Receding Horizon. With the previous method, the k-th timed transition of a run of length n is sampled according to DPD and PDF that depend on k and n. This dependency on k and n is not suitable for large n as it requires storage of as many polynomials as the length of the run to generate n. Also, one might wish to randomly generate arbitrarily long runs without a prescribed bound on the length. To take the kth timed transition in the recursive method for uniform sampling, we use DPD and PDF that depend on v_{n-k}, that is, on the volume measure of the possible $(n-k)$ step future. The idea of the following method is to replace $(n-k)$ by a fixed $m \ll n$ at every step of the sampling. The constant m can be seen as a receding horizon used in control theory [17]. At each step we consider only the possible m step future to generate the next timed transition.

Method 2 (Random Sampling with Finite Receding Horizon m). *Given a split DTA, $n \in \mathbb{N}$ and $m \in \mathbb{N}$, precompute the volume functions $v_0 = 1, \ldots, v_m = \Psi^m(1)$ (see Proposition 1), then sample n-length timed words in linear time using the same DPD $\omega(v_m, s)$ and PDF $\varphi(v_m, s, \delta)$ for every $k = 1..n$.*

The precomputation is polynomial in m. Hence this methods is more efficient than Method 1 when $m \ll n$, but it does not yield exact uniform sampling.

Table 2. Table for Example 4.

m	$(C^+/C^-)-1$	$n_{0.01}$	m	$(C^+/C^-)-1$	$n_{0.01}$	m	$(C^+/C^-)-1$	$n_{0.01}$
0	3	1	4	3.272×10^{-4}	35	8	2.364×10^{-7}	42 098
1	0.3229	2	5	8.431×10^{-5}	124	9	2.520×10^{-8}	394 801
2	1.659×10^{-2}	3	6	9.308×10^{-6}	1 076	10	5.304×10^{-9}	1.8760×10^{6}
3	4.444×10^{-3}	6	7	1.409×10^{-6}	7 069	11	4.487×10^{-10}	2.2178×10^{7}

Quasi-Uniform Random Sampling. We now present a trade-off between exact uniform sampling (Method 1) and the finite receding horizon sampling (Method 2). We give bounds on the distance to uniformity for this method, which we conjecture to be small in practice for small horizon m. This conjecture is supported by theoretical results of previous works [3,5,7] and by practical experiments (notably in Example 4 below).

Method 3 (Switching Method for Quasi-uniform Sampling). *Given a split DTA, $n \in \mathbb{N}$ and $m \in \mathbb{N}$, precompute the volume functions $v_0 = 1, \ldots, v_m = \Psi^m(1)$ (see Proposition 1), then generate the $n - m$ first letters as in* Method 2 *and use* Method 1 *from the current state for the last m steps.*

This method ensures quasi-uniform sampling in the following sense.

Theorem 2. *If in* Method 3 *there exist constants $C^-, C^+ \in \mathbb{R}_{>0}$ such that $C^- v_{m+1} \leq v_m \leq C^+ v_{m+1}$, then the weight of every timed word lies in the interval $[(1 - \varepsilon_{m,n})/v_n(s_0), (1 + \varepsilon_{m,n})/v_n(s_0)]$, with $\varepsilon_{m,n} = (C^+/C^-)^{(n-m-1)} - 1$.*

Example 4. For the running example (Example 1) we determine the tightest constraints $C^- \overset{\text{def}}{=} \inf_{s \in \mathbb{S}} v_m(s)/v_{m+1}(s)$ and $C^+ \overset{\text{def}}{=} \sup_{s \in \mathbb{S}} v_m(s)/v_{m+1}(s)$ for $m = 0..11$. We observe empirically that C^+/C^- tends to 1 exponentially fast when m grows (see Table 2). Given a maximal tolerated error of ε, one can determine for every m the maximal n, called n_ε, such that $\varepsilon_{m,n} \leq \varepsilon$ for every $n \leq n_\varepsilon$; formally, $n_\varepsilon \overset{\text{def}}{=} m + 1 + \lfloor \log_{(C^+/C^-)}(1 + \varepsilon) \rfloor$. First values of $n_{0.01}$ as a function of m are given in Table 2; for instance, using receding horizon for $m = 11$ one can generate timed words of length $20,000,000$ with a divergence to uniformity less than 1%.

Our sampling method requires the computation of a complete zone graph, as opposed to on-the-fly techniques used in state-of-the-art statistical model checkers; this is the price we pay for statistical evaluation of quantities of timed words in complex sub-languages as described in the next section.

5   Applications and Experiments

5.1   Tackling General Timed Languages

It is well known that language inclusion for languages recognised by nondeterministic TAs (NTAs) is undecidable, even when a robust semantics is considered [14]. The situation is even worse for stopwatch automata, hybrid automata, etc., for which the reachability problem is undecidable. However, we can handle a statistical variant of the inclusion problem when, first, an overapproximation of the language described by a DTA is known and, second, the languages admit decision procedures for the membership problem defined as: given a language $\mathcal{L}$ and a word w, is $w \in \mathcal{L}$? Our method is based on statistical volume estimation that relies on the quasi-uniform random sampling developed in the previous section. The complexity results given below are expressed in terms of the number of membership queries one has to solve.

Application 1 (Statistical Volume Estimation). *Given a timed language $\mathcal{L}$, $n \in \mathbb{N}$, a confidence level θ, an error bound ε, and an over-approximation of the language recognised by a DTA $\mathcal{C}$, that is, $\mathcal{L}_n \subset L_n(\mathcal{C})$, define $N \geq (1/\varepsilon^2) \log (\theta/2)$ (Chernoff-Hoeffding bound); draw N samples uniformly at random in $L_n(\mathcal{C})$ and answer N queries for membership in $\mathcal{L}$ to return a value p such that $\mathrm{Vol}(\mathcal{L}_n)/\mathrm{Vol}(L_n(\mathcal{C}))$ lies in $[p - \varepsilon, p + \varepsilon]$ with confidence $1 - \theta$.*

Application 2 (Inclusion Measurement). *Given two timed languages $\mathcal{L}'$, $\mathcal{L}''$ and an over-approximation of the two languages recognised by a DTA $\mathcal{C}$ one can use the previous application with $\mathcal{L} = \mathcal{L}' \setminus \mathcal{L}''$ to evaluate the volume $\mathrm{Vol}(\mathcal{L}'_n \setminus \mathcal{L}''_n)$. If a positive value is returned, a timed word in $\mathcal{L}'_n \setminus \mathcal{L}''_n$ has been detected and one can surely claim $\mathcal{L}'_n \nsubseteq \mathcal{L}''_n$. Otherwise, a null value allows one to claim with confidence $1 - \theta$ that either the inclusion holds or the difference of the two languages is smaller than $\varepsilon \mathrm{Vol}(L_n(\mathcal{C}))$.*

Application 3 (Uniform Sampling). *Given a timed language $\mathcal{L}$ and $n \in \mathbb{N}$, and an over-approximation of the language recognised by a DTA $\mathcal{C}$, that is, $\mathcal{L}_n \subseteq L_n(\mathcal{C})$, draw samples uniformly at random in $L_n(\mathcal{C})$ until one falls in $\mathcal{L}_n$.*

The sampling is uniform: every timed word of $\mathcal{L}_n$ has the same density of probability to be output. The expected number of samplings in $L_n(\mathcal{C})$ to sample one timed word in $\mathcal{L}_n$ is $\mathrm{Vol}(\mathcal{L}_n)/\mathrm{Vol}(L_n(\mathcal{C}))$. The choice of $\mathcal{C}$ is crucial, since if $L_n(\mathcal{C})$ is a too coarse approximation of $\mathcal{L}_n$ the probability of a sample from $L_n(\mathcal{C})$ to be in $\mathcal{L}_n$ is small and the methods become inefficient. We leave as future work the design of heuristics that, given a general timed language $\mathcal{L}$, automatically generate a DTA that recognises a good over-approximation of $\mathcal{L}$.

5.2   Implementation and Experiments

We implemented the techniques using three tools: PRISM [16], SageMath [20] and COSMOS [4]. The workflow is depicted in Fig. 5. We modify the tools to meet our needs. We adapted PRISM's forward reachability algorithm to implement the splitting algorithm of Sect. 3. We also export the split zone graph in a file format easy to read for SageMath. We use SageMath to compute distributions and weights of transitions as rational functions of clock valuations, which are exported and read by COSMOS in the form of a Stochastic Petri Net with general distributions. COSMOS then samples trajectories of this model, checks the membership of

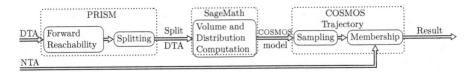

Fig. 5. Tool workflow. For the running example (Example 1), the DTA is the automaton in Fig. 2 (Left), the zone graph is the automaton in Fig. 3, the COSMOS model is the zone graph annotated with probability distributions as described in Example 3, and examples of trajectories are depicted in Fig. 4.

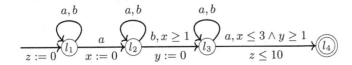

Fig. 6. The NTA $\mathcal{B}$ for Example 5. Every transition has a guard $z \leq 10$ omitted.

the language of a given NTA, and returns the probability. We have modified COS-MOS to handle distributions given by arbitrary rational functions and to compute the membership of a timed word in an NTA. Our implementation can be found at http://www.prismmodelchecker.org/files/qest16.

Example 5. Let $\mathcal{A}$ be the DTA of the running example (Example 1). The NTA $\mathcal{B}$ of Fig. 6 recognises the timed words that contain *aba* as a subword within the first 10 time units, where the latter *a* occurs at most 3 time units after the former and there is at least 1 time unit between *b* and both *as*. We have estimated $\mathtt{Vol}(L_{10}(\mathcal{A}) \cap L_{10}(\mathcal{B}))/\mathtt{Vol}(L_{10}(\mathcal{A}))$ by implementing Application 1. Sampling was performed using Method 2 with $m = 5$. The result is in the interval $[0.679, 0.688]$ with confidence level 0.99; 58,000 simulations were used in $5s$.

A Case Study. We additionally consider a larger case study of a failure and repair system modelled as an NTA (see [5] for more details). We consider a model with K machines that need to be fully repaired for the overall system to work properly. Each machine contains N levels of failure and can fail at most n_b times between two full repairs. The model is implemented by an NTA $\mathcal{A}$ with Nn_b locations and $K + 1$ clocks. The property we are interested in is encoded in another NTA $\mathcal{B}$ with 4 locations and 2 clocks. We apply our method by over-approximating the NTA $\mathcal{A}$ with a DTA $\mathcal{C}$ with $R \stackrel{\text{def}}{=} KN$ locations and 2 clocks. The results are reported in Table 3. We use our approach to sample timed words of length 50 of the DTA $\mathcal{C}$ and check their membership in $L_{50}(\mathcal{A})$ and $L_{50}(\mathcal{B})$. We compare receding horizon sampling to isotropic sampling. We observe that for isotropic sampling the probability for a timed word in $L_{50}(\mathcal{A})$ to be in $L_{50}(\mathcal{B})$ (denoted by $P_{50}(\mathcal{B}|\mathcal{A})$) tends to 1 quickly when R increases, which, for large values of R, might be interpreted as an inclusion of the languages. On the other hand, with the receding horizon sampling the same probability ($P_{50}(\mathcal{B}|\mathcal{A})$) tends to zero, which shows that the model does not satisfy the property. This result demonstrates the necessity of (quasi)-uniform

Table 3. Result of receding horizon sampling compared to isotropic sampling for the case study with two machines ($K = 2$). "Pre Time" is the pre-computation time, "Sim Time" is the simulation time. The meaning of R, $P_{50}(\mathcal{A}|\mathcal{C})$ and $P_{50}(\mathcal{B}|\mathcal{A})$ is described in the text. The receding horizon is $8 + R$. The number of samples is $100,000$.

R	Receding horizon					Isotropic		
	Pre Time	#Zones	Sim Time	$P_{50}(\mathcal{B}|\mathcal{A})$	$P_{50}(\mathcal{A}|\mathcal{C})$	Sim Time	$P_{50}(\mathcal{B}|\mathcal{A})$	$P_{50}(\mathcal{A}|\mathcal{C})$
4	45s	380	133s	0.999977	0.86539	36s	0.990439	0.03347
6	99s	581	369s	0.997717	0.58701	39s	0.975795	0.05123
8	219s	783	5005s	0.930944	0.06111	56s	0.995179	0.07052
10	417s	985	5773s	0.509091	0.00275	55s	0.999893	0.09325
12	745s	1187	7954s	0.0344828	0.00029	64s	1	0.1019

sampling to explore the behaviour of the model, since the results of isotropic simulation significantly diverge from those of (quasi)-uniform simulation, and thus do not yield reliable information about the system.

We also observe that the probability for timed words in the over-approximation $L_{50}(\mathcal{C})$ to fall in $L_{50}(\mathcal{A})$ (denoted by $P_{50}(\mathcal{A}|\mathcal{C})$) tends to zero, meaning that it becomes too crude for large values of R. Thus, tight over-approximations are important to obtain efficient simulation of an NTA through a DTA.

The time required for receding horizon simulation is high compared to isotropic, since it requires sampling of complex distributions involving many polynomials.

6   Conclusion and Further Work

We have developed the foundations for the practical application of xsvolumetry of timed languages to quantitative and statistical verification of complex properties for TAs. We implemented our work in a tool chain and provide first experiments.

On the theoretical side, we want to show that constants in Method 3 and Theorem 2 can be chosen to guarantee arbitrarily small divergence from exact uniform sampling and consider extending the theory to probabilistic TAs. We would also like to implement membership checking in COSMOS for generaltimed languages (e.g. recognised by stopwatch automata, LHA, etc.). We also plan to use our random sampling algorithms to detect forgetful cycles described in [3], which are needed to synthesise controllers robust to timing imprecision [18].

References

1. Alur, R., Dill, D.L.: A theory of timed automata. Theoret. Comput. Sci. **126**, 183–235 (1994)
2. Asarin, E., Basset, N., Béal, M.-P., Degorre, A., Perrin, D.: Toward a timed theory of channel coding. In: Jurdziński, M., Ničković, D. (eds.) FORMATS 2012. LNCS, vol. 7595, pp. 27–42. Springer, Heidelberg (2012)
3. Asarin, E., Basset, N., Degorre, A.: Entropy of regular timed languages. Inf. Comput. **241**, 142–176 (2015)

4. Ballarini, P., Barbot, B., Duflot, M., Haddad, S., Pekergin, N.: HASL: a new approach for performance evaluation and model checking from concepts to experimentation. Perform. Eval. **90**, 53–77 (2015)
5. Barbot, B., Basset, N., Beunardeau, M., Kwiatkowska, M.: Uniform sampling for timed automata with application to language inclusion measurement. Technical report CS-RR-16-04, University of Oxford (2016)
6. Basset, N.: Counting and generating permutations using timed languages. In: Pardo, A., Viola, A. (eds.) LATIN 2014. LNCS, vol. 8392, pp. 502–513. Springer, Heidelberg (2014)
7. Basset, N.: A maximal entropy stochastic process for a timed automaton. In: Fomin, F.V., Freivalds, R., Kwiatkowska, M., Peleg, D. (eds.) ICALP 2013, Part II. LNCS, vol. 7966, pp. 61–73. Springer, Heidelberg (2013)
8. Bengtsson, J.E., Yi, W.: Timed automata: semantics, algorithms and tools. In: Desel, J., Reisig, W., Rozenberg, G. (eds.) Lectures on Concurrency and Petri Nets. LNCS, vol. 3098, pp. 87–124. Springer, Heidelberg (2004)
9. Bohlender, D., Bruintjes, H., Junges, S., Katelaan, J., Nguyen, V.Y., Noll, T.: A review of statistical model checking pitfalls on real-time stochastic models. In: Margaria, T., Steffen, B. (eds.) ISoLA 2014, Part II. LNCS, vol. 8803, pp. 177–192. Springer, Heidelberg (2014)
10. David, A., Larsen, K.G., Legay, A., Mikucionis, M., Poulsen, D.B.: UPPAAL SMC tutorial. STTT **17**(4), 397–415 (2015)
11. Denise, A., Gaudel, M.-C., Gouraud, S.-D., Lassaigne, R., Oudinet, J., Peyronnet, S.: Coverage-biased random exploration of large models and application to testing. STTT **14**(1), 73–93 (2012)
12. Flajolet, P., Zimmerman, P., Van Cutsem, B.: A calculus for the random generation of labelled combinatorial structures. Theoret. Comput. Sci. **132**(1), 1–35 (1994)
13. Grosu, R., Smolka, S.A.: Monte Carlo model checking. In: Halbwachs, N., Zuck, L.D. (eds.) TACAS 2005. LNCS, vol. 3440, pp. 271–286. Springer, Heidelberg (2005)
14. Henzinger, T.A., Raskin, J.-F.: Robust undecidability of timed and hybrid systems. In: Lynch, N.A., Krogh, B.H. (eds.) HSCC 2000. LNCS, vol. 1790, pp. 145–159. Springer, Heidelberg (2000)
15. Jaynes, E.T.: Information theory and statistical mechanics II. Phys. Rev. Online Arch. (PROLA) **108**(2), 171–190 (1957)
16. Kwiatkowska, M., Norman, G., Parker, D.: PRISM 4.0: verification of probabilistic real-time systems. In: Gopalakrishnan, G., Qadeer, S. (eds.) CAV 2011. LNCS, vol. 6806, pp. 585–591. Springer, Heidelberg (2011)
17. Murray, R.M., Hauser, J., Jadbabaie, A., Milam, M.B., Petit, N., Dunbar, W.B., Franz, R.: Online control customization via optimization-based control. In: Software-Enabled Control: Information Technology for Dynamical Systems, p. 149 (2003)
18. Oualhadj, Y., Reynier, P.-A., Sankur, O.: Probabilistic robust timed games. In: Baldan, P., Gorla, D. (eds.) CONCUR 2014. LNCS, vol. 8704, pp. 203–217. Springer, Heidelberg (2014)
19. Oudinet, J., Denise, A., Gaudel, M.-C., Lassaigne, R., Peyronnet, S.: Uniform monte-carlo model checking. In: Giannakopoulou, D., Orejas, F. (eds.) FASE 2011. LNCS, vol. 6603, pp. 127–140. Springer, Heidelberg (2011)
20. Stein, W.A., et al.: Sage Mathematics Software (Version 6.9). The Sage Development Team (2015). http://www.sagemath.org
21. Younes, H.L.S., Simmons, R.G.: Statistical probabilistic model checking with a focus on time-bounded properties. Inf. Comput. **204**(9), 1368–1409 (2006)

Inferring Covariances for Probabilistic Programs

Benjamin Lucien Kaminski[✉], Joost-Pieter Katoen[✉],
and Christoph Matheja[✉]

Software Modeling and Verification Group, RWTH Aachen University,
Aachen, Germany
{benjamin.kaminski,katoen,matheja}@cs.rwth-aachen.de

Abstract. We study weakest precondition reasoning about the (co) variance of outcomes and the variance of run–times of probabilistic programs with conditioning. For outcomes, we show that approximating (co)variances is computationally more difficult than approximating expected values. In particular, we prove that computing both lower and upper bounds for (co)variances is Σ_2^0–complete. As a consequence, neither lower nor upper bounds are computably enumerable. We therefore present invariant–based techniques that *do* enable enumeration of both upper and lower bounds, once appropriate invariants are found. Finally, we extend this approach to reasoning about run–time variances.

Keywords: Probabilistic programs · Covariance · Run–time

1 Introduction

Probabilistic programs describe manipulations on uncertain data in a succinct way. They are normal–looking programs describing how to obtain a distribution over the outputs. Using mostly standard programming language constructs, a probabilistic program transforms a prior distribution into a posterior distribution. Probabilistic programs provide a structured means to describe e.g., Bayesian networks (from AI), random encryption (from security), or predator–prey models (from biology) [5] succinctly.

The posterior distribution of a program is mostly determined by approximate means such as Markov Chain Monte Carlo (MCMC) sampling using (variants of) the well–known Metropolis–Hasting approach. This yields estimates for various measures of interest, such as expected values, second moments, variances, covariances, and the like. Such estimates typically come with weak guarantees in the form of confidence intervals, asserting that with a certain confidence the measure has a certain value. In contrast to these weak guarantees, we aim at the *exact* inference of such measures and their bounds. We hereby focus both on correctness and on run–time analysis of probabilistic programs. Put shortly, we

This work was supported by the Excellence Initiative of the German federal and state government.

G. Agha and B. Van Houdt (Eds.): QEST 2016, LNCS 9826, pp. 191–206, 2016.
DOI: 10.1007/978-3-319-43425-4_14

are interested in obtaining *quantitative* statements about the possible outcomes of programs well as their run times.

This paper studies reasoning about the (co)variance of outcomes and the variance of run–times of probabilistic programs. Our programs support sampling from discrete probability distributions, conditioning on the outcomes of experiments by observations [5], and unbounded while–loops[1]. In the first part of the paper, we study the *theoretical complexity* of obtaining (co)variances on outcomes. We show that obtaining bounds on (co)variances is computationally more difficult than for expected values. In particular, we prove that computing both upper *and* lower bounds for (co)variances of program outcomes is Σ_2^0–complete, thus *not recursively enumerable*. This contrasts the case for expected values where lower bounds *are recursively enumerable*, while only upper bounds are Σ_2^0–complete [7]. We also show that determining the precise values of (co)variances as well as checking whether the (co)variance is infinite are both Π_2^0–complete. These results rule out analysis techniques based on finite loop–unrollings as complete approaches for reasoning about the covariances of outcomes of probabilistic programs.

In the second part of the paper, we therefore develop a weakest precondition reasoning technique for obtaining covariances on outcomes and variances on run–times. As with deductive reasoning for ordinary sequential programs, the crux is to find suitable loop–invariants. We present a couple of invariant–based proof rules that provide a sound and complete method to computably enumerate both upper and lower bounds on covariances, once appropriate invariants are found. We establish similar results for variances of the run–time of programs. The results of this paper extend McIver and Morgans approach for obtaining expectations of probabilistic programs [11], recent techniques for expected run–time analysis [9], and complement results on termination analysis [4,7].

Some proofs had to be omitted due to lack of space. They can be found in an extended version of this paper [8].

2   Preliminaries

We study approximating the covariance of two random variables (ranging over program states) after successful termination of a probabilistic program on a given input state. Our development builds upon the *conditional probabilistic guarded command language (cpGCL)* [6]—an extension of Dijkstra's guarded command language [3] endowed with probabilistic choice and conditioning constructs.

Definition 1 (cpGCL [6]). *Let $\mathbb{V}$ be a finite set of program variables[2]. Then the set of programs in cpGCL, denoted $\mathbb{P}$, adheres to the grammar*

$$\mathbb{P} ::= \texttt{skip} \mid \texttt{empty} \mid \texttt{diverge} \mid \texttt{halt} \mid x := E \mid \mathbb{P}; \mathbb{P} \mid \texttt{if } (B) \; \{\mathbb{P}\} \texttt{ else } \{\mathbb{P}\}$$
$$\mid \{\mathbb{P}\} \; [p] \; \{\mathbb{P}\} \mid \texttt{while } (B) \; \{\mathbb{P}\} \mid \texttt{observe } B \; ,$$

---

[1] This contrasts MCMC–based analysis, as this is restricted to bounded programs.

[2] We restrict ourselves to a finite set of program variables for reasons of cleanness of the presentation. In principle, a countable set of program variables could be allowed.

where $x \in \mathbb{V}$, E is an arithmetical expression over $\mathbb{V}$, $p \in [0, 1] \cap \mathbb{Q}$ is a rational probability, and B is a Boolean expression over arithmetic expressions over $\mathbb{V}$.

If a program C contains neither a probabilistic choice $\{C'\} \, [p] \, \{C''\}$ nor an observe*–statement, we say that C is* non–probabilistic.

We briefly go over the meaning of the language constructs. Furthermore, we assign each statement an execution time in order to reason about the *run–time* of programs. skip (empty) does nothing—i.e. does not alter the current variable valuations—and consumes one (no) unit of time. diverge is syntactic sugar for the certainly non–terminating program while (true) {skip}. halt consumes no unit of time and halts program execution immediately (even when encountered inside a loop). It represents an *improper* termination of the program. $x := E$, C_1; C_2, if (B) $\{C_1\}$ else $\{C_2\}$, and while (B) $\{C'\}$ are standard variable assignment, sequential composition, conditional choice, and while–loop constructs. Assignments and guard evaluations consume one unit of time.

$\{C_1\} \, [p] \, \{C_2\}$ is a probabilistic choice construct: With probability p the program C_1 is executed and with probability $1 - p$ the program C_2 is executed. Flipping the p–coin itself consumes one unit of time. observe B is the conditioning construct. Whenever in the execution of a program, an observe B is encountered, such that the current variable valuation satisfies the guard B, nothing happens except that one unit of time is being consumed. If, however, an observe B is encountered along an execution trace that occurs with probability q, such that B is *not* satisfied, this trace is blocked as it is considered an *undesired execution*. The probabilities of the remaining execution traces are then conditioned to the fact that this undesired trace was not encountered, i.e. the probabilities of the remaining execution traces are renormalized by $1 - q$. We refer to encountering such an undesired execution as an *observation violation*. For more details on conditioning and its semantics, see [6].

Notice that we do not include non–deterministic choice constructs (as opposed to probabilistic choice construct) in our language, as we would then run into similar problems as in [6, Sect. 6] in the presence of conditioning.

Example 1 (Conditioning Inside a Loop). Consider the following loop:

```
while (c = 1){ {c := 0} [0.5] {x := x + 1}; observe c = 1 ∨ x is odd }
```

Without the observe–statement, this loop would generate a geometric distribution on x. By considering the observe–statement, this distribution is conditioned to the fact that after termination x is odd.   △

Given a probabilistic program C, an initial state σ, and a random variable f mapping program states to positive reals, we could now ask: What is the *conditional* expected value of f after proper termination of program C on input σ, *given that no observation was violated during the execution*? An answer to this question is given by the conditional weakest pre–expectation calculus introduced in [6]. For summarizing this calculus, we first formally characterize the random variables f, commonly called *expectations* [11]:

Definition 2 (Expectations [6,11]**).** *Let* $\mathbb{S} = \{\sigma \mid \sigma : \mathbb{V} \to \mathbb{Q}\}$, *where* $\mathbb{Q}$ *is the set of rational numbers, be the* set of program states.[3] *Then the* set of expectations *is defined as* $\mathbb{E} = \{f \mid f : \mathbb{S} \to \mathbb{R}_{\geq 0}^{\infty}\}$, *and the* set of bounded expectations *is defined as* $\mathbb{E}_{\leq 1} = \{f \mid f : \mathbb{S} \to [0, 1]\}$. *A complete partial order* $\preceq$ *on both* $\mathbb{E}$ *and* $\mathbb{E}_{\leq 1}$ *is given by* $f_1 \preceq f_2$ *iff* $\forall \sigma \in \mathbb{S} : f_1(\sigma) \leq f_2(\sigma)$.

The *weakest (liberal) pre–expectation transformer* wp $: \mathbb{P} \to (\mathbb{E} \to \mathbb{E})$ (wlp $: \mathbb{P} \to (\mathbb{E}_{\leq 1} \to \mathbb{E}_{\leq 1})$) is defined according to Table 1 (middle column). By means of these two transformers, we can give an answer to the question posed above: Namely, the fraction $\text{wp}[C](f)(\sigma)/\text{wlp}[C](1)(\sigma)$ is indeed the conditional expected value of f after termination of C on input σ, given that no observation was violated during C's execution [6]. Consequently, we define:

Table 1. Definition of wp, wlp, and rt. $[x/E]$ is a syntactic replacement with $f[x/E](\sigma) = f(\sigma[x \mapsto \sigma(E)])$. $[B]$ is the indicator function of B with $[B](\sigma) = 1$ if $\sigma \models B$, and $[B](\sigma) = 0$ otherwise. $F \circ H(f)$ is the functional composition of F and H applied to f. lfp $X.\ F(X)$ (gfp $X.\ F(X)$) is the least (greatest) fixed point of F with respect to $\preceq$. Definitions of wlp for the other language constructs are as for wp and thus omitted.

C	wp $[C]\,(f)$	rt $[C]\,(t)$
skip	f	$t[\tau/\tau + 1]$
empty	f	t
diverge	0	∞
halt	0	0
$x := E$	$f[x/E]$	$t[x/E,\ \tau/\tau + 1]$
$C_1;\ C_2$	$\text{wp}[C_1] \circ \text{wp}[C_2]\,(f)$	$\text{rt}[C_1] \circ \text{rt}[C_2]\,(t)$
if $(B)\ \{C_1\}$ else $\{C_2\}$	$[B] \cdot \text{wp}[C_1]\,(f)$ $+[\neg B] \cdot \text{wp}[C_2]\,(f)$	$([B] \cdot \text{rt}[C_1]\,(t)$ $+[\neg B] \cdot \text{rt}[C_2]\,(t))[\tau/\tau + 1]$
$\{C_1\}\ [p]\ \{C_2\}$	$p \cdot \text{wp}[C_1]\,(f)$ $+(1 - p) \cdot \text{wp}[C_2]\,(f)$	$(p \cdot \text{rt}[C_1]\,(t)$ $+(1 - p) \cdot \text{rt}[C_2]\,(t))[\tau/\tau + 1]$
while $(B)\ \{C'\}$	lfp $X.\ [\neg B] \cdot f$ $+[B] \cdot \text{wp}[C']\,(X)$	lfp $X.\ ([\neg B] \cdot t$ $+[B] \cdot \text{rt}[C']\,(X))[\tau/\tau + 1]$
observe B	$[B] \cdot f$	$[B] \cdot t[\tau/\tau + 1]$

C	wlp $[C]\,(f)$
diverge	1
halt	1
while $(B)\ \{C'\}$	gfp $X.\ [\neg B] \cdot f + [B] \cdot \text{wlp}[C']\,(X)$

---

[3] Notice that $\mathbb{S}$ is countable and computably enumerable as $\mathbb{V}$ is finite.

Definition 3 (Conditional Expected Values [6]). *Let $C \in \mathbb{P}$, $\sigma \in \mathbb{S}$, and $f \in \mathbb{E}$. Then the* conditional expected value *of f after executing C on input σ given that no observation was violated is defined as*[4]

$$\mathsf{E}_{[\![C]\!](\sigma)}(f) \;=\; \frac{\mathsf{wp}\,[C]\,(f)\,(\sigma)}{\mathsf{wlp}\,[C]\,(1)\,(\sigma)} \;.$$

Having the definition for conditional expected values readily available, we can now turn towards defining the conditional (co)variance of a (two) random variables. We simply translate the textbook definition to our setting:

Definition 4 (Conditional (Co)variances). *Let $C \in \mathbb{P}$, $\sigma \in \mathbb{S}$, and $f, g \in \mathbb{E}$. Then the* conditional covariance *of the two random variables f and g after executing C on input σ, given that no observation was violated is defined as*

$$\mathsf{Cov}_{[\![C]\!](\sigma)}(f, g) \;=\; \mathsf{E}_{[\![C]\!](\sigma)}(f \cdot g) - \mathsf{E}_{[\![C]\!](\sigma)}(f) \cdot \mathsf{E}_{[\![C]\!](\sigma)}(g) \;.$$

The conditional variance *of the single random variable f after executing C on input σ, given that no observation was violated is defined as the conditional covariance of f with itself, i.e. $\mathsf{Var}_{[\![C]\!](\sigma)}(f) = \mathsf{Cov}_{[\![C]\!](\sigma)}(f, f)$.*

3   Computational Hardness of Computing (Co)variances

In this section, we will investigate the computational hardness of computing upper and lower bounds for conditional (co)variances. The results will be stated in terms of levels in the arithmetical hierarchy—a concept we first briefly recall:

Definition 5 (The Arithmetical Hierarchy [10,12]). *For every $n \in \mathbb{N}$, the class Σ_n^0 is defined as $\Sigma_n^0 = \{\mathcal{A} \mid \mathcal{A} = \{x \mid \exists y_1 \forall y_2 \exists y_3 \cdots \exists/\forall y_n : (x, y_1, y_2, y_3, \ldots, y_n) \in \mathcal{R}\}$, $\mathcal{R}$ is a decidable relation$\}$ and the class Π_n^0 is defined as $\Pi_n^0 = \{\mathcal{A} \mid \mathcal{A} = \{x \mid \forall y_1 \exists y_2 \forall y_3 \cdots \exists/\forall y_n : (x, y_1, y_2, y_3, \ldots, y_n) \in \mathcal{R}\}$, $\mathcal{R}$ is a decidable relation$\}$. Note that we require the values of variables to be drawn from a computable domain. Multiple consecutive quantifiers of the same type can be contracted to one quantifier of that type, so the number n really refers to the number of necessary quantifier alternations. A set $\mathcal{A}$ is called* arithmetical, *iff $\mathcal{A} \in \Gamma_n^0$, for $\Gamma \in \{\Sigma, \Pi\}$ and $n \in \mathbb{N}$. The arithmetical sets form a strict hierarchy, i.e. $\Gamma_n^0 \subset \Gamma_{n+1}^0$ holds for $\Gamma \in \{\Sigma, \Pi\}$ and $n \geq 0$. Furthermore, note that $\Sigma_0^0 = \Pi_0^0$ is exactly the class of the decidable sets and Σ_1^0 is exactly the class of the computably enumerable sets.*

---

[4] We make use of the convention that $\frac{0}{0} = 0$. Note that since our probabilistic choice is a discrete choice and our language does not support sampling from continuous distributions, the problematic case of "$\frac{0}{0}$" can only occur if executing C on input σ will result in a violation of an observation with probability 1.

Next, we recall the concept of many–one reducibility and completeness:

Definition 6 (Many–One Reducibility and Completeness [2,12,14]**).**
Let $\mathcal{A}$, $\mathcal{B}$ be arithmetical sets and let X be some appropriate universe such that $\mathcal{A}, \mathcal{B} \subseteq X$. $\mathcal{A}$ is called many–one reducible *(or simply* reducible*) to $\mathcal{B}$, denoted $\mathcal{A} \leq_m \mathcal{B}$, iff there exists a computable function $r : X \to X$, such that $\forall\, x \in X :$ $\left(x \in \mathcal{A} \iff r(x) \in \mathcal{B}\right)$. If r is a function such that r reduces $\mathcal{A}$ to $\mathcal{B}$, we denote this by $r : \mathcal{A} \leq_m \mathcal{B}$. Note that $\leq_m$ is transitive.*

$\mathcal{A}$ is called Γ_n^0–complete, for $\Gamma \in \{\Sigma, \Pi\}$, iff both $\mathcal{A} \in \Gamma_n^0$ and $\mathcal{A}$ is Γ_n^0–hard, meaning $\mathcal{C} \leq_m \mathcal{A}$, for any set $\mathcal{C} \in \Gamma_n^0$. Note that if $\mathcal{B} \in \Gamma_n^0$ and $\mathcal{A} \leq_m \mathcal{B}$, then $\mathcal{A} \in \Gamma_n^0$, too. Furthermore, note that if $\mathcal{A}$ is Γ_n^0–complete and $\mathcal{A} \leq_m \mathcal{B}$, then $\mathcal{B}$ is necessarily Γ_n^0–hard. Lastly, note that if $\mathcal{A}$ is Σ_n^0–complete, then $\mathcal{A} \in \Sigma_n^0 \setminus \Pi_n^0$. Analogously, if $\mathcal{A}$ is Π_n^0–complete, then $\mathcal{A} \in \Pi_n^0 \setminus \Sigma_n^0$.

In the following, we study the hardness of obtaining covariance approximations both from above and from below. Furthermore, we are interested in exact values of covariances as well as in deciding whether the covariance is infinite. In order to formally investigate the arithmetical complexity of these problems, we define four problem sets which relate to upper and lower bounds for covariances and to the question whether the covariance is infinite:

Definition 7 (Approximation Problems for Covariances). *We define the following decision problems:*

$$(C, \sigma, f, g, q) \in \mathcal{LCOVAR} \iff \mathsf{Cov}_{[\![C]\!](\sigma)}(f, g) > q$$
$$(C, \sigma, f, g, q) \in \mathcal{RCOVAR} \iff \mathsf{Cov}_{[\![C]\!](\sigma)}(f, g) < q$$
$$(C, \sigma, f, g, q) \in \mathcal{COVAR} \iff \mathsf{Cov}_{[\![C]\!](\sigma)}(f, g) = q$$
$$(C, \sigma, f, g) \in {}^{\infty}\mathcal{COVAR} \iff \mathsf{Cov}_{[\![C]\!](\sigma)}(f, g) \in \{-\infty, +\infty\}$$

where $C \in \mathbb{P}$, $\sigma \in \mathbb{S}$, $f, g \in \mathbb{E}$, and $q \in \mathbb{Q}$.[5]

The first fact we establish about the hardness of computing upper and lower bounds of covariances is that this is at most Σ_2^0–hard, thus not harder than deciding whether a non–probabilistic program, i.e. a program without observations and probabilistic choice, does *not* terminate on all inputs, or deciding whether a probabilistic program terminates after an expected finite number of steps [7,13]. Formally, we establish the following results:

Lemma 1. *$\mathcal{LCOVAR}$ and $\mathcal{RCOVAR}$ are both in Σ_2^0.*

For proving Lemma 1, we revert to a fact established in [7]: All lower bounds for expected outcomes are computably enumerable. As a consequence, there exists a computable function $\mathsf{wp}^k\,[C]\,(f)\,(\sigma)$ that is ascending in k, such that for given $C \in \mathbb{P}$, $\sigma \in \mathbb{S}$, and $f \in \mathbb{E}$, we have

$$\forall\, k \in \mathbb{N} : \mathsf{wp}^k\,[C]\,(f)\,(\sigma) \;\leq\; \mathsf{wp}\,[C]\,(f)\,(\sigma), \quad \text{and}$$

$$\sup_{k \in \mathbb{N}} \mathsf{wp}^k\,[C]\,(f)\,(\sigma) \;=\; \mathsf{wp}\,[C]\,(f)\,(\sigma) .$$

---

[5] Note that, for obvious reasons, we restrict to *computable* expectations f, g only.

Intuitively, for every $k \in \mathbb{N}$ the function $\mathsf{wp}^k [C] (f) (\sigma)$ outputs a lower bound of $\mathsf{wp} [C] (f) (\sigma)$ in ascending order.

Similarly, lower bounds for $\mathsf{wlp} [C] (1) (\sigma)$ can be enumerated. To see this, note that $\mathsf{wp} [C] (1) (\sigma) = 1$ for any observe–free program C and any state σ. $\mathsf{wp} [C] (1) (\sigma)$ can only be decreased by violation of an observation. Informally,

$$\mathsf{wp} [C] (1) (\sigma) = 1 - \text{``Probability of } C \text{ violating an observation''} \ .$$

Lower bounds for the latter probability can be enumerated by successively exploring the computation tree of C on input σ and accumulating the probability mass of all execution traces that lead to a violation of an observation. As a consequence, there must exist a computable function $\mathsf{wlp}^k [C] (1) (\sigma)$ that is descending in k, such that for given $C \in \mathbb{P}$ and $\sigma \in \mathbb{S}$,

$$\forall k \in \mathbb{N} : \mathsf{wlp} [C] (1) (\sigma) \leq \mathsf{wlp}^k [C] (1) (\sigma), \quad \text{and}$$
$$\mathsf{wlp} [C] (1) (\sigma) = \inf_{k \in \mathbb{N}} \mathsf{wlp}^k [C] (1) (\sigma) \ .$$

Since $\mathsf{wp}^k [C] (f) (\sigma)$ is ascending and $\mathsf{wlp}^k [C] (1) (\sigma)$ is descending in k, the quotient $\mathsf{wp}^k[C](f)(\sigma)/\mathsf{wlp}^k[C](1)(\sigma)$ is ascending in k. We can now prove Lemma 1:

Proof (Lemma 1). For $\mathcal{LCOVAR} \in \Sigma_2^0$, consider $(C, \sigma, f, g, q) \in \mathcal{LCOVAR}$ iff

$$\exists k \, \forall \ell : \ \frac{\mathsf{wp}^k [C] (f \cdot g) (\sigma)}{\mathsf{wlp}^k [C] (1) (\sigma)} - \frac{\mathsf{wp}^\ell [C] (f) (\sigma) \cdot \mathsf{wp}^\ell [C] (g) (\sigma)}{\mathsf{wlp}^\ell [C] (1) (\sigma)^2} > q \ .$$

For the proof for $\mathcal{RCOVAR}$, see [8] □

Regarding the hardness of deciding whether a given rational is equal to the covariance and the hardness of deciding non–finiteness of covariances, we establish that this is at most Π_2^0–hard, thus not harder than deciding whether a non–probabilistic program terminates on all inputs, or deciding whether a probabilistic program does *not* terminate after an expected finite number of steps [7,13]. Formally, we establish the following:

Lemma 2. $\mathcal{COVAR}$ and $^\infty\mathcal{COVAR}$ are both in Π_2^0.

So far we provided upper bounds for the computational hardness of solving approximation problems for covariances. We now show that these bounds are tight in the sense that these problems are *complete* for their respective level of the arithmetical hierarchy. For that we need a Σ_2^0– and a Π_2^0–hard problem in order to perform the necessary reductions for proving the hardness results. Adequate problems are the problem of almost–sure termination and its complement:

Theorem 1 (Hardness of the Almost–Sure Termination Problem [7]). *Let $C \in \mathbb{P}$ be observe–free. Then C terminates almost–surely on input $\sigma \in \mathbb{S}$, iff it does so with probability 1. The problem set $\mathcal{AST}$ is defined as $(C, \sigma) \in \mathcal{AST}$ iff C terminates almost–surely on input σ. We denote the complement of $\mathcal{AST}$ by $\overline{\mathcal{AST}}$.[6] $\mathcal{AST}$ is Π_2^0–complete and $\overline{\mathcal{AST}}$ is Σ_2^0–complete.*

---

[6] Note that by "complement" we mean not exactly a set theoretic complement but rather all pairs (C, σ) such that C does not terminate almost–surely on σ.

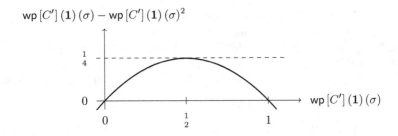

Fig. 1. Plot of the termination probability of a program against the resulting variance.

By reduction from $\overline{\mathcal{AST}}$ we now establish the following hardness results:

Lemma 3. $\mathcal{LCOVAR}$ and $\mathcal{RCOVAR}$ are both Σ_2^0-hard.

Proof. For proving the Σ_2^0-hardness of $\mathcal{LCOVAR}$, consider the reduction function $r_{\mathcal{L}}(C, \sigma) = (C', \sigma, v, v, 0)^7$, with $C' = v := 0; \{\text{skip}\}\,[1/2]\,\{C\}; v := 1$, where variable v does not occur in C. Now consider the following:

$$
\begin{aligned}
\mathsf{Cov}_{[\![C']\!](\sigma)}(v, v) &= \frac{\mathsf{wp}\,[C']\,(v^2)\,(\sigma)}{\mathsf{wlp}\,[C']\,(1)\,(\sigma)} - \frac{\mathsf{wp}\,[C']\,(v)\,(\sigma)^2}{\mathsf{wlp}\,[C']\,(1)\,(\sigma)^2} \\
&= \frac{\mathsf{wp}\,[C']\,(v^2)\,(\sigma)}{1} - \frac{\mathsf{wp}\,[C']\,(v)\,(\sigma)^2}{1^2} \quad (C' \text{ is observe-free}) \\
&= \mathsf{wp}\,[C']\,(v^2)\,(\sigma) - \mathsf{wp}\,[C']\,(v)\,(\sigma)^2
\end{aligned}
$$

Since v does not occur in C and v is set from 0 to 1 if and only if C' has terminated, this is equal to:

$$
\begin{aligned}
&= \mathsf{wp}\,[C']\,(1^2)\,(\sigma) - \mathsf{wp}\,[C']\,(1)\,(\sigma)^2 \\
&= \mathsf{wp}\,[C']\,(1)\,(\sigma) - \mathsf{wp}\,[C']\,(1)\,(\sigma)^2
\end{aligned}
$$

Note that $\mathsf{wp}\,[C']\,(1)\,(\sigma)$ is exactly the probability of C' terminating on input σ. A plot of this termination probability against the resulting variance is given in Fig. 1. We observe that $\mathsf{Cov}_{[\![C']\!](\sigma)}(v, v) = \mathsf{wp}\,[C']\,(1)\,(\sigma) - \mathsf{wp}\,[C']\,(1)\,(\sigma)^2 > 0$ iff C' terminates *neither* with probability 0 *nor* with probability 1. Since, however, C' terminates by construction *at least* with probability $1/2$, we obtain that $\mathsf{Cov}_{[\![C']\!](\sigma)}(v, v) > 0$ iff C' terminates with probability less than 1, which is the case iff C terminates with probability less than 1. Thus $r_{\mathcal{L}}(C, \sigma) = (C', \sigma, v, v, 0) \in \mathcal{LCOVAR}$ iff $(C, \sigma) \in \overline{\mathcal{AST}}$. Thus, $r_{\mathcal{L}} : \overline{\mathcal{AST}} \leq_m \mathcal{LCOVAR}$. Since $\overline{\mathcal{AST}}$ is Σ_2^0-complete, if follows that $\mathcal{LCOVAR}$ is Σ_2^0-hard.

For the the proof for $\mathcal{RCOVAR}$, see [8]. $\qquad\square$

A hardness results for $\mathcal{COVAR}$ is obtained by reduction from $\mathcal{AST}$.

---

[7] We write v for the expectation that in state σ returns $\sigma(v)$.

Lemma 4. $\mathcal{COVAR}$ *is Π_2^0–hard.*

Proof. Similar to Lemma 3 using $r_V(C, \sigma) = \big(C', \sigma, v, v, \frac{1}{4}\big)$, with $C' = v := 0$; $\{\texttt{diverge}\}$ $[1/2]$ $\{C\}$; $v := 1$. For details, see [8]. □

For a hardness result on $^\infty\mathcal{COVAR}$ we use the universal halting problem for non–probabilistic programs.

Theorem 2 (Hardness of the Universal Halting Problem [13]). *Let C be a non–probabilistic program. The universal halting problem is the problem of deciding whether C terminates on all inputs. Let $\mathcal{UH}$ denote the problem set, defined as $C \in \mathcal{UH}$ iff $\forall \sigma \in \mathbb{S} : C$ terminates on input σ. $\mathcal{UH}$ is Π_2^0–complete.*

We now establish by reduction from $\mathcal{UH}$ the remaining hardness result:

Lemma 5. $^\infty\mathcal{COVAR}$ *is Π_2^0–hard.*

Proof. For proving the Π_2^0–hardness of $^\infty\mathcal{COVAR}$ we use the reduction function $r_\infty(C) = (C', \sigma, v, v)$, where σ is arbitrary but fixed and C' is the program

$c := 1;\ i := 0;\ x := 0;\ v := 0;\ term := 0;\ InitC;$
$\texttt{while } (c \neq 0)\{$
    $StepC;\ \texttt{if } (term\ =\ 1)\{\ v := 2^x;\ i := i+1;\ term := 0;\ InitC\ \};$
    $\{c := 0\}\ [0.5]\ \{c := 1\};\ x := x+1\ \}\ ,$

where $InitC$ is a non–probabilistic program that initializes a simulation of the program C on input $e(i)$ (where $e : \mathbb{N} \to \mathbb{S}$ is some computable enumeration of $\mathbb{S}$), and $StepC$ is a non–probabilistic program that does one single (further) step of that simulation and sets $term$ to 1 if that step has led to termination of C.

Intuitively, the program C' starts by simulating C on input $e(0)$. During the simulation, it—figuratively speaking—gradually looses interest in further simulating C by tossing a coin after each simulation step to decide whether to continue the simulation or not. If eventually C' finds that C has terminated on input $e(0)$, it sets the variable v to a number exponential in the number of coin tosses that were made so far, namely to 2^x. C' then continues with the same procedure for the next input $e(1)$, and so on.

The variable x keeps track of the number of loop iterations (starting from 1 as the first loop iteration will definitely take place), which equals the number of coin tosses. The x–th loop iteration takes place with probability $1/2^x$. The expected value $\mathsf{E}_{[\![C']\!](\sigma)}(v)$ is thus given by a series of the form $S = \sum_{i=1}^\infty v_i/2^i$, where $v_i = 2^j$ for some $j \in \mathbb{N}$. Two cases arise:

(1) $C \in \mathcal{UH}$, i.e. C terminates on every input. In that case, v will infinitely often be updated to 2^x. Therefore, summands of the form $2^i/2^i$ will appear infinitely often in S and so S diverges. Hence, the expected value of v is infinity and therefore, the variance of v must be infinite as well. Thus, $(C', \sigma, v, v) \in {}^\infty\mathcal{COVAR}$.

(2) $C \notin \mathcal{UH}$, i.e. there exists some input σ' with minimal $i \in \mathbb{N}$ such that $e(i) = \sigma'$ on which C does not terminate. In that case, the numerator of all summands of S is upper bounded by some constant 2^j and thus S converges. Boundedness of the v_i's implies that the series $\sum_{i=1}^\infty v_i^2/2^i = \mathsf{E}_{[\![C']\!](\sigma)}(v^2)$ also converges. Hence, the variance of v is finite and $(C', \sigma, v, v) \notin {}^\infty\mathcal{COVAR}$. □

Lemmas 1 to 5 together directly yield the following completeness results:

Theorem 3 (The Hardness of Approximating Covariances).

1. $\mathcal{LCOVAR}$ and $\mathcal{RCOVAR}$ are both Σ_2^0-complete.
2. $\mathcal{COVAR}$ and $\infty\mathcal{COVAR}$ are both Π_2^0-complete.

Remark 1 (The Hardness of Approximating Variances). It can be shown that *variance* approximation is not easier than covariance approximation: exactly the same completeness results as in Theorem 3 hold for analogous variance approximation problems. In fact, we have always reduced to approximating a variance for obtaining our hardness results on covariances.   △

As an immediate consequence of Theorem 3, computing both upper and lower bounds for covariances is equally difficult. This is *contrary to the case for expected values*: While computing upper bounds for expected values is also Σ_2^0-complete, computing lower bounds is Σ_1^0-complete, thus lower bounds are computably enumerable [7]. Therefore we can computably enumerate an ascending sequence that converges to the sought–after expected value. By Theorem 3 this is *not possible* for a covariance as Σ_2^0-sets are in general not computably enumerable.

  Theorem 3 rules out techniques based on finite loop–unrollings as *complete* approaches for reasoning about the covariances of outcomes of probabilistic programs. As this is a rather sobering insight, in the next section we will investigate invariant–aided techniques that are complete and can be applied to tackle these approximation problems.

4   Invariant–Aided Reasoning on Outcome Covariances

For straight–line (i.e. loop–free) programs, upper and lower bounds for covariances are obviously computable, e.g. by using the decompositions from Definitions 3 and 4, and the inference rules from Table 1. Problems do arise, however, for loops. We have seen in the previous section that neither upper nor lower bounds are computably enumerable. In this section we therefore present an invariant–aided approach for enumerating bounds on covariances of loops. The underlying principle of such techniques is quite commonly a result due to Park:

Theorem 4 (Park's Lemma [15]). *Let $(D, \sqsubseteq)$ be a complete partial order and $F : D \to D$ be continuous. Then, for all $d \in D$, it holds that $F(d) \sqsubseteq d$ implies $\mathsf{lfp}\, F \sqsubseteq d$, and $d \sqsubseteq F(d)$ implies $d \sqsubseteq \mathsf{gfp}\, F$.*

Using this theorem, we can verify in a relatively easy fashion that some element is an over–approximation of the least fixed point or an under–approximation of the greatest fixed point of a continuous mapping on a complete partial order. In the following, let $C = \mathtt{while}\ (B)\ \{C'\}$. In order to exploit Park's Lemma for enumerating bounds on covariances for this while–loop, recall

$$\mathsf{Cov}_{\llbracket C \rrbracket(\sigma)}\,(f,\,g)\ =\ \mathsf{E}_{\llbracket C \rrbracket(\sigma)}\,(f \cdot g) - \mathsf{E}_{\llbracket C \rrbracket(\sigma)}\,(f) \cdot \mathsf{E}_{\llbracket C \rrbracket(\sigma)}\,(g)$$

$$=\ \frac{\mathsf{wp}\,[C]\,(f \cdot g)\,(\sigma)}{\mathsf{wlp}\,[C]\,(1)\,(\sigma)} - \frac{\mathsf{wp}\,[C]\,(f)\,(\sigma) \cdot \mathsf{wp}\,[C]\,(g)\,(\sigma)}{\mathsf{wlp}\,[C]\,(1)\,(\sigma)^2}\ .$$

By inspection of the last line, we can see that for obtaining an over–approximation of $\mathsf{Cov}_{[\![C]\!](\sigma)}(f, g)$, it suffices to over–approximate $\mathsf{wp}[C'](f \cdot g)(\sigma)/\mathsf{wlp}[C'](1)(\sigma)$, which can be done by over–approximating $\mathsf{wp}[C'](f \cdot g)(\sigma)$ and under–approximating $\mathsf{wlp}[C'](1)(\sigma)$. Since wp (wlp) of a loop is defined in terms of a least (greatest) fixed point, we can apply Park's Lemma for over–approximating this fraction. This leads us to the following proof rule:

Theorem 5 (Invariant–Aided Over–Approximation of Covariances).
Let $C = \mathtt{while}\ (B)\ \{C'\}$, $\sigma \in \mathbb{S}$, $f, g \in \mathbb{E}$, $F_h(X) = [\neg B] \cdot h + [B] \cdot \mathsf{wp}[C'](X)$, for any $h \in \mathbb{E}$, and $G(Y) = [\neg B] + [B] \cdot \mathsf{wlp}[C'](Y)$. Furthermore, let $\widehat{X} \in \mathbb{E}$ and $\widehat{Y} \in \mathbb{E}_{\leq 1}$, such that $F_{f \cdot g}(\widehat{X}) \preceq \widehat{X}$, $\widehat{Y} \preceq G(\widehat{Y})$, and $\widehat{Y}(\sigma) > 0$. Then for all $k \in \mathbb{N}$ it holds that[8]

$$\mathsf{Cov}_{[\![C]\!](\sigma)}(f, g) \ \leq\ \frac{\widehat{X}(\sigma)}{\widehat{Y}(\sigma)} - \frac{F_f^k(\mathbf{0})(\sigma) \cdot F_g^k(\mathbf{0})(\sigma)}{G^k(\mathbf{1})(\sigma)^2} .$$

By this method we can computably enumerate upper bounds for covariances once appropriate invariants are found. The catch is that if we choose the invariants, such that $F_{f \cdot g}(\widehat{X})(\sigma) < \widehat{X}(\sigma)$ or $\widehat{Y}(\sigma) < G(\widehat{Y})(\sigma)$, then the enumeration will *not* get arbitrarily close to the actual covariance. Note, however, that our method is complete since we could have chosen $\widehat{X} = \mathsf{lfp}\ F_{f \cdot g}$ and $\widehat{Y} = \mathsf{gfp}\ G$:

Corollary 1 (Completeness of Theorem 5). *Let $C = \mathtt{while}\ (B)\ \{C'\}$, $\sigma \in \mathbb{S}$, $f, g \in \mathbb{E}$. Then there exist $\widehat{X} \in \mathbb{E}$ and $\widehat{Y} \in \mathbb{E}_{\leq 1}$, such that*

$$\inf_{k \in \mathbb{N}} \frac{\widehat{X}(\sigma)}{\widehat{Y}(\sigma)} - \frac{F_f^k(\mathbf{0})(\sigma) \cdot F_g^k(\mathbf{0})(\sigma)}{G^k(\mathbf{1})(\sigma)^2} \ =\ \mathsf{Cov}_{[\![C]\!](\sigma)}(f, g) .$$

By considerations analogous to the ones above, we can formulate dual results for lower bounds. For details, see [8].

Example 2 (Application of Theorem 5). Reconsider the loop from Example 1. For reasoning about the variance of x, we pick the invariants

$$\widehat{X} \ =\ [c \neq 0] \cdot x^2 + [c = 1] \cdot ([x \text{ is even}] \cdot 1/27 \left(9x^2 + 30x + 41\right)$$
$$+ [x \text{ is odd}] \cdot 2/27 \left(9x^2 + 12x + 20\right)), \quad \text{and}$$
$$\widehat{Y} \ =\ [c \neq 0] + [c = 1] \cdot ([x \text{ is even}] \cdot 1/3 + [x \text{ is odd}] \cdot 2/3) ,$$

which satisfy the preconditions of Theorem 5. If we enter the loop in a state σ with $\sigma(c) = 1$ and $\sigma(x) = 0$, we have $\widehat{X}(\sigma)/\widehat{Y}(\sigma) = 41/9$ which is our first upper bound. We can now enumerate further bounds by doing fixed point iteration on $F_x(X) = [c \neq 1] \cdot x + [c = 1] \cdot \mathsf{wp}[\text{loop body}](X) = [c \neq 1] \cdot$

---

[8] Here $F_h^k(X)$ stands for k–fold application of F_h to X.

$x + [c = 1] \cdot \frac{1}{2}([x \text{ is odd}] \cdot X[c/0] + X[x/x+1])$ and $G(Y) = [c \neq 1] + [c = 1] \cdot \text{wlp}\,[\text{loop body}]\,(Y) = [c \neq 1] + [c = 1] \cdot \frac{1}{2}([x \text{ is odd}] \cdot Y[c/0] + Y[x/x+1])$:

$$\frac{41}{9} - \frac{F_x^1(0)(\sigma)^2}{G^1(1)(\sigma)^2} = \frac{41}{9} - \frac{F_x^2(0)(\sigma)^2}{G^2(1)(\sigma)^2} = \frac{41}{9}, \qquad \frac{41}{9} - \frac{F_x^3(0)(\sigma)^2}{G^3(1)(\sigma)^2} = \frac{37}{9}, \qquad \cdots$$

Finally, this sequence converges to $41/9 - 25/9 = 16/9$ as the variance of x. $\triangle$

5   Reasoning About Run–Time Variances

In addition to the (co)variance of outcomes we are interested in the variance of the program's *run–time*. Intuitively, the run–time of a program corresponds to its number of executed operations, where each operation is weighted according to some run–time model. For simplicity, our run–time model assumes skip, guard evaluations and assignments to consume one unit of time. Other statements are assumed to consume no time at all. More elaborated run–time models, e.g. in which the run–time of assignments depends on the size of a given expression, are possible design choices that can easily be integrated in our formalization.

We describe the run–time variance in terms of an operational model Markov Chain (MC) with rewards. The model is similar to the ones studied in [6,9], but additionally keeps track of the run–time in a dedicated variable τ which is *not accessible by the program*, but may occur in expectations.

Definition 8 (Run–Time Expectations). *Let* $\mathbb{S}_\tau = \{\sigma \mid \mathbb{V} \uplus \{\tau\} \to \mathbb{Q}\}$*. The set of run–time expectations is then defined as* $\mathbb{E}_\tau = \{t \mid t : \mathbb{S}_\tau \to \mathbb{R}_{\geq 0}^\infty\}$*.*

A corresponding wp–style calculus to reason about expected run–times and variances of probabilistic programs is presented afterwards.

We first briefly recall some necessary notions about MCs and refer to [1, Chap. 10] for a comprehensive introduction. A *Markov Chain* is a tuple $\mathcal{M} = (\mathcal{S}, \mathbf{P}, s_I, rew)$, where $\mathcal{S}$ is a countable set of *states*, $s_I \in \mathcal{S}$ is the initial state, $\mathbf{P} : \mathcal{S} \times \mathcal{S} \to [0,1]$ is the *transition probability function* such that for each state $s \in \mathcal{S}$, $\sum_{s' \in \mathcal{S}} \mathbf{P}(s, s') \in \{0,1\}$, and $rew : \mathcal{S} \to \mathbb{R}_{\geq 0}$ is a *reward function*. Instead of $\mathbf{P}(s, s') = p$, we often write $s \xrightarrow{p} s'$. A *path* in $\mathcal{M}$ is a finite or infinite sequence $\pi = s_0 s_1 \ldots$ such that $s_i \in S$ and $\mathbf{P}(s_i, s_{i+1}) > 0$ for each $i \geq 0$ (where we tacitly assume $\mathbf{P}(s_i, s_{i+1}) = 0$ if π is a finite path of length n and $i \geq n$). The *cumulative reward* and the probability of a finite path $\hat{\pi} = s_0 \ldots s_n$ are given by $rew(\hat{\pi}) = \sum_{k=0}^{n-1} rew(s_k)$ and $\text{Pr}^{\mathcal{M}}\{\hat{\pi}\} = \prod_{k=0}^{n-1} \mathbf{P}(s_k, s_{k+1})$. These notions are lifted to infinite paths by the standard cylinder set construction (cf. [1]).

Given a set of target states $T \subseteq \mathcal{S}$, $\Diamond T$ denotes the set of all paths in $\mathcal{M}$ reaching a state in T from initial state s_I. Analogously, all paths starting in s_I that never reach a state in T are denoted by $\neg\Diamond T$. The *expected reward* that $\mathcal{M}$ eventually reaches T from a state $s \in \mathcal{S}$ is defined as follows:

$$\text{ExpRew}^{\mathcal{M}}(\Diamond T) = \begin{cases} \sum_{\pi \in \Diamond T} \text{Pr}^{\mathcal{M}}\{\pi\} \cdot rew(\pi) & \text{if } \sum_{\pi \in \Diamond T} \text{Pr}^{\mathcal{M}}\{\pi\} = 1 \\ \infty & \text{if } \sum_{\pi \in \Diamond T} \text{Pr}^{\mathcal{M}}\{\pi\} < 1. \end{cases}$$

Moreover, the *conditional expected reward* of $\mathcal{M}$ reaching T from s under the condition that a set of undesired states $U \subseteq \mathcal{S}$ is never reached is given by[9]

$$\mathsf{CExpRew}^{\mathcal{M}}(\Diamond T \mid \neg \Diamond U) \;=\; \frac{\mathsf{ExpRew}^{\mathcal{M}}(\Diamond T \cap \neg \Diamond U)}{\mathsf{Pr}^{\mathcal{M}}\{\neg \Diamond U\}}.$$

We are now in a position to define an operational model for our probabilistic programming language $\mathbb{P}$. Let $\downarrow$ and $\mathcal{f}$ be two special symbols denoting successful termination of a program and failure of an observation, respectively.

Definition 9 (The Operational MC of a $\mathbb{P}$-Program). *Given a program $C \in \mathbb{P}$, an initial program state $\sigma_0 \in \mathbb{S}_\tau$ and a post–run–time $t \in \mathbb{E}$, the according MC is given by $\mathcal{M}_{\sigma_0}^t[C] = (\mathcal{S}, \mathbf{P}, s_I, rew)$, where*

- $\mathcal{S} = ((\mathbb{P} \cup \{\downarrow\} \cup \{\downarrow; C \mid C \in \mathbb{P}\}) \times \mathbb{S}_\tau) \cup \{\langle \mathrm{sink} \rangle, \langle \mathcal{f} \rangle\}$,
- *the transition probability function $\mathbf{P}$ is given by the rules in Fig. 2,*
- $s_I = \langle C, \sigma_0 \rangle$, *and*
- $rew : \mathcal{S} \to \mathbb{R}_{\geq 0}$ *is the reward function defined by $rew(s) = t(\sigma)$ if $s = \langle \downarrow, \sigma \rangle$ for some $\sigma \in \mathbb{S}_\tau$ and $rew(s) = 0$, otherwise.*

In this construction, $\sigma_0(\tau)$ represents the *post–execution time* of a program, i.e. the run–time that is added after a program finishes its execution. Hence, τ precisely captures the run–time of a program if $\sigma_0(\tau) = 0$. The rules presented in Fig. 2 defining the transition probability function are mostly self–explanatory. Since we assume guard evaluations, probabilistic choices, assignments and the statement `skip` to consume one unit of time. Hence, τ is incremented accordingly for each of these statements and remains untouched otherwise.

Figure 3 sketches the structure of the operational MC $\mathcal{M}_\sigma^t[C]$. Here, clouds represent a set of states and squiggly arrows indicate that a set of states is reachable by one or more paths. Each run either terminates successfully (i.e. it visits some state $\langle \downarrow, \sigma' \rangle$), or violates an observation (i.e. it visits $\langle \mathcal{f} \rangle$), or diverges. In the first two cases each run eventually ends up in the $\langle \mathrm{sink} \rangle$ state. Note that states of the form $\langle \downarrow, \sigma' \rangle$ are the only ones that may have a positive reward. Furthermore, each of the auxiliary states of the form $\langle \downarrow, \sigma' \rangle$, $\langle \mathcal{f} \rangle$ and $\langle \mathrm{sink} \rangle$ is needed to properly deal with `diverge`, `halt` and `observe` B.

Since τ precisely captures the run–time of a program if τ is initially set to 0, the *expected run–time* of executing $C \in \mathbb{P}$ on input $\sigma \in \mathbb{S}_\tau$ with $\sigma(\tau) = 0$ is given by the conditional expected reward of $\mathcal{M}_\sigma^\tau[C]$ reaching $\langle \mathrm{sink} \rangle$, given that no observation fails, i.e. $\mathsf{E}_{[\![C]\!](\sigma)}(\tau) = \mathsf{CExpRew}^{\mathcal{M}_\sigma^\tau[C]}(\Diamond \langle \mathrm{sink} \rangle \mid \neg \Diamond \langle \mathcal{f} \rangle)$. Then, in compliance with Definition 4, the *run–time variance* $\mathsf{RTVar}_{[\![C]\!](\sigma)}$ of $C \in \mathbb{P}$ in state $\sigma \in \mathbb{S}_\tau$ with $\sigma(\tau) = 0$ is given by $\mathsf{E}_{[\![C]\!](\sigma)}(\tau^2) - \left(\mathsf{E}_{[\![C]\!](\sigma)}(\tau) \right)^2$ which is

$$\mathsf{CExpRew}^{\mathcal{M}_\sigma^{\tau^2}[C]}(\Diamond \langle \mathrm{sink} \rangle \mid \neg \Diamond \langle \mathcal{f} \rangle) - \left(\mathsf{CExpRew}^{\mathcal{M}_\sigma^\tau[C]}(\Diamond \langle \mathrm{sink} \rangle \mid \neg \Diamond \langle \mathcal{f} \rangle) \right)^2 .$$

In the following we provide a corresponding wp–style calculus to reason about expected run–times and run–time variances of probabilistic programs. A formal

---

[9] Again, we stick to the convention that $\frac{0}{0} = 0$.

$$\frac{}{\langle \downarrow, \sigma \rangle \xrightarrow{1} \langle \text{sink} \rangle} \text{[terminated]} \qquad \frac{}{\langle \text{sink} \rangle \xrightarrow{1} \langle \text{sink} \rangle} \text{[sink]}$$

$$\frac{}{\langle \text{empty}, \sigma \rangle \xrightarrow{1} \langle \downarrow, \sigma \rangle} \text{[empty]} \qquad \frac{}{\langle \text{skip}, \sigma \rangle \xrightarrow{1} \langle \downarrow, \sigma[\tau/\tau+1] \rangle} \text{[skip]}$$

$$\frac{}{\langle \text{halt}, \sigma \rangle \xrightarrow{1} \langle \text{sink} \rangle} \text{[halt]} \qquad \frac{}{\langle x := E, \sigma \rangle \xrightarrow{1} \langle \downarrow, \sigma[x/E, \tau/\tau+1] \rangle} \text{[assgn]}$$

$$\frac{\langle C_1, \sigma \rangle \xrightarrow{p} \langle C_1', \sigma' \rangle \quad 0 < p \le 1}{\langle C_1; C_2, \sigma \rangle \xrightarrow{p} \langle C_1'; C_2, \sigma' \rangle} \text{[seq-1]} \qquad \frac{}{\langle \downarrow; C_2, \sigma \rangle \xrightarrow{1} \langle C_2, \sigma \rangle} \text{[seq-2]}$$

$$\frac{}{\langle \{C_1\} [p] \{C_2\}, \sigma \rangle \xrightarrow{p} \langle C_1, \sigma[\tau/\tau+1] \rangle} \text{[pc-1]}$$

$$\frac{}{\langle \{C_1\} [p] \{C_2\}, \sigma \rangle \xrightarrow{1-p} \langle C_2, \sigma[\tau/\tau+1] \rangle} \text{[pc-2]}$$

$$\frac{[B](\sigma) = 1}{\langle \text{if } (B) \{C_1\} \text{ else } \{C_2\}, \sigma \rangle \xrightarrow{1} \langle C_1, \sigma[\tau/\tau+1] \rangle} \text{[if-true]}$$

$$\frac{[B](\sigma) = 0}{\langle \text{if } (B) \{C_1\} \text{ else } \{C_2\}, \sigma \rangle \xrightarrow{1} \langle C_2, \sigma[\tau/\tau+1] \rangle} \text{[if-false]}$$

$$\frac{}{\langle \text{while } (B) \{C\}, \sigma \rangle \xrightarrow{1} \langle \text{if } (B) \{C; \text{ while } (B) \{C\}\} \text{ else } \{\text{empty}\}, \sigma \rangle} \text{[while]}$$

$$\frac{}{\langle \text{diverge}, \sigma \rangle \xrightarrow{1} \langle \text{diverge}, \sigma \rangle} \text{[diverge]}$$

$$\frac{[B](\sigma) = 1}{\langle \text{observe } B, \sigma \rangle \xrightarrow{1} \langle \downarrow, \sigma[\tau/\tau+1] \rangle} \text{[observe-true]}$$

$$\frac{[B](\sigma) = 0}{\langle \text{observe } B, \sigma \rangle \xrightarrow{1} \langle \text{E} \rangle} \text{[observe-false]} \qquad \frac{}{\langle \text{E} \rangle \xrightarrow{1} \langle \text{sink} \rangle} \text{[observe-failed]}$$

Fig. 2. Rules for defining the transition probability function of the MC of a $\mathbb{P}$–program.

definition of the *run–time transformer* rt : $\mathbb{P} \to (\mathbb{E}_\tau \to \mathbb{E}_\tau)$ is provided in Table 1 (rightmost column). Intuitively, it behaves like wp except that a *dedicated run–time variable* τ is updated accordingly for each program statement that consumes time. In [9], a transformer for expected run–times without the need for an additional variable τ is studied. However, this approach fails when reasoning about run–time variances since it fails to capture expected squared run–times. The run–time transformer rt precisely captures the notion of expected run–time of our operational model.

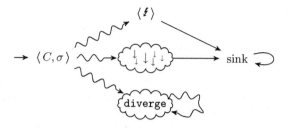

Fig. 3. Schematic depiction of the structure of the operational MC $\mathcal{M}_\sigma^t [C]$.

Theorem 6 (Operational–Denotational Correspondence). *Let $C \in \mathbb{P}$, $t \in \mathbb{E}_\tau$, and $\sigma \in \mathbb{S}_\tau$. Then*

$$\mathsf{CExpRew}^{\mathcal{M}_\sigma^t [C]} (\Diamond \langle \mathrm{sink} \rangle \mid \neg \Diamond \langle \mathit{f} \rangle) \;=\; \frac{\mathsf{rt}\,[C]\,(t)\,(\sigma)}{\mathsf{wlp}\,[C]\,(1)\,(\sigma)}.$$

As a result of Theorem 6 we immediately obtain a formal definition of the run–time variance of probabilistic programs in terms of rt and wlp. Formally, the *run–time variance* of $C \in \mathbb{P}$ in state $\sigma \in \mathbb{S}_\tau$ with $\sigma(\tau) = 0$ is given by

$$
\begin{aligned}
\mathsf{RTVar}_{[C]}(\sigma) \;&=\; \mathsf{CExpRew}^{\mathcal{M}_\sigma^{\tau^2} [C]} (\Diamond \langle \mathrm{sink} \rangle \mid \neg \Diamond \langle \mathit{f} \rangle) \\
&\quad - \left(\mathsf{CExpRew}^{\mathcal{M}_\sigma^{\tau} [C]} (\Diamond \langle \mathrm{sink} \rangle \mid \neg \Diamond \langle \mathit{f} \rangle) \right)^2 \\
&=\; \frac{\mathsf{rt}\,[C]\,(\tau^2)\,(\sigma)}{\mathsf{wlp}\,[C]\,(1)\,(\sigma)} - \frac{(\mathsf{rt}\,[C]\,(\tau)\,(\sigma))^2}{(\mathsf{wlp}\,[C]\,(1)\,(\sigma))^2}.
\end{aligned}
$$

Since rt is continuous (cf. [8] for a formal proof), the invariant–aided approach based on Park's Lemma (Theorem 4) presented in Sect. 4 is applicable to approximate run–time variances as well. We present the result for approximating upper bounds only. The dual result for lower bounds is obtained analogously.

Theorem 7 (Invariant–Aided Over–Approximation of Run–Time Variances). *Let $C =$ while (B) $\{C'\}$ and $\sigma \in \mathbb{S}_\tau$ with $\sigma(\tau) = 0$. Moreover, let $F_h(X) = [\neg B] \cdot h + [B] \cdot \mathsf{rt}\,[C']\,(X)$, and $G(Y) = [\neg B] + [B] \cdot \mathsf{wlp}\,[C']\,(Y)$. Furthermore, let $\widehat{X} \in \mathbb{E}_\tau$ and $\widehat{Y} \in \mathbb{E}_{\leq 1}$, such that $F_{\tau^2}(\widehat{X}) \preceq \widehat{X}, \widehat{Y} \preceq G(\widehat{Y})$, and $\widehat{Y}(\sigma) > 0$. Then for each $k \in \mathbb{N}$, it holds*

$$\mathsf{RTVar}_{[C]}(\sigma) \;\leq\; \frac{\widehat{X}(\sigma)}{\widehat{Y}(\sigma)} - \left(\frac{F_\tau^k(0)(\sigma)}{G^k(1)(\sigma)} \right)^2.$$

The proof of Theorem 7 is analogous to the proof of Theorem 5. Again, since it is always possible to choose $\widehat{X} = \mathsf{lfp}\,F_{\tau^2}$ and $\widehat{Y} = \mathsf{gfp}\,G$, Theorem 7 is complete, i.e. there exist $\widehat{X} \in \mathbb{E}_\tau$ and $\widehat{Y} \in \mathbb{E}_{\leq 1}$ such that

$$\inf_{k \in \mathbb{N}} \frac{\widehat{X}(\sigma)}{\widehat{Y}(\sigma)} - \left(\frac{F_\tau^k(0)(\sigma)}{G^k(1)(\sigma)} \right)^2 \;=\; \mathsf{RTVar}_{[C]}(\sigma).$$

6   Conclusion

We have studied the computational hardness of obtaining both upper and lower bounds on (co)variance of outcomes and established that this is Σ_2^0–complete. Thus neither upper nor lower bounds are computably enumerable. Furthermore, we have established that deciding whether the (co)variance equals a given rational and deciding whether the covariance is infinite is Π_2^0–complete.

In the second part of the paper, we continued by presenting a sound and complete invariant–aided approach which allows to computably enumerate upper and lower bounds on (co)variances of while–loops, once appropriate loop–invariants are found. Finally, we have shown how this approach can be extended to reason about the variance of run–times.

References

1. Baier, C., Katoen, J.P.: Principles of Model Checking. MIT Press, Cambridge (2008)
2. Davis, M.D.: Computability, Complexity, and Languages: Fundamentals of Theoretical Computer Science. Academic Press, Cambridge (1994)
3. Dijkstra, E.W.: A Discipline of Programming. Prentice-Hall, Englewood Cliffs (1976)
4. Fioriti, L.M.F., Hermanns, H.: Probabilistic termination: soundness, completeness, and compositionality. In: POPL 2015, pp. 489–501. ACM (2015)
5. Gordon, A.D., Henzinger, T.A., Nori, A.V., Rajamani, S.K.: Probabilistic programming. In: Future of Software Engineering (FOSE), pp. 167–181. ACM (2014)
6. Jansen, N., Kaminski, B.L., Katoen, J.P., Olmedo, F., Gretz, F., McIver, A.: Conditioning in probabilistic programming. ENTCS **319**, 199–216 (2015)
7. Kaminski, B.L., Katoen, J.-P.: On the hardness of almost–sure termination. In: Italiano, G.F., Pighizzini, G., Sannella, D.T. (eds.) MFCS 2015. LNCS, vol. 9234, pp. 307–318. Springer, Heidelberg (2015)
8. Kaminski, B.L., Katoen, J.P., Christoph, M.: Inferring covariances for probabilistic programs. ArXiv e-prints, June 2016
9. Kaminski, B.L., Katoen, J.-P., Matheja, C., Olmedo, F.: Weakest precondition reasoning for expected run–times of probabilistic programs. In: Thiemann, P. (ed.) ESOP 2016. LNCS, vol. 9632, pp. 364–389. Springer, Heidelberg (2016). doi:10. 1007/978-3-662-49498-1_15
10. Kleene, S.C.: Recursive predicates and quantifiers. Trans. AMS **53**(1), 41–73 (1943)
11. McIver, A., Morgan, C.: Abstraction, Refinement and Proof for Probabilistic Systems. Springer, Heidelberg (2004)
12. Odifreddi, P.: Classical Recursion Theory: The Theory of Functions and Sets of Natural Numbers. Elsevier, Amsterdam (1992)
13. Odifreddi, P.: Classical Recursion Theory, vol. II. Elsevier, Amsterdam (1999)
14. Post, E.L.: Recursively enumerable sets of positive integers and their decision problems. Bull. AMS **50**(5), 284–316 (1944)
15. Wechler, W.: Universal Algebra for Computer Scientists. EATCS Monographs on Theoretical Computer Science, vol. 25. Springer, Heidelberg (1992)

Should Network Calculus Relocate?
An Assessment of Current Algebraic
and Optimization-Based Analyses

Steffen Bondorf[(⊠)] and Jens B. Schmitt

Distributed Computer Systems (DISCO) Lab,
University of Kaiserslautern, Kaiserslautern, Germany
{bondorf,jschmitt}@cs.uni-kl.de

Abstract. Network calculus (NC) offers a framework for worst-case analysis of queueing networks. It enables to derive deterministic bounds on flow delay and server backlog. The continuous evolution of NC led to a set of different analyses. In fact, it even resulted in two entirely different branches of the methodology. Both start with a common network description based on bounding functions on flow arrivals and forwarding service. Anything that follows, i.e., the actual analysis leading to a worst-case performance bound, vastly differs. For long, there was only the algebraic NC, the formalism created as a system theory for communication networks. It matured and eventually seemed to have reached its limits regarding the accuracy of bounds. The problems preventing it from attaining tight bounds in feed-forward networks were overcome with optimization-based analysis. However, this approach was proven NP-hard without an efficient analysis algorithm known for it. Therefore, it was proposed to confine to a less complex optimization-based analysis instead. Like algebraic NC analyses, it derives tight bounds for some networks and valid bounds with varying accuracy for other networks. In this paper, we investigate the consequences of this tradeoff and identify a new and crucial analysis principle that allows us to compare both NC branches more comprehensively than simply ranking delay bounds.

Keywords: Network calculus · Algebraic analysis · Optimization

1 Introduction

Network calculus (NC) is a methodology for the worst-case analysis of queueing systems. It provides different analysis procedures to derive deterministic bounds on buffer requirements and flow delays. Thus, it can be employed in the verification of real-time systems. In fact, *algebraic* network calculus has seen application in avionics [8,11,12]. E.g., the Airbus A380's backbone AFDX network (Avionics Full-Duplex Ethernet) has been certified using network calculus. Additionally, tool support is available for algebraic network calculus: open-source [2], commercial [9] as well as an internal tool of a company manufacturing AFDX

© Springer International Publishing Switzerland 2016
G. Agha and B. Van Houdt (Eds.): QEST 2016, LNCS 9826, pp. 207–223, 2016.
DOI: 10.1007/978-3-319-43425-4_15

switches (Rockwell Collins: ConfGen [13]) and others (Hirschmann Automation [17], SIEMENS [14]). Moreover, algebraic NC has continuously seen improvements, e.g., w.r.t. the procedure of a feed-forward analysis and the computational effort [3,4], as well as attainable features [3] and accuracy of results [4,5].

All these developments took place despite the introduction of an entirely different analysis approach based on the NC system description: optimization-based network calculus. Initially developed by [19] to prove a property of algebraic NC analysis that can inhibit deriving tight bounds, it was further developed into an alternative feed-forward analysis, the LP analysis, in [7]. The LP analysis is able to derive tight bounds, yet, it is also NP-hard with no efficient algorithm known for it. As this insight obviously prevents network calculus to relocate to optimization, an accurate (not necessarily tight) ULP analysis was proposed based on the new optimization approach. It is of course not NP-hard, however, neither was it benchmarked comprehensively against the current algebraic NC analyses. Therefore, the question of switching the fundamental analysis approach of network calculus has not been answered yet.

In this paper, we derive a new analysis principle for feed-forward networks – similar to the existing principles – that pinpoints the problem of algebraic NC, a problem theoretically solved with the LP analysis of [7]. With this principle at hand, we can provide an in-depth NC analysis evaluation. This treatment of NC also allows us to reveal the weaknesses of the ULP. We comprehensively compare it to current algebraic network calculus analyses, foremost the PMOO analysis and its extensions, and finally provide an evaluation that gives insight on the gap between both NC branches. Our work helps to better assess the severity of the shortfall of current algebraic NC and thus tool implementations. Moreover, our insights provide a guide to future work w.r.t. improving algebraic NC accuracy.

The remainder of this paper is structured as follows: Sect. 2 presents some background on NC: The system description common to both branches as well as the algebraic operations. In Sect. 3, we examine the two NC branches regarding their approaches, strengths and weaknesses when analyzing a network. This allows us to assess accuracy of the currently employed four NC analyses (SFA, PMOO, LP, and ULP) in Sect. 4. Section 5 concludes the paper.

2   Network Calculus Background

2.1   The System Description

Data Arrivals and Forwarding Service. Flows are characterized by functions cumulatively counting their data. They belong to the set $\mathcal{F}_0$ of nonnegative, wide-sense increasing functions:

$$\mathcal{F}_0 = \left\{ f : \mathbb{R} \to \mathbb{R}_\infty^+ \mid f(0) = 0, \ \forall s \leq t \ : \ f(s) \leq f(t) \right\}, \ \mathbb{R}_\infty^+ := [0, +\infty) \cup \{+\infty\}.$$

We are particularly interested in the functions $A(t)$ and $A'(t)$ cumulatively counting a flow's data put into a server s and put out from s, both up until time t. These functions allow for a straight-forward derivation of flow delays.

Definition 1. *(Flow Delay)* Assume a flow with input A crosses a server s and results in the output A'. The (virtual) delay for a data unit arriving at time t is

$$D(t) = \inf \{\tau \geq 0 \mid A(t) \leq A'(t + \tau)\}.$$

Note, that the order of data within the flow needs to be retained for the (virtual) delay calculation [18].

Network calculus operates in the interval time domain, i.e., its functions of $\mathcal{F}_0$ bound the maximum data arrivals of a flow during any duration of length d.

Definition 2. *(Arrival Curve)* Given a flow with input A, a function $\alpha \in \mathcal{F}_0$ is an arrival curve for A iff

$$\forall t \, \forall d \; 0 \leq d \leq t \; : \; A(t) - A(t - d) \leq \alpha(d).$$

For example, periodic traffic with a maximum packet size b and a maximum arrival rate r can be bounded by token-bucket curves $\mathcal{F}_{\text{TB}} = \{\gamma_{r,b} \mid \gamma_{r,b}(0) = 0,$ $\forall d > 0 \; : \; \gamma_{r,b}(d) = b + r \cdot d\} \subseteq \mathcal{F}_0$.

Scheduling and buffering leading to the output function $A'(t)$ depend on a server's forwarding service. It is lower bounded in interval time as well.

Definition 3. *(Service Curve)* If the service provided by a server s for a given input A results in an output A', then s offers a service curve $\beta \in \mathcal{F}_0$ iff

$$\forall t \; : \; A'(t) \geq \inf_{0 \leq d \leq t} \{A(t - d) + \beta(d)\}.$$

For instance, service offered by Ethernet connections can be described by rate-latency curves $\mathcal{F}_{\text{RL}} = \{\beta_{R,T} \mid \beta_{R,T}(d) = \max\{0, R \cdot (d - T)\} \subseteq \mathcal{F}_0$.

A number of servers fulfill a stricter definition of service curves that guarantees a higher output during periods of queued data (backlogged periods).

Definition 4. *(Strict Service Curve)* Let $\beta \in \mathcal{F}_0$. Server s offers a strict service curve β to a flow iff, during any backlogged period of duration d, the output of the flow is at least equal to $\beta(d)$.

2.2 Algebraic Network Calculus

Network calculus was cast in a (min, +)-algebraic framework in [10,15]. We will first depict the basic operations and then present their combination for flow analysis.

Table 1. Network calculus notation for flows, arrivals and service.

Quantifier	Definition
foi	Flow of interest, the flow under analysis
$\langle s_x, \ldots, s_y \rangle$	Tandem of consecutive servers s_x to s_y
$P(f)$	Path of flow f (a tandem of servers)
α^f, α_s^f	Arrival curve of flow f, arrival bound at server s
β_s	Service curve of server s
$\beta_s^{\text{l.o.} f}, \beta_{\langle s_x, \ldots, s_y \rangle}^{\text{l.o.} f}$	Left-over service curve for f at server s, on tandem $\langle s_x, \ldots, s_y \rangle$

(min,+)-Operations. The following operations allow to manipulate arrival and service curves while retaining their worst-case semantic.

Definition 5. *((min,+)-Operations)* The (min,+)-aggregation, -convolution and -deconvolution of two functions $f, g \in \mathcal{F}_0$ are defined as

$$aggregation: (f + g)(t) = f(t) + g(t),$$
$$convolution: (f \otimes g)(t) = \inf_{0 \leq s \leq t} \{f(t - s) + g(s)\},$$
$$deconvolution: (f \oslash g)(t) = \sup_{u \geq 0} \{f(t + u) - g(u)\}.$$

The system description's service curve definition then translates to $A' \geq A \otimes \beta$, the arrival curve definition to $A \otimes \alpha \geq A$, and performance characteristics can be bounded using the deconvolution $\alpha \oslash \beta$:

Theorem 1. (Performance Bounds) *Consider a server s that offers a service curve β. Assume a flow f with arrival curve α traverses the server. Then we obtain the following performance bounds for f:*

$$delay\ bound: \forall t \in \mathbb{R}^+: \ D(t) \leq \inf \{d \geq 0 \,|\, (\alpha \oslash \beta)(-d) \leq 0\},$$
$$output\ bound: \forall d \in \mathbb{R}^+: \ \alpha'(d) = (\alpha \oslash \beta)(d),$$

where the delay bound holds independent of t and α' is an arrival curve for A'.

Analyzing an entire flow with cross-traffic on its path is enabled by the following theorems. Table 1 depicts the notation we use for the network analysis.

Theorem 2. (Concatenation of Servers) *Consider a single flow f crossing a tandem of servers $s_1, \ldots, s_n$ where each s_i offers a service curve β_{s_i}. The overall service curve for f is their concatenation by convolution*

$$\beta_{s_1} \otimes \ldots \otimes \beta_{s_n} = \bigotimes_{i=1}^{n} \beta_{s_i}.$$

Theorem 3. (Left-Over Service Curve) *Consider a server s that offers a strict service curve β and that serves two input flows, f_1 and f_2 with arrival curves*

α^{f_1} and α^{f_2}, respectively. The minimum service f_1 is guaranteed to receive is lower bounded by the so-called left-over service curve

$$\beta_s^{l.o.f_1} = \beta_s \ominus \alpha^{f_0},$$

with $(\beta \ominus \alpha)(d) := \sup_{0 \le u \le d}\{(\beta - \alpha)(u)\}$ denoting the non-decreasing upper closure of $(\beta - \alpha)(d)$.

3   Network Calculus Feed-Forward Analyses (FFA)

3.1   Algebraic Network Calculus Analysis

An algebraic network calculus FFA computes the end-to-end delay bound for a specific flow interest (foi). Conceptually, the analysis proceeds in two steps [2,3]:

1. The analysis abstracts from the feed-forward network to the analyzed flow's path (tandem of servers). This step is enabled by recursively backtracking flows, decomposing the network into tandems [2] along their paths and bounding output arrivals of cross-traffic with Theorem 1, the output bound. Then, arrival curves that bound the worst-case shape of cross-flows are known at the location of interference with the foi.
2. The foi's end-to-end delay bound in the feed-forward network can now be calculated with a less complex *tandem analysis*. The flow's end-to-end service curve is derived and the delay bound computed.

The second step of the algebraic feed-forward analysis procedure has seen much treatment in the literature. Effort focused on improving the ability to capture flow scheduling and cross-traffic multiplexing effects in a tandem analysis. This effort resulted in two basic principles for left-over service curve derivation of tandems that improve algebraic NC's accuracy:

The PBOO-Principle and the Separate Flow Analysis [15]: The SFA is a straight-forward, hop-by-hop application of Theorems 3 and 2: First subtract cross-traffic arrivals and then concatenate the left-over service curves. Deriving the delay bound with a single, end-to-end left-over service curve will consider the flow of interest's burst term only once. This principle is therefore called Pay Bursts Only Once (PBOO). However, for cross-flows present at multiple consecutive hops, bursts impact the derivation multiple times.

The PMOO-Principle and -Analysis [20]: The PMOO analysis provides an alternative derivation containing each burst term only once. Its left-over service curve derivation reverses the operations, i.e., it convolves the tandem of servers before subtracting cross-traffic. Due to this end-to-end approach for all flows on the analyzed tandem, the PMOO analysis was considered superior to SFA. Yet, [19] shows that the SFA can arbitrarily outperform a PMOO tandem analysis. Both algebraic analyses thus complement each other.

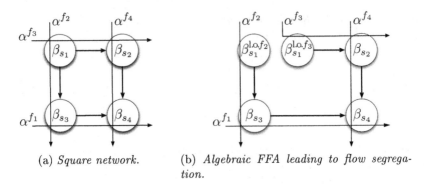

(a) *Square network.*   (b) *Algebraic FFA leading to flow segregation.*

Fig. 1. The square network and the result of the algebraic FFA applied to it if $f_1(\alpha^{f_1})$ is the analyzed flow of interest.

3.2   A New Principle for Feed-Forward Analysis: PSOO

Artifacts of Algebraic FFA. In [4], the expression cross-flow segregation was coined for the situation where a network calculus analysis considers cross-flows to be mutually interfering in the worst case. It was identified in a procedure for the FFA step 1 that derives an individual arrival bound for each cross-flow [6]. While this situation could be avoided easily, we found further problems in the FFA, so-called *analysis artifacts*, enforcing segregation nonetheless.

Artifact 1: The FFA Procedure. During the backtracking in step 1, several tandem analyses are executed such that the FFA can re-compose tandem-local results. This imposes a fundamental problem: Each tandem analysis is an independent instance of SFA or PMOO analysis, operating on its own worst-case assumptions that retain the overall FFA's worst-case modeling. The worst case is, in fact, a segregation of flows. Problems arise if flows share a server, demultiplex and rejoin again – either each other directly or indirectly via another flow. For an example see Fig. 1: The flows at server s_1 (service β_{s_1}) are treated independently of each other, i.e., they assume worst-case mutual interference that is not attainable in a real network. This independence is expressed by the segregation of service β_{s_1} into $\beta_{s_1}^{l.o.f_2}$ and $\beta_{s_1}^{l.o.f_3}$, both with a latency for the respective flow that is greater than β_{s_1}'s latency. Thus, they cannot sum up to the full service β_{s_1}.

Artifact 2: Interdependence Between FFA Steps. We also found the need for segregation in tandem networks – although this topology does not allow flows to demultiplexing, take different paths and rejoin again. This second artifact is illustrated in the non-nested tandem with cross-traffic arrival bounding in Fig. 2. Applying a PMOO FFA, i.e., using PMOO in the FFA step 2, requires knowledge about each cross-flow's arrival bounds individually. This can only be achieved by segregately executing the FFA step 1 for each cross-flow xf_1 and xf_2. This artifact thus leads to two independently computed left-over service

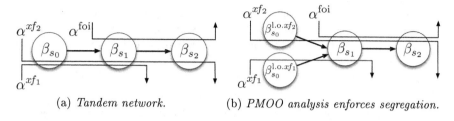

(a) *Tandem network.*   (b) *PMOO analysis enforces segregation.*

Fig. 2. A non-nested tandem and the result of the PMOO FFA applied to it.

curves $\beta_{s_0}^{\mathrm{l.o.}xf_1}$ and $\beta_{s_0}^{\mathrm{l.o.}xf_2}$ that do not add up to β_{s_0}, i.e., the overly pessimistic analysis does not consider usage of the entire forwarding resources.

Effect on the Analysis Result. The pessimism of independent left-over service curve derivations inevitable results in less accurate delay bounds. We illustrate this situation for the tandem network of Fig. 2. In the following, we compare the PMOO delay bound for f_1 delay bound with PMOO FFA (left side, Fig. 2b) against a derivation using the entire service offered by s_1 (right side, Fig. 2a):

$$\beta_{s_0}^{\mathrm{l.o.}xf_1} + \beta_{s_0}^{\mathrm{l.o.}xf_2} < \beta_{s_0}$$

$$\Rightarrow \underbrace{\left(\alpha^{xf_1} \oslash \beta_{s_0}^{\mathrm{l.o.}xf_1}\right)}_{=:\alpha_{s_1}^{xf_1}} + \underbrace{\left(\alpha^{xf_2} \oslash \beta_{s_0}^{\mathrm{l.o.}xf_2}\right)}_{=:\alpha_{s_1}^{xf_2}} > \underbrace{\left(\alpha^{xf_1} + \alpha^{xf_2}\right) \oslash \beta_{s_0}}_{=:\alpha_{s_1}^{[xf_1,xf_2]}}$$

$$\Rightarrow \beta_{s_1} \ominus \left(\alpha_{s_1}^{xf_1} + \alpha_{s_1}^{xf_2}\right) < \beta_{s_1} \ominus \alpha_{s_1}^{[xf_1,xf_2]}$$

$$\Rightarrow \beta_{\langle s_1,s_2\rangle}^{\mathrm{l.o.segr\,foi}} < \beta_{\langle s_1,s_2\rangle}^{\mathrm{l.o.foi}}$$

$$\Rightarrow D_{\mathrm{segr}}^{\mathrm{foi}} > D^{\mathrm{foi}}$$

As the segregated left-over service curves do not sum up to the original service curve, they cause larger output bounds for xf_1 and xf_2. This, in turn, results in a smaller left-over service curve for the foi at server s_1. Therefore, the end-to-end left-over service curve under flow segregation, $\beta_{\langle s_1,s_2\rangle}^{\mathrm{l.o.segr\,foi}}$, is smaller than the one without flow segregation, $\beta_{\langle s_1,s_2\rangle}^{\mathrm{l.o.foi}}$. This results in a larger delay bound.

All of the segregated flows consider each other with isolated, local worst-case assumptions that retain the global worst case. This leads to an overall situation assumed by the analysis that cannot be attained in a realistic system. We capture this problem in a novel principle to be strived for in a feed-forward network calculus analysis, formulating it similar to PBOO and PMOO (Sect. 3.1).

The Pay Segregation Only Once (PSOO) Principle: If the arrivals of two flows have to be bounded segregately in the feed-forward analysis and these flows both cross the same server before interfering with the flow of interest, then they should not be segregated in a way that imposes worst-case mutual interference assumptions on both. Segregation of cross-flows should only be paid for once by the ensemble of the two flows.

For instance, in the algebraic FFA equation, segregated flows should not have to consider each other fully in their respective arrival bounding. Although this leads to valid intermediate bounds on arrivals and left-over service, the according behavior is not attainable by a realistic system and thus the eventual performance bound cannot be tight.

Mitigation with by Aggregation. Having derived these artifacts of algebraic FFA, we can mitigate them in different ways. On the one hand, it is possible to prevent their occurrence by routing restrictions. However, adapting the network to be analyzed may not be justifiable. Thus, we strife for a different mitigation strategy: flow aggregation. Yet, aggregation is not universally practical in algebraic NC. Therefore, we depict the state-of-the-art optimization-based NC that does not suffer from these artifacts next. Then, we benchmark it against implementations of our mitigation strategy and evaluate accuracy loss in networks where we cannot prevent algebraic NC from violating the PSOO principle.

3.3   Optimization-Based Network Calculus Analysis and PSOO

An optimization-based FFA was proposed in [7]. It transforms the NC system description of Sect. 2.1 into a set of linear programs as follows:

1. Starting from the foi's sink server, flows and their cross-flows are recursively backtracked. For every link traversed backwards, the start of backlogged periods at the connected servers is related. This results in a partial order where there is no given order relation for servers on parallel paths.
2. Next, the partial order is extended to the set of all compatible total orders. In contrast to algebraic FFA, the backtracking result is not directly used to derive performance bounds. The extension enumerates all potential relations of backlogged periods on parallel paths to attain all potential entanglement in the network. Total orders of particular interest are those assuming an equal (start of) backlogged period for flows that later are demultiplexed. I.e., in Fig. 1's network, the flows f_2 and f_3 are related with a common start of the backlogged period at server s_1 in order to implement PSOO.
3. Based on the network calculus model, each total order is converted into one linear program. Strict service curves, arrival curves, non-decreasing functions, non-negativity and the flow constraint derive LP constraints.
4. The set of LPs represents all potential entanglements in the network, not only worst-case ones. Therefore, all linear programs must be solved in the final step. The maximum among their solutions is a valid worst-case delay for the foi.

The LP FFA was, however, shown to be NP-hard due to step 2. Therefore, the authors propose to confine to a less complex, accurate analysis by skipping the extension of the partial order. This analysis is known as the *unique* LP (ULP). The ULP is, of course, less constrained than any single linear program of the LP; the constraints are chosen such that it guarantees to derive an upper bound on delay and backlog.

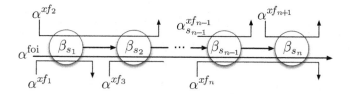

Fig. 3. Tandem network with non-nested cross-traffic arrivals.

4 Accuracy Evaluation of Network Calculus Analyses

The LP analysis and the ULP analysis both implement the PBOO and the PMOO principle. Yet, the ULP skips the LP's step 2 that is crucial for the PSOO principle. Therefore, it must operate on worst-case assumptions when bounding the arrivals of cross-traffic. As arrival bounding is not distinguishable from the foi analysis during an optimization-based analysis, we aim to gain knowledge about the attained worst case by evaluations. In contrast to the literature, we add the PMOO analysis that implements PBOO and PMOO principles but not PSOO in order to observe the impact of LP's PSOO implementation on the accuracy of NC delay bounds.

As there is no comprehensive tool support for the LP or the ULP available, we need to rely on the tooling provided by the authors of these analyses[1]. It allows to analyze arbitrary tandem networks. Additionally, LpSolve lp files for the square network are available.

In this paper, we benchmark instances of these networks against the advanced algebraic NC that has been implemented in the DiscoDNC [2] in the meantime. Additionally, we provide the equations created by different analyses where it is helpful for illustration of analysis principle violations. We are the first to contribute PMOO delay bounds to a comparison of the two NC branches.

4.1 The Non-nested Tandem with Overlapping Interference

In [7], the so-called non-nested tandem was analyzed first. It consists of a sequence of servers crossed by the flow of interest and overlapping interference of cross-flows such that there are three flows at every server (see Fig. 3). This is a classic example in network calculus; it was already used when introducing the PMOO principle and analysis [20]. Two evaluations are carried out: one that investigates the impact of the tandem's length and another one varying the utilization for the tandem of length 20. Arrival curves and service curves are taken from [7]: Both evaluations assign rate-latency service $\beta_{R,T} = \beta_{10,\frac{1}{10}}$. The evaluation with varying utilization $u \in \{0.1, \ldots, 0.9\}$ assigns token-bucket arrival curves of $\alpha = \gamma_{r,b} = \gamma_{\frac{10u}{3},1}$ where $\frac{10u}{3}$ is rounded to two decimal digits. The evaluation of tandem length impact is carried out at a utilization of 0.2, i.e., all arrivals are shaped to $\alpha = \gamma_{0.67,1}$.

---

[1] http://perso.bretagne.ens-cachan.fr/~bouillar/NCbounds/.

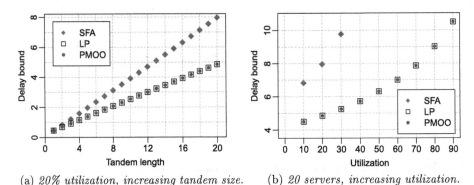

(a) *20% utilization, increasing tandem size.*   (b) *20 servers, increasing utilization.*

Fig. 4. Delay bounds in the non-nested tandem with overlapping interference.

Separate Flow Analysis (SFA). First, we derive the flow of interest's end-to-end left-over service curve (FFA step 2):

$$\beta_{P(\text{foi})}^{\text{l.o.foi}} = \beta_{s_1}^{\text{l.o.foi}} \otimes \beta_{s_2}^{\text{l.o.foi}} \otimes \ldots \otimes \beta_{s_n}^{\text{l.o.foi}}$$

$$= \left(\beta_{s_1} \ominus \left(\alpha_{s_1}^{xf_1} + \alpha_{s_1}^{xf_2}\right)\right) \otimes \left(\beta_{s_2} \ominus \left(\alpha_{s_2}^{xf_2} + \alpha_{s_2}^{xf_3}\right)\right) \otimes \ldots \otimes \left(\beta_{s_n} \ominus \left(\alpha_{s_n}^{xf_n} + \alpha_{s_n}^{xf_{n+1}}\right)\right)$$

$$= \left(\beta_{s_1} \ominus \left(\alpha^{xf_1} + \alpha^{xf_2}\right)\right) \otimes \left(\beta_{s_2} \ominus \left(\alpha_{s_2}^{xf_2} + \alpha_{s_2}^{xf_3}\right)\right) \otimes \ldots \otimes \left(\beta_{s_n} \ominus \left(\alpha_{s_n}^{xf_n} + \alpha^{xf_{n+1}}\right)\right) \tag{1}$$

In the equation, we can see that the PMOO property is not fulfilled. We need to pay for xf_m arrivals, $m \in \{2, \ldots n\}$, at both servers the cross-flow shares with the foi. Thus, we need to compute the xf_m arrivals at the respective second server of interference (FFA step 1; Eq. 1; αs with server indices). For each server s_i, $i \in \{2, \ldots, n\}$, we get:

$$\beta_{s_i} \ominus \left(\alpha_{s_i}^{xf_i} + \alpha_{s_i}^{xf_{i+i}}\right)$$

$$= \beta_{s_i} \ominus \left(\left(\alpha^{xf_i} \oslash \left(\beta_{s_{i-1}} \ominus \left(\alpha^{xf_{i-2}} \oslash \left(\ldots \left(\beta_{s_1} \ominus \alpha^{xf_1}\right) \ldots\right)\right)\right)\right) + \alpha^{xf_{i+1}}\right) \tag{2}$$

Note, that cross-traffic arrival bounding demands to recursively backtrack cross-flows of cross-flows at every server, yet, this backtracking never leaves the foi's path. The recursion terminates when the sources of flows are reached. See Eq. 2 where αs do not have server indices as we know $\alpha_{s_1}^{xf_1} = \alpha^{xf_1}$, $\alpha_{s_n}^{xf_{n+1}} = \alpha^{xf_{n+1}}$ and $\alpha_{s_{m-1}}^{xf_m} = \alpha^{xf_m}$ for $m \in \{2, \ldots n\}$ from the network description.

Pay Multiplexing Only Once (PMOO) Analysis. PMOO was specifically designed to counteract the burstiness increase being considered in the analysis multiple times. I.e, each xf_m's burstiness only appears once in its $\beta_{P(\text{foi})}^{\text{l.o.foi}}$-derivation [20]. The PMOO analysis does not demand the cross-flow arrival bounding (Eq. 2) and thus performs better on this non-nested tandem.

Linear Programming (LP) Analysis and Comparison. The results for the LP analysis are depicted alongside PMOO and SFA in Fig. 4. They all scale

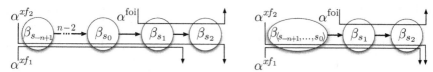

(a) *Tandem network, n PSOO violations.* (b) *Flow extension counteracts segregation.*

Fig. 5. Tandem with algebraic FFA Artifact 2 and mitigation by flow extension.

linearly with the length of the tandem (Figs. 4a) – the PBOO principle is responsible for this behavior. The second evaluation shows that all delay bounds grow super-linearly when increasing the utilization. Yet, the PMOO principle implemented by the eponymous analysis and the LP analysis leads to much smaller, better scaling delay bounds compared to the SFA (Fig. 4b).

For both parts of this evaluation, the PMOO analysis that was omitted in [7] performs equal to the LP analysis.

4.2 The Non-nested Tandem with Cross-Traffic Arrival Bounding

In this section, we will evaluate a tandem network with arbitrarily many PSOO violations due to algebraic FFA Artifact 2. To do so, we generalize the network of Fig. 2a by adding a variable number of servers to be traversed by the two cross-flows (see Fig. 5a). Each server constitutes one PSOO violation. Parameters are chosen according to the previous tandem evaluation: We assign rate-latency service $\beta_{R,T} = \beta_{10,\frac{1}{10}}$ and token-bucket arrival curves $\alpha = \gamma_{r,b}$. These are either fixed to $\gamma_{2,1}$ for a network utilization of 60% or kept variable at $\gamma_{\frac{10u}{3},1}$, $u \in \{0.1, \ldots, 0.9\}$, where $\frac{10u}{3}$ is rounded to two decimal digits.

Separate Flow Analysis (SFA). While SFA can aggregately bound both cross-flows xf_1 and xf_2 at server s_1, its left-over service curve derivation at s_2 requires segregation in order to only consider xf_2 there. [5] presents a new countermeasure to this artifact called Burst Reduction (BR). It caps xf_2's burstiness at s_2 with the backlog bound aggregately caused by both flows at server s_1.

Pay Multiplexing Only Once (PMOO) Analysis. The PMOO analysis enforces PSOO violations in this network, too (see Sect. 3.2). This can be mitigated by a preceding step to the analysis called Flow Extension (FE) [1]. This step transforms the analyzed network to a different one that is worse form the foi's point of view: Interference of cross-flow xf_1 is extended to server s_2 as depicted in Fig. 5b. While this reduces the service available to the foi at s_2 in the network, the PMOO FFA does not suffer from Artifact 2 anymore. It can now handle the cross-flows aggregately and convolve the servers they traverse. Therefore, it computes a better delay bound in the worse network setting. We are the first to show FE's benefits in arbitrary multiplexing networks.

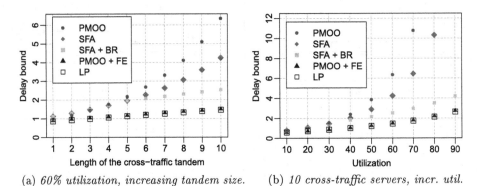

(a) *60% utilization, increasing tandem size.*   (b) *10 cross-traffic servers, incr. util.*

Fig. 6. Delays in the non-nested tandem with cross-traffic arrival bounding.

Linear Programming (LP) Analysis and Comparison. Equal to the previous tandem network, we carry out two evaluations: one that increases the length of the cross-traffic tandem $s_{-n+1}, \ldots, s_0$ and thus the amount of PSOO violations and another one that increases the maximum utilization in the network.

The scaling with respect to the tandem length is depicted in Fig. 6a. We fixed the network utilization, defined by server s_1 crossed by three flows, at 60%. SFA and PMOO analysis both scale super-linearly with the length of the cross-traffic tandem, i.e., with the amount of PSOO violations. Burst reduction (BR) significantly decreases the delay bound of the SFA for long tandems, however, it is outperformed by PMOO with flow extension (FE). The difference between the LP analysis and PMOO + FE remains small and steady for all tandem lengths – it is the penalty of assuming xf_1 crossing server s_2, too.

For the second evaluation, we fixed the tandem length at 12 servers, i.e., 10 servers with PSOO violations. Again, plain analyses (SFA and PMOO) scale super-linearly and the deviation from LP delays becomes large with growing utilization. The two analyses amended with aggregation-based countermeasures perform considerably better; PMOO + FE derives nearly identical bounds to the LP analysis. The difference does not remain steady, yet, it does not grow much. In Fig. 6b, PMOO + FE's triangles stay within LP's squares, i.e., current algebraic NC analyses counteract the PSOO violations with small amendments and their results are highly competitive with the ones derived by optimization.

4.3   The Square Network

Next, we evaluate the square network from [7] (see Fig. 1). This network is evaluated for varying utilizations: service curves remain $\beta_{s_i} = \beta_{R,T} = \beta_{10,\frac{1}{10}}$, $i \in \{1, 2, 3, 4\}$ and arrival curves are adapted to the utilization $u \in \{0.1, \ldots, 0.9\}$. As there are only two flows per server, this setting translates to $\alpha^{f_i} = \gamma_{r,b} = \gamma_{\frac{10u}{2},1}$.

Separate Flow Analysis (SFA). We start with the SFA left-over service curve derivation steps shared by every utilization's analysis:

$$\beta^{\text{l.o.}f_1}_{\langle s_3, s_4 \rangle} = \beta^{\text{l.o.}f_1}_{s_3} \otimes \beta^{\text{l.o.}f_1}_{s_4} = \left(\beta_{s_3} \ominus \alpha^{f_2}_{s_3} \right) \otimes \left(\beta_{s_4} \ominus \alpha^{f_4}_{s_4} \right) \tag{3}$$

with the following cross-traffic arrival boundings:

$$\alpha^{f_2}_{s_3} = \alpha^{f_2} \oslash \beta^{\text{l.o.}f_2}_{s_1} = \alpha^{f_2} \oslash \left(\beta_{s_1} \ominus \alpha^{f_3} \right) \tag{4}$$

$$\alpha^{f_4}_{s_4} = \alpha^{f_4} \oslash \beta^{\text{l.o.}f_4}_{s_2} = \alpha^{f_4} \oslash \left(\beta_{s_2} \ominus \left(\alpha^{f_3} \oslash \left(\beta_{s_1} \ominus \alpha^{f_2} \right) \right) \right) \tag{5}$$

The cross-traffic arrival boundings show mutual interference assumptions of Fig. 1: It computes $\beta_{s_1} \ominus \alpha^{f_3}$ and $\beta_{s_1} \ominus \alpha^{f_2}$, both will be in the SFA left-over service curve derivation of Eq. 3 – the PSOO principle is not implemented.

This derivation illustrates the different approaches applied by algebraic NC to model the flow of interest's worst-case scenario in a feed-forward network. On the foi's path (i.e., the foi tandem analysis part of step 2), the left-over service curve derivation assumes lowest priority for the flow of interest (cf. Theorem 3). In the arrival bounding, each flow's worst-case interference is modeled individually. Therefore, at server s_1, flows f_2 and f_3 are considered as mutual interference: $\beta^{\text{l.o.}f_2}_{s_1} = \left(\beta_{s_1} \ominus \alpha^{f_3} \right)$ and $\beta^{\text{l.o.}f_3}_{s_1} = \left(\beta_{s_1} \ominus \alpha^{f_2} \right)$. Note, that we cannot exploit burst reduction because flows only interfere with each other on single servers.

Pay Multiplexing Only Once (PMOO) Analysis. The PMOO left-over service curve [20] is derived as follows:

$$\beta^{\text{l.o.}f_1}_{\langle s_3, s_4 \rangle} = \beta_{R^{\text{l.o.}f_1}_{\langle s_3, s_4 \rangle}, T^{\text{l.o.}f_1}_{\langle s_3, s_4 \rangle}} \tag{6}$$

$$R^{\text{l.o.}f_1}_{\langle s_3, s_4 \rangle} = \left(R_{s_3} - r^{f_2} \right) \wedge \left(R_{s_4} - r^{f_4} \right) \tag{7}$$

$$T^{\text{l.o.}f_1}_{\langle s_3, s_4 \rangle} = T_{s_3} + T_{s_4} + \frac{b^{f_2}_{s_3} + b^{f_4}_{s_4} + r^{f_2}_{s_3} \cdot T_{s_3} + r^{f_4}_{s_4} \cdot T_{s_4}}{R^{\text{l.o.}f_1}_{\langle s_3, s_4 \rangle}} \tag{8}$$

The mutual interference modeling persists. It can be seen in the cross-flow burst terms $b^{f_2}_{s_3}$ and $b^{f_4}_{s_4}$ in Eq. 8 that equal the one of the according arrival bounds $\alpha^{f_2}_{s_3}$ and $\alpha^{f_4}_{s_4}$ used in the SFA. Cross-traffic is bounded with Eqs. 4 and 5, again, as it belongs to FFA step 1. I.e., the PMOO left-over service curve derivation of Eq. 6 will also suffer from the mutual interference problem depicted in Fig. 1 due to Eq. 8. In this network, flow extension is not beneficial as both cross-flows of f_1 arrive from different links.

Linear Programming (LP, ULP) Analysis and Comparison. Based on the derivation of algebraic analysis of the non-nested tandems and the square network (Eqs. 1 to 8) as well as the observed properties and delay bounds, we can predict the relative performance of NC analyses in the square network:

Recall that the LP approach enumerates all potential entanglements of flows by extending the partial order (defined by consecutive hops of flows) to the set

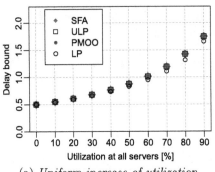 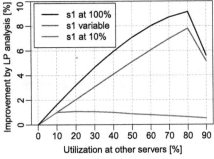

(a) *Uniform increase of utilization.*   (b) *Relative improvement of bounds.*

Fig. 7. Delay bounds in the square network.

of all compatible total orders – the benefit is pointed out in Sect. 3.3, step 2. The ULP, in contrast, is solely based on the partial order, the extension step is omitted due to its combinatorial explosion. This results in a smaller set of ULP constraints, especially the potential entanglements of f_2 and f_3 at s_1 are not considered anymore. Instead, they are assumed to constitute the respective flow's worst case. Based on this insight, it is not surprising that the ULP actually does not beat the PMOO analysis in the square network (see Fig. 7a). In fact, even the SFA yields the same delays. The reason for SFA's accuracy is the lack of multi-hop interference, the effect captured with the PMOO principle does not manifest in this network.

Our evaluation shows that the ULP models the worst-case interference between f_2 and f_3 in the same way as the algebraic analyses do.

The square network also illustrates the case where we cannot mitigate the PSOO violation in algebraic NC. Thus, we conclude our evaluation by investigating the potential superiority of LP delay bounds over PMOO delay bounds. Figure 7b depicts three different settings:

- First, we fixed the utilization of server s_1 at 100 %. For the mutual interference assumptions violating the PSOO principle, this leads to the maximum burstiness increase attainable for flows f_2 and f_3. As depicted in Sect. 3, the burstiness will propagate through the analysis and eventually result in a loose delay bound. However, the final impact also depends on the utilization of servers that forward the flows with overly pessimistic burstiness. We evaluated utilizations of the remaining servers s_2, s_3 and s_4 ranging from 10 % to 90 % (see [7]). As expected, we can first observe an amplification of the PSOO violation's effect that lets the LP analysis outperform the algebraic ones by an increasing margin. Yet, it reaches its maximum of 9.15 % at 80 % utilization. After this peak, its impact declines to a level below the 40 % network utilization.
- A similar decline can be observed in the second evaluation, taken from [7] and depicted in Fig. 7a, where the rate at s_1 varies with the other servers.

– We added another utilization level to the evaluation to confirm this observation. With a fixed utilization of 10 % at s_1, the peak is already reached at a network utilization of 10 %.

Thus, our results show that the benefit of implementing the PSOO principle is bounded as the network utilization becomes more impactful. Compared to the impact of previous analysis principles, the PSOO did not lead to a vast decrease of delay bounds. Moreover, implementing it requires huge effort. There are already eleven linear programs derives from this small square network alone.

4.4 Outlook

Evaluation of Larger, More Involved Networks. In this paper, we restricted our evaluation to tandem networks and the square network. This restriction is caused by a practical problem of the LP analysis. Besides being NP-hard, the effort to derive its linear programs them scales super-exponentially with the network size. The authors of this analysis point out two particular problems [6]:

1. The LPs relate the start of backlogged periods at servers. The amount of these dates to be related to each other grows exponentially with the network size. This problem directly impact the second one.
2. Dates are not totally ordered. This problem corresponds to step 2 given in Sect. 3.3: extension of a partial order to the set of all compatible total orders. Even for rather small networks, this step suffers from combinatorial explosion [16] and known algorithms for linear extensions like the Varol-Rotem algorithm [21] do not allow for the analysis of larger networks. Computational effort becomes prohibitive.

For these reasons, network calculus lacks comprehensive tool support for the only analysis implementing the PSOO principle as of today. In tandem networks, flows cannot take parallel paths and thus there is only one total order. This fact is exploited by the tool we used for the evaluations in Sects. 4.1 and 4.2. For the square network evaluation of Sect. 4.3, the authors of the LP analysis provide the required linear programs. Analyzing larger networks remains an open issue.

Improving the Accuracy of the ULP Analysis. The ULP constitutes the return to accurate, yet, untight bounds in general feed-forward networks. However, for some special networks it derives tight bounds nonetheless. The special case holds in tandem networks (see Fig. 1) where there is only one order. For more involved networks, we only know that the ULP has less constraints than any of the LPs. Improving the ULP's result can only be achieved by adding more constraints to it. The addition of constraints found in a total order can, however, result in a linear program that produces an invalid bound. Identifying constraints that improve the derived bound while guaranteeing to retain its validity is an open research topic of optimization-based NC.

5 Conclusion

In this paper, we assessed both branches of network calculus, the algebraic and the optimization-based analysis branch. We aimed at a more comprehensive comparison of those generally incomparable alternatives to derive delay bounds from a NC system description. A new principle for feed-forward analysis, the Pay Segregation Only Once (PSOO), enabled us to derive new insights on both branches of NC such that we were able to predict the relative results between analyses. Moreover, we provide evidence that the PSOO principle that is only implemented by the NP-hard LP analysis does not lead to vastly improved delay bounds when it is compared to the PMOO analysis of algebraic network calculus. Thus, there is currently no clear proof for the necessity to relocate network calculus by abandoning the idea of an algebraic system theory. In current NC, the LP analysis is rather a tool to benchmark algebraic NC analysis in small networks and as such it helps to find algebraic NC's weak spots, e.g., the violation of the PSOO principle we present and evaluate in this work.

References

1. Bisti, L., Lenzini, L., Mingozzi, E., Stea, G.: Estimating the worst-case delay in FIFO tandems using network calculus. In: Proceedings of ValueTools (2008)
2. Bondorf, S., Schmitt, J.B.: The DiscoDNC v2 – a comprehensive tool for deterministic network calculus. In: Proceedings of ValueTools (2014)
3. Bondorf, S., Schmitt, J.B.: Boosting sensor network calculus by thoroughly bounding cross-traffic. In: Proceedings of IEEE INFOCOM (2015)
4. Bondorf, S., Schmitt, J.B.: Calculating accurate end-to-end delay bounds - you better know your cross-traffic. In: Proceedings of ValueTools (2015)
5. Bondorf, S., Schmitt, J.B.: Improving cross-traffic bounds in feed-forward networks - there is a job for everyone. In: Remke, A., Haverkort, B.R. (eds.) MMB & DFT 2016. LNCS, vol. 9629, pp. 9–24. Springer, Heidelberg (2016)
6. Bouillard, A.: Algorithms and efficiency of network calculus. Habilitation thesis, ENS (2014)
7. Bouillard, A., Jouhet, L., Thierry, E.: Tight performance bounds in the worst-case analysis of feed-forward networks. In: Proceedings of IEEE INFOCOM (2010)
8. Boyer, M., Fraboul, C.: Tightening end-to-end delay upper bound for AFDX network calculus with rate latency FIFO servers using network calculus. In: Proceedings of IEEE WFCS (2008)
9. Boyer, M., Navet, N., Olive, X., Thierry, E.: The PEGASE project: precise and scalable temporal analysis for aerospace communication systems with network calculus. In: Margaria, T., Steffen, B. (eds.) ISoLA 2010, Part I. LNCS, vol. 6415, pp. 122–136. Springer, Heidelberg (2010)
10. Chang, C.-S.: Performance Guarantees in Communication Networks. Springer, Heidelberg (2000)
11. Frances, F., Fraboul, C., Grieu, J.: Using Network Calculus to Optimize AFDX Network. In: ERTS (2006)
12. Geyer, F., Carle, G.: Network engineering for real-time networks: comparison of automotive and aeronautic industries approaches. IEEE Commun. Mag. 54(2), 106–112 (2016)

13. Grieu, J.: Analyse et évaluation de techniques de commutation Ethernet pour l'interconnexion des systèmes avioniques. Ph.D. thesis, INPT (2004)
14. Kerschbaum, S., Hielscher, K.-S.J., Klehmet, U., German, R.: A framework for establishing performance guarantees in industrialautomation networks. In: Fischbach, K., Krieger, U.R. (eds.) MMB & DFT 2014. LNCS, vol. 8376, pp. 177–191. Springer, Heidelberg (2014)
15. Le Boudec, J.-Y., Thiran, P.: Network calculus. In: Thiran, P., Boudec, J.-Y. (eds.) Network Calculus. LNCS, vol. 2050, pp. 3–81. Springer, Heidelberg (2001)
16. Ruskey, F.: Combinatorial Generation. CRC Press, Boca Raton (2003)
17. Schmidt, M., Veith, S., Menth, M., Kehrer, S.: DelayLyzer: a tool for analyzing delay bounds in industrial ethernet networks. In: Fischbach, K., Krieger, U.R. (eds.) MMB & DFT 2014. LNCS, vol. 8376, pp. 260–263. Springer, Heidelberg (2014)
18. Schmitt, J.B., Gollan, N., Bondorf, S., Martinovic, I.: Pay bursts only once holds for (some) non-FIFO Systems. In: Proceedings of IEEE INFOCOM (2011)
19. Schmitt, J.B., Zdarsky, F.A., Fidler, M.: Delay bounds under arbitrary multiplexing: when network calculus leaves you in the lurch... In: Proceedings of IEEE INFOCOM (2008)
20. Schmitt, J.B., Zdarsky, F.A., Martinovic, I.: Improving performance bounds in feed-forward networks by paying multiplexing only once. In: Proceedings of GI/ITG MMB (2008)
21. Varol, Y.L., Rotem, D.: An algorithm to generate all topological sorting arrangements. Comput. J. **24**(1), 83–84 (1981)

Markov Decision Processes
and Markovian Analysis

Verification of General Markov Decision Processes by Approximate Similarity Relations and Policy Refinement

Sofie Haesaert[1]([⊠]), Alessandro Abate[2], and Paul M.J. Van den Hof[1]

[1] Eindhoven University of Technology, Eindhoven, The Netherlands
s.haesaert@tue.nl
[2] University of Oxford, Oxford, UK
alessandro.abate@cs.ox.ac.uk

Abstract. In this work we introduce new approximate similarity relations that are shown to be key for policy (or control) synthesis over general Markov decision processes. The models of interest are discrete-time Markov decision processes, endowed with uncountably-infinite state spaces and metric output (or observation) spaces. The new relations, underpinned by the use of metrics, allow in particular for a useful trade-off between deviations over probability distributions on states, and distances between model outputs. We show that the new probabilistic similarity relations can be effectively employed over general Markov decision processes for verification purposes, and specifically for control refinement from abstract models.

1 Introduction

The formal verification of computer systems allows for the quantification of their properties and for their correct functioning. Whilst verification has classically focused on finite-state models, with the ever more ubiquitous embedding of digital components into physical systems richer models are needed, and correct functioning can only be expressed over the combined behaviour of both a digital computer and its surrounding physical system. It is in particular of interest to synthesise the part of the computer software that controls or interacts with the physical system automatically, with low likelihood of malfunctioning. Quite importantly, when computers interact with physical systems such as biological processes, power networks, and smart-grids, stochastic models are key.

Systems with uncertainty and non-determinism can be naturally modelled as Markov decision processes (MDP). In this work, we focus on general Markov decision processes (gMDP) with uncountable state spaces as well as metric output spaces. The characterisation of properties or the synthesis of policies over such processes can in general not be attained analytically [4], so an alternative is the approximation of the original (concrete) models by simpler (abstract) models that are prone to be analysed or algorithmically verified [12], such as finite-state MDP [11]. Clearly, it is then paramount to provide formal guarantees on this approximation step.

In this work we develop a new notion of approximate similarity relation to assist in the computationally efficient controller synthesis of gMDP. The use of

© Springer International Publishing Switzerland 2016
G. Agha and B. Van Houdt (Eds.): QEST 2016, LNCS 9826, pp. 227–243, 2016.
DOI: 10.1007/978-3-319-43425-4_16

similarity relations on *finite-state* probabilistic models has been broadly investigated, either via exact notions of probabilistic simulation and bisimulation
relations [17,21], or via approximate notions [9,10]. On the other hand, similar
notions over general, uncountable-state spaces have been only recently studied: available relations either hinge on stability requirements on model outputs
[16,24] (established via martingale theory or contractivity analysis), or alternatively enforce structural abstractions of a model [8] by exploiting continuity
conditions on its probability laws [1,3].

In this work, we want to quantify properties with a certified precision *both*
in the deviation of the probability laws for finite-time events (as in the classical
notion of probabilistic bisimulation) and of the output trajectories (as studied
for dynamical models). To this end, we generalise the exact probabilistic simulation and bisimulation relations to allow for errors in the probability laws
and deviations over the output space (Sect. 4). A case study on smart buildings
(Sect. 5) is used to evaluate this new approximate similarity relations, which are
specifically tailored to perform control synthesis. The new approximate similarity relation generalises notions of probabilistic simulation relations [17,21], and
their approximate versions [9,10].

Key to this work, we further show that a control strategy for a gMDP can be
obtained as a refinement of a strategy synthesised for an abstract model, at the
expense of bounded deviations in transition probabilities and outputs as defined
by their similarity relation.

In view of space, details on measurability properties and precise derivations
of proofs of the statements are relegated to an extended version [14], which also
contains a more detailed comparison with literature.

2   Verification of General Markov Decision Processes

2.1   Preliminaries and Notations

For two sets A and B a relation $\mathcal{R} \subset A \times B$ is a subset of their Cartesian
product that relates elements $x \in A$ with elements $y \in B$, denoted as $x\mathcal{R}y$. We
use the following notation for the mappings $\mathcal{R}(\tilde{A}) := \{y : x\mathcal{R}y, \ x \in \tilde{A}\}$ and
$\mathcal{R}^{-1}(\tilde{B}) := \{x : x\mathcal{R}y, \ y \in \tilde{B}\}$ for $\tilde{A} \subseteq A$ and $\tilde{B} \subseteq B$. A relation over a set
defines a preorder if it is reflexive, $\forall x \in A : x\mathcal{R}x$; and transitive, $\forall x, y, z \in A$: if
$x\mathcal{R}y$ and $y\mathcal{R}z$ then $x\mathcal{R}z$. A relation $\mathcal{R} \subseteq A \times A$ is an equivalence relation if it is
reflexive, transitive and symmetric, $\forall x, y \in A$: if $x\mathcal{R}y$ then $y\mathcal{R}x$.

A measurable space is a pair $(\mathbb{X}, \mathcal{F})$ with sample space $\mathbb{X}$ and σ-algebra $\mathcal{F}$
defined over $\mathbb{X}$, which is equipped with a topology. As a specific instance of
$\mathcal{F}$ consider the Borel measurable space $(\mathbb{X}, \mathcal{B}(\mathbb{X}))$. In this work, we restrict our
attention to Polish spaces and generally consider the Borel σ-field [6]. Recall
that a Polish space is a separable and completely metrisable topological space.
A simple example of such a space is the real line.

A probability measure $\mathbb{P}(\cdot)$ for $(\mathbb{X}, \mathcal{F})$ is a non-negative map, $\mathbb{P}(\cdot) : \mathcal{F} \to$
$[0,1]$ such that $\mathbb{P}(\mathbb{X}) = 1$ and such that for all countable collections $\{A_i\}_{i=1}^{\infty}$ of
pairwise disjoint sets in $\mathcal{F}$, it holds that $\mathbb{P}(\bigcup_i A_i) = \sum_i \mathbb{P}(A_i)$. Together with

the measurable space, such a probability measure $\mathbb{P}$ defines the probability space, which is denoted as $(\mathbb{X}, \mathcal{F}, \mathbb{P})$ and has realisations $x \sim \mathbb{P}$. Let us further denote the set of all probability measures for a given measurable pair $(\mathbb{X}, \mathcal{F})$ as $\mathcal{P}(\mathbb{X}, \mathcal{F})$. For a probability space[1] $(\mathbb{X}, \mathcal{F}_{\mathbb{X}}, \mathbb{P})$ and a measurable space $(\mathbb{Y}, \mathcal{F}_{\mathbb{Y}})$, a $(\mathbb{Y}, \mathcal{F}_{\mathbb{Y}})$-valued *random variable* is a function $y : \mathbb{X} \rightarrow \mathbb{Y}$ that is $(\mathcal{F}_{\mathbb{X}}, \mathcal{F}_{\mathbb{Y}})$-measurable, and which induces the probability measure $y_* \mathbb{P}$ in $\mathcal{P}(\mathbb{Y}, \mathcal{F}_{\mathbb{Y}})$. For a given set $\mathbb{X}$ a metric or distance function $\mathbf{d}_{\mathbb{X}}$ is a function $\mathbf{d}_{\mathbb{X}} : \mathbb{X} \times \mathbb{X} \rightarrow \mathbb{R}_0^+$.

2.2 gMDP Models - Syntax and Semantics

General Markov decision processes are related to control Markov processes [3] and Markov decision processes [5,20], and are formalised as follows.

Definition 1 (Markov Decision Process (MDP)). *A discrete-time MDP* $\mathbf{M} = (\mathbb{X}, \pi, \mathbb{T}, \mathbb{U})$ *is defined over an uncountable state space* $\mathbb{X}$, *and characterised by* $\mathbb{T}$, *a conditional stochastic kernel that assigns to each point* $x \in \mathbb{X}$ *and control* $u \in \mathbb{U}$ *a probability measure* $\mathbb{T}(\cdot \,|\, x, u)$ *over* $(\mathbb{X}, \mathcal{B}(\mathbb{X}))$. *For any set* $A \in \mathcal{B}(\mathbb{X})$, $\mathbb{P}_{x,u}(x(t+1) \in A) = \mathbb{T}(A \,|\, x(t) = x, u)$, *where* $\mathbb{P}_{x,u}$ *denotes the conditional probability* $\mathbb{P}(\cdot \,|\, x, u)$. *The initial probability distribution is* $\pi : \mathcal{B}(\mathbb{X}) \rightarrow [0,1]$.

At every state the state transition depends non-deterministically on the choice of $u \in \mathbb{U}$. When chosen according to a distribution $\mu_u : \mathcal{B}(\mathbb{U}) \rightarrow [0,1]$, we refer to the stochastic control input as μ_u. Moreover the transition kernel is denoted as $\mathbb{T}(\cdot \,|\, x, \mu_u) = \int_{\mathbb{U}} \mathbb{T}(\cdot \,|\, x, u) \mu_u(du) \in \mathcal{P}(\mathbb{X}, \mathcal{B}(\mathbb{X}))$. Given a string of inputs (possibly randomised) $u(0), u(1), \ldots, u(N)$, over a finite time horizon $\{0, 1, \ldots, N\}$, and an initial condition x_0 (sampled from distribution π), the state at the $(t+1)$-st time instant, $x(t+1)$, is obtained as a realisation of the controlled Borel-measurable stochastic kernel $\mathbb{T}(\cdot \,|\, x(t), u(t))$ – these semantics induce paths (or executions) of the MDP.

Definition 2 (General Markov Decision Process (gMDP)). *A discrete-time gMDP* $\mathbf{M} = (\mathbb{X}, \pi, \mathbb{T}, \mathbb{U}, h, \mathbb{Y})$ *is an MDP combined with a metric output space* $(\mathbb{Y}, \mathbf{d}_{\mathbb{Y}})$, *and a measurable output mapping* $h : \mathbb{X} \rightarrow \mathbb{Y}$.

The gMDP semantics are directly inherited from those of the MDP. Further, output traces of gMDP are obtained as mappings of MDP paths, namely $\{y(t)\}_{0:N} := y(0), y(1), \ldots, y(N)$, where $y(t) = h\big(x(t)\big)$. Denote the class of all gMDP with the metric output space $\mathbb{Y}$ as $\mathcal{M}_{\mathbb{Y}}$. Note that gMDP can be regarded as a super-class of the known labelled Markov processes (LMP) [8] as elucidated in [1].

Example 1. Consider a stochastic process defined as the solution of the stochastic difference equation

$$\mathbf{M} : x(t+1) = f(x(t), u(t)) + e(t), \qquad y(t) = h(x(t)) \in \mathbb{Y},$$

---

[1] The index $\mathbb{X}$ in $\mathcal{F}_{\mathbb{X}}$ distinguishes the given σ-algebra on $\mathbb{X}$ from that on $\mathbb{Y}$, which is denoted as $\mathcal{F}_{\mathbb{Y}}$. Whenever possible this index will be dropped.

with variables $x(t), u(t), e(t)$, taking values in $\mathbb{R}^n$, representing the state, control input (external non-determinism), and noise terms respectively. The process is initialised as $x(0) \sim \pi$, and driven by $e(t)$, a white noise sequence with zero-mean normal distributions and variance Σ_e. This stochastic process, defined as a dynamical model with dynamics characterised by the stochastic difference equation above, is a gMDP characterised by a tuple $(\mathbb{R}^n, \pi, \mathbb{T}, \mathbb{R}^n, h, \mathbb{Y})$, where the conditional transition kernel is defined as $\mathbb{T}(\cdot\,|x, u) = \mathcal{N}(\cdot\,|f(x(t), u(t)), \Sigma_e)$, a normal probability distribution with mean $f(x(t), u(t))$ and variance Σ_e.   □

A policy is a selection of control inputs based on the past history of states and actions. We allow controls to be selected via universally measurable maps [5] from the state to the control space, so that time-bounded properties such as safety can be maximised [12]. When the selected controls are only dependent on the current states and thus conditionally independent of history (or memoryless), the policy is referred to as Markov. A Markov policy μ for a gMDP $\mathbf{M} = (\mathbb{X}, \pi, \mathbb{T}, \mathbb{U}, h, \mathbb{Y})$ is a sequence $\mu = (\mu_1, \mu_2, \mu_3, \ldots)$ of universally measurable maps $\mu_t : \mathbb{X} \to \mathcal{P}(\mathbb{U}, \mathcal{B}(\mathbb{U}))$ $t = 0, 1, 2, \ldots$, from the state space $\mathbb{X}$ to the set of controls. Recall that a function $f : \mathbb{Z}_1 \to \mathbb{Z}_2$ is universally measurable if the inverse image of every Borel set is measurable with respect to every complete probability measure on $\mathbb{Z}_1$ that measures all Borel subsets of $\mathbb{Z}_1$.

The execution $\{x(t), t \in [0, N]\}$ initialised by $x_0 \in \mathbb{X}$ and controlled with Markov policy μ is a stochastic process defined on the canonical sample space $\Omega := \mathbb{X}^{N+1}$ endowed with its product topology $\mathcal{B}(\Omega)$. This stochastic process has a probability measure $\mathbb{P}$ uniquely defined by the transition kernel $\mathbb{T}$, policy μ, and initial distribution π [5, Prop. 7.45].

Of interest to us are time-dependent properties such as those expressed as specifications in a temporal logic of choice. This leads to problems where one maximises the probability that a sequence of labelled sets is reached within a time limit and in the right order. One can intuitively understand that in general the optimal policy leading to the maximal probability is not a Markov (memoryless) policy. We introduce the notion of a control strategy, and define it as a broader, memory-dependent version of the Markov policy above. Such a strategy for controlling a gMDP is formulated next as a Markov process that takes the state of the gMDP as input.

Definition 3 (Control Strategy). *A control strategy* $\mathbf{C} = (\mathbb{X}_\mathbf{C}, x_{\mathbf{C}0}, \mathbb{X}, \mathbb{T}_\mathbf{C}^t, h_\mathbf{C}^t)$ *for a gMDP* $\mathbf{M}$ *with state space* $\mathbb{X}$ *and control space* $\mathbb{U}$ *over the time horizon* $t = 0, 1, 2, \ldots, N$ *is an inhomogenous Markov process with state space* $\mathbb{X}_\mathbf{C}$; *an initial state* $x_{\mathbf{C}0}$; *inputs* $x \in \mathbb{X}$; *time-dependent, universally measurable kernels* $\mathbb{T}_\mathbf{C}^t$, $t = 0, 1, \ldots, N$; *and with universally measurable output maps* $h_\mathbf{C}^t : \mathbb{X}_\mathbf{C} \to \mathcal{P}(\mathbb{U}, \mathcal{B}(\mathbb{U}))$, $t = 1, \ldots, N$, *with elements* $\mu \in \mathcal{P}(\mathbb{U}, \mathcal{B}(\mathbb{U}))$.   □

Unlike a Markov policy, the control strategy is in general dependent on the history, as it has an internal state that can be used to remember relevant past events. Note that the first control $u(0)$ is selected by drawing $x_\mathbf{C}(1)$ according to $\mathbb{T}_\mathbf{C}^0(\cdot\,|x_\mathbf{C}(0), x(0))$, where $x_\mathbf{C}(0) = x_{\mathbf{C}0}$, and selecting $u(0)$ from measure $\mu_\mathbf{C}^0 = h_\mathbf{C}^0(x_\mathbf{C}(1))$. This is then repeated at every time step, when the controller selects a control $u(t)$ by updating its internal state $\mathbb{T}_{\mathbf{C}}^t(\cdot \,|x_{\mathbf{C}}(t), x(t))$ and then selecting $u(t)$ according to $\mu_{\mathbf{C}}^t = h_{\mathbf{C}}^t(x_{\mathbf{C}}(t+1))$. The control strategy applied to $\mathbf{M}$ can be both stochastic (it is a realisation of $\mathbb{T}_{\mathbf{C}}^t(\cdot \,|\, x_{\mathbf{C}}(t), x(t))$), a function of the initial state $x(0)$, and of time.

The execution $\{(x(t), x_{\mathbf{C}}(t)), t \in [0, N]\}$ of a gMDP $\mathbf{M}$ controlled with strategy $\mathbf{C}$, is defined on the canonical sample space $\Omega := (\mathbb{X} \times \mathbb{X}_{\mathbf{C}})^{N+1}$ endowed with its product topology $\mathcal{B}(\Omega)$. This stochastic process is associated to a unique probability measure $\mathbb{P}_{\mathbf{C}\times\mathbf{M}}$, since the stochastic kernels $\mathbb{T}_{\mathbf{C}}^t$ and $\mathbb{T}$ are Borel measurable and composed via universally measurable policies [5, Prop. 7.45].

2.3   gMDP Verification and Strategy Refinement: The Idea

We qualitatively anticipate the main result of this work. We intend to provide a general framework to synthesise control policies over a formal abstraction $\tilde{\mathbf{M}}$ of a concrete complex model $\mathbf{M}$, with the understanding that $\tilde{\mathbf{M}}$ is much simpler to be manipulated (analytically or computationally) than $\mathbf{M}$ is. We define a simulation relation under which a policy $\tilde{\mathbf{C}}$ for the abstract Markov process $\tilde{\mathbf{M}}$ implies the existence of a policy $\mathbf{C}$ for $\mathbf{M}$, so that we can quantify differences in the stochastic transition kernels and in the output trajectories for the two closed-loop models. This allows us to derive bounds on the probability of satisfaction of a specification for $\mathbf{M}\times\mathbf{C}$ from the satisfaction probability of modified specifications for $\tilde{\mathbf{M}} \times \tilde{\mathbf{C}}$. This setup allows dealing with finite-horizon temporal properties, including safety verification as a relevant instance.

The results in this paper are to be used in parallel with optimisation, both for selecting the control refinement and for synthesising a policy on the abstract model. It has been shown in [5] that stochastic optimal control, even for a system on a "basic" state space, can lead to measurability issues: in order to avoid these issues we follow [5,9] and the developed theory for Polish spaces and Borel (or universally) measurable notions. Throughout the paper we will give as clarifying examples Markov processes evolving, as in Example 1, over Euclidean spaces which are a special instances of Polish spaces.

3   Exact (Bi-)simulation Relations Based on Lifting

In this section we define probabilistic simulation and bisimulation relations that are, respectively, a preorder and an equivalence relation on $\mathcal{M}_{\mathbb{Y}}$. Before introducing these relations, we first extend Segala's notion [21] of *lifting* to uncountable state spaces, which allows us to equate the transition kernels of two given gMDPs. Thereafter, we leverage liftings to define (bi-)simulation relations over $\mathcal{M}_{\mathbb{Y}}$, which characterise the similarity in the controllable behaviours of the two gMDPs. Subsequently we show that these similarity relations also imply controller refinement, i.e., within the similarity relation a control strategy for a given gMDP can be refined to a controller for another gMDP. In the next section, we show that this exact notion of similarity allows a more general notion of approximate probabilistic simulation. The new notions of similarity relations extend the known exact notions in [17], and the approximate notions of [9,10].

3.1   Lifting for General Markov Decision Processes

Consider two gMDP $\mathbf{M}_1, \mathbf{M}_2 \in \mathcal{M}_{\mathbb{Y}}$ mapping to a common output space $\mathbb{Y}$ with metric $\mathbf{d}_{\mathbb{Y}}$. For $\mathbf{M}_1 = (\mathbb{X}_1, \pi_1, \mathbb{T}_1, \mathbb{U}_1, h_1, \mathbb{Y})$ and $\mathbf{M}_2 = (\mathbb{X}_2, \pi_2, \mathbb{T}_2, \mathbb{U}_2, h_2, \mathbb{Y})$ at given state-action pairs $x_1 \in \mathbb{X}_1, u_1 \in \mathbb{U}_1$ and $x_2 \in \mathbb{X}_2, u_2 \in \mathbb{U}_2$, respectively, we want to relate the corresponding transition kernels, namely the probability measures $\mathbb{T}_1(\cdot | x_1, u_1) \in \mathcal{P}(\mathbb{X}_1, \mathcal{B}(\mathbb{X}_1))$ and $\mathbb{T}_2(\cdot | x_2, u_2) \in \mathcal{P}(\mathbb{X}_2, \mathcal{B}(\mathbb{X}_2))$.

Similar to the coupling of measures in $\mathcal{P}(\mathbb{X}, \mathcal{F})$ [2,18], consider the *coupling* of two arbitrary probability spaces $(\mathbb{X}_1, \mathcal{F}_1, \mathbb{P}_1)$ and $(\mathbb{X}_2, \mathcal{F}_2, \mathbb{P}_2)$ (cf. [22]). A probability measure $\mathbb{P}_c$ defined on $(\mathbb{X}_1 \times \mathbb{X}_2, \mathcal{F})$ *couples* the two spaces if the projections p_1, p_2, with $x_1 = p_1(x_1, x_2)$ and $x_2 = p_2(x_1, x_2)$, define respectively an $(\mathbb{X}_1, \mathcal{F}_1)$- and an $(\mathbb{X}_2, \mathcal{F}_2)$-valued random variable, such that $\mathbb{P}_1 = p_{1*}\mathbb{P}_c$ and $\mathbb{P}_2 = p_{2*}\mathbb{P}_c$. For *finite- or countably infinite-state* stochastic processes a closely-related concept has been introduced in [21] and referred to as *lifting*: the transition probabilities are coupled using a weight function in a way that respects a given relation over the combined state spaces. Rather than using weight functions over a countable or finite domain [21], we introduce lifting as a coupling of measures over Polish spaces.

Since we assume that the state spaces are Polish and have a corresponding Borel σ-field for the given probability spaces $(\mathbb{X}_1, \mathcal{B}(\mathbb{X}_1), \mathbb{P}_1)$ and $(\mathbb{X}_2, \mathcal{B}(\mathbb{X}_2), \mathbb{P}_2)$ with $\mathbb{P}_1 := \mathbb{T}_1(\cdot | x_1, u_1)$ and $\mathbb{P}_2 := \mathbb{T}_2(\cdot | x_2, u_2)$, the natural choice for the σ-algebra becomes[2] $\mathcal{B}(\mathbb{X}_1 \times \mathbb{X}_2) = \mathcal{B}(\mathbb{X}_1) \otimes \mathcal{B}(\mathbb{X}_2)$ and the question of finding a coupling can be reduced to finding a probability measure in $\mathcal{P}(\mathbb{X}_1 \times \mathbb{X}_2, \mathcal{B}(\mathbb{X}_1 \times \mathbb{X}_2))$.

Definition 4 (Lifting for General State Spaces). *Let $\mathbb{X}_1, \mathbb{X}_2$ be two sets with associated measurable spaces $(\mathbb{X}_1, \mathcal{B}(\mathbb{X}_1))$ and $(\mathbb{X}_2, \mathcal{B}(\mathbb{X}_2))$ and let the Borel measurable set $\mathcal{R} \subseteq \mathbb{X}_1 \times \mathbb{X}_2$ be a relation. We denote by $\bar{\mathcal{R}} \subseteq \mathcal{P}(\mathbb{X}_1, \mathcal{B}(\mathbb{X}_1)) \times \mathcal{P}(\mathbb{X}_2, \mathcal{B}(\mathbb{X}_2))$ the corresponding lifted relation, so that $\Delta \bar{\mathcal{R}} \Theta$ holds if there exists a probability space $(\mathbb{X}_1 \times \mathbb{X}_2, \mathcal{B}(\mathbb{X}_1 \times \mathbb{X}_2), \mathbb{W})$ (equivalently, a lifting $\mathbb{W}$) satisfying*

1. *for all $X_1 \in \mathcal{B}(\mathbb{X}_1)$: $\mathbb{W}(X_1 \times \mathbb{X}_2) = \Delta(X_1)$;*
2. *for all $X_2 \in \mathcal{B}(\mathbb{X}_2)$: $\mathbb{W}(\mathbb{X}_1 \times X_2) = \Theta(X_2)$;*
3. *for the probability space $(\mathbb{X}_1 \times \mathbb{X}_2, \mathcal{B}(\mathbb{X}_1 \times \mathbb{X}_2), \mathbb{W})$ it holds that $s\mathcal{R}t$ with probability 1, or equivalently that $\mathbb{W}(\mathcal{R}) = 1$.*

Remark 1. We have implicitly required that the σ-algebra $\mathcal{B}(\mathbb{X}_1 \times \mathbb{X}_2)$ contains not only sets of the form $X_1 \times \mathbb{X}_2$ and $\mathbb{X}_1 \times X_2$, but also specifically the sets that characterise the relation $\mathcal{R}$. Since the spaces $\mathbb{X}_1$ and $\mathbb{X}_2$ have been assumed to be Polish, it holds that every open (closed) set in $\mathbb{X}_1 \times \mathbb{X}_2$ belongs to $\mathcal{B}(\mathbb{X}_1) \otimes \mathcal{B}(\mathbb{X}_2) = \mathcal{B}(\mathbb{X}_1 \times \mathbb{X}_2)$ [6, Lemma 6.4.2]. As an example also consider the diagonal relation $\mathcal{R}_{diag} := \{(x, x) : x \in \mathbb{X}\}$ over $\mathbb{X} \times \mathbb{X}$, of importance for some examples introduced later. This is a Borel measurable set [6, Theorem 6.5.7].   □

3.2   Exact Probabilistic (Bi-)simulation Relations via Lifting

Similar to the alternating notions for probabilistic game structures in [25], we provide a simulation that relates any input chosen for the abstract process with

---

[2] $\mathcal{B}(\mathbb{X}_1) \otimes \mathcal{B}(\mathbb{X}_2)$ denotes the product σ-algebra of $\mathcal{B}(\mathbb{X}_1)$ and $\mathcal{B}(\mathbb{X}_2)$.

one for the concrete process. We aim to compare the models behaviour with respect to how they can be controlled, and thus allow for more elaborate handling of the inputs than in the probabilistic simulation relations of [9,10,21], paving the way to controller refinement. We introduce the notion of *interface function* in order to connect the controllable behaviour of the two gMDP:

$$\mathcal{U}_v : \mathbb{U}_1 \times \mathbb{X}_1 \times \mathbb{X}_2 \to \mathcal{P}(\mathbb{U}_2, \mathcal{B}(\mathbb{U}_2)),$$

where we require that $\mathcal{U}_v$ is a Borel measurable function. This means that $\mathcal{U}_v$ induces a Borel measurable stochastic kernel, denoted by $\mathcal{U}_v$, over $\mathbb{U}_2$ given $\mathbb{U}_1 \times \mathbb{X}_1 \times \mathbb{X}_2$. The notion of interface function is known in the context of correct-by-design controller synthesis and of hierarchical controller refinement [13,23]. The lifting of the transition kernels for the chosen interface generates a stochastic kernel $\mathbb{W}_\mathbb{T}$ conditioned on the inputs $\mathbb{U}_1$ and $\mathbb{X}_1 \times \mathbb{X}_2$. Let us trivially extend the interface function to $\mathcal{U}_v(\mu_1, x_1, x_2) := \int_{\mathbb{U}_1} \mathcal{U}_v(u_1, x_1, x_2)\mu_1(du_1)$.

Definition 5 (Probabilistic Simulation). *Consider two gMDP* $\mathbf{M}_i, i = 1, 2$, $\mathbf{M}_i = (\mathbb{X}_i, \pi_i, \mathbb{T}_i, \mathbb{U}_i, h_i, \mathbb{Y})$. *The gMDP* $\mathbf{M}_1$ *is stochastically simulated by* $\mathbf{M}_2$ *if there exists an interface function* $\mathcal{U}_v$ *and relation* $\mathcal{R} \subseteq \mathbb{X}_1 \times \mathbb{X}_2 \in \mathcal{B}(\mathbb{X}_1 \times \mathbb{X}_2)$, *for which there exists a Borel measurable stochastic kernel* $\mathbb{W}_\mathbb{T}(\cdot | u_1, x_1, x_2)$ *on* $\mathbb{X}_1 \times \mathbb{X}_2$ *given* $\mathbb{U}_1 \times \mathbb{X}_1 \times \mathbb{X}_2$, *such that* $\forall(x_1, x_2) \in \mathcal{R}$:

1. $h_1(x_1) = h_2(x_2)$;
2. $\forall u_1 \in \mathbb{U}_1$, $\mathbb{T}_1(\cdot | x_1, u_1)\, \bar{\mathcal{R}}\, \mathbb{T}_2(\cdot | x_1, \mathcal{U}_v(u_1, x_1, x_2))$, *with lifted probability measure* $\mathbb{W}_\mathbb{T}(\cdot | u_1, x_1, x_2)$;
3. $\pi_1 \bar{\mathcal{R}} \pi_2$.

The relationship between the two models is denoted as $\mathbf{M}_1 \preceq \mathbf{M}_2$.

Definition 6 (Probabilistic Bisimulation). *Under the same conditions as above,* $\mathbf{M}_1$ *is a probabilistic bisimulation of* $\mathbf{M}_2$ *if there exists a relation* $\mathcal{R} \subseteq \mathbb{X}_1 \times \mathbb{X}_2$ *such that* $\mathbf{M}_1 \preceq \mathbf{M}_2$ *w.r.t.* $\mathcal{R}$ *and* $\mathbf{M}_2 \preceq \mathbf{M}_1$ *w.r.t. the inverse relation* $\mathcal{R}^{-1} \subseteq \mathbb{X}_2 \times \mathbb{X}_1$. $\mathbf{M}_1$ *and* $\mathbf{M}_2$ *are said to be probabilistically bisimilar, which is denoted* $\mathbf{M}_1 \approx \mathbf{M}_2$.

For every gMDP $\mathbf{M}$: $\mathbf{M} \preceq \mathbf{M}$ and $\mathbf{M} \approx \mathbf{M}$. This can be seen by considering the diagonal relation $\mathcal{R}_{diag} = \{(x_1, x_2) \in \mathbb{X} \times \mathbb{X} \mid x_1 = x_2\}$ and selecting equal inputs for the associated interfaces. The resulting equal transition kernels $\mathbb{T}(\cdot | x, u)\bar{\mathcal{R}}_{diag}\mathbb{T}(\cdot | x, u)$ are lifted by the measure $\mathbb{W}_\mathbb{T}(dx'_1 \times dx'_2 | u, x_1, x_2) = \delta_{x'_1}(dx'_2)\mathbb{T}(dx'_1 | x_1, u)$ where δ denotes the Dirac distribution.

Example 2 (Lifting for Diagonal Relations). Consider the specific case of the gMDP ($\mathbf{M}_1$) introduced in Example 1, and a slight variation of it ($\mathbf{M}_2$), both given as stochastic dynamic processes as

$$\mathbf{M}_1 : x(t+1) = ax(t) + bu(t) + e(t) \in \mathbb{R}, \qquad y(t) = h(x(t)) \in \mathbb{R},$$
$$\mathbf{M}_2 : x(t+1) = ax(t) + bu(t) + \tilde{e}(t) + \tilde{u}(t) \in \mathbb{R}, \quad y(t) = h(x(t)) \in \mathbb{R},$$

with variables $x(t), x(t+1), u(t), \tilde{u}(t), e(t), \tilde{e}(t)$ and constants a, b taking values in $\mathbb{R}$, and with dynamics initialised with the same probability distribution at

$t = 0$ and driven by white noise sequences $e(t), \tilde{e}(t)$, both with zero-mean normal distributions and with variance equal to 1 and 1.25, respectively. $\mathbf{M}_1 \preceq \mathbf{M}_2$. For every action u_1 chosen for $\mathbf{M}_1$, select the control input pair $(u_2, \tilde{u}_2) \in \mathbb{U}_2 = \mathbb{R}^2$ as $u_2 = u_1$, and $\tilde{u}_2$ according to the zero-mean normal distribution with variance 0.25, then the associated *interface* is $\mathcal{U}_v(\cdot|u_1, x_1, x_2) = \delta_{u_1}(du_2)\mathcal{N}(d\tilde{u}_2|0, 0.25)$. For this interface the stochastic dynamics of the two processes are equal, and can be lifted with $\mathcal{R}_{diag}$, namely $\mathbb{T}_1(\cdot|x, u)\bar{\mathcal{R}}_{diag}\mathbb{T}_2(\cdot|x, \mathcal{U}_v)$.   □

Remark 2. Over $\mathcal{M}_{\mathbb{Y}}$, the class of gMDP with a shared output space, the relation $\preceq$ is a preorder, as it is reflexive (see Example 2) and transitive (see Corollary 6). Moreover $\approx$ is an equivalence relation as it is also symmetric (Corollary 6).   □

3.3   Controller Refinement via Probabilistic Simulation Relations

The ideas underlying the controller refinement are first discussed, after which it is shown that the refined controller induces a strategy as per Definition 3. Finally the equivalence of properties defined over the closed-loop gMDPs is shown.

Consider two gMDP $\mathbf{M}_i = (\mathbb{X}_i, \pi_i, \mathbb{T}_i, \mathbb{U}, h_i, \mathbb{Y})$ $i = 1, 2$ with $\mathbf{M}_1 \preceq \mathbf{M}_2$. Given the entities $\mathcal{U}_v$ and $\mathbb{W}_{\mathbb{T}}$ associated to $\mathbf{M}_1 \preceq \mathbf{M}_2$, the distribution of the next state x_2' of $\mathbf{M}_2$ is given as $\mathbb{T}_2(\cdot|x_2, \mathcal{U}_v(u_1, x_1, x_2))$, and is equivalently defined via the lifted measure as the marginal of $\mathbb{W}_{\mathbb{T}}(\cdot|u_1, x_1, x_2)$ on $\mathbb{X}_2$. Therefore, the distribution of the combined next state (x_1', x_2'), defined as $\mathbb{W}_{\mathbb{T}}(\cdot|u_1, x_1, x_2)$, can be expressed as

$$\mathbb{W}_{\mathbb{T}}(dx_1' \times dx_2'|u_1, x_1, x_2) = \mathbb{W}_{\mathbb{T}}(dx_1'|x_2', u_1, x_1, x_2)\mathbb{T}_2(dx_2'|x_2, \mathcal{U}_v(u_1, x_1, x_2)),$$

where $\mathbb{W}_{\mathbb{T}}(dx_1'|x_2', u_1, x_1, x_2)$ is referred to as the conditional probability given x_2' (c.f. [7, Corollary 3.1.2]). Similarly, the conditional measure for the initialisation $\mathbb{W}_\pi$ is denoted as $\mathbb{W}_\pi(dx_1(0) \times dx_2(0)) = \mathbb{W}_\pi(dx_1(0)|x_2(0))\pi_2(dx_2(0))$.

Now suppose that we have a control strategy for $\mathbf{M}_1$, referred to as $\mathbf{C}_1$, and we want to construct the refined control strategy $\mathbf{C}_2$ for $\mathbf{M}_2$, which is such that events defined over the output space have equal probability. This refinement procedure follows directly from the interface and the conditional probability distributions, and is described in Algorithm 1. The above execution algorithm is separated into the refined control strategy $\mathbf{C}_2$ and its gMDP $\mathbf{M}_2$. $\mathbf{C}_2$ is composed of $\mathbf{C}_1$, the stochastic kernel $\mathbb{W}_{\mathbb{T}}$, and the interface $\mathcal{U}_v$, and it remembers the previous state of $\mathbf{M}_2$.

Theorem 1 (Refined Control Strategy). *Let gMDP $\mathbf{M}_1$ and $\mathbf{M}_2$ be related as $\mathbf{M}_1 \preceq \mathbf{M}_2$, and consider the control strategy $\mathbf{C}_1 = (\mathbb{X}_{\mathbf{C}_1}, x_{\mathbf{C}_10}, \mathbb{X}_1, \mathbb{T}_{\mathbf{C}_1}^t, h_{\mathbf{C}_1}^t)$ for $\mathbf{M}_1$ as given. Then there exists at least one refined control strategy $\mathbf{C}_2 = (\mathbb{X}_{\mathbf{C}_2}, x_{\mathbf{C}_20}, \mathbb{X}_2, \mathbb{T}_{\mathbf{C}_2}^t, h_{\mathbf{C}_2}^t)$ as defined in Definition 3, with*

- *state space $\mathbb{X}_{\mathbf{C}_2} := \mathbb{X}_{\mathbf{C}_1} \times \mathbb{X}_1 \times \mathbb{X}_2$, with elements $x_{\mathbf{C}_2} = (x_{\mathbf{C}_1}, x_1, x_2)$;*
- *initial state $x_{\mathbf{C}_20} := (x_{\mathbf{C}_10}, 0, 0)$;*
- *input variable $x_2 \in \mathbb{X}_2$, namely the state variable of $\mathbf{M}_2$;*

– *time-dependent stochastic kernels* $\mathbb{T}_{\mathbf{C}_2}^t$, *defined as*

$$\mathbb{T}_{\mathbf{C}_2}^0(dx_{\mathbf{C}_2}|x_{\mathbf{C}_2 0}, x_2(0)) := \mathbb{T}_{\mathbf{C}_1}^0(dx_{\mathbf{C}_1}|x_{\mathbf{C}_1 0}, x_1)\mathbb{W}_\pi(dx_1|x_2)\delta_{x_2(0)}(dx_2) \text{ and}$$
$$\mathbb{T}_{\mathbf{C}_2}^t(dx'_{\mathbf{C}_2}|x_{\mathbf{C}_2}(t), x_2(t)) := \mathbb{T}_{\mathbf{C}_1}^t(dx'_{\mathbf{C}_1}|x_{\mathbf{C}_1}, x'_1)$$
$$\mathbb{W}_\mathbb{T}(dx'_1|x'_2, h_{\mathbf{C}_1}^t(x_{\mathbf{C}_1}), x_2, x_1)\delta_{x_2(t)}(dx'_2) \text{ for } t \in [1, N];$$

– *measurable output maps* $h_{\mathbf{C}_2}^t(x_{\mathbf{C}_1}, \tilde{x}_1, x_2) := \mathcal{U}_v(h_{\mathbf{C}_1}^t(x_{\mathbf{C}_1}), x_1, x_2).$ □

Algorithm 1: Refinement of Control Strategy $\mathbf{C}_1$ as $\mathbf{C}_2$

Given the interface function $\mathcal{U}_v$, and the (conditional) stochastic kernels $\mathbb{W}_\mathbb{T}(dx'_1|x'_2, u_1, x_1, x_2)$ and $\mathbb{W}_\pi(dx_1(0)|x_2(0))$.
Initialise by drawing
– the initial state $x_2(0)$ from π_2, and
– the initial state $x_1(0)$ from $\mathbb{W}_\pi(\cdot|x_2(0))$.
Run starting at $t = 0$,
1. given $x_1(t)$, select $u_1(t)$ according $\mathbf{C}_1$,
2. choose randomised input $\mu_{2t} = \mathcal{U}_v(u_1(t), x_1(t), x_2(t))$,
   draw $x_2(t+1)$ from $\mathbb{T}_2(\cdot|x_2(t), \mu_{2t})$,
3. draw $x_1(t+1)$ from $\mathbb{W}_\mathbb{T}(\cdot|x_2(t+1), u_1(t), x_1(t), x_2(t))$,
4. set $t := t + 1$, return.

Both the time-dependent stochastic kernels $\mathbb{T}_{\mathbf{C}_2}^t$ and the output maps $h_{\mathbf{C}_2}^t$, for $t \in [0, N]$, are universally measurable, since Borel measurable maps are universally measurable and the latter are closed under composition [5, Chapter 7].

Since, by the above construction of $\mathbf{C}_2$, traces in the output spaces of the closed loop systems $\mathbf{C}_1 \times \mathbf{M}_1$ and $\mathbf{C}_2 \times \mathbf{M}_2$ have equal distribution, it follows that measurable events have equal probability, as stated next.

Theorem 2. *If* $\mathbf{M}_1 \preceq \mathbf{M}_2$, *then for all control strategies* $\mathbf{C}_1$ *there exists a control strategy* $\mathbf{C}_2$ *such that, for all measurable events* $A \in \mathcal{B}(\mathbb{Y}^{N+1})$,

$$\mathbb{P}_{\mathbf{C}_1 \times \mathbf{M}_1}(\{y_1(t)\}_{0:N} \in A) = \mathbb{P}_{\mathbf{C}_2 \times \mathbf{M}_2}(\{y_2(t)\}_{0:N} \in A),$$

with respective output traces $\{y_1(t)\}_{0:N}$ *and* $\{y_2(t)\}_{0:N}$ *of* $\mathbf{C}_1 \times \mathbf{M}_1$ *and* $\mathbf{C}_2 \times \mathbf{M}_2$.

4   New Approximate (Bi-)simulation Relations via Lifting

The requirement on an exact simulation relation between two models is evidently restrictive. This is also shown in the following example of gMDPs.

Example 3 (Models with a Shared Noise Source). Consider an output space $\mathbb{Y} := \mathbb{R}^d$, with a metric $\mathbf{d}_\mathbb{Y}(x, y) := \|x - y\|$ (the Euclidean norm), and two gMDP expressed as noisy dynamic processes:

$$\mathbf{M}_1 : x_1(t+1) = f(x_1(t), u_1(t)) + e_1(t), \qquad y_1(t) = h(x_1(t)),$$
$$\mathbf{M}_2 : x_2(t+1) = f(x_2(t), u_2(t)) + e_2(t), \qquad y_2(t) = h(x_2(t)),$$

where f and h are both globally Lipschitz. Namely, there is an $0 < L < 1$ such that $\|f(x_1, u) - f(x_2, u)\| \leq L\|x_1 - x_2\|$ for all $x_1, x_2 \in \mathbb{R}^n$ and for all

u, and in addition an $0 < H$ such that $\|h(x_1) - h(x_2)\| \leq H\|x_1 - x_2\|$. Suppose the probability distributions of the random variable e_1 and of e_2 can be coupled with distribution $\mathbb{P}_{e_1 \times e_2}$, and that there exists a value $c \in \mathbb{R}$, such that $\mathbb{P}_{e_1 \times e_2}[\|e_1 - e_2\| < c] = 1$. Then for every pair of states $x_1(t)$ and $x_2(t)$ of $\mathbf{M}_1$ and $\mathbf{M}_2$ respectively, the difference between state transitions is bounded as $\|x_1(t+1) - x_2(t+1)\| \leq L\|x_1(t) - x_2(t)\| + c$ with probability 1. Therefore, we know that if $\|x_1(0) - x_2(0)\| \leq \frac{c}{1-L}$, then for all $t \geq 0$, $\|x_1(t) - x_2(t)\| \leq \frac{c}{1-L}$, and $\|y_1(t) - y_2(t)\| \leq \frac{cH}{1-L}$.

Even though the difference in the output of the two models is bounded with probability 1, it is impossible to provide an approximation error using either the method in [16] (hinging on stochastic stability assumptions), or using (approximate) relations as in [9,10]: with the former approach, for the same input sequence $u(t)$ the output trajectories of $\mathbf{M}_1$ and $\mathbf{M}_2$ have bounded difference, but do not converge to each other; with the latter approach, the relation defined via a normed difference cannot satisfy the required notion of transitivity.   □

As mentioned before and highlighted in the previous Example 3, we are interested in introducing a new approximate version of the notion of probabilistic simulation relation, which allows for both δ-differences in the stochastic transition kernels, and ϵ-differences in the output trajectories. For the former prerequisite, we relax the requirements on the lifting in Definition 4.

Definition 7 (δ-Lifting for General State Spaces). *Let $\mathbb{X}_1, \mathbb{X}_2$ be two sets with associated measurable spaces $(\mathbb{X}_1, \mathcal{B}(\mathbb{X}_1)), (\mathbb{X}_2, \mathcal{B}(\mathbb{X}_2))$, and let $\mathcal{R} \subseteq \mathbb{X}_1 \times \mathbb{X}_2$ be a relation for which $\mathcal{R} \in \mathcal{B}(\mathbb{X}_1 \times \mathbb{X}_2)$. We denote by $\bar{\mathcal{R}}_\delta \subseteq \mathcal{P}(\mathbb{X}_1, \mathcal{B}(\mathbb{X}_1)) \times \mathcal{P}(\mathbb{X}_2, \mathcal{B}(\mathbb{X}_2))$ the corresponding lifted relation (acting on $\Delta\bar{\mathcal{R}}_\delta\Theta$), if there exists a probability space $(\mathbb{X}_1 \times \mathbb{X}_2, \mathcal{B}(\mathbb{X}_1 \times \mathbb{X}_2), \mathbb{W})$ satisfying*

1. *for all $X_1 \in \mathcal{B}(\mathbb{X}_1)$: $\mathbb{W}(X_1 \times \mathbb{X}_2) = \Delta(X_1)$;*
2. *for all $X_2 \in \mathcal{B}(\mathbb{X}_2)$: $\mathbb{W}(\mathbb{X}_1 \times X_2) = \Theta(X_2)$;*
3. *for the probability space $(\mathbb{X}_1 \times \mathbb{X}_2, \mathcal{B}(\mathbb{X}_1 \times \mathbb{X}_2), \mathbb{W})$ it holds that $s\mathcal{R}t$ with probability at least $1 - \delta$, or equivalently that $\mathbb{W}(\mathcal{R}) \geq 1 - \delta$.*

We leverage Definition 7 to introduce a new approximate similarity relation that encompasses both approximation requirements, obtaining the following ϵ, δ-approximate probabilistic simulation.

Definition 8 (ϵ, δ-Approximate Probabilistic Simulation). *Consider gMDP $\mathbf{M}_i = (\mathbb{X}_i, \pi_i, \mathbb{T}_i, \mathbb{U}_i, h_i, \mathbb{Y}), i = 1, 2$, over a shared metric output space $(\mathbb{Y}, \mathbf{d}_\mathbb{Y})$. $\mathbf{M}_1$ is ϵ, δ-stochastically simulated by $\mathbf{M}_2$ if there exists an interface function $\mathcal{U}_v$ and a relation $\mathcal{R} \subseteq \mathbb{X}_1 \times \mathbb{X}_2$, for which there exists a Borel measurable stochastic kernel $\mathbb{W}_\mathbb{T}(\cdot|u_1, x_1, x_2)$ on $\mathbb{X}_1 \times \mathbb{X}_2$ given $\mathbb{U}_1 \times \mathbb{X}_1 \times \mathbb{X}_2$, such that $\forall(x_1, x_2) \in \mathcal{R}$:*

1. *$\mathbf{d}_\mathbb{Y}(h_1(x_1), h_2(x_2)) \leq \epsilon$;*
2. *$\forall u_1 \in \mathbb{U}_1$, $\mathbb{T}_1(\cdot|x_1, u_1) \bar{\mathcal{R}}_\delta \mathbb{T}_2(\cdot|x_2, \mathcal{U}_v(u_1, x_1, x_2))$, with lifted probability measure $\mathbb{W}_\mathbb{T}(\cdot|u_1, x_1, x_2)$;*
3. *$\pi_1 \bar{\mathcal{R}}_\delta \pi_2$.*

The simulation relation is denoted as $\mathbf{M}_1 \preceq_\epsilon^\delta \mathbf{M}_2$.

Definition 9 (ϵ, δ-Approximate Probabilistic Bisimulation). *Under the same conditions as before* $\mathbf{M}_1$ *is an* ϵ, δ-*probabilistic bisimulation of* $\mathbf{M}_2$ *if there exists a relation* $\mathcal{R} \subseteq \mathbb{X}_1 \times \mathbb{X}_2$ *such that* $\mathbf{M}_1 \preceq_\epsilon^\delta \mathbf{M}_2$ *w.r.t.* $\mathcal{R}$ *and* $\mathbf{M}_1 \preceq_\epsilon^\delta \mathbf{M}_2$ *w.r.t.* $\mathcal{R}^{-1} \subset \mathbb{X}_2 \times \mathbb{X}_1$. $\mathbf{M}_1$ *and* $\mathbf{M}_2$ *are said to be* ϵ, δ-*probabilistically bisimilar, denoted as* $\mathbf{M}_1 \approx_\epsilon^\delta \mathbf{M}_2$.

In the next section we use the introduced similarity relations to quantify the probability of events of a gMDP via its abstraction and to refine controllers.

4.1   Controller Refinement via Approximate Simulation Relations

Consider two gMDP $\mathbf{M}_1$ and $\mathbf{M}_2$ for which $\mathbf{M}_1$ is the abstraction of the concrete model $\mathbf{M}_2$. The following result is an approximate version of Theorem 2, and provides the main result of this paper, i.e., approximate equivalence of properties defined over the gMDP $\mathbf{M}_1$ and $\mathbf{M}_2$.

Theorem 3. *If* $\mathbf{M}_1 \preceq_\epsilon^\delta \mathbf{M}_2$, *then for all control strategies* $\mathbf{C}_1$ *there exists a control strategy* $\mathbf{C}_2$ *such that for the output traces* $\{y_1(t)\}_{0:N}$ *and* $\{y_2(t)\}_{0:N}$ *of* $\mathbf{C}_1 \times \mathbf{M}_1$ *and* $\mathbf{C}_2 \times \mathbf{M}_2$, *it holds that for all measurable events* $A \subset \mathbb{Y}^{N+1}$,

$$\mathbb{P}_{\mathbf{C}_1 \times \mathbf{M}_1}\left(\{y_1(t)\}_{0:N} \in A_{-\epsilon}\right) - \gamma \leq \mathbb{P}_{\mathbf{C}_2 \times \mathbf{M}_2}\left(\{y_2(t)\}_{0:N} \in A\right) \leq \mathbb{P}_{\mathbf{C}_1 \times \mathbf{M}_1}\left(\{y_1(t)\}_{0:N} \in A_\epsilon\right) + \gamma,$$

with constant $1 - \gamma := (1 - \delta)^{N+1}$, *and with the* ϵ-*expansion of* A *defined as*

$$A_\epsilon := \left\{\{y_\epsilon(t)\}_{0:N} | \exists \{y(t)\}_{0:N} \in A : \max_{t \in [0,N]} \mathbf{d}_\mathbb{Y}(y_\epsilon(t), y(t)) \leq \epsilon\right\}$$

and similarly the ϵ-*contraction defined as* $A_{-\epsilon} := \{\{y(t)\}_{0:N} | \{\{y(t)\}_{0:N}\}_\epsilon \subset A\}$ *where* $\{\{y(t)\}_{0:N}\}_\epsilon$ *is the point-wise* ϵ-*expansion of* $\{y(t)\}_{0:N}$.

Key to show this result is the existence of a refined control strategy $\mathbf{C}_2$, which we detail next. Given a control strategy $\mathbf{C}_1$ over the time horizon $t \in \{0, \ldots, N\}$, there is a control strategy $\mathbf{C}_2$ that refines $\mathbf{C}_1$ over $\mathbf{M}_2$. The control strategy is conceptually given in Algorithm 2. Whilst the state (x_1, x_2) of $\mathbf{C}_2$ is in $\mathcal{R}$, the control refinement from $\mathbf{C}_1$ follows in the same way as for the exact case of Sect. 3.3. Hence, similar to the control refinement for exact probabilistic simulations, the *basic ingredients* of $\mathbf{C}_2$ are the states x_1 and x_2, whose stochastic transition to the pair (x_1', x_2') is governed firstly by a point distribution $\delta_{x_2(t)}(dx_2')$ based on the measured state $x_2(t)$ of $\mathbf{M}_2$; and, subsequently, by the lifted probability measure $\mathbb{W}_\mathbb{T}(dx_1' \mid x_2', u_1, x_2, x_1)$, *conditioned on* x_2'.

On the other hand, whenever the state (x_1, x_2) leaves $\mathcal{R}$ the control chosen by strategy $\mathbf{C}_1$ cannot be refined to $\mathbf{M}_2$ and fails. A new control strategy $\mathbf{C}_{rec}$, referred to as *recovery*, can be used to control the residual trajectory of $\mathbf{M}_2$. The choice is of no importance to the result in Theorem 3, as it bounds errors on probabilistic events based on the event that the states stay in the relation.

Theorem 4 (Refined Control Strategy). *Let gMDP* $\mathbf{M}_1$ *and* $\mathbf{M}_2$, *with* $\mathbf{M}_1 \preceq_\epsilon^\delta \mathbf{M}_2$, *and control strategy* $\mathbf{C}_1 = (\mathbb{X}_{\mathbf{C}_1}, x_{\mathbf{C}_1 0}, \mathbb{X}_1, \mathbb{T}_{\mathbf{C}_1}^t, h_{\mathbf{C}_1}^t)$ *for* $\mathbf{M}_1$ *be* *given. Then for every recovery control strategy* $\mathbf{C}_{rec}$, *a refined control strategy* $\mathbf{C}_2 = (\mathbb{X}_{\mathbf{C}_2}, x_{\mathbf{C}_2 0}, \mathbb{X}_2, \mathbb{T}^t_{\mathbf{C}_2}, h^t_{\mathbf{C}_2})$ *is obtained as an* inhomogenous Markov process *with two discrete modes of operation,* {refinement} *and* {recovery}, *based on Algorithm 2.*

By dividing the execution in Algorithm 2 into a control strategy and a gMDP $\mathbf{M}_2$, we again obtain a refined control strategy with tuple $(\mathbb{X}_{\mathbf{C}_2}, x_{\mathbf{C}_2 0}, \mathbb{X}_2, \mathbb{T}^t_{\mathbf{C}_2}, h^t_{\mathbf{C}_2})$.

Algorithm 2: Refinement of $\mathbf{C}_1$ as $\mathbf{C}_2$

Given the interface function $\mathcal{U}_v$, the (conditional) stochastic kernels $\mathbb{W}_{\mathbb{T}}(dx'_1|x'_2, u_1, x_1, x_2)$ and $\mathbb{W}_\pi(dx_1(0)|x_2(0))$, and the chosen recovery strategy $\mathbf{C}_{rec}$.

Initialise by drawing

- the initial state $x_2(0)$ from π_2, and
- the initial state $x_1(0)$ from $\mathbb{W}_\pi(\cdot\,|x_2(0))$.

Run starting at $t = 0$, **while** $t \leq N$

1. **if** $(x_1(t), x_2(t)) \in \mathcal{R}$ **go to** 2. **else skip to** 6.
2. given $x_1(t)$, select $u_1(t)$ from $\mathbf{C}_1$,   {refine}
3. choose randomised input $\mu_{2t} = \mathcal{U}_v(u_1(t), x_1(t), x_2(t))$,
   draw $x_2(t+1)$ from $\mathbb{T}_2(\cdot\,|x_2(t), \mu_{2t})$,
4. draw $x_1(t+1)$ from $\mathbb{W}_{\mathbb{T}}(\cdot\,|x_2(t+1), u_1(t), x_1(t), x_2(t))$,
5. set $t := t+1$, go to 1.
6. given $x_2(t)$, compute μ_t (from $\mathbf{C}_{rec}$),   {recover}
7. draw $x_2(t+1)$ from $\mathbb{T}_2(\cdot\,|x_2(t), \mu_t)$,
8. set $t := t+1$, go to 6.

4.2 Examples and Properties

Example 4 (Models with a Shared Noise Source – Contin'd from Above). Based on the relation $\mathcal{R} := \{(x_1, x_2) : \|x_1 - x_2\| \leq \frac{c}{1-L}\}$ it can be shown that $\mathbf{M}_1 \approx^0_\epsilon \mathbf{M}_2$ with $\epsilon = \frac{Hc}{1-L}$, since, firstly, it holds that $\mathbf{d}_{\mathbb{Y}}(h(x_1), h(x_2)) \leq \epsilon$ for all $(x_1, x_2) \in \mathcal{R}$, with $\mathbf{d}_{\mathbb{Y}} = \|h(x_1) - h(x_2)\|$. Additionally, for all $(x_1, x_2) \in \mathcal{R}$ and for any input u_1 the selection $u_2 = u_1$ is such that $\mathbb{T}_1(\cdot\,|x_1, u_1)\bar{\mathcal{R}}_0\mathbb{T}_2(\cdot\,|x_2, u_1)$, note that $\bar{\mathcal{R}}_0$ is equal to $\bar{\mathcal{R}}$ (the lifted relation from $\mathcal{R}$). The lifted stochastic kernel is $\mathbb{W}_{\mathbb{T}}(dx'_1 \times dx'_2|u_1, x_1, x_2) := \int_\omega \delta_{f(x_1, u_1)+g_1(\omega)}(dx'_1)\delta_{f(x_2, u)+g_2(\omega)}(dx'_2)\mathbb{P}_\omega(d\omega)$, this stochastic kernel is Borel measurable if $f(x_1, u_1) + g_1(\omega)$ and $f(x_2, u) + g_2(\omega)$ are Borel measurable mappings. The identity interface is Borel measurable.   □

Example 5 (Relationship to Model with Truncated Noise). Consider the stochastic dynamical process $\mathbf{M}_1 : x(t+1) = f(x(t), u(t)) + e(t)$ with output mapping $y(t) = h(x(t))$, operating over the Euclidean state space $\mathbb{R}^n$, and driven by a white noise sequence $e(t) \in \mathbb{R}^n$ with distribution $\mathbb{P}_e$. The output space $y \in \mathbb{Y} \subseteq \mathbb{R}^d$ is endowed with the Euclidean norm $\mathbf{d}_{\mathbb{Y}} = \|\cdot\|$. Select a domain

$D \subset \mathbb{R}^n$ so that, at any given time instant t, $e(t) \in D$ with probability $1 - \delta$. Then define a truncated white noise sequence $\tilde{e}(t)$, with distribution $\mathbb{P}_e (\cdot | D)$. The resulting model $\mathbf{M}_2$ driven by $\tilde{e}(t)$ is $\mathbf{M}_2 : x(t + 1) = f(x(t), u(t)) + \tilde{e}(t)$, with the same output mapping $y(t) = h(x(t))$. We show that $\mathbf{M}_2$ is a $0, \delta$-approximate probabilistic bisimulation of $\mathbf{M}_1$, i.e. $\mathbf{M}_1 \approx_0^\delta \mathbf{M}_2$. Select $\mathcal{R} := \{(x_1, x_2) \text{ for } x_1, x_2 \in \mathbb{R}^n | x_1 = x_2\}$, and choose as interface the identity function, i.e., $\mathcal{U}_v(u_1, x_1, x_2) = u_1$. Denote $t_1(e) = f(x_1, u_1) + e$ and $t_2(\tilde{e}) = f(x_2, u_1) + \tilde{e}$, then a lifting measure depending on $x_1, x_2 \in \mathcal{R}$ and u_1, is

$$\mathbb{W}_\mathbb{T}(dx_1' \times dx_2' | u_1, x_1, x_2) := \int_{e \in D} \delta_{x_1'}(dx_2')\delta_{t_1(e)}(dx_1')\mathbb{P}_e(de) \qquad (1)$$
$$+ \int_{e \in \mathbb{R}^n \setminus D} \delta_{t_1(e)}(dx_1')\mathbb{P}_e(de) \int_{\tilde{e}} \delta_{t_2(\tilde{e})}(dx_2')\mathbb{P}_e(d\tilde{e}|D).$$

□

Example 6 (Relationship Between Noiseless and Truncated-Noise Models). Continuing with Example 5, consider the model with truncated noise $\mathbf{M}_2$ as defined before. In what sense is $\mathbf{M}_2$ approximated by its noiseless version $\mathbf{M}_3$, namely $\mathbf{M}_3 : x(t + 1) = f(x(t), u(t))$, with $y(t) = h(x(t))$? Under requirements on the Lipschitz continuity $\|f(x_1, u) - f(x_2, y)\| \leq L\|x_1 - x_2\|$ $0 < L < 1$, $\|h(x_1) - h(x_2)\| \leq H\|x_1 - x_2\|$, and on the boundedness of D and of $c = \max_{d \in D} \|d\|$, Example 3 can be leveraged by concluding that $\mathbf{M}_2 \approx_\epsilon^0 \mathbf{M}_3$, with $\epsilon = \frac{Hc}{1-L}$.[3]   □

In the Examples 5 and 6 we have that $\mathbf{M}_1$ is approximated by $\mathbf{M}_2$, which is subsequently approximated by $\mathbf{M}_3$. The following theorem and corollary attains a quantitative answer on the question whether $\mathbf{M}_1$ is approximated by $\mathbf{M}_3$.

Theorem 5 (Transitivity of $\preceq_\epsilon^\delta$). *Consider three gMDP $\mathbf{M}_i$, $i = 1, 2, 3$, defined by tuples $(\mathbb{X}_i, \pi_i, \mathbb{T}_i, \mathbb{U}_i, h_i, \mathbb{Y})$, with shared output space.*

$$\text{If } \mathbf{M}_1 \preceq_{\epsilon_a}^{\delta_a} \mathbf{M}_2 \text{ and } \mathbf{M}_2 \preceq_{\epsilon_b}^{\delta_b} \mathbf{M}_3, \text{ then } \mathbf{M}_1 \preceq_{\epsilon_a + \epsilon_b}^{\delta_a + \delta_b} \mathbf{M}_3.$$

Next, as a corollary of this theorem, we discuss further transitivity properties for simulation and bisimulation relations.

Corollary 6 (Transitivity Properties). *Following Theorem 5, it holds that*

- *if $\mathbf{M}_1 \approx_{\epsilon_a}^{\delta_a} \mathbf{M}_2$ and $\mathbf{M}_2 \approx_{\epsilon_b}^{\delta_b} \mathbf{M}_3$, then $\mathbf{M}_1 \approx_{\epsilon_a + \epsilon_b}^{\delta_a + \delta_b} \mathbf{M}_3$, and*
- *if $\mathbf{M}_1 \preceq \mathbf{M}_2$ and $\mathbf{M}_2 \preceq \mathbf{M}_3$, then $\mathbf{M}_1 \preceq \mathbf{M}_3$, and*
- *if $\mathbf{M}_1 \approx \mathbf{M}_2$ and $\mathbf{M}_2 \approx \mathbf{M}_3$, then $\mathbf{M}_1 \approx \mathbf{M}_3$.*

Example 7 (Combination of Examples 5 and 6 via Corollary 6). For the models in Examples 5 and 6 we can conclude that $\mathbf{M}_1 \approx_\epsilon^\delta \mathbf{M}_3$. This means that a stochastic system as in $\mathbf{M}_1$ in Example 5 can be approximated via its deterministic counterpart, and that the approximation error can be expressed via the probability (i.e. amount of truncation cf. Example 5) and the output error (i.e. Example 6). This allows for explicit trading off between output deviation and deviation in probability.   □

---

[3] Alternatively, if $\mathbf{M}_2$ with non-deterministic input $\tilde{e} \in D$ is an ϵ_a- alternating bisimulation [23] of $\mathbf{M}_3$ then $\mathbf{M}_2 \approx_{\epsilon_a}^0 \mathbf{M}_3$.

5   Case Study: Energy Management in Smart Buildings

We are interested in developing advanced solutions for the energy management of smart buildings. We consider a simple building that is divided in two connected zones, each with a radiator affecting the heat exchange in that zone. The temperature fluctuations in the two zones and the ambient temperature dynamics are modelled via $\mathbf{M}$ as a Gaussian process [15]:

$$\mathbf{M} : x(t+1) = Ax(t) + Bu(t) + Fe(t), \qquad y(t) = \begin{bmatrix} 1 & 0 & 0 \\ 0 & 1 & 0 \end{bmatrix} x(t), \quad (2)$$

with stable dynamics characterised by matrices

$$A = \begin{bmatrix} 0.8725 & 0.0625 & 0.0375 \\ 0.0625 & 0.8775 & 0.0250 \\ 0 & 0 & 0.9900 \end{bmatrix}, \quad B = \begin{bmatrix} 0.0650 & 0 \\ 0 & 0.60 \\ 0 & 0 \end{bmatrix}, \quad F = \begin{bmatrix} 0.05 & -0.02 & 0 \\ -0.02 & 0.05 & 0 \\ 0 & 0 & 0.1 \end{bmatrix},$$

where $x_{1,2}(t)$ are the temperatures in zone 1 and 2, respectively; $x_3(t)$ is the deviation of the ambient temperature from its mean; and $u(t) \in \mathbb{R}^2$ is the control input. The state variables are initiated as $x(0) = [16\ 14\ -5]^T$. The disturbance $e(t)$ is a sequence of independent distributed standard Gaussian distributions, for all $t \in \mathbb{R}^+$. This stochastic process can be written as a gMDP as detailed in Example 1. For the model abstraction, we select the controllable dynamics of the mean of the state variables, and consequently omit the ambient temperature:

$$\tilde{\mathbf{M}} : \begin{cases} \tilde{x}(t+1) = \tilde{A}\tilde{x}(t) + \tilde{B}\tilde{u}(t) \in \mathbb{R}^2, & \text{with } \tilde{A} := \begin{bmatrix} 0.8725 & 0.0625 \\ 0.0625 & 0.8775 \end{bmatrix}, \\ \tilde{y}(t) = \begin{bmatrix} 1 & 0 \\ 0 & 1 \end{bmatrix} \tilde{x}(t), & \tilde{B} := \begin{bmatrix} 0.0650 & 0 \\ 0 & 0.60 \end{bmatrix}. \end{cases} \quad (3)$$

We then obtain that, as intuitive, $\tilde{\mathbf{M}} \preceq_\epsilon^\delta \mathbf{M}$.

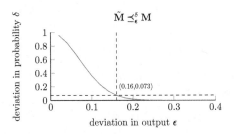

Fig. 1. Figure of trade-off between the output error ϵ and the probability error δ for the δ, ϵ-approximate probabilistic simulation $\tilde{\mathbf{M}} \preceq_\epsilon^\delta \mathbf{M}$. We have selected the pair $(\epsilon, \delta) = (0.16, 0.073)$ as an ideal trade-off.

In order to compute specific values of ϵ and δ, we select the relation $\mathcal{R} := \{(\tilde{x}, x) \in \mathbb{R}^2 \times \mathbb{R}^3 \mid \sqrt{(\tilde{x}_1 - x_1)^2 + (\tilde{x}_2 - x_2)^2} \le \epsilon\}$ and the interface function $\mathcal{U}_v(\tilde{u}, \tilde{x}, x) = \tilde{u} + \tilde{B}^{-1}(\tilde{A}\tilde{x} - \bar{A}x)$, with $\bar{A} = \begin{bmatrix} 0.8725 & 0.0625 & 0.0375 \\ 0.0625 & 0.8775 & 0.0250 \end{bmatrix}$. A stochastic kernel $\mathbb{W}_\mathbb{T}$ for the lifting is $\mathbb{W}_\mathbb{T}(d\tilde{x}' \times dx' \mid \tilde{u}, \tilde{x}, x) = \int_e \delta_{\tilde{f}}(d\tilde{x}')\,\delta_{f(e)}(dx')\mathcal{N}(de \mid 0, I)$, with $\tilde{f} = \tilde{A}\tilde{x} + \tilde{B}\tilde{u}$ and $f(e) = Ax + B\mathcal{U}_v(\tilde{u}, \tilde{x}, x) + Fe$. The lower bound on $\mathbb{W}_\mathbb{T}(\mathcal{R} \mid \tilde{u}, \tilde{x}, x) \le 1 - \delta$ has been computed and traded off against the output deviation in Fig. 1.

We are interested in the goal, expressed for the model $\mathbf{M}$, of increasing the likelihood of reaching the target set $T = [20.5,\ 21]^2$ and staying there thereafter. For the abstract model we have developed a strategy, as in [15], satisfying by construction the property expressed in LTL-like notation with the formula $\varphi = \Diamond\Box T$ and shrunken to $\varphi_{-\epsilon}$ (as per Theorem 3). This strategy is synthesised as a correct-by-construction controller using PESSOA [19], where the discrete-time dynamics are further discretised over state and action spaces: we have selected a state quantisation of 0.05 over the range $[15, 25]^2$, and an input quantisation of 0.05 over the set $[10, 30]^2$. It can be observed that the controller regulates the abstract model $\tilde{\mathbf{M}}$ to eventually remain within the target region, as shown in Fig. 2. We now want to verify that indeed, when refined to the concrete stochastic model, this strategy implies the reaching and staying in the safe set up to some probabilistic error. The refined strategy is obtained from this control strategy as discussed in Sect. 4.1, and recovers from exits out of the relation $\mathcal{R}$ by resetting the abstract states in the relation. A simulation study is given in Fig. 2: as predicted, the behaviour of the controlled concrete model $\mathbf{M}$ stays close to that of $\tilde{\mathbf{M}}$. Over a time horizon of 200 steps the output error exceeds the level $\epsilon = 0.16$ only a few (four) times. Indeed, the probability that the concrete state leaves the relation with the abstract model ($\leq \delta$, with $\delta = 0.073$) leads, over N time steps, to a bound on the probability that it does not satisfy the LTL property: Theorem 3 ensures that this probability is provably less than $1 - (1-\delta)^N \approx N\delta$. In practice, whenever state exits the relation, then the controller recovers by resetting the state of the abstract model and re-applying the strategy again, and thanks to the ϵ-contraction $\varphi_{-\epsilon}$ of the concrete specification, $\mathbf{M}$ will abide by φ with a high confidence.

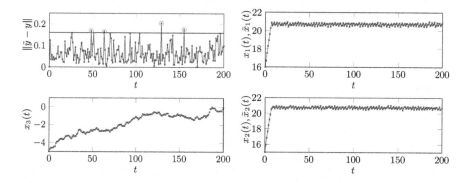

Fig. 2. Refined control for deterministic model applied to $\mathbf{M}$. The figure (above left) evaluates the accuracy of the approximation, and gives with red circles the instances in which the relation is left. The plot (below left) gives the ambient temperature. The plots on the right give the temperature inside the rooms. The (very small) blue crosses give the actual temperature in the rooms (x_1, x_2) and cover the deterministic simulation of $(\tilde{x}_1, \tilde{x}_2)$ drawn in black. (Color figure online)

6   Conclusions

In this work we have discussed new approximate similarity relations for general control Markov processes, and shown that they can be effectively employed for abstraction-based verification and controller refinement. The new relations in particular allow for a useful trade-off over deviations between probability distributions on the states and distances between model outputs.

Alongside practical applications of the developed notions, current research efforts focus on further generalisation of Theorem 3 to specific quantitative properties expressed via temporal logics. We are moreover interested in expanding on the properties of the similarity relations.

Acknowledgement. The research of Sofie Haesaert is supported by DISC through a personal grant from the NWO graduate program.

References

1. Abate, A., Kwiatkowska, M., Norman, G., Parker, D.: Probabilistic model checking of labelled markov processes via finite approximate bisimulations. In: van Breugel, F., Kashefi, E., Palamidessi, C., Rutten, J. (eds.) Horizons of the Mind. LNCS, vol. 8464, pp. 40–58. Springer, Heidelberg (2014)

2. Abate, A., Redig, F., Tkachev, I.: On the effect of perturbation of conditional probabilities in total variation. Stat. Probab. Lett. **88**, 1–8 (2014)

3. Abate, A.: Approximation metrics based on probabilistic bisimulations for general state-space Markov processes: a survey. Electron. Notes Theor. Comput. Sci. **297**, 3–25 (2013)

4. Abate, A., Prandini, M., Lygeros, J., Sastry, S.: Probabilistic reachability and safety for controlled discrete time stochastic hybrid systems. Automatica **44**(11), 2724–2734 (2008)

5. Bertsekas, D.P., Shreve, S.E.: Stochastic Optimal Control: The Discrete Time Case. Athena Scientific, Belmont (1996)

6. Bogachev, V.I.: Measure Theory. Springer Science & Business Media, Heidelberg (2007)

7. Borkar, V.S.: Probability Theory: An Advanced Course. Springer Science & Business Media, Heidelberg (2012)

8. Desharnais, J., Gupta, V., Jagadeesan, R., Panangaden, P.: Metrics for labelled Markov processes. Theor. Comput. Sci. **318**(3), 323–354 (2004)

9. Desharnais, J., Laviolette, F., Tracol, M.: Approximate analysis of probabilistic processes: logic, simulation and games. In: QEST, pp. 264–273, September 2008

10. D'Innocenzo, A., Abate, A., Katoen, J.P.: Robust PCTL model checking. In: Proceedings of the 15th ACM International Conference on Hybrid Systems: Computation and Control, pp. 275–285 (2012)

11. Esmaeil Zadeh Soudjani, S., Gevaerts, C., Abate, A.: FAUST^2: Formal Abstractions of Uncountable-STate STochastic processes. In: Baier, C., Tinelli, C. (eds.) TACAS 2015. LNCS, vol. 9035, pp. 272–286. Springer, Heidelberg (2015)

12. Esmaeil Zadeh Soudjani, S., Abate, A.: Adaptive and sequential gridding procedures for the abstraction and verification of stochastic processes. SIAM J. Appl. Dyn. Syst. **12**(2), 921–956 (2013)

13. Girard, A., Pappas, G.J.: Hierarchical control system design using approximate simulation. Automatica **45**(2), 566–571 (2009)
14. Haesaert, S., Esmaeil Zadeh Soudjani, S., Abate, A.: Verification of general Markov decision processes by approximate similarity relations and policy refinement, May 2016. arXiv preprint arXiv:1605.09557
15. Haesaert, S., Abate, A., Van den Hof, P.M.J.: Correct-by-design output feedback of LTI systems. In: Conference on Decision and Control, pp. 6159–6164 (2015)
16. Julius, A.A., Pappas, G.J.: Approximations of stochastic hybrid systems. IEEE Trans. Autom. Control **54**(6), 1193–1203 (2009)
17. Larsen, K.G., Skou, A.: Bisimulation through probabilistic testing. Inf. Comput. **94**(1), 1–28 (1991)
18. Lindvall, T.: Lectures on the Coupling Method. Courier Corporation, North Chelmsford (2002)
19. Mazo Jr., M., Davitian, A., Tabuada, P.: PESSOA: a tool for embedded controller synthesis. In: Touili, T., Cook, B., Jackson, P. (eds.) CAV 2010. LNCS, vol. 6174, pp. 566–569. Springer, Heidelberg (2010)
20. Meyn, S.P., Tweedie, R.L.: Markov Chains and Stochastic Stability. Communications and Control Engineering Series. Springer, London (1993)
21. Segala, R.: Modeling and verification of randomized distributed real-time systems. Ph.D. thesis, Massachusetts Institute of Technology (1995)
22. Skala, H.J.: The existence of probability measures with given marginals. Ann. Probab. **21**, 136–142 (1993)
23. Tabuada, P.: Verification and Control of Hybrid Systems. Springer, Heidelberg (2009)
24. Zamani, M., Esfahani, P.M., Majumdar, R., Abate, A., Lygeros, J.: Symbolic control of stochastic systems via approximately bisimilar finite abstractions. IEEE Trans. Autom. Control **59**(12), 3135–3150 (2014)
25. Zhang, C., Pang, J.: On probabilistic alternating simulations. In: Calude, C.S., Sassone, V. (eds.) TCS 2010. IFIP AICT, vol. 323, pp. 71–85. Springer, Heidelberg (2010)

Policy Learning for Time-Bounded Reachability in Continuous-Time Markov Decision Processes via Doubly-Stochastic Gradient Ascent

Ezio Bartocci[1], Luca Bortolussi[2,3,4], Tomáš Brázdil[5], Dimitrios Milios[6(✉)],
and Guido Sanguinetti[6,7]

[1] Faculty of Informatics, Vienna University of Technology, Vienna, Austria
[2] Department of Maths and Geosciences, University of Trieste, Trieste, Italy
[3] CNR/ISTI, Pisa, Italy
[4] Modelling and Simulation Group, Saarland University, Saarbrücken, Germany
[5] Faculty of Informatics, Masaryk University, Brno, Czech Republic
[6] School of Informatics, University of Edinburgh, Edinburgh, UK
dmilios@inf.ed.ac.uk
[7] SynthSys, Centre for Synthetic and Systems Biology,
University of Edinburgh, Edinburgh, UK

Abstract. Continuous-time Markov decision processes are an important class of models in a wide range of applications, ranging from cyber-physical systems to synthetic biology. A central problem is how to devise a policy to control the system in order to maximise the probability of satisfying a set of temporal logic specifications. Here we present a novel approach based on statistical model checking and an unbiased estimation of a functional gradient in the space of possible policies. The statistical approach has several advantages over conventional approaches based on uniformisation, as it can also be applied when the model is replaced by a black box, and does not suffer from state-space explosion. The use of a stochastic gradient to guide our search considerably improves the efficiency of learning policies. We demonstrate the method on a proof-of-principle non-linear population model, showing strong performance in a non-trivial task.

1 Introduction

Continuous-time Markov Decision Processes (CTMDPs) [2] are a very powerful mathematical framework to solve control and dependability problems in real-time systems featuring both probabilistic and nondeterministic behaviours. Examples include applications such as the control of epidemic processes [15,20], power management [27], queueing systems [32] and cyber-physical systems [22]. A CTMDP extends a continuous-time Markov chain (CTMC) by introducing a decision maker (also called *scheduler*) that can perform actions with an associated cost or reward. CTMDPs are particularly useful modelling tools to address important problems such as *model checking* [1] and *planning*.

Model checking aims to verify if a CTMDP satisfies a desired requirement for a given class of schedulers or for all possible schedulers. The requirement of

G. Agha and B. Van Houdt (Eds.): QEST 2016, LNCS 9826, pp. 244–259, 2016.
DOI: 10.1007/978-3-319-43425-4_17

interest is usually expressed in terms of the *min/max* probability for a CTMDP to satisfy the temporal logic property [1] of interest. In particular, the main target of the current quantitative model checking techniques for CTMDPs is the *time-bounded reachability* [2,13,25,28,29], a property that requires a CTMDP to reach a particular set of states within a time bound.

Planning or *scheduling* is an orthogonal problem w.r.t. model checking. It consists in devising the optimal sequence of actions (or *policy*) to control the system in order to maximise the probability to satisfy a temporal logic specification such as the aforementioned time-bounded reachability. In the case of CTMDP the optimal scheduling can be either *timed* or *untimed* depending on whether or not the scheduler is aware of the passing of time. Timed optimal scheduling can be further classified in *late* or *early* depending on whether the decision of choosing an action can change while the time passes in a state or it remains unchanged.

In this paper we present a novel statistical approach to compute lower bounds on the maximum reachability probability of a CTMDP. Our method uses a basis-function regression approach to compactly encode schedulers and effectively search for an optimal one. We consider here *randomised time-dependent early schedulers*, and focus on population models, where the state space of the CTMDP is represented by a set of integer-valued variables counting how many entities of each kind are in the system. This is a large class of models: queueing and performance models [13], epidemic scenarios, biological systems are all members of this class. Population models, despite being so common, suffer severely from state space explosion, with the number of states growing exponentially with the number of variables. This reflects on the size of the schedulers: in principle, we would need to store a function of time for each state of the CTMDP, which is unfeasible. This paper contains two main novel insights. First, we leverage the structure of the state space, which can be embedded as a discrete grid in real space, to obtain a continuous relaxation of the problem and consider schedulers defined on such a continuous space. The advantage now is that we can treat time and space uniformly, representing schedulers as continuous functions. This opens up the use of machine learning methods to represent continuous functions as combinations of basis functions, and allows us to define the optimisation problem as a search in such a continuous function space. The second main contribution of the work is to set up an efficient stochastic gradient ascent search algorithm, which considerably speeds up the search in the space of functions. This is based on a novel algorithm using Gaussian Processes (GPs) and statistical model checking to sample in an unbiased manner the gradient of the functional associating a reachability probability with a randomized scheduler. This method allows us to effectively learn schedulers that maximise (locally) the reachability probability.

Organisation of the Paper. In Sect. 2 we present the related work and in Sect. 3 we provide the necessary formal background on CTMDPs. In Sect. 4 we present our algorithm to learn optimal policies using stochastic functional gradient ascent techniques. In Sect. 5 we demonstrate our algorithm on an epidemiology case study. Finally, we draw our conclusion in Sect. 6.

2   Related Work

Symbolic model checking algorithms for discrete-time Markov decision processes have been intensively investigated in [3,7] and implemented in popular tools such as PRISM [19]. In the area of CTMDPs, the problem of time optimal planning has been first considered from a theoretical point of view in [23]. In the last decade there has been a great effort on developing practical model checking techniques for CTMDPs [2,13,25,28,29] (i.e., based on uniformization [2]) with the introduction of efficient approximation algorithms that provide also formal error bounds. Generally, all these techniques rely on the a-priori knowledge of the CTMDP model under investigation and they suffer the state-explosion problem.

In this light, methods based on statistical model checking are particularly attractive, even though they may suffer when the property to be verified is a rare-event. In [16] the authors presented a statistical model checking algorithm for the discrete-time case; their approach was however based on random search combined with a greedy selection criterion, which is difficult to analyse in terms of convergence properties, and may be practically difficult to tune. The availability of an unbiased estimate of the (functional) gradient allows us to improve on the efficiency, and to leverage a rich theory on the convergence of stochastic gradient ascent algorithms. Our approach relies on using Gaussian Processes (GPs), a probability distribution over the space of functions which universally approximates continuous functions. This ability of GPs to provide efficient approximations to intractable functions has been recently exploited in a formal modelling context in a number of publications [5,9,10].

Our work is closely related to research in the area of machine learning, where much research has gone on defining good local search methods to learn effective randomised schedulers, for different criteria like time bounded reward, time unbounded discounted reward, receding horizon. These approaches combine simulation with efficient exploration schemes, like gradient ascent [6,31], path integral policy improvement [33], or the cross entropy method [21], see [34] for a survey. Our approach differs in two main directions: firstly, we are interested in complex rewards associated with trajectories of the system, i.e. reachability probabilities. Secondly, we work directly in continuous time, which prevents the use of simple finite-dimensional gradient ascent methods. In particular, the GP-based method of defining a stochastic gradient ascent algorithm is novel, to the best of our knowledge.

3   Preliminaries

Definition 1. *A continuous-time Markov decision process (CTMDP) is a tuple* $\mathcal{M} = (S, \mathcal{A}, R, s_0)$, *where* S *is a finite set of states,* $\mathcal{A}$ *is a finite set of actions,* $R : S \times \mathcal{A} \times S \to \mathbb{R}_{\geq 0}$ *is the* rate function, *and* $s_0 \in S$ *is the initial state.*

An action $a \in \mathcal{A}$ is *enabled* in a state $s \in S$ if there is a state $s' \in S$ such that $R(s, a, s') > 0$. We call $\mathcal{A}(s)$ the set of enabled actions in s. A *continuous-time Markov chain (CTMC)* is a CTMDP where every $\mathcal{A}(s)$ is a singleton.

We define $E(s,a) = \sum_{s'} R(s,a,s')$ the *exit* rate from a state s when an action a is chosen. We also let $P(s,a,s') = R(s,a,s')/E(s,a)$ be the probability of jumping from s to s' if a is selected.

Intuitively, a run of CTMDP starts in a state s_0 and proceeds as follows: Assume that the CTMDP is currently in a state s_i. First, an action a_i is selected, then the CTMDP waits for a delay t_i randomly chosen according to an exponential distribution with the exit rate $E(s_i, a_i)$, and then a next state s_{i+1} is chosen randomly with the probability $P(s_i, a_i, s_{i+1})$. This produces a run $s_0 a_0 t_0 s_1 a_1 t_1 \cdots$.

In order to obtain a complete semantics, we need to specify how the actions are selected in every step. Obviously, in CTMC, only a single action is enabled in each state. In CTMDP, actions need to be chosen by a scheduler defined as follows.

Definition 2. *An* (early timed) *scheduler is a function* $\sigma : \mathbb{R}_{\geq 0} \times S \times \mathcal{A} \to [0,1]$ *which to every* $t \in \mathbb{R}_{\geq 0}$, $s \in S$ *and* $a \in \mathcal{A}$ *assigns a probability measure* $\sigma(t,s,a)$ *that the action* a *is chosen in* s *at time* t.

A scheduler σ is *deterministic* if for every $t \in \mathbb{R}_{\geq 0}$, $s \in S$ and $a \in \mathcal{A}$ we have that $\sigma(t,s,a) \in \{0,1\}$. We denote by Σ and Σ_D the sets of all schedulers and all deterministic schedulers, respectively.

Remark 1. An early scheduler has the following property: whenever an execution of the CTMDP enters into a state s at time t, the scheduler chooses an action and commits to it. It cannot be changed while the system remains in state s, in contrast with late schedulers, that can change action while in a state.

Once a scheduler σ and an initial state s is fixed, we obtain the unique probability measure $\mathbb{P}_\sigma^{\mathcal{M},s}$ over the space of all runs initiated in s using standard definitions [26].

Time-Bounded Reachability. Let $G \subset S$ be a set of goal states and let $I = [t_1, t_2] \subseteq [0, \infty)$ be a closed interval. Denote by $\mathbb{P}_\sigma^{\mathcal{M},s}(\diamond_I G)$ the probability that G is reached from s within the time interval I using the scheduler σ. Our goal is to maximize $\mathbb{P}_\sigma^{\mathcal{M},s}(\diamond_I G)$, i.e. compute a scheduler σ^* satisfying

$$\mathbb{P}_{\sigma^*}^{\mathcal{M},s}(\diamond_I G) = \sup_{\sigma \in \Sigma} \mathbb{P}_\sigma^{\mathcal{M},s}(\diamond_I G)$$

We say that such a scheduler σ^* is *optimal*.

Proposition 1 [26]. *There always exists an optimal scheduler.*

A proof for the proposition above can be found in the archive version of the paper [4].

When dealing with time-bounded reachability, we may safely assume that schedulers are defined only on the interval $[0,T]$, i.e., on a compact set. An equivalent problem is to maximise a time-bounded safety property $\square_I G$, requiring the CTMDP to remain in a region G during the time-interval I. In this case, we have that $\mathbb{P}_{\sigma^*}^{\mathcal{M},s}(\square_I G) = \mathbb{P}_{\sigma^*}^{\mathcal{M},s}(\neg \diamond_I S \setminus G) = \inf_{\sigma \in \Sigma} \mathbb{P}_\sigma^{\mathcal{M},s}(\diamond_I S \setminus G)$.

Population CTMDPs. In this work, we will consider CTMDPs modelled in a special way, reminiscent of population processes which are very common in performance modelling, epidemiology, systems biology. The basic idea is that we will have populations of agents, belonging to one or more classes, that can interact together and thus evolve in time. Individual agents are typically indistinguishable, hence the state of the system can be described by a set of variables counting the amount of agents of each kind in the system. A non-deterministic action in this context typically represents an action of a global controller, enforcing a policy controlling the system, or effects on the environment.

More formally, we will describe a Population CTMDP (PCTMDP), extending population processes [8,17], as a tuple $(\boldsymbol{X}, \mathcal{T}, \mathcal{A}, \boldsymbol{s}_0)$, where:

- $\boldsymbol{X} = X_1, \ldots, X_n$ is a vector of population variables, $X_i \in \mathbb{N}$, which we assume take values on $S = \mathbb{N}^n \cap E$, where E is a compact subset of $\mathbb{R}^n$ (hence S is finite);
- $\boldsymbol{s}_0 \in S$ is the initial state;
- $\tau \in \mathcal{T}$ is the set of transitions, of the form $(a, \boldsymbol{v}, f(\boldsymbol{X}))$, where a is an action from the set $\mathcal{A}$, $\boldsymbol{v}$ is an update vector, specifying that the state after the execution of a transition in state $\boldsymbol{s}$ is $\boldsymbol{s} + \boldsymbol{v}$, and $f(\boldsymbol{X})$ is the state-dependent rate function.

The idea of this model is that in each state an action a is chosen, and then the model evolves by a race condition between transitions guarded by the action a. If a transition is enabled by all possible actions, we can either specify a copy of it guarded by each model action a, or use the notation $(*, \boldsymbol{v}, f(\boldsymbol{X}))$. The CTMDP $\mathcal{M} = (S, \mathcal{A}, R)$ associated with a PCTMDP $(\boldsymbol{X}, \mathcal{T}, \mathcal{A}, \boldsymbol{x}_0)$ is defined by specifying the state space $S = \mathbb{N}^n \cap E$ and the rate function R as

$$R(\boldsymbol{s}, a, \boldsymbol{s}') = \sum \{ f_\tau(\boldsymbol{s}) \mid \tau = (a, \boldsymbol{v}, f(\boldsymbol{s})) \wedge \boldsymbol{s}' = \boldsymbol{s} + \boldsymbol{v} \}.$$

It is easy to observe, modulo the introduction of enough variables and actions, that the expressive power of PCTMDPs is the same as that of CTMDPs introduced earlier.

4   Learning Optimal Policies via Stochastic Functional Gradient Ascent

In this section we give a variational formulation of the control problem of determining the optimal scheduler for a CTMDP. We show how to approximate statistically in an unbiased way the functional gradient of the time-bounded reachability probability, and give a convergent algorithm to achieve this.

4.1   Reachability Probability as a Functional

As defined in Sect. 3, a scheduler is a way of resolving non-determinism by associating a (time-dependent) probability to each action/ state pair. We will realise a scheduler as a vector $\mathbf{f}$ of functions $f_\alpha : E \times [0,T] \to \mathbb{R}$, one for each action $\alpha \in \mathcal{A}$, where E is the compact subset of $\mathbb{R}^n$ used to define S for the PCTMDP formalism. The corresponding probability of an action α at a state $\mathbf{X}$ can be retrieved using the soft-max (logistic) transform as follows:

$$p_X(\alpha \mid t) \equiv \sigma(t, \alpha, \mathbf{X}) = \frac{\exp(f_\alpha(\mathbf{X}, t))}{\sum_{\alpha' \in \mathcal{A}} \exp(f_{\alpha'}(\mathbf{X}, t))}, \qquad \mathbf{X} \in S, t \in [0, T] \quad (1)$$

Given a scheduler σ, a CTMDP is reduced to a CTMC $\mathcal{M}_\sigma$, and the problem of estimating the probability of a reachability property $\phi = \diamond_I G$ can be reduced to the computation of a transient probability for $\mathcal{M}_\sigma$ by standard techniques [1]. The satisfaction probability can be therefore viewed as a *functional*

$$Q \colon \mathcal{F} \to \mathbb{R}$$

where $\mathcal{F}$ is the set of all possible scheduler functions. The functional is defined explicitly as follows: consider a sample trajectory $\{s, a, t\}_n \equiv s_0 \xrightarrow{\alpha_0, t_0} s_1 \xrightarrow{\alpha_1, t_1} \ldots s_n \xrightarrow{\alpha_n, t_n} s_{n+1}$ from the CTMC $\mathcal{M}_\sigma$ obtained from the CTMDP by selecting a scheduler. Let $\phi = \diamond_I G$, $I = [t_1, t_2]$ be a reachability property, and denote by $\{s, a, t\}_n \models \phi$ the fact that the trajectory reaches G within the specified time bound. We can encode it in the following indicator function:

$$I_\phi(\{s, a, t\}_n) = \begin{cases} 1, & \{s, a, t\}_n \models \phi \\ 0, & \text{otherwise.} \end{cases} \quad (2)$$

Then the expected reachability value associated with the scheduler σ, represented by the vector of functions $\mathbf{f} = \{f_\alpha\}_{\alpha \in \mathcal{A}}$, is defined as follows:

$$Q\left[\mathbf{f}(\mathbf{X}, t)\right] = E_{\mathcal{M}_\sigma}\left[I_\phi(\{s, a, t\}_n)\right], \quad (3)$$

where expectation is taken with respect to the distribution on trajectories of $\mathcal{M}_\sigma$. Notice that in general it is computationally very hard to analytically compute the r.h.s. in the above equation, as it amounts to transient analysis for a time-inhomogeneous CTMC; we therefore need to resort to statistical model checking methods [18,35] to approximate in a Monte Carlo way the expectation in equation (3).

To formulate the continuous time control problem of determining the optimal scheduler, we need to define the concept of functional derivative.

Definition 3. *Let $Q \colon \mathcal{F} \to \mathbb{R}$ be a functional defined on a space of functions $\mathcal{F}$. The functional derivative of Q at $f \in \mathcal{F}$ along a function $g \in \mathcal{F}$, denoted by $\frac{\delta Q}{\delta f}$, is defined by*

$$\int \frac{\delta Q}{\delta f}(\mathbf{X}, t)\, g(\mathbf{X}, t)\, ds\, dt = \lim_{\epsilon \to 0} \frac{Q[f(\mathbf{X}, t) + \epsilon g(\mathbf{X}, t)] - Q[f(\mathbf{X}, t)]}{\epsilon} \quad (4)$$

whenever the limit on the r.h.s. exists.

Notice that if we restrict ourselves to piecewise constant functions on a grid, the definition above returns the standard definition of gradient of a finite-dimensional function. We can now give a variational definition of optimal scheduler

Lemma 1. *An optimal scheduler σ is associated with a function f such that*

$$max_{g \in \mathcal{F}} \left\| \int \frac{\delta Q}{\delta f}(\boldsymbol{X}, t) \, g(\boldsymbol{X}, t) \, dsdt \right\|_2 = 0 \tag{5}$$

where $\| \cdot \|_2$ denotes the L^2 norm on functions.

The variational formulation above allows us to attack the problem via direct optimisation through a gradient ascent algorithm, as we will see below.

4.2 Stochastic Estimation of the Functional Gradient

It is well-known that a gradient ascent approach is guaranteed to find the global optimum of a convex objective function. Gradient ascent starts from an initial solution which is updated iteratively towards the direction that induces the steepest change in the objective function; that direction is given by the gradient of the function. For a functional $Q[f]$ the concept of gradient is captured by the functional derivative $\frac{\delta Q}{\delta f}$, which is a function of $\boldsymbol{X}, t$ that dictates the rate of change of the functional Q when f is perturbed at the point $(\boldsymbol{X}, t)$. In the case of functional optimisation, the gradient ascent update will have the form:

$$f' = f + \gamma \frac{\delta Q}{\delta f} \tag{6}$$

where γ is the learning rate which controls the effect of each update, and $\frac{\delta Q}{\delta f}$ is the functional derivative of Q. Unfortunately, an analytic expression for the functional derivative of the functional defined in (3) is usually not available.

We can however obtain an unbiased estimate of the functional derivative by using the infinite-dimensional generalisation of this simple lemma.

Lemma 2. *Let $q \colon \mathbb{R}^n \to \mathbb{R}$ be a smooth function, and let $\nabla q(\mathbf{v})$ be its gradient at a point $\mathbf{v}$. Let $\mathbf{w}$ be a random vector from an isotropic, zero mean distribution $p(\mathbf{w})$. For $\epsilon \ll 1$, define*

$$\hat{\mathbf{w}} = \begin{cases} \mathbf{w}, & \text{if } q(\mathbf{v} + \epsilon\mathbf{w}) - q(\mathbf{v}) > 0 \\ -\mathbf{w}, & \text{otherwise.} \end{cases} \tag{7}$$

Then

$$E_p[\epsilon\hat{\mathbf{w}}] \propto \nabla q(\mathbf{v}) + O(\epsilon^2).$$

Proof. The tangent space of $\mathbb{R}^n$ at the point $\mathbf{v}$ is naturally decomposed in the orthogonal direct sum of a subspace of dimension 1 parallel to the gradient, and a subspace of dimension $n-1$ tangent to the level surfaces of the function q. For small ϵ, any change in the value of the function q will be due to movement in the gradient direction. As the distribution p is isotropic, every direction is equally likely in $\mathbf{w}$; however, the flipping operation in the definition of $\hat{\mathbf{w}}$ in (7) ensures that the component of $\hat{\mathbf{w}}$ along the gradient $\nabla q(\mathbf{v})$ is always positive, while it does not affect the orthogonal components. Therefore, in expectation, $\hat{\mathbf{w}}$ returns the direction of the functional gradient.

4.3   Scheduler Representation in Terms of Basis Functions

In order to obtain an unbiased estimate of a functional gradient, we need to define a zero-mean isotropic distribution on a suitable space of functions. To do so, we introduce the concept of Gaussian Process, a generalisation of the multivariate Gaussian distribution to infinite dimensional spaces of functions (see, e.g. [30]).

Definition 4. *A Gaussian Process (GP) over an input space $\mathcal{X}$ is an infinite-dimensional family of real-valued random variables indexed by $x \in \mathcal{X}$ such that, for every finite subset $X \subset \mathcal{X}$, the finite dimensional marginal obtained by restricting the GP to X follows a multi-variate normal distribution.*

Thus, a GP can be thought as a distribution over functions $f \colon \mathcal{X} \to \mathbb{R}$ such that, whenever the function is evaluated at a finite number of points, the resulting random vector is normally distributed. In the following, we will only consider $\mathcal{X} = \mathbb{R}^d$ for some integer d.

Just as the Gaussian distribution is characterised by two parameters, a GP is characterised by two functions, the *mean* and *covariance* function. The mean function plays a relatively minor role, as one can always add a deterministic mean function, without loss of generality; in our case, since we are interested in obtaining small perturbations, we will set it to zero. The covariance function, which captures the correlations between function values at different inputs, instead plays a vital role, as it defines the type of functions which can be sampled from a GP. We will use the *Radial Basis Function* (RBF) covariance, defined as follows:

$$\mathrm{cov}(f(x_1), f(x_2)) = k(x_1, x_2) = \alpha^2 \exp\left[-\frac{\|x_1 - x_2\|^2}{\lambda^2}\right]. \tag{8}$$

where α and λ are the amplitude and length-scale parameters of the covariance function. To gain insight into the geometry of the space of functions associated with a GP with RBF covariance, we report without proof the following lemma (see e.g. Rasmussen and Williams, Ch 4.2.1 [30]).

Lemma 3. *Let $\mathcal{F}_N$ be the space of random functions $f = \sum_{j=1}^{N} w_j \phi_j(x)$ generated by taking linear combinations of basis functions $\phi_j(x) = \exp\left[-\frac{\|x - \mu_j\|^2}{\lambda^2}\right]$, with $\mu_j \in \mathbb{R}$ and independent Gaussian coefficients $w_j \sim \mathcal{N}(0, \alpha^2/N)$. The sample space of a GP with RBF covariance defined by (8) is the infinite union of the spaces $\mathcal{F}_N$.*

We refer to the basis functions entering in the constructive definition of GPs given in Lemma 3 as *kernel functions*. Two immediate consequences of the previous Lemma are important for us:

- A GP with RBF covariance defines an *isotropic* distribution in its sample space (this follows immediately from the i.i.d. definition of the weights in Lemma 3);
- The sample space of a GP with RBF covariance is a dense subset of the space of all continuous functions (see also [9] and references therein).

GPs therefore provide us with a convenient way of extending the procedure described in Lemma 2 to the infinite dimensional setting. In particular, Lemma 3 implies that any scheduler function $f \in \mathcal{F}$ that is a sample from a GP (with RBF covariance) can be approximated to arbitrary accuracy in terms of basis functions as follows:

$$f(\boldsymbol{X}, t) = \sum_{j=1}^{N} w_j \exp \left[-0.5([\boldsymbol{X}, t]^\top - \mu_j)^\top \Lambda^{-1}([\boldsymbol{X}, t]^\top - \mu_j) \right] \qquad (9)$$

where $\mu_j \in \mathbb{R}^n \times [0, T]$ is the centre of a Gaussian kernel function, Λ is a diagonal matrix that contains $n + 1$ squared length-scale parameters of the kernel functions, and n is the dimensionality of the state-space. This formulation allows describing functions (aka points in an infinitely dimensional Hilbert space) as points in the finite vector space spanned by the weights $\mathbf{w}$. Note that the proposed basis function representation implies relaxation of the population variables to the continuous domain, though in practice we are only interested in evaluating $f(\boldsymbol{X}, t)$ for integer-valued $\boldsymbol{X}$.

The advantage of the kernel representation is that we do not need to account for all states $\boldsymbol{X} \in S$, but only for N Gaussian kernels with centres μ_j for $1 \le j \le N$. Therefore, the value of the scheduler at a particular state $\boldsymbol{X}$ will be determined as a linear combination of the kernel functions, with proximal kernels contributing more due to the exponential decay of the kernel functions. This method offers a compact representation of the scheduler, and essentially does not suffer from state-space explosion, as we treat states as continuous. Moreover, we do not lose accuracy, as every function on S can be extended to a continuous function on E by interpolation. On the practical side, we consider that the kernel functions are spread evenly across the joint space (state space & time), and the length-scale for each dimension is considered to be equal to the distance of two successive kernels.[1]

4.4   A Stochastic Gradient Ascent Algorithm

Given a scheduler σ, we first evaluate the reachability probability via statistical model checking. We then perturb the corresponding functions f_α by adding a draw from a zero-mean GP with marginal variance scaled by $\epsilon \ll 1$, and evaluate

---

[1] Kernel functions typically also have an amplitude parameter, which we consider to be equal to 1.

again by statistical model checking the probability of the perturbed scheduler. If this is increased, we take a step in the perturbed direction, otherwise we take a step in the opposite direction. Notice that this procedure can be repeated for multiple independent perturbation functions to obtain a more robust estimate. The whole procedure is described in Algorithm 1, which produces an estimate for the gradient of the functional Q at a vector $\mathbf{f}$ of functions f_α by considering the average of k random directions. We are now ready to state our main result:

Algorithm 1. Estimate the functional gradient of $Q[\mathbf{f}]$

Require: Vector $\mathbf{f}$ of functions f_α, scaling factor ϵ, batch size k
Ensure: An estimate of the functional derivative (gradient) $\nabla Q \equiv \frac{\delta Q}{\delta \mathbf{f}}$
   Set gradient $\nabla Q = 0$
   Evaluate $Q[\mathbf{f}]$ via statistical model checking
   for $i = 1$ to k **do**
      Consider random direction $\mathbf{g}$ such that $\forall \alpha \in \mathcal{A}$, we have:

$$g_a \sim \mathcal{N}(0, 1)$$

      Evaluate $Q[\mathbf{f} + \epsilon \mathbf{g}]$
      Estimate the directional derivative:

$$\nabla_{\mathbf{g}} Q = \frac{Q[\mathbf{f} + \epsilon \mathbf{g}] - Q[\mathbf{f}]}{\epsilon}$$

   if $\nabla_{\mathbf{g}} Q > 0$ **then**
      $\nabla Q \leftarrow \nabla Q + \frac{1}{k} \mathbf{g}$
   else
      $\nabla Q \leftarrow \nabla Q - \frac{1}{k} \mathbf{g}$
   end if
   end for

Theorem 1. *Algorithm 1 gives an unbiased estimate of the functional gradient of the functional $Q[f_\alpha]$.*

Proof. Since both the statistical model checking estimation and the gradient estimation are unbiased and independent of each other, this follows.

   Therefore, we can use this stochastic estimate of the functional gradient to devise a stochastic gradient ascent algorithm which directly solves the variational problem in equation (5). This is summarised in Algorithm 2, which requires as input an initial vector of functions $\mathbf{f}_0$, and a learning rate γ_0. The effects of the learning rate on the convergence properties of the method have been extensively studied in the literature. In particular, for a decreasing learning rate convergence is guaranteed in the strictly convex scenario, if the following conditions are satisfied: $\sum_n \gamma_n = \infty$ and $\sum_n \gamma_n^2 < \infty$ [11,24], suggesting a $\Theta(n^{-1})$ decrease for the learning rate. In non-convex problems, such as the ones considered in this work,

---

Algorithm 2. Stochastic gradient ascent for $Q[\mathbf{f}]$

Require: Initial function vector $\mathbf{f}_0$, learning rate γ_0, $n_{\max}$ iterations
Ensure: A function vector $\mathbf{f}$ that approximates a local optimum of Q
    for $n \leftarrow 1$ **to** $n_{\max}$ **do**
        Estimate the functional gradient ∇Q by using Algorithm 1
        Update: $\mathbf{f}_n \leftarrow \mathbf{f}_{n-1} + \gamma_{n-1} \nabla Q$
    end for

---

the $\Theta(n^{-1})$ decrease is generally too aggressive, leading to vulnerability to local optima. Following the recommendations of [12], we adopt a more conservative strategy:

$$\gamma_n = \gamma_0 \, n^{-1/2} \tag{10}$$

where γ_0 is an initial value for the learning rate, which is problem dependent.

5 Example

We demonstrate the stochastic gradient ascent algorithm on a simple epidemiology that features no permanent recovery, also known as the SIS model. The system is modelled as a PCTMDP, in which the state is described by two variables denoting the population of susceptible (X_S) and infected individuals (X_I). We assume that no immunity to the infection is gained upon recovery. The objective is to monitor how infection progresses over time, given that there is a non-deterministic choice at each step among actions in $\mathcal{A} = \{no\ treatment, treatment\}$, indicating whether an external action is taken to deal with the infection.

This non-deterministic choice will affect the dynamics of the system, which are represented by a list of transitions together with their rate functions, in the biochemical notation style (see e.g. [14]):

infection (*): $S + I \xrightarrow{k_i} I + I$, with rate function $k_i \, X_S \, X_I$;

slow recovery (*no treatment*): $I \xrightarrow{k_r} S$, with rate function $k_r \, X_I$;

self-infection (*no treatment*): $S \xrightarrow{k_i} I$, with rate function $k_i \, X_S/2$;

fast recovery (*treatment*): $I \xrightarrow{k_r} S$, with rate function $\alpha k_r \, X_I$;

death (*treatment*): $I \xrightarrow{k_r} \emptyset$, with rate function $k_d \, X_I$;

death (*treatment*): $S \xrightarrow{k_r} \emptyset$, with rate function $k_d \, X_S$;

Among the transitions above, only *infection* has the same rate regardless of any non-deterministic choice. If the *no treatment* action is chosen, infected individuals recover slowly as prescribed by the *slow recovery* transition, while there is a small chance of self-infection. If treatment is applied, the recovery rate is increased by a factor $\alpha > 1$, and the chance of spontaneous infection is eliminated. We assume however that the treatment is associated with some very negative side-effects that result in a small probability of death, either for healthy of infected individuals.

In this example, we seek to construct a scheduler that maximises the probability of having no deaths and no infected individuals during the time interval $[t_1, t_2]$, i.e. maximising the safety property

$$\Box_{[t_1, t_2]} G \qquad G = \{S = N\} \tag{11}$$

The application of treatment contributes in accelerating the extinction of the infected population, but it also introduces a possibility of death. Therefore a policy of constantly applying treatment cannot be optimal with respect to the satisfiability of the property considered. Moreover, maximising the satisfaction probability requires a time-dependent scheduler, as the treatment application has to be appropriately timed so that it has effect in the time-interval $[t_1, t_2]$.

In the experiments that follow, we illustrate how the stochastic gradient ascent algorithm converges to solutions that maximise this probability. We consider a system with total population $N = 100$, and initial populations $X_{S_0} = 90$ and $X_{I_0} = 10$. The rate constants are $k_i = 0.0012$ for infection, $k_r = 0.1$ for recovery, $k_d = 0.0002$ for the death event, while the increase in the recovery rate due to treatment is fixed to $\alpha = 10$. The time bounds for the safety property considered are $t_1 = 50$ and $t_2 = 60$. Regarding the stochastic gradient ascent parameters, the learning rate at the n-th step is $\gamma_n = \gamma_0/\sqrt{n}$, where $\gamma_0 = 5$. For the numerical estimation of the directional derivatives, we consider $\epsilon = 0.1$ and the batch size for the gradient estimation was fixed to $k = 5$. For each estimation of the Q function, we have used 1000 simulation runs. In all cases, the algorithm was run for 100 iterations, meaning that a total of 600000 simulation runs were used for each experiment.

We first present an example that illustrates the importance of time in the satisfaction of the time-bounded property in (11). Figure 1 reports a scheduler which is given as a solution by the stochastic gradient ascent approach. The scheduler is presented as a multivariate function that takes values in $[0, 1]$, indicating the probability of selecting the *no treatment* action for different values of state and time. In particular, we have a series of surface plots, each of which summarises the probability of *no treatment* as function of the 2-dimensional state-space for a different time-point. The white colour denotes that *no treatment* is selected with probability 1, while the black colour implies that *treatment* is used instead. We can see that *treatment* is only preferable for a particular time window and for certain parts of the state-space, that is $X_S > 80$ and $X_I < 20$. This makes sense, as the probability of achieving full recovery from a state with more than 20 infected is too small to justify the risks connected with treatment. More specifically, *treatment* is selected with high probability for $t \in [33.75, 52.5]$, which precedes with a very small overlap the time interval if interest, which is $[50, 60]$. Intuitively, to maximise the probability that all of the population is recovered over the course of a particular interval, the *treatment* action should be engaged just before. In a different case, there is an increased risk of death, as a consequence of the negative effects of prolonged treatment.

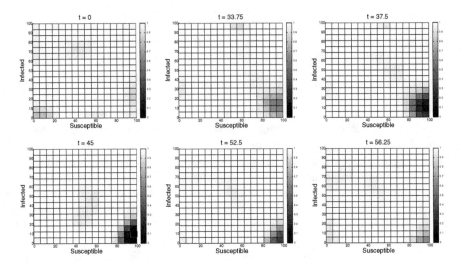

Fig. 1. Example of scheduler that (locally) maximises the probability of $\mathbf{G}_{[t_1,t_2]}S = N$. The white area indicates high probability of choosing the *no treatment* action; the dark area indicates high probability of choosing *treatment*.

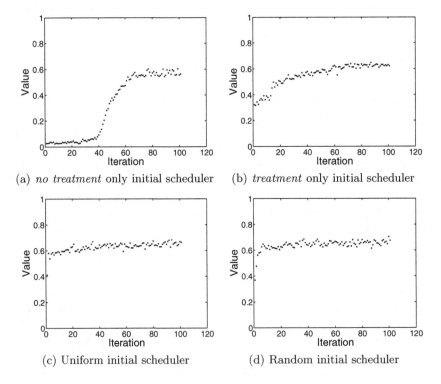

(a) *no treatment* only initial scheduler

(b) *treatment* only initial scheduler

(c) Uniform initial scheduler

(d) Random initial scheduler

Fig. 2. Stochastic gradient ascent starting from different initial schedulers

We next investigate how the algorithm responds to different initial schedulers. In Fig. 2, we monitor how the value of the functional Q as function of the scheduler evolves during the course of the algorithm, starting from different initial solutions. More specifically, Fig. 2(a) depicts the evolution of Q values starting from a scheduler where *no treatment* is globally selected as an action. The initial satisfaction probability is very small, but after a number of iterations it converges to values above 0.6. Figure 2(b) summarises the results where the initial solution selects *treatment* everywhere; apparently this initial solution has been closer to the local optimum and the convergence rate had been significantly faster in this case. Convergence is even faster in Fig. 2(c), where a uniform initial solution was used; that is that each of the two possible actions has equal probability $\forall s \in S$ and $\forall t \in T$. Finally, in Fig. 2(d) we report the Q values for a run starting from a randomly initialised scheduler. In the last two instances, the starting point has had Q values at around 0.4, which is closer to the maximum; therefore the algorithm naturally required fewer iterations to converge to a good solution. Although the convergence rate is apparently dependent on the initial solution, the experiments considered resulted in solutions of similar value, which obtain satisfaction probabilities at around 0.65. It is important to note however that there is no guarantee that the algorithm will converge to the global maximum, since the problem considered in not convex in the general case.

6   Conclusions

Continuous time Markov Decision processes play an important role in many applications, yet they are relatively understudied in the formal methods literature. Part of the problem resides in the difficulty to provide effective characterisations of time-varying schedulers. Recent methodologies [13] have focussed on iterative algorithms based on uniformisation over an increasingly fine time discretisation. While such methods have the ability to compute exactly (up to numerical precision) the objective function (reachability probability), their scalability to large systems is significantly hampered by the state-space explosion problem. Furthermore, such approaches rely on the availability of a mathematical description of the systems, and are therefore not applicable to control black-box systems where a reliable model is not available.

Our approach is suitable instead when the model of the system we want to control is not available a-priori. Our algorithm relies on using GPs, a probability distribution over the space of functions which universally approximates continuous functions.

A potentially significant limitation of our approach is its vulnerability to locally optimal choices. This is a common problem in optimisation, where global convergence in the non-convex case is well known to be hard. Theoretically, this means that our approach can only provide a lower-bound on the reachability probability; nevertheless, this can still be a very valuable result in practical scenarios. Empirically, we observed that the algorithm had excellent performance in a challenging test set; its computational efficiency also means that practical strategies to avoid local optima, such as multiple restarts, can be feasibly employed.

Acknowledgements. L.B. acknowledges partial support from the EU-FET project QUANTICOL (nr. 600708) and by FRA-UniTS. G.S. and D.M. acknowledge support from the European Reasearch Council under grant MLCS306999. T.B. is supported by the Czech Science Foundation, grant No. 15-17564S. E.B. acknowledges the partial support of the Austrian National Research Network S 11405-N23 (RiSE/SHiNE) of the Austrian Science Fund (FWF), the ICT COST Action IC1402 Runtime Verification beyond Monitoring (ARVI) and the IKT der Zukunft of Austrian FFG project HARMONIA (nr. 845631).

References

1. Baier, C., Haverkort, B., Hermanns, H., Katoen, J.-P.: Model-checking algorithms for continuous-time Markov chains. IEEE Trans. Softw. Eng. **29**(6), 524–541 (2003)
2. Baier, C., Hermanns, H., Katoen, J.-P., Haverkort, B.R.: Efficient computation of time-bounded reachability probabilities in uniform continuous-time Markov decision processes. Theor. Comput. Sci. **345**(1), 2–26 (2005)
3. Baier, C., Kwiatkowska, M.Z.: Model checking for a probabilistic branching time logic with fairness. Distrib. Comput. **11**, 125–155 (1998)
4. Bartocci, E., Bortolussi, L., Brázdil, T., Milios, D., Sanguinetti, G.: Policy learning for time-bounded reachability in continuous-time Markov decision processes via doubly-stochastic gradient ascent (2016). CoRR ArXiv, abs/1605.09703
5. Bartocci, E., Bortolussi, L., Nenzi, L., Sanguinetti, G.: System design of stochastic models using robustness of temporal properties. Theor. Comput. Sci. **587**, 3–25 (2015)
6. Baxter, J., Bartlett, P.L., Weaver, L.: Experiments with infinite-horizon, policy-gradient estimation. J. Artif. Int. Res. **15**(1), 351–381 (2011)
7. Bianco, A., de Alfaro, L.: Model checking of probabilistic and nondeterministic systems. In: Thiagarajan, P.S. (ed.) Foundations of Software Technology and Theoretical Computer Science. LNCS, vol. 1026, pp. 499–513. Springer, Heidelberg (1995)
8. Bortolussi, L., Hillston, J., Latella, D., Massink, M.: Continuous aproximation of collective systems behaviour: a tutorial. Perform. Eval. **70**(5), 317–349 (2013)
9. Bortolussi, L., Milios, D., Sanguinetti, G.: Smoothed model checking for uncertain continuous time Markov chains. Inf. Comput. **247**, 235–253 (2016)
10. Bortolussi, L., Sanguinetti, G.: Learning and designing stochastic processes from logical constraints. In: Joshi, K., Siegle, M., Stoelinga, M., D'Argenio, P.R. (eds.) QEST 2013. LNCS, vol. 8054, pp. 89–105. Springer, Heidelberg (2013)
11. Bottou, L.: Large-scale machine learning with stochastic gradient descent. In: Proceedings of COMPSTAT, pp. 177–186. Physica-Verlag HD (2010)
12. Bottou, L.: Stochastic gradient descent tricks. In: Montavon, G., Orr, G.B., Müller, K.-R. (eds.) Neural Networks: Tricks of the Trade, 2nd edn. LNCS, vol. 7700, 2nd edn, pp. 421–436. Springer, Heidelberg (2012)
13. Butkova, Y., Hatefi, H., Hermanns, H., Krcál, J.: Optimal continuous time Markov decisions. In: Finkbeiner, B., Pu, G., Zhang, L. (eds.) ATVA 2015. LNCS, vol. 9364, pp. 166–182. Springer, Heidelberg (2015)
14. Gillespie, D.T.: Exact stochastic simulation of coupled chemical reactions. J. Phys. Chem. **81**(25), 2340–2361 (1977)
15. Guo, X., Hernández-Lerma, O., Prieto-Rumeau, T., Cao, X.-R., Zhang, J., Hu, Q., Lewis, M.E., Vélez, R.: A survey of recent results on continuous-time Markov decision processes. TOP **14**(2), 177–261 (2006)

16. Henriques, D., Martins, J., Zuliani, P., Platzer, A., Clarke, E.M.: Statistical model checking for Markov decision processes. In: Proceedings of QEST, pp. 84–93. IEEE Computer Society (2012)
17. Henzinger, T., Jobstmann, B., Wolf, V.: Formalisms for specifying Markovian population models. Int. J. Found. Comput. Sci. **22**(04), 823–841 (2011)
18. Jha, S.K., Clarke, E.M., Langmead, C.J., Legay, A., Platzer, A., Zuliani, P.: A Bayesian approach to model checking biological systems. In: Degano, P., Gorrieri, R. (eds.) CMSB 2009. LNCS, vol. 5688, pp. 218–234. Springer, Heidelberg (2009)
19. Kwiatkowska, M., Norman, G., Parker, D.: PRISM 4.0: verification of probabilistic real-time systems. In: Gopalakrishnan, G., Qadeer, S. (eds.) CAV 2011. LNCS, vol. 6806, pp. 585–591. Springer, Heidelberg (2011)
20. Lefevre, C.: Optimal control of a birth and death epidemic process. Oper. Res. **29**(5), 971–982 (1981)
21. Mannor, S., Rubinstein, R.Y., Gat, Y.: The cross entropy method for fast policy search. In: ICML, pp. 512–519 (2003)
22. Medina Ayala, A.I., Andersson, S.B., Belta, C.: Probabilistic control from time-bounded temporal logic specifications in dynamic environments. In: Proceedings of ICRA 2012, pp. 4705–4710. IEEE (2012)
23. Miller, B.: Finite state continuous time Markov decision processes with an infinite planning horizon. J. Math. Anal. Appl. **22**(3), 552–569 (1968)
24. Murata, N.: A statistical study of on-line learning. In: On-Line Learning in Neural Networks, pp. 63–92. Cambridge University Press, Cambridge (1998)
25. Neuhaeusser, M.R., Zhang, L.: Time-bounded reachability probabilities in continuous-time Markov decision processes. In: Proceedings of QEST, pp. 209–218. IEEE (2010)
26. Neuhäußer, M.R.: Model checking nondeterministic and randomly timed systems. Ph.D. thesis, RWTH Aachen University (2010)
27. Qiu, Q., Wu, Q., Pedram, M.: Stochastic modeling of a power-managed system-construction and optimization. IEEE Trans. Comput. Aided Des. Integr. Circ. Syst. **20**(10), 1200–1217 (2001)
28. Rabe, M.N., Schewe, S.: Finite optimal control for time-bounded reachability in CTMDPs and continuous-time Markov games. Acta Inform. **48**, 291–315 (2011)
29. Rabe, M.N., Schewe, S.: Optimal time-abstract schedulers for CTMDPs and continuous-time Markov games. Theor. Comput. Sci. **467**, 53–67 (2013)
30. Rasmussen, C.E., Williams, C.K.I.: Gaussian Processes for Machine Learning. MIT Press, Cambridge (2006)
31. Rosenstein, M., Barto, A.G.: Robot weightlifting by direct policy search. In: Proceedings of IJCAI, vol. 17, pp. 839–846 (2001)
32. Sennott, L.I.: Stochastic Dynamic Programming and the Control of Queueing Systems. Wiley, New York (1998)
33. Stulp, F., Sigaud, O.: Path integral policy improvement with covariance matrix adaptation (2012). CoRR ArXiv, arXiv:1206.4621
34. Stulp, F., Sigaud, O.: Policy improvement methods: between black-box optimization and episodic reinforcement learning (2012)
35. Younes, H.L.S., Simmons, R.G.: Statistical probabilistic model checking with a focus on time-bounded properties. Inf. Comput. **204**(9), 1368–1409 (2006)

Compact Representation of Solution Vectors in Kronecker-Based Markovian Analysis

Peter Buchholz[1], Tuğrul Dayar[2(✉)], Jan Kriege[1], and M. Can Orhan[2]

[1] Informatik IV, Technical University of Dortmund, 44221 Dortmund, Germany
{peter.buchholz,jan.kriege}@cs.tu-dortmund.de
[2] Department of Computer Engineering, Bilkent University,
TR-06800 Bilkent, Ankara, Turkey
{tugrul,morhan}@cs.bilkent.edu.tr

Abstract. It is well known that the infinitesimal generator underlying a multi-dimensional Markov chain with a relatively large reachable state space can be represented compactly on a computer in the form of a block matrix in which each nonzero block is expressed as a sum of Kronecker products of smaller matrices. Nevertheless, solution vectors used in the analysis of such Kronecker-based Markovian representations still require memory proportional to the size of the reachable state space, and this becomes a bigger problem as the number of dimensions increases. The current paper shows that it is possible to use the hierarchical Tucker decomposition (HTD) to store the solution vectors during Kronecker-based Markovian analysis relatively compactly and still carry out the basic operation of vector-matrix multiplication in Kronecker form relatively efficiently. Numerical experiments on two different problems of varying sizes indicate that larger memory savings are obtained with the HTD approach as the number of dimensions increases.

Keywords: Markov chains · Kronecker products · Hierarchical Tucker decomposition · Reachable state space · Compact vectors

1 Introduction

Modelling and analysis of multi-dimensional Markov chains (MC) on high end desk-top computers is an area of research with ongoing interest. When a discrete-event dynamic system is composed of interacting subsystems, it may be possible to provide a state-based mathematical model for its behaviour as a multi-dimensional MC with each dimension of the MC representing a different subsystem and a number of events that trigger state changes at certain transition rates. In this kind of model, subsystems can change state locally by themselves, that is, independently of states the other subsystems are in, or they can change state synchronously with some or all the other subsystems depending on their local states. The state space of such a model is therefore determined by the combination of states the subsystems can be in under the operational semantics of the system. Hence, a subset of the Cartesian product of the subsystem state

© Springer International Publishing Switzerland 2016
G. Agha and B. Van Houdt (Eds.): QEST 2016, LNCS 9826, pp. 260–276, 2016.
DOI: 10.1007/978-3-319-43425-4_18

spaces forms the so called reachable state space. Usually not all states from the Cartesian product are reachable because synchronized transitions prohibit some specific combinations of subsystem states to be reachable [3,6]. It is important to be able to represent this reachable state space and the transitions among its states compactly and then analyse the steady-state or transient behaviour of the underlying system as accurately and as efficiently as possible.

When the reachable state space at hand is relatively large but finite, the infinitesimal generator underlying the MC can be represented as a block matrix in which each nonzero block is expressed as a sum of Kronecker products of smaller rectangular matrices [7]. This is the form of the Kronecker representation in hierarchical Markovian models [3], where rectangularity of the smaller matrices is possible due to the product state space of the modelled system being larger than its reachable state space [9]. When the product state space is equal to the reachable state space, the smaller matrices turn out to be square as in stochastic automata networks [19,20].

For Kronecker-based Markovian representations, analysis methods employ vector-Kronecker product multiplication as the basic operation [21]. Therein, the challenge is to perform this operation in as little of memory and as fast as possible. When the factors in the Kronecker product terms are relatively dense, the operation can be performed efficiently by the shuffle algorithm [10]. When the factors are relatively sparse, it may be more efficient to obtain nonzeros of the generator in Kronecker form on the fly and multiply them with corresponding elements of the vector [6]. Recently, the shuffle algorithm has been modified so that relevant elements of the vector are multiplied with submatrices of factors in which zero rows and columns are omitted [8]. This approach is shown to avoid unnecessary floating-point operations (flops) that evaluate to zero during the course of the multiplication and possibly reduces the amount of memory used. In many cases, a smaller number of flops than the shuffle algorithm and the algorithm that generates nonzeros on the fly is possible. Nevertheless, the memory allocated for the vectors in all mentioned algorithms is still proportional to the size of the reachable state space, and this size increases rapidly with the number of dimensions.

The current paper takes a different approach and attempts to reduce the amount of memory allocated to solution vectors in Kronecker-based Markovian analysis by using the hierarchical Tucker decomposition (HTD) [14,15]. HTD is originally conceived in the context of providing a compact approximate representation for dense multi-dimensional data [12] in a manner similar to the tensor-train decomposition [18], but is somewhat more suitable to our requirements in that the decomposition is available through a tree data structure with logarithmic depth in the number of dimensions. Both decompositions have the special feature of possessing approximation errors that can be user controlled, and hence, approximations accurate to machine precision are computable using them. Clearly, with such decompositions it is always possible to trade quality of approximation for compactness of representation, and how compact the solution vector in HTD format remains throughout the solution process is an interesting question to investigate. The tensor train decomposition has been applied in [13] to approximate the solution vector for models where the product space is reachable using an alternating least squares approach. HTD has, to the best of our knowledge, not been applied to structured Markov chains yet.

Here, we show that a compact solution vector in HTD format can be multiplied with a sum of Kronecker products to yield another compact solution vector in HTD format. In doing this, we note that the multiplication of the compact solution vector in HTD format with a Kronecker product term does not increase the memory requirements of the compact vector, but the addition of two compact vectors does, which necessitates some kind of truncation, hence, approximation, to be introduced to the addition operation only. Then, starting from an initial solution, the compact vector in HTD format is iteratively multiplied with the uniformized generator matrix of a given MC in Kronecker form until a predetermined stopping criterion is met. Indeed, we are interested in observing how the memory requirements of the compact solution vector in HTD format changes over the course of iterations due to the sequence of multiply, add, and truncate operations in each iteration, together with the average time it takes to perform the iteration and the influence of the approximation error on the quality of the solution. The same numerical experiment is performed with a flat solution vector as long as the reachable state space size using the modified shuffle algorithm. The two approaches are compared for their memory and timing requirements, leading us to the conclusion that compact vectors in HTD format become relatively more memory efficient as the number of dimensions increases.

In passing to the organization of the paper, we remark that compact representations for solution vectors in Markovian analysis have also been considered from the perspective of binary decision diagrams [5,16]. The proposed compact structures therein have not been time-wise competitive, whereas the approach investigated in this paper seems to be a step forward. The organization of the paper is as follows. In Sect. 2, we provide background information on HTD and the related algorithms that are be used in our Kronecker setting. In Sect. 3, we discuss implementation issues associated with using HTD within the NSolve package of the Abstract Petri Net Notation (APNN) toolbox [1,2]. In Sect. 4, we present results of numerical experiments on two different problems of varying sizes and having transitions that take place at different time scales. Section 5 concludes the paper.

2   Compact Vectors in Kronecker Setting

Let us consider a d-dimensional Markovian system, where $\mathcal{S}_h$ denotes the state space of the hth ($h = 1, \ldots, d$) component in the d-dimensional MC, and assume that $\mathcal{S}_h$ are defined on consecutive nonnegative integers starting from 0. We denote the reachable state space of the system by $\mathcal{S} \subseteq \times_{h=1}^{d} \mathcal{S}_h$, where $\times_{h=1}^{d} \mathcal{S}_h$ is the product state space. Now, let $\mathcal{S}^{(i)} = \times_{h=1}^{d} \mathcal{S}_h^{(i)}$, where $\mathcal{S}_h^{(i)}$ is a partition of $\mathcal{S}_h$ in the form of consecutive integers for $i = 1, \ldots, J$. Then $\mathcal{S}^{(1)}, \ldots, \mathcal{S}^{(J)}$ is a Cartesian product partitioning of $\mathcal{S}$ if $\mathcal{S} = \cup_{i=1}^{J} \mathcal{S}^{(i)}$ and $\mathcal{S}^{(i)} \cap \mathcal{S}^{(j)} = \emptyset$ for $i \neq j$ and $i, j = 1, \ldots, J$ [9].

The infinitesimal generator $\mathbf{Q}$ underlying the MC can be viewed as a $(J \times J)$ block matrix induced by the Cartesian product partitioning of $\mathcal{S}$ as in [7,9]

$$\mathbf{Q} = \begin{bmatrix} \mathbf{Q}^{(1,1)} & \cdots & \mathbf{Q}^{(1,J)} \\ \vdots & \ddots & \vdots \\ \mathbf{Q}^{(J,1)} & \cdots & \mathbf{Q}^{(J,J)} \end{bmatrix}.$$

Block (i,j) of $\mathbf{Q}$ for $i,j = 1, \ldots, J$ is given by

$$\mathbf{Q}^{(i,j)} = \begin{cases} \sum_{k \in \mathcal{K}^{(i,j)}} \mathbf{Q}_k^{(i,j)} + \mathbf{Q}_D^{(i)} & \text{if } i = j, \\ \sum_{k \in \mathcal{K}^{(i,j)}} \mathbf{Q}_k^{(i,j)} & \text{otherwise,} \end{cases}$$

where

$$\mathbf{Q}_k^{(i,j)} = \alpha_k \bigotimes_{h=1}^{d} \mathbf{Q}_{k,h}^{(i,j)}, \quad \mathbf{Q}_D^{(i)} = -\sum_{j=1}^{J} \sum_{k \in \mathcal{K}^{(i,j)}} \alpha_k \bigotimes_{h=1}^{d} \mathrm{diag}(\mathbf{Q}_{k,h}^{(i,j)} \mathbf{e}),$$

$\otimes$ is the Kronecker product operator, α_k is the rate associated with continuous-time transition k, $\mathcal{K}^{(i,j)}$ is the set of transitions in block (i,j), $\mathbf{e}$ represents a column vector of ones, $\mathrm{diag}(\mathbf{a})$ denotes the diagonal matrix with the entries of vector $\mathbf{a}$ along its diagonal, and $\mathbf{Q}_{k,h}^{(i,j)}$ is the submatrix of the transition matrix $\mathbf{Q}_{k,h}$ whose row and column state spaces are $\mathcal{S}_h^{(i)}$ and $\mathcal{S}_h^{(j)}$, respectively [3]. In practice, the matrices $\mathbf{Q}_{k,h}$ are sparse [7] and held in sparse row format since the nonzeros in each of its rows indicate the possible transitions from the state with that row index. The advantage of partitioning the reachable state space is the elimination of unreachable states from the set of rows and columns of the generator to avoid unnecessary flops due to unreachable states. We also remark that the continuous-time transition rate of a Kronecker product term, α_k, can be eliminated by scaling one factor in the term with that rate.

To simplify the discussion and the notation, we consider the multiplication of a single block of $\mathbf{Q}$ from the left with a (sub)vector, and therefore, omit the indices (i,j) and write the index k associated with the transition as a superscript in parentheses above the matrices forming the block. Hence, we concentrate on the operation

$$\mathbf{y}^T := \mathbf{x}^T \sum_{k=1}^{K} \bigotimes_{h=1}^{d} \mathbf{Q}_h^{(k)},$$

where $\mathbf{Q}_h^{(k)}$ is a $(m_h \times n_h)$ matrix, implying $\bigotimes_{h=1}^{d} \mathbf{Q}_h^{(k)}$ is a $(\prod_{h=1}^{d} m_h \times \prod_{h=1}^{d} n_h)$ matrix, and $\mathbf{x}$ is a $(\prod_{h=1}^{d} m_h \times 1)$ vector. K is equal to the number of terms in the sum, i.e., $|\mathcal{K}^{(i,j)}|$ if we consider block (i,j). Observe that this is the operation that takes place when each block of a block matrix in Kronecker form such as $\mathbf{Q}$ gets multiplied on the left by an iteration subvector. In fact, the same subvector multiplies all blocks in a row of the matrix in Kronecker form.

To be consistent with the literature, we consider in the following multiplications of Kronecker products $\otimes_{h=1}^{d} \mathbf{A}_h^{(k)}$ with column vector $\mathbf{x}$ and their summation in the usual matrix-vector form

$$\mathbf{y} := \sum_{k=1}^{K} \left(\bigotimes_{h=1}^{d} \mathbf{A}_h^{(k)} \right) \mathbf{x},$$

where $\mathbf{A}_h^{(k)}$ is the transpose of $\mathbf{Q}_h^{(k)}$ and of size $(n_h \times m_h)$. In particular, we are interested in its implementation as

$$\mathbf{y}^{(1)} := \mathbf{0}, \quad \mathbf{x}^{(k)} := \left(\bigotimes_{h=1}^{d} \mathbf{A}_h^{(k)} \right) \mathbf{x}, \quad \mathbf{y}^{(k+1)} := \mathbf{y}^{(k)} + \mathbf{x}^{(k)} \quad \text{for } k = 1, \ldots, K,$$

and $\mathbf{y} := \mathbf{y}^{(K+1)}$, where $\mathbf{0}$ is a column vector of 0's. Now, we turn to the HTD format.

2.1 HTD Format

Assuming without loss of generality that d is a power of 2, the $(\prod_{h=1}^{d} m_h \times 1)$ vector $\mathbf{x}$ in (orthogonalized) HTD format can be expressed as

$$\mathbf{x} = (\mathbf{U}_1 \otimes \cdots \otimes \mathbf{U}_d)\mathbf{c},$$

where $\mathbf{U}_h$ for $h = 1, \ldots, d$ are $(m_h \times r_h)$ orthogonal basis matrices for the different dimensions in the model and

$$\mathbf{c} = (\mathbf{B}_{1,2} \otimes \cdots \otimes \mathbf{B}_{d-1,d}) \cdots (\mathbf{B}_{1,\ldots,d/2} \otimes \mathbf{B}_{d/2+1,\ldots,d})\mathbf{B}_{1,\ldots,d}$$

is a $(\prod_{h=1}^{d} r_h \times 1)$ vector in the form of a product of $\log_2 d$ matrices each of which except the last is a Kronecker product of a number of transfer matrices $\mathbf{B}_t$ related to each other as in the full binary tree of Fig. 1. The transfer matrix

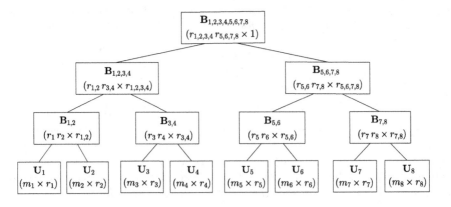

Fig. 1. Matrices forming $\mathbf{x}$ in HTD format for $d = 8$.

$\mathbf{B}_t$ is of size $(r_{t_l} r_{t_r} \times r_t)$ with the node index t defined as $t := t_l, t_r$, and $r_{1,\ldots,d} = 1$ since $\mathbf{B}_{1,\ldots,d}$ is at the root of the tree [14, pp. 5–6].

The $(d-1)$ intermediate nodes of the binary tree in Fig. 1 store the transfer matrices $\mathbf{B}_t$ and its leaves store the basis matrices $\mathbf{U}_h$ so that each intermediate node has two children. In orthogonalized HTD format of $\mathbf{x}$, one can also conceive of orthogonal basis matrices $\mathbf{U}_t = (\mathbf{U}_{t_l} \otimes \mathbf{U}_{t_r})\mathbf{B}_t$, at intermediate nodes with r_t columns that relate the orthogonal basis matrices $\mathbf{U}_{t_l}$ and $\mathbf{U}_{t_r}$ for the two children of transfer matrix $\mathbf{B}_t$ with the transfer matrix itself. In fact, the orthogonal matrix $\mathbf{U}_t$ has in its columns the singular vectors associated with the largest r_t singular values [11, pp. 76–79] of the matrix obtained by taking index t as row index, the remaining indices in order as column index of the d-dimensional data at hand (i.e., with a slight abuse of notation, $\mathbf{x}(t, \{1, \ldots, d\} - t)$). Hence, we have the concepts of "hierarchy of matricizations" and "higher-order singular value decomposition (HOSVD)", and r_t is the rank of the truncated HOSVD. More detailed information regarding this can be found in [12,14]. We remark that $\mathbf{B}_t$ may also be viewed as a 3-dimensional array of size $(r_{t_l} \times r_{t_r} \times r_t)$ having as many indices in each of its three dimensions as the number of columns in the matrices in its two children and itself, respectively. The number of transfer matrices in the lth factor forming $\mathbf{c}$ is the Kronecker product of $2^{\log_2 d - l}$ transfer matrices for $l = 1, \ldots, \log_2 d - 1$. In fact, $\mathbf{c}$ is a product of Kronecker products, and so is $\mathbf{x}$, but neither has to be formed explicitly.

When d is not a power of 2, it is still useful to keep the tree in a balanced form, for instance, as in Fig. 2 for which

$$\mathbf{x} = (((\mathbf{U}_1 \otimes \mathbf{U}_2)\mathbf{B}_{1,2}) \otimes \mathbf{U}_3 \otimes \mathbf{U}_4 \otimes \mathbf{U}_5)(\mathbf{B}_{1,2,3} \otimes \mathbf{B}_{4,5})\mathbf{B}_{1,2,3,4,5}.$$

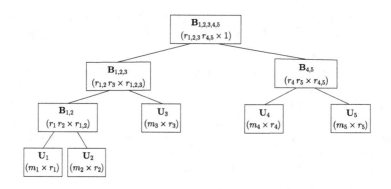

Fig. 2. Matrices forming $\mathbf{x}$ in HTD format for $d = 5$.

Assuming that $r_{\max} = \max_t(r_t)$ and $m_{\max} = \max(m_1, \ldots, m_d)$, memory requirement for matrices in the binary tree associated with HTD format is bounded by $dm_{\max}r_{\max}$ at the leaves, $r_{\max}^2$ at the root, and $(d-2)r_{\max}^3$ at other intermediate nodes, thus, totally $dm_{\max}r_{\max} + (d-2)r_{\max}^3 + r_{\max}^2$. In the next subsection, we show how a particular rank-1 vector can be represented in HTD format.

2.2   Uniform Distribution in HTD Format

Let $\mathbf{x} = \mathbf{e}/m$ be the $(m \times 1)$ uniform distribution vector, where $m = \prod_{h=1}^{d} m_h$. Then $\mathbf{x}$ may be represented in HTD format with all matrices having rank-1 for which the basis matrices given by $\mathbf{U}_h = \mathbf{e}/\sqrt{m_h}$ are of size $(m_h \times 1)$ for $h = 1, \ldots, d$ and the transfer matrices given by

$$\mathbf{B}_t = \begin{cases} (\prod_{h=1}^{d} \sqrt{m_h})/m & \text{if } t \text{ corresponds to root} \\ 1 & \text{otherwise} \end{cases}$$

are (1×1). Note that memory taken up by flat representation of $\mathbf{x}$ is m nonzeros, whereas that with HTD format is $d - 1 + \sum_{h=1}^{d} m_h$ nonzeros since the $(d-1)$ transfer matrices are all scalars equal to 1 except the one corresponding to the root. In passing to the multiplication of a compact vector with a Kronecker product, we remark that each basis matrix $\mathbf{U}_h$ for the uniform distribution has only a single column and that column is unit 2-norm, implying all $\mathbf{U}_h$ are orthogonal.

2.3   Multiplication of Vector in HTD Format with Kronecker Product

Assuming that $\mathbf{x}$ is in HTD format with orthogonal basis matrices $\mathbf{U}_h$ and transfer matrices $\mathbf{B}_t$ forming vector $\mathbf{c}$, the operation

$$\mathbf{x}^{(k)} := \left(\bigotimes_{h=1}^{d} \mathbf{A}_h^{(k)} \right) \mathbf{x} \quad \text{is equivalent to performing} \quad \mathbf{x}^{(k)} := \left(\bigotimes_{h=1}^{d} \mathbf{A}_h^{(k)} \mathbf{U}_h \right) \mathbf{c}$$

since $\mathbf{x} = (\otimes_{h=1}^{d} \mathbf{U}_h)\mathbf{c}$. Hence, the only thing that needs to be done to carry out the computation of $\mathbf{x}^{(k)}$ in HTD format is to multiply the $(n_h \times m_h)$ Kronecker factor $\mathbf{A}_h^{(k)}$ with the corresponding $(m_h \times r_h)$ orthogonal basis matrix $\mathbf{U}_h$ for $h = 1, \ldots, d$. Clearly, the $(n_h \times r_h)$ product matrix $\mathbf{A}_h^{(k)} \mathbf{U}_h$ need not be orthogonal. But this does not pose much of a problem, since $\mathbf{x}^{(k)}$ can be transformed into orthogonalized HTD format if the need arises by computing the QR decomposition [11, pp. 246–250] of $\mathbf{A}_h^{(k)} \mathbf{U}_h = \tilde{\mathbf{U}}_h \mathbf{R}_h$ for $h = 1, \ldots, d$, propagating the triangular factors $\mathbf{R}_h$ into the transfer matrices, and orthogonalizing the updated transfer matrices at intermediate nodes in a similar manner up to the root as in Algorithm 1 in [14, p. 12]. However, the situation is not as good for the addition of two compact vectors.

2.4   Addition of Two Vectors in HTD Format and Truncation

Addition of two matrices $\mathbf{Y}$ and $\mathbf{X}$ with given singular value decompositions (SVDs) [11, pp. 76–79]

$$\mathbf{Y} = \mathbf{U_Y} \mathbf{\Sigma_Y} \mathbf{V_Y}^T \quad \text{and} \quad \mathbf{X} = \mathbf{U_X} \mathbf{\Sigma_X} \mathbf{V_X}^T$$

results in

$$\mathbf{Y} + \mathbf{X} := (\mathbf{U_Y} \ \mathbf{U_X}) \begin{pmatrix} \mathbf{\Sigma_Y} & \\ & \mathbf{\Sigma_X} \end{pmatrix} (\mathbf{V_Y} \ \mathbf{V_X})^T .$$

Here, $\mathbf{\Sigma_Y}, \mathbf{\Sigma_X}$ are diagonal matrices of singular values, whereas $\mathbf{U_Y}, \mathbf{U_X}$ and $\mathbf{V_Y}, \mathbf{V_X}$ are orthogonal matrices of left and right singular (row) vectors associated with matrices $\mathbf{Y}, \mathbf{X}$, respectively. SVD is a rank revealing factorization in that the number of nonzero singular values of a matrix corresponds to its column rank. This implies that the sum $(\mathbf{Y} + \mathbf{X})$ has a rank equal to the sum of the ranks of the two matrices that are added.

The situation for the sum $\mathbf{y}^{(k+1)}$ of the two vectors $\mathbf{y}^{(k)}$ and $\mathbf{x}^{(k)}$ in HTD format is no different if one replaces the SVD with HOSVD. This is conveniently illustrated for $d = 4$ by Fig. 5 in [14, p. 11]. For the following steps performing the addition and representing the resulting vector in HTD format, we exploit the algorithms presented in [14]. Among three alternative approaches that have been investigated therein for computing $\mathbf{y}$, the best seems to be to multiply, add and then truncate K times as demonstrated in Fig. 11 of [14]. This approach is coded in Algorithm 7 of [14, p. 23] which works by calling Algorithm 3 that takes care of the reduced Gramians computations of a compact vector in non-orthogonalized HTD format. Recall that the compact vector $\mathbf{x}^{(k)}$ obtained after multiplication does not need to be in orthogonal HTD format even though $\mathbf{x}$ might have been. Once Algorithm 3 is executed, Algorithm 7 takes over and computes the truncated HOSVD for the sum of two vectors $\mathbf{y}^{(k)}$ and $\mathbf{x}^{(k)}$ in HTD format without initial orthogonalization. The output $\mathbf{y}^{(k+1)}$ of Algorithm 7 is a truncated compact vector in orthogonalized HTD format and this operation is repeated K times until $\mathbf{y}$ is obtained. The number of flops executed by Algorithm 7 is $O(dK^2 r_{max}^2 (n_{max} + r_{max}^2 + K r_{max}))$, where $n_{max} = \max(n_1, \dots, n_d)$. The significance of this algorithm is that one can impose an accuracy of ϵ on the truncated HOSVD by choosing rank r_t in node t based on dropping the smallest singular values whose squared sum is less than or equal to $\epsilon^2/(2d-3)$ [14, pp. 18–19]. This is a very nice result but also implies that the truncation leads to an approximate solution vector.

2.5 Computing the 2-Norm of a Vector in HTD Format

Normally, it is more relevant to compute the maximum (i.e., infinity) norm of a solution vector in probabilistic analysis even though all norms are known to be equivalent [11, pp. 68–70]. However, the computation of the maximum value (in magnitude) of the elements of a compact vector requires being able to know which indexed value is the largest and also its value, which seems to be costly for a compact vector in HTD format. Therefore, we consider the computation of the 2-norm of vector $\mathbf{y}$ given by $||\mathbf{y}||_2 = \sqrt{\mathbf{y}^T \mathbf{y}}$.

Fortunately, $||\mathbf{y}||_2$ can be obtained using Algorithm 2 in [14, p. 14], which computes inner products of two compact vectors in HTD format. Here, the only difference is that the two vectors are the same vector $\mathbf{y}$. The algorithm starts from the leaves of the binary tree and moves towards the root, requiring the same sequence of operations in the first part of the computation of reduced Gramians in Algorithm 3 in [14, p. 17]. But, this has already been discussed in the previous subsection.

Now, we can move to implementation issues regarding compact solution vectors in HTD format for Kronecker-based Markovian representations.

3   Implementation Issues

The implementation is done within the NSolve package of the APNN Toolbox [1,2]. The binary tree data structure accompanying the HTD format is allocated at the outset depending on the value of d. It is stored in the form of an array of tree nodes from root to leaves level by level so that accessing the children of a parent node or the parent of a child node becomes relatively easy. In a tree node t, there are pointers to matrices $\mathbf{U}_t$ for leaves and $\mathbf{B}_t$ for intermediate nodes which we have seen and accounted for before, but also pointers to matrices $\mathbf{R}_t$ and, as we explain shortly, (2×2) block matrices $\mathbf{M}_t$ and $\mathbf{G}_t$ for each node. Since we expect solution vectors to be dense, the matrices in the compact representation are stored as full matrices including those corresponding to the blocks of $\mathbf{M}_t$ and $\mathbf{G}_t$. The nonzero elements of the full matrices are kept in a one-dimensional real array so that relevant LAPACK methods available at [17] can be called without having to copy vectors. We choose to store transposes of the matrices representing the compact solution vector in row sparse format (meaning they are stored by columns) so that relevant LAPACK methods can be called without having to transpose the input matrices.

The multiplication of the sparse Kronecker factors $\mathbf{A}_h^{(k)}$ with the orthogonal basis matrices $\mathbf{U}_h$ in $\mathbf{x}^{(k)} := \left(\bigotimes_{h=1}^d \mathbf{A}_h^{(k)} \mathbf{U}_h \right) \mathbf{c}$ is implemented using straightforward sparse matrix-vector multiplication. After the compact vector $\mathbf{x}^{(k)}$ is computed, the tree nodes of $\mathbf{y}^{(k)}$ are visited and its respective fields are updated so that we have $\mathbf{y}^{(k+1)}$ at hand. Efficient computation of the reduced Gramian matrices $\mathbf{G}_t$ as in Algorithm 3 of [14, p. 17] for $\mathbf{y}^{(k+1)}$ requires exploiting the block structure of the new transfer matrices $\mathbf{B}_t$ whose blocks are already available in the corresponding tree nodes of $\mathbf{y}^{(k+1)}$ after the addition operation. Clearly, there is no need to generate block matrices (or a cubic blocks as in Fig. 5 of [14, p. 11]) with these blocks explicitly. We prefer to store $\mathbf{M}_t$ and $\mathbf{G}_t$ as (2×2) block matrices because of the add a term and then truncate approach followed. Let us next elaborate on this.

Assuming that $r_t(\mathbf{y}^{(k)})$ and $r_t(\mathbf{x}^{(k)})$ denote the ranks of matrices in compact representations of the two vectors that are summed up in node t, $\mathbf{M}_t$ and $\mathbf{G}_t$ become $(r_t(\mathbf{y}^{(k)})r_t(\mathbf{x}^{(k)}) \times r_t(\mathbf{y}^{(k)})r_t(\mathbf{x}^{(k)}))$ matrices, where the first diagonal block is $(r_t(\mathbf{y}^{(k)}) \times r_t(\mathbf{y}^{(k)}))$ and the second diagonal block is $(r_t(\mathbf{x}^{(k)}) \times r_t(\mathbf{x}^{(k)}))$. Then the computation $\mathbf{M}_t := \mathbf{U}_t^T \mathbf{U}_t$ for leaf nodes can be formulated in (2×2) block manner as

$$\mathbf{M}_t^{(i,j)} := (\mathbf{U}_t^{(i)})^T (\mathbf{U}_t^{(j)}) \quad \text{for } i, j = 1, 2,$$

where $\mathbf{U}_t^{(1)}$ and $\mathbf{U}_t^{(2)}$ denote basis matrices of $\mathbf{y}^{(k)}$ and $\mathbf{x}^{(k)}$ at leaf node t, respectively. This computation requires multiplying two full matrices for which the DGEMM routine of LAPACK may be used. On the other hand, the computation

$\mathbf{M}_t := \mathbf{B}_t^T (\mathbf{M}_{t_l} \otimes \mathbf{M}_{t_r}) \mathbf{B}_t$ for intermediate nodes can be formulated from the bottom of the tree to the root in (2×2) block manner as

$$\mathbf{M}_t^{(i,j)} := (\mathbf{B}_t^{(i)})^T (\mathbf{M}_{t_l}^{(i,j)} \otimes \mathbf{M}_{t_r}^{(i,j)})(\mathbf{B}_t^{(j)}) \quad \text{for } i,j = 1,2,$$

where $\mathbf{B}_t^{(1)}$ and $\mathbf{B}_t^{(2)}$ denote transfer matrices of $\mathbf{y}^{(k)}$ and $\mathbf{x}^{(k)}$ at node t, respectively.

Similarly, we have reduced Gramian computations, but in opposite direction from root to leaves, that can be formulated in (2×2) block manner for $i,j = 1,2$ as $\mathbf{G}_t^{(i,j)} := 1$ when t corresponds to root; otherwise,

$$\mathbf{G}_{t_l}^{(i,j)} := (\mathbf{B}_{t:2,3}^{(i)})^T (\mathbf{M}_{t_r}^{(i,j)} \otimes \mathbf{G}_t^{(i,j)}) \mathbf{B}_{t:2,3}^{(j)}$$

and

$$\mathbf{G}_{t_r}^{(i,j)} := (\mathbf{B}_{t:1,3}^{(i)})^T (\mathbf{M}_{t_l}^{(i,j)} \otimes \mathbf{G}_t^{(i,j)}) \mathbf{B}_{t:1,3}^{(j)},$$

where $\mathbf{B}_{t:2,3}^{(1)}$ and $\mathbf{B}_{t:1,3}^{(1)}$ are transfer matrices $\mathbf{B}_t^{(1)}$ of $\mathbf{y}^{(k)}$ organized respectively as $(r_{t_r}(\mathbf{y}^{(k)})r_t(\mathbf{y}^{(k)}) \times r_{t_l}(\mathbf{y}^{(k)}))$ and $(r_{t_l}(\mathbf{y}^{(k)})r_t(\mathbf{y}^{(k)}) \times r_{t_r}(\mathbf{y}^{(k)}))$ matrices and $\mathbf{B}_{t:2,3}^{(2)}$ and $\mathbf{B}_{t:1,3}^{(2)}$ are transfer matrices $\mathbf{B}_t^{(2)}$ of $\mathbf{x}^{(k)}$ organized respectively as $(r_{t_r}(\mathbf{x}^{(k)})r_t(\mathbf{x}^{(k)}) \times r_{t_l}(\mathbf{x}^{(k)}))$ and $(r_{t_l}(\mathbf{x}^{(k)})r_t(\mathbf{x}^{(k)}) \times r_{t_r}(\mathbf{x}^{(k)}))$ matrices. Such matrices are called matricizations of the given matrix (in this case, the transfer matrix $\mathbf{B}_t^{(1)}$ or $\mathbf{B}_t^{(2)}$ along specific dimensions), and therefore, represent different organizations of the same data. We remark that the off-diagonal blocks of $\mathbf{M}_t$ and $\mathbf{G}_t$ respectively satisfy the relationships $\mathbf{M}_t^{(i,j)} = (\mathbf{M}_t^{(j,i)})^T$ and $\mathbf{G}_t^{(i,j)} = (\mathbf{G}_t^{(j,i)})^T$. Therefore, only one off-diagonal block for these two matrices in each node needs to be computed. The computation of the three blocks of $\mathbf{M}_t$ and $\mathbf{G}_t$ requires multiplications using DGEMM with matricizations and contraction of multi-dimensional data involving $\mathbf{B}_t^{(i)}$ matrices for $i = 1,2$ as discussed in [14, pp. 9–10, 12–13]. We use two auxiliary vectors of length $\max_{t_l, t_r, t}(r_{t_l} r_{t_r} r_t)$ to implement these operations. The disadvantage of not storing $\mathbf{M}_t$ and $\mathbf{G}_t$ as (2×2) block matrices is that longer auxiliary vectors would need to be allocated.

Truncation of a compact vector requires QR and singular value decompositions [11, pp. 76–79, 246–250] as in Algorithm 7 of [14, p. 23] to be performed. In order to compute these decompositions, DGEQRF and DGESDD routines of LAPACK are used. Since we expect input matrices to be dense, we do not call routines expecting sparse matrices. For a leaf node t, the $(m_t \times (r_t(\mathbf{y}^{(k)}) + r_t(\mathbf{x}^{(k)})))$ input matrix $\mathbf{U}_t$ maybe obtained by concatenating the matrices $\mathbf{U}_t^{(1)}$ and $\mathbf{U}_t^{(2)}$ corresponding to $\mathbf{y}^{(k)}$ and $\mathbf{x}^{(k)}$, respectively. Since the input matrix is also an output matrix, the upper-triangular factor $\mathbf{R}_t$ of the QR decomposition is returned from DGEQRF in the upper-triangular part of the input matrix in which the lower-triangular part has the Householder reflections amounting to the orthogonal factor $\mathbf{Q}_t$. After $\mathbf{R}_t$ is obtained, $\mathbf{R}_t \mathbf{G}_t \mathbf{R}_t^T$ needs to be formed. To this end, we first transform the block matrix $\mathbf{G}_t$ to a dense matrix (with a single block) and multiply this new matrix held as a one-dimensional array from left and right using the DTRMM routine of LAPACK. Note that DTRMM does not accept a trapezoid

$\mathbf{R}_t$; however, this case can be handled by multiplying triangular and rectangular parts of $\mathbf{R}_t$ separately using DTRMM and DGEMM. Hence, there is no need to copy the output of DGEQRF to another matrix including $\mathbf{R}_t$. Once $\mathbf{R}_t\mathbf{G}_t\mathbf{R}_t^T$ is formed, it needs to be decomposed for its singular values and vectors. To this end, we prefer to use the DGESDD routine over the DGESVD routine since it is said to be faster [17]. We remark that this routine computes singular values through the symmetric eigenvalue decomposition, and the singular vectors are truncated at a certain number or possibly by omitting some corresponding to the smaller singular values based on an error tolerance. $\mathbf{S}_t$ ends up being the matrix holding the r_t singular vectors. Then the orthogonal basis matrix $\mathbf{U}_t = \mathbf{Q}_t\mathbf{S}_t$ is computed using the DORMQR routine. In order to avoid storing $\mathbf{S}_t$, we prefer to update $\mathbf{R}_t$ with $\mathbf{S}_t^T$ as in the htucker package [15].

The same sequence of operations are carried out level by level from the parents of the leaves to the top of the tree excluding the root. The product $\mathbf{S}_t^T\mathbf{R}_t$ is computed using DTRMM (also possibly with an additional call to DGEMM when $\mathbf{R}_t$ is trapezoid) and stored in the matrix that was allocated for $\mathbf{R}_t$. Note that $\mathbf{S}_t^T\mathbf{R}_t = (\mathbf{F}_t^{(1)} \ \mathbf{F}_t^{(2)})$ is an $(r_t \times (r_t(\mathbf{y}^{(k)}) + r_t(\mathbf{x}^{(k)})))$ matrix with the two blocks $\mathbf{F}_t^{(l)}$ for $l = 1, 2$, where r_t is the rank of node t after truncation. Then for a non-leaf node t, the QR factorization of $\sum_{i=1}^2 (\mathbf{F}_{t_l}^{(i)} \otimes \mathbf{F}_{t_l}^{(i)})\mathbf{B}_t^{(i)}$ needs to be computed. This computation requires multiplications using DGEMM with matricizations of multi-dimensional data involving $\mathbf{B}_t^{(i)}$ matrices for $i = 1, 2$ as discussed in [14, pp. 9–10]. Finally, the transfer matrix $\mathbf{B}_t = \mathbf{Q}_t\mathbf{S}_t$ is computed using DORMQR.

4   Results of Numerical Experiments

In this section, we consider two example models that have been used as benchmarks in [4]. The first is an availability model with d subsystems in which different time scales occur. Each subsystem models a processing node with 2 processors, one acting as a cold spare, a bus and two memory modules. Time to failure is exponentially distributed with rate 5×10^{-4} for processors, 4×10^{-4} for buses and 10^{-4} for memory modules. Components are repaired by a global repair facility with preemptive priority such that components from subsystem 1 have the highest priority and components from subsystem d have the least priority. Furthermore, the repair of the bus has priority over the repair of the processor which has priority over the repair of the memory module. The repair times of components are exponentially distributed. The repair rates of a processor, a bus, and a memory from subsystem 1 are given respectively as 1, 2, and 4. The same rates for other subsystems are given respectively as 0.1, 0.2, and 0.4. For this model, the reachable state space is equal to the product state space and contains 12^d states. We consider availability models with $d = 3, 4, 5, 6, 7, 8$.

The second example is a model of a polling system of two servers serving customers from d finite capacity queues, which are cyclically visited by the servers. Customers arrive to the system according to a Poisson process with rate 1.5 and are distributed with queue specific probabilities among the queues each of which is assumed to have a capacity of 10. If a server visits a nonempty queue, it serves one customer and then travels to the next queue. On the other hand, a server arriving at an empty queue, skips the queue and travels to the next queue. Service and travelling times of servers are exponentially distributed respectively with rates 1 and 10. Each subsystem in the model describes one queue, and the J partitions of the reachable state space for this model are defined according to the number of servers serving customers at a queue or travelling to the next queue. For each subsystem we obtain 62 states partitioned into 3 subsets. The reachable state space of the complete model has $J = \binom{d+1}{2}$ partitions, and we consider polling system models with $d = 3, 4, 5, 6, 7$.

Table 1. Properties of availability and polling models

	Availability		Polling	
d	J	$|\mathcal{S}|$	J	$|\mathcal{S}|$
3	1	1,728	6	25,443
4	1	20,736	10	479,886
5	1	248,832	15	8,065,860
6	1	2,985,984	21	125,839,395
7	1	35,831,808	28	1,863,521,121
8	1	429,981,696		

The goal of this paper is to compare the memory and timing requirements for a vector-matrix product computation using the full vector and the HTD format approaches. Furthermore, we have to evaluate the accuracy of the computation if truncation is performed in the HTD format. Therefore, we consider in the following iteration steps of the Power method. This is not the most efficient solution method for steady-state analysis, but similar iteration steps can be applied in more advanced iterative techniques and they can be directly used in uniformization for transient analysis. For each model, the solution vector $\boldsymbol{\pi}^{(it)}$ at iteration it is multiplied with

$$\mathbf{P} := \mathbf{I} + \Delta\mathbf{Q}, \quad \text{where } \Delta := 0.999/\max_{s\in\mathcal{S}} |q_{s,s}|,$$

starting with the uniform distribution in $\boldsymbol{\pi}^{(0)}$, so that we have

$$\boldsymbol{\pi}^{(it)} := \boldsymbol{\pi}^{(it-1)}\mathbf{P} \quad \text{for } it = 1, 2, \ldots, \texttt{maxit}$$

with the associated error vector $\mathbf{e}^{(it)} := \boldsymbol{\pi}^{(it)} - \boldsymbol{\pi}^{(it-1)}$. Note that $\mathbf{e}^{(it)} = \Delta\boldsymbol{\pi}^{(it-1)}\mathbf{Q}$, the scaled residual vector corresponding to the previous iteration vector. Here, $\texttt{maxit}$ is the maximum number of iterations and we set $\texttt{maxit} := 1,000$. The numerical experiments are performed on an an Intel Core i7 Quad-Core 3.6 GHz processor with 32 GB of main memory.

Table 2 contains the results for the availability model. Time is in seconds and Memory indicates the number of allocated real array elements. For the chosen truncation accuracy of $\epsilon \in [10^{-9}, 10^{-7}]$, the norm of the final error vector is the same for the full and HTD representations. Due to the reduced memory requirements of the vector, the compact representation results even in smaller iteration times when d increases. It should be mentioned that the model is not symmetric due to the priority repair strategy but, as it is common in availability models, the probability distribution becomes unbalanced because repair rates are higher than failure rates.

Table 2. Numerical results for availability models

	Full			Compact			
d	Time	Memory	$\|e^{(\texttt{maxit})}\|_2$	ϵ	Time	Memory	$\|e^{(\texttt{maxit})}\|_2$
3	0	5,391	5×10^{-7}	10^{-7}	1	2,102	7×10^{-7}
				10^{-8}	1	2,298	5×10^{-7}
				10^{-9}	2	3,400	5×10^{-7}
4	0	62,598	3×10^{-6}	10^{-7}	3	2,809	2×10^{-6}
				10^{-8}	4	4,107	3×10^{-6}
				10^{-9}	6	6,938	3×10^{-6}
5	2	747,132	1×10^{-5}	10^{-7}	6	3,719	9×10^{-6}
				10^{-8}	9	5,327	1×10^{-5}
				10^{-9}	13	9,278	1×10^{-5}
6	38	8,958,897	3×10^{-5}	10^{-7}	13	6,756	3×10^{-5}
				10^{-8}	18	9,398	3×10^{-5}
				10^{-9}	28	14,120	3×10^{-5}
7	513	107,496,741	7×10^{-5}	10^{-7}	15	6,726	7×10^{-5}
				10^{-8}	28	10,381	7×10^{-5}
				10^{-9}	43	16,150	7×10^{-5}
8	6,329	1,289,946,786	2×10^{-6}	10^{-7}	22	9,078	9×10^{-5}
				10^{-8}	37	12,340	9×10^{-6}
				10^{-9}	66	26,041	3×10^{-6}

The situation is more ambiguous for the polling example whose results are given in Table 3. For the larger configurations, we obtain savings in memory by several orders of magnitude even with the smallest truncation accuracy of ϵ. Time-wise the conventional approach is faster for small configurations, but it is outperformed by the compact representation for larger state spaces (i.e. $d = 6$), if ϵ is not too small. The largest configuration with $d = 7$ can only be analysed with the compact vector representation.

In Fig. 3 the ranks of the different matrices forming the HTD are shown for a truncation accuracy of $\epsilon = 10^{-7}$. It can be seen that the ranks remain moderate. The matrices are fairly dense such that a sparse storage of the matrices for the vector representation is not necessary which can be seen in Fig. 4.

Table 3. Numerical results for polling models

d	Full			Compact			
	Time	Memory	$\|\|e^{(\mathrm{maxit})}\|\|_2$	ϵ	Time	Memory	$\|\|e^{(\mathrm{maxit})}\|\|_2$
3	0	82,599	4×10^{-6}	10^{-7}	50	49,297	4×10^{-6}
				10^{-8}	83	72,281	4×10^{-6}
				10^{-9}	108	89,257	4×10^{-6}
4	5	1,496,563	5×10^{-6}	10^{-7}	285	143,436	5×10^{-6}
				10^{-8}	1,175	397,349	5×10^{-6}
				10^{-9}	3,272	774,834	5×10^{-6}
5	103	24,791,966	5×10^{-6}	10^{-7}	409	221,850	5×10^{-6}
				10^{-8}	2,522	800,742	5×10^{-6}
				10^{-9}	10,951	2,136,401	5×10^{-6}
6	1,896	383,988,648	3×10^{-6}	10^{-7}	254	177,534	3×10^{-6}
				10^{-8}	2,448	883,360	3×10^{-6}
				10^{-9}	19,423	3,564,320	3×10^{-6}
7	n/a	5,661,610,381	n/a	10^{-7}	196	217,254	3×10^{-6}
				10^{-8}	1,831	900,220	2×10^{-6}
				10^{-9}	21,668	5,037,050	2×10^{-6}

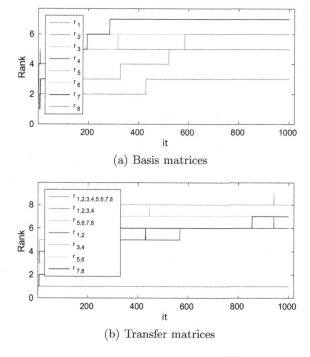

(a) Basis matrices

(b) Transfer matrices

Fig. 3. Ranks of basis and transfer matrices forming $\pi^{(it)}$, availability $d = 8$ (Color figure online)

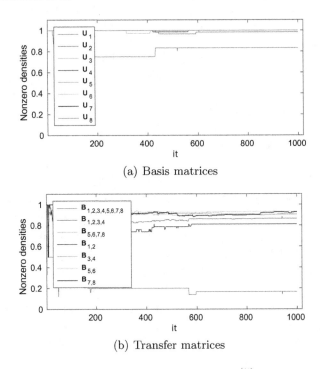

(a) Basis matrices

(b) Transfer matrices

Fig. 4. Densities of basis and transfer matrices forming $\pi^{(it)}$, availability $d = 8$ (Color figure online)

5   Conclusion

We present in this paper a compact representation for the iteration vector of large structured Markov models which has been adopted from numerical analysis where the techniques have been developed in the recent years. It is shown that this vector representation can be combined naturally with a hierarchical Kronecker representation of generator matrices of structured Markov models. The basic step of iterative numerical algorithms to compute transient or steady-state solutions can be conveniently combined with the compact vector representation. Our first examples indicate that in contrast to previously tried compact representations for the vector (e.g., [5,16]), the new approach is memory and also relatively time efficient such that it bears the potential to increase the size of solvable models on a given computer significantly.

There are several things to be done. In particular, more experiments are necessary to confirm our results. The vector-matrix multiplications have to be embedded in more advanced solution techniques like projection or multi-level solution techniques. However, this can be easily done with the available software environment.

Acknowledgement. This work is supported by the Alexander von Humboldt Foundation through the Research Group Linkage Programme. The research of the last author is supported by The Scientific and Technological Research Council of Turkey.

References

1. APNN-Toolbox. Abstract Petri Net Notation Toolbox. http://www4.cs. uni-dortmund.de/APNN-TOOLBOX
2. Bause, F., Buchholz, P., Kemper, P.: A toolbox for functional and quantitative analysis of DEDS. In: Puigjaner, R., Savino, N.N., Serra, B. (eds.) TOOLS 1998. LNCS, vol. 1469, pp. 356–359. Springer, Heidelberg (1998)
3. Buchholz, P.: Hierarchical structuring of superposed GSPNs. IEEE Trans. Softw. Eng. **25**(2), 166–181 (1999)
4. Buchholz, P., Dayar, T.: On the convergence of a class of multilevel methods for large, sparse Markov chains. SIAM J. Matrix Anal. Appl. **29**(3), 1025–1049 (2007)
5. Buchholz, P., Kemper, P.: Compact representations of probability distributions in the analysis of superposed GSPNs. In: Proceedings of the 9th International Workshop on Petri Nets and Performance Models, Aachen, Germany, pp. 81–90. IEEE Press, New York, September 2001
6. Buchholz, P., Ciardo, G., Donatelli, S., Kemper, P.: Complexity of memory-efficient Kronecker operations with applications to the solution of Markov models. INFORMS J. Comput. **12**(3), 203–222 (2000)
7. Dayar, T.: Analyzing Markov Chains using Kronecker Products: Theory and Applications. Springer, New York (2012)
8. Dayar, T., Orhan, M.C.: On vector-Kronecker product multiplication with rectangular factors. SIAM J. Sci. Comput. **37**(5), S526–S543 (2015)
9. Dayar, T., Orhan, M.C.: Cartesian product partitioning of multi-dimensional reachable state spaces. Probab. Eng. Inf. Sci. **30**(3), 413–430 (2016)
10. Fernandes, P., Plateau, B., Stewart, W.J.: Efficient descriptor-vector multiplications in stochastic automata networks. J. ACM **45**(3), 381–414 (1998)
11. Golub, G.H., Van Loan, C.F.: Matrix Computations, 4th edn. Johns Hopkins University Press, Baltimore (2012)
12. Hackbusch, W.: Tensor Spaces and Numerical Tensor Calculus. Springer, Heidelberg (2012)
13. Kressner, D., Macedo, F.: Low-rank tensor methods for communicating Markov processes. In: Norman, G., Sanders, W. (eds.) QEST 2014. LNCS, vol. 8657, pp. 25–40. Springer, Heidelberg (2014)
14. Kressner, D., Tobler, C.: htucker — A Matlab toolbox for tensors in hierarchical Tucker format. Technical report 2012-02, Mathematics Institute of Computational Science and Engineering, Lausanne, Switzerland, August 2012. http://anchp.epfl. ch/htucker
15. Kressner, D., Tobler, C.: Algorithm 941: htucker — a matlab toolbox for tensors in hierarchical Tucker format. ACM Trans. Math. Softw. **40**(3), 22 (2014)
16. Kwiatkowska, M., Mehmood, R., Norman, G., Parker, D.: A symbolic out-of-core solution method for Markov models. Electron. Notes Theor. Comput. Sci. **68**(4), 589–604 (2002)
17. Netlib, A.: Collection of Mathematical Software, Papers, and Databases. http:// www.netlib.org
18. Oseledets, I.V.: Tensor-train decomposition. SIAM J. Sci. Comput. **33**(5), 2295–2317 (2011)

19. Plateau, B.: On the stochastic structure of parallelism and synchronization models for distributed algorithms. Perform. Eval. Rev. **13**(2), 147–154 (1985)
20. Plateau, B., Fourneau, J.-M.: A methodology for solving Markov models of parallel systems. J. Parallel Distrib. Comput. **12**(4), 370–837 (1991)
21. Stewart, W.J.: Introduction to the Numerical Solution of Markov Chains. Princeton University Press, Princeton (1994)

Networks

A Comparison of Different Intrusion Detection Approaches in an Advanced Metering Infrastructure Network Using ADVISE

Michael Rausch[1]([✉]), Brett Feddersen[1], Ken Keefe[1], and William H. Sanders[2]

[1] Information Trust Institute, University of Illinois at Urbana-Champaign,
Urbana, IL, USA
{mjrausc2,bfeddrsn,kjkeefe}@illinois.edu
[2] Department of Electrical and Computer Engineering,
University of Illinois at Urbana-Champaign, Urbana, IL, USA
whs@illinois.edu

Abstract. Utilities responsible for Advanced Metering Infrastructure (AMI) networks must be able to defend themselves from a variety of potential attacks so they may achieve the goals of delivering power to consumers and maintaining the integrity of their equipment and data. Intrusion detection systems (IDSes) can play an important part in the defense of such networks. Utilities should carefully consider the strengths and weaknesses of different IDS deployment strategies to choose the most cost-effective solution. Models of adversary behavior in the presence of different IDS deployments can help with making this decision as we demonstrate through a case study that uses a model created in the ADversary VIew Security Evaluation (ADVISE) formalism (which calculates metrics used to compare different IDSes). We show how these metrics give valuable insight into the selection of the appropriate IDS architecture for an AMI network.

Keywords: Advanced Metering Infrastructure (AMI) · Smart grid · ADversary VIew Security Evaluation (ADVISE) · Security modeling · Intrusion Detection Systems (IDS)

1 Introduction

Many utility companies are creating Advanced Metering Infrastructure (AMI) networks, which incorporate smart meters and other intelligent components into the power grid. The added functionality allows utilities to monitor and control their smart grid with more precision than was previously possible. As an example, a utility company can use an AMI infrastructure to remotely collect more frequent meter readings, which allows them to respond more accurately to fluctuations in power demand.

Unfortunately, AMI networks increase the attack surface of a power grid. For example, an unscrupulous customer may compromise a single smart meter

© Springer International Publishing Switzerland 2016
G. Agha and B. Van Houdt (Eds.): QEST 2016, LNCS 9826, pp. 279–294, 2016.
DOI: 10.1007/978-3-319-43425-4_19

so that it sends false data to under-report electricity consumption, resulting in a lower bill. Distributed denial of service attacks, traffic injection attacks, and Byzantine attacks are examples of new threats to these cyber-enhanced power grids. As utility companies build and maintain AMI infrastructures they should be aware of the possibility of these attacks, and work to create a cost-efficient architecture that minimizes the expected damage.

One obvious way of limiting the potential damage of an attack is to detect and respond to the attack before it can cause much harm. An intrusion detection system (IDS) can help a utility company detect an attack. There are several different IDS architectures that can be deployed by a utility company as a defensive precaution. Each architecture has a different cost and degree of effectiveness. A utility company must decide whether its application warrants an IDS, and if so, which would give the best protection for the best price.

One approach for informing this critical design decision is to build a sound, state-based stochastic model of the system and the possible IDS architectures that can be applied to it. Quantitative metrics can be calculated on the models to determine which configuration provides the best cost/security balance.

Our approach is to study a multi-layered power grid example and the potential IDS implementations that can be applied to this grid. We used the ADversary VIew Security Evaluation (ADVISE) [6] modeling formalism in the Möbius modeling tool [10] for this work. We consider several different adversaries interested in attacking such a system and calculate useful and relevant security metrics. Using our approach, a utility company can make a more informed decision about how to implement an IDS on its grid.

To make an informed decision regarding the selection and implementation of various IDS approaches, it is necessary to know the probability that an adversary would successfully attack a system, given its type of IDS architecture. Given a particular adversary and IDS approach, it would be useful to know the estimated probability of detecting the adversary, the estimated damage to the utility due to activity of the adversary, and the type of attack chosen by the adversary. We create a model of the adversary behavior that is detailed enough to give insight into these metrics. We do not claim that the quantitative metrics generated by the model are accurate in any absolute sense. However, we do believe that they may be very useful when comparing the relative strengths and weaknesses of modeled systems. A model that gives quantitative security metrics will give a system designer another approach to supplement the advice and intuition of security experts.

The remainder of this paper is organized as follows. Section 2 provides a concise overview of AMI networks, IDS systems on AMI networks, and the ADVISE formalism. Section 3 offers a description of system we modeled. Section 4 offers a detailed explanation of the ADVISE model that was constructed, including the adversary profiles that were considered and the metrics that were defined on the system. Section 5 shows our quantitative results and our interpretation of them. Section 6 discusses previous work that seeks to examine power grid security using a variety of methods. Finally, Sect. 7 concludes the paper.

2   Background

2.1   AMI Overview

An AMI gives a utility company the ability to remotely communicate with the electric meters in its grid. There are many possible network hierarchies. For example, some smart meters connect to the utility company through the consumer's Internet connection. However, our example system uses a hierarchy of communication gateways that rely on the utility's own network infrastructure, as depicted in Fig. 1.

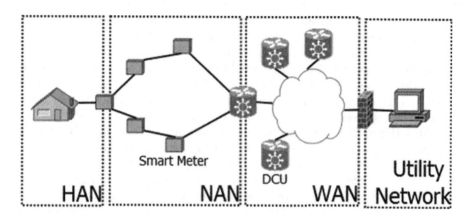

Fig. 1. Example system diagram.

At the bottom level of our hierarchy, a smart meter forms the core of a *home area network* (HAN). The HAN may include other components, such as smart appliances designed to draw less power during times of high demand and more power during times of low demand. If there are multiple devices in the HAN, the smart meter itself may act as a network gateway for the other devices. Multiple HANs, and data collection unit (DCU) gateways serving them, are connected together to form a *neighborhood area network* (NAN). The DCU gateways in NANs are connected to the utility via a *wide area network* (WAN).

Many different communication technologies can be used in an AMI. The WAN usually utilizes higher bandwidth, long-range communication technologies such as long-range wireless, satellite, or power line communication. The NANs don't have the same bandwidth or range requirements, and can use shorter-range wireless. We assume in our case study that the NAN uses a wireless mesh network.

2.2   IDS Overview

Intrusion detection systems are intended to monitor a system for suspicious activity, to raise an alert if a security event occurs, and to log information to determine how an attack proceeded through the system. A number of different IDS deployment strategies are possible in an AMI. In this paper, we consider centralized IDSes, dedicated distributed IDSes, and embedded distributed IDSes.

A centralized IDS deployment scheme would place an IDS at the top of the network hierarchy, in the utility company's network. The IDS would monitor all traffic flowing into and out of the utility company's command and control center LAN network, and raise an alert if it detected anything suspicious. However, it would be completely unaware of inter-meter traffic, since that would not pass through the top level of the hierarchy. Alternatively, a utility could deploy a distributed set of IDSes to monitor inter-meter communication. This approach would still require a central node to coordinate the monitoring, so it would have all of the benefits of a centralized IDS approach, with some additional installation and maintenance costs associated with the additional IDSes.

We studied two main varieties of distributed IDSes: dedicated and embedded. A dedicated device IDS deployment would have the same components as a centralized deployment, and in addition it would have a number of geographically distributed dedicated IDS devices in wireless communication with the smart meters. These IDSes would monitor all AMI traffic within wireless range. An embedded IDS deployment incorporates intrusion detection directly into the smart meter. Like the dedicated IDS architecture, the embedded IDSes communicate and cooperate with the central IDS device.

There are a number of trade-offs to consider when evaluating these IDS designs. A centralized IDS would potentially miss large families of attacks because it is unaware of inter-meter communication. It is, however, the cheapest IDS option. An embedded or dedicated IDS scheme would be able to observe inter-meter communication, possibly allowing it to detect a larger set of attacks than a centralized scheme, but would cost more. A dedicated architecture would cost more because many additional devices would have to be purchased and maintained, and separate permits and location sites would have to be acquired to install these devices. However, the device would be able to monitor inter-meter communication in the NAN. One dedicated device could serve multiple smart meters. An embedded system would not require separate building sites or permits, but every single meter would cost slightly more because of the added IDS capability. Given the large number of meters involved, even a small increase in price for an individual meter would potentially be very costly for a utility company. In addition to monitoring inter-meter communication, an embedded IDS architecture would be able to detect attacks on the meter itself. This means the embedded IDS option provides the greatest coverage against possible attacks.

Clearly, a utility company should seek the most cost-effective solution. The choice can be made and justified with metrics derived from the analysis of mathematical models, such as the one we developed using the ADVISE modeling formalism.

2.3   ADVISE Overview

The ADversary VIew Security Evaluation (ADVISE) method [6] is used to calculate quantitative security metrics via executable models of adversary behavior in a system [5]. At a high level, a modeler creates an Attack Execution Graph (AEG), which is similar to a standard attack tree, but incorporates additional details about each attack's properties, such as its cost, time to completion, and probability of success. The AEG also contains nodes that track the state of the model, such as the prerequisites and goals held by the adversary at a particular discrete point in time. Different adversaries may exhibit very different behaviors when attacking the same system, since their initial foothold in the system, knowledge, skills, and goals of interest may differ dramatically. An adversary's preference for avoiding cost, avoiding detection, and earning reward also plays a pivotal role in the approach taken when attacking a system. A modeler is given the ability to create different adversary profiles before executing the model in ADVISE to reflect this reality.

An AEG is defined by the tuple

$$< A, R, K, S, G, C >$$

where A is the set of attack steps, R is the set of access domains available to the adversary, K is the set of information that can be known by the adversary, S is the set of skills possessed by the adversary, and G is the set of goals that the adversary attempts to achieve. The relation C defines the set of directed connecting arcs from $e \in R \cup K \cup S \cup G$ to $a \in A$, where e is a prerequisite element needed in order to attempt a. This relation also defines the set of directed connecting arcs from $a \in A$ to $e \in R \cup K \cup S \cup G$, where e is an affected element that may be changed by the performance of a. The elements R, K, S, and G are state variables that hold an integer value that usually represents whether the element represented by the state variable has been obtained (1 or 0).

An attack step is defined by the tuple

$$< B, T, C, O >$$

where B is a Boolean precondition that indicates whether or not the attack step is currently enabled, T is the timing distribution that is sampled to determine the time it takes to complete the attack step, C is the cost to the adversary for attempting the attack, and O is the set of outcomes of the attack (such as success or failure). Each outcome contains a Pr, D, and E, which are the probability the outcome will be selected from an attack step's O, the probability of being detected for that outcome, and the effect of that outcome on the state of the model, respectively. An adversary uses the solution of a Competitive Markov Decision Process [1] as described in [6] to select the best attack step given the adversary's characteristics, limitations and preferences.

System metrics are then defined using rate- and event-based performance variables [8]. Reliability of a device, preferred paths of attack for an adversary, and expected costs for the adversary and defender are all examples of possible metrics. Discrete event simulation is used to generate a statistically sound estimate of the defined metrics.

3   Power Grid Description

In this case study, we consider a hypothetical utility company with an urban deployment of an AMI network, as shown in Fig. 1 and described in Sect. 2.1. We have based our system on the system described in [2], following it in detail whenever possible. In this network, zero or more smart appliances connect to a smart meter at each home and together form a HAN. Multiple HANs are connected to one another and one or more gateways via a wireless mesh network to form a NAN. Multiple NANs are connected to one another and to the utility command and control center network via a WAN.

The utility wishes to supply power to consumers, protect their equipment, ensure the integrity of communication in the AMI network, and ensure the confidentiality of communication in the AMI network.

The utility company in this scenario is primarily concerned about attacks from three classes of adversaries: unscrupulous customers who wish to under-report their electricity consumption to unfairly lower their bill, disgruntled insider employees who wish to cause as much monetary damage as possible in retribution for a perceived wrong, and sophisticated, well-funded terrorist organizations or nation-states who wish to interrupt the delivery of power and cause as much damage as possible. The utility company estimates that over a 20-year period, there will be 1,000 attempts to under-report electricity consumption, a 0.1 % chance that a disgruntled employee will attempt a massively damaging attack, and a 0.01 % chance of being attacked by a terrorist organization. An adversary may choose from a variety of attacks to achieve a goal. We utilized the literature search conducted in [3] to compile a list of attacks for inclusion in our model.

The utility company wishes to compare the cost-effectiveness of various proposed IDS architectures. In particular it wishes to compare the centralized IDS solution with the two distributed IDS solutions: embedded and dedicated. The utility can easily obtain the estimated installation and maintenance costs of an IDS from vendors. However, estimating the expected benefit of implementing the IDS is much more difficult. We attempt to make such an estimate with an ADVISE model.

4   ADVISE Model

We used the ADVISE formalism as implemented in Möbius to construct an Attack Execution Graph to gain insight into the adversary behavior. We created a model that was detailed enough to calculate the quantitative security metrics of interest, while minimizing the number of assumptions that a more detailed model would have forced us to make. We were primarily interested in three

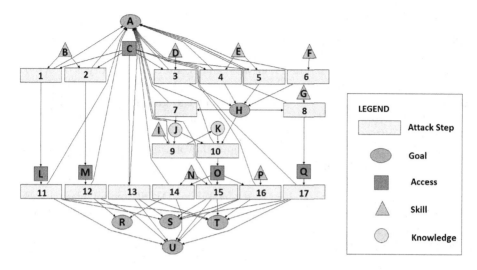

Fig. 2. Attack Execution Graph of ADVISE model.

key metrics: the estimated probability of detecting an adversary, the estimated damage to the utility company due to adversary behavior, and the attack used to damage the utility company. To calculate these metrics, we developed a model of the attacks against the system and a model of the adversaries that could execute the attacks.

4.1 Attack Execution Graph Model

The Attack Execution Graph, which is shown in Fig. 2, contains three main adversary goals; three auxiliary goals; seventeen attack steps that an adversary may attempt when trying to reach a goal; a number of supporting access, knowledge, and skill state variables that may help an adversary satisfy the preconditions for attempting a particular attack; and arcs that connect the attack steps to particular accesses, knowledge, skills, and goals and signify the relationships between them. The access, skill, knowledge, and goal state variables in our model hold a value of zero if they are not held by the adversary, and a positive integer otherwise.

The set of goals desired by the adversary drives his or her behavior and are therefore one of the most important components of the model. Cheating the company by under-reporting electricity consumption, interrupting the delivery of electric power, and damaging the utility's equipment are the three most important goals for the adversary in this study. Those goals are represented in Fig. 2 by Goals R, S, and T, respectively. In addition, the adversary wishes to remain undetected; this goal is represented by Goal A, the *Undetected* goal. Goal H is the supporting goal of acquiring compromised smart meters. This is an important prerequisite for several attack steps, but is also a goal in its own right for some adversaries. Finally, Goal U represents the goal of achieving at least one of the three primary goals described previously. Usually goal state variables initially hold a value of zero; and the value is incremented on the successful conclusion of an attack. The notable exception is the *Undetected* goal, which initially holds a value of 1 and is decremented to 0 if an attack fails and the adversary is detected.

We assume that the adversary may not continue to attack after being detected. For this reason, every attack step in the graph is connected by an arc to the *Undetected* goal, which is a prerequisite for attempting every attack step. If the *Undetected* goal holds a value of zero, the adversary may not attempt any attack, with the exception of the unique *DoNothing* attack step [6]. Attack steps can be attempted by an adversary either to achieve a goal directly or to change the model state to make it easier to achieve a goal later. An attack step must have at least one outcome. In this particular ADVISE model every attack step outcome results either in the certain detection of an adversary, or the adversary remaining undetected. In other words, if p_o is the probability of detection associated with an outcome o, then $p_o = 0$ or $p_o = 1$, but $p_o \notin (0, 1)$.

Attack Step 1 in the AEG diagram is an *Install Long Range Jammer Attack.* This attack step requires the adversary to be undetected, to be in reasonably close proximity to the smart meters, and to have skill in installing wireless jammers. It would result in the adversary's having access to a long-range wireless jammer that incapacitates the wireless mesh network in the NAN. The adversary must hold several prerequisites to attempt this attack, including the *NodeInstallationSkill*, represented by Skill B, the *PhysicalSmartMeterAccess*, represented by Access C, and the *Undetected* goal. At the successful conclusion of the attack, the adversary gains the *LongRangeJammerAccess*, Access L, whose value is incremented from 0 to 1. Attack Step 2, *Install Short Range Jammer*, is very similar, but its purpose is to gain access to a short-range wireless jammer that blocks communication in a HAN rather than a NAN, so Attack 2 is connected to *ShortRangeJammerAccess*, Access M, rather than Attack 1's *LongRangeJammerAccess*. Attack 2's prerequisites are identical to Attack 1's prerequisites.

Any one of Attack Steps 3, 4, 5, and 6 may be attempted by an adversary in an effort to obtain the *NumCompromisedSmartMeters* goal (Goal H), which would give the adversary control of smart meters in the AMI network. Attack Step 3, *InstallMaliciousSmartMeter*, aims to accomplish this goal by installing a new meter (controlled by the adversary) that tricks the AMI network into accepting it as one of its own smart meters. Attack Step 3 requires physical access to the AMI network and skill in installation as prerequisites, and so is connected to the *PhysicalAccess* access and the *SmartMeterInstallationSkill* skill, shown as Access C and Skill D, respectively. Attack Step 4, *PhysicalSmartMeterExploit*, represents an adversary attempt to physically tamper with the smart meter to gain control of it. An adversary must have physical access to the smart meters and skill in this exploit to attempt the attack step, so Attack Step 4 is connected to Access C and Skill E, which are the *PhysicalAccess* access and the *PhysicalSmartMeterExploitSkill* skill, respectively. Attack Step 5, *MassMeterCompromise*, is very similar to Attack Step 4, with the major difference being that 50 smart meters are compromised if this attack step is achieved instead of just one.

Finally, Attack Step 6 also compromises 50 smart meters, but it requires the adversary to have the appropriate skill (*RemoteSmartMeterExploitSkill*, shown as Skill F) and does not require the adversary to have physical access.

Attack Steps 7, 9, and 10 are related because their sequence leads to Access O, the *RoutingCapability* access, which is a prerequisite for Attack Steps 14, 15, and 16. Attack Step 7, *CollectCryptoKeys*, represents the adversary's attempt to collect cryptographic keys from the compromised smart meters. The adversary must have access to compromised smart meters to attempt the attack, and if the attack step is successful, it leads to the acquisition of knowledge of the cryptographic keys, which is represented by Knowledge Item J. The *AnalyzeTraffic* attack step (Attack Step 9) requires the adversary to hold Knowledge Item J and Skill I (I being the *TrafficAnalysis* skill) in order to attempt the attack step. If successful, the adversary gains sufficient knowledge of the traffic in the network to launch sophisticated routing and Byzantine attacks. This knowledge is represented by the knowledge item *TrafficKnowledge*, which is Knowledge Item K. Finally, the adversary may attempt Attack Step 10, *GainRoutingCapability*, if he or she has knowledge of the keys and traffic and at least one compromised smart meter. If the prerequisites have been satisfied, the adversary will successfully execute the attack step and gain the *RoutingCapability* access.

Attack Step 8, *CreateBotnet*, gives the adversary the *BotnetAccess* access, depicted as Access Q, which is a prerequisite for launching resource exhaustion attacks such as DDoS attacks. To attempt the attack step, the adversary must hold Skill G, the *BotnetShepherd* skill, as well as at least 50 smart meters, represented by a value greater than or equal to 50 in Goal H.

There are seven attack steps that directly achieve at least one of the three most significant goals (Attacks 11–17). First, Attack Step 11, the *Major Jamming Attack*, requires the adversary to have access to a long range jammer; it results in a significant interruption of service in the NAN, and also damages equipment, since important commands for coordinating the network are not delivered. Attack Step 12, which is the *Minor Jamming Attack*, requires access to a short-range jammer: it does not result in loss of power or damage to equipment but may be utilized to help an unscrupulous customer give a false power reading. Attack Step 13, *PhysicalAttack*, represents a major physical, non-cyber attack on the equipment of the utility company, e.g. shooting one or more transformers. This attack requires only physical access to the equipment, and causes a significant blackout and major damage to the equipment. It has a relatively high probability of detection, but requires only minimal prerequisites to attempt. Attack Steps 14 and 15, *MinorRoutingAttack* and *MajorRoutingAttack*, respectively, are similar in that they have the same prerequisites, *RoutingAttack* skill and *RoutingCapability* access (Skill N and Access O, respectively), but have different intended goals. The *MinorRoutingAttack* underreports the electricity consumption of one customer. The *MajorRoutingAttack*, in contrast, leads to interrupted service and damage to the AMI network equipment. Attack Step 16, *ByzantineAttack*, requires that the adversary hold the *RoutingCapability* access and the *ByzantineAttack* skill (Access O and Skill P, respectively), and a successful outcome for the adversary leads to damaged equipment and interrupted service. Finally, Attack Step 17, the *ResourceExhaustion* attack, requires *BotnetAccess* and results in damaged equipment and interrupted service.

In addition, there is one implied attack step not shown in the diagram, the *DoNothing* attack step, which an adversary may attempt at any time and has no effect on the model state, costs nothing, and will never lead to the detection of the adversary. This attack step may be attempted by an adversary when the payoff for attempting any other attack step does not justify the risk of detection and the cost of attempting the attack step.

Each attack step contains detailed information about the probability of success, probability of detection, cost to attempt, effects on the system, duration, and other information. Space considerations prevent us from explaining the details of every attack step in this model, but we discuss one attack step as an example. The *CreateBotnet* attack, Attack Step 8 in the diagram, is assumed to cost the adversary $1,000 to attempt, and to take 8 h to complete. If the attack is to be attempted, the *Undetected* goal state variable and the *BotnetShepherdSkill* skill state variable must both contain a positive value, and the *NumCompromisedSmartMeters* goal state variable must hold a value greater than or equal to 50. If these conditions are not met, the attack step cannot be attempted. If the attack step is attempted, one of three outcomes, *FailureUndetected*, *FailureDetected*, or *Success*, is randomly chosen according to their probabilities of occurrence. The *FailureUndetected* outcome represents the event in which the adversary attempts the attack and fails, but remains undetected. It has no effect on the state of the model, and has a probability of 0.05. The *FailureDetected* outcome represents the event in which the adversary attempts the attack, fails, and is detected. If this outcome is randomly selected by the simulation, it modifies the model state by changing the value of the *Undetected* goal from 1 to 0, disabling any future attack. This outcome is also assumed to have a probability of 0.05. Finally, the *Success* outcome represents the successful completion of the attack. It has the effect of giving the adversary access to a botnet of smart meters, which is represented by changing the value of the *BotnetAccess*, State Variable Q, from 0 to 1. This outcome has a 0.9 probability of being selected if the attack step is attempted. All the other attack steps in the model have a similar level of detail.

The probability that an attack step will lead to a successful outcome for an adversary, as well as the effect an outcome will have on the system, may be adjusted based on the IDS approach being modeled.

4.2   Attacker Model

In addition to a model of attacks against the system, we need a model of the adversary, since different adversaries have different goals, and different initial access, skills, and knowledge related to the system. Even adversaries with identical goals may weigh these goals differently. These differences can lead to very different behaviors when the attackers are confronted by the same system. Table 1 shows the state variables initially held by each adversary, and corresponds to an initial configuration of state variables in the AEG (Fig. 1). As can be seen from the table, the

Table 1. Initial state values and parameters for adversaries.

Initial state	Customer	Insider	Terrorist
BotnetAccess		X	
RoutingCapability		X	
PhysicalAccess	X	X	X
CryptoKeys		X	
TrafficKnowledge		X	
RoutingAttackSkill	X	X	X
NodeInstallationSkill	X	X	X
SmartMeterInstallationSkill		X	X
TrafficAnalysisSkill	X	X	X
BotnetShepherdSkill		X	X
ByzantineAttackSkill	X	X	X
PhysicalSmartMeterExploit	X	X	X
NumCompromisedMeters		51	
Undetected	X	X	X

customer adversary is assumed to have access to a physical smart meter (his or her own) and some skill in various attacks, perhaps obtained via compromises published on the Internet. The customer wants to achieve the goal of cheating the power company by under-reporting electricity consumption. The insider adversary is in some ways the most powerful adversary, because the insider starts with the most access, knowledge, and skills of any adversary considered, and in addition is the only adversary assumed to start with a number of compromised smart meters. However, this adversary is constrained by a relatively high desire to avoid detection, which is expressed in the model by placing a high payoff on maintaining the *Undetected* goal. The insider wishes to cause as much monetary damage as possible to the utility company without being detected. Finally, the terrorist adversary has fewer initial forms of access, knowledge, and skill than the insider, but wants to achieve the same goal of causing the utility company as much monetary damage as possible by interrupting the delivery of power and damaging equipment. The terrorist is assumed to be less concerned than the insider with the possibility of being detected and apprehended (expressed in the model by a relatively low payoff on maintaining the *Undetected* goal), which means the terrorist is much more likely to try risky attacks.

4.3 Metrics

We use the ADVISE model described above to calculate three metrics. All three are determined through the creation of performance variables [8] calculated by simulation in Möbius. We took the cross-product of the adversaries {Insider, Customer, Terrorist} and the IDS approaches {None, Central, Dedicated, Embedded}, and ran a simulation for every element of this set. We estimated the mean of every performance variable with a 0.95 confidence level and a 0.1 confidence interval.

The first metric is qualitative, it is the attack that the adversary attempts that leads to one of the three major goals (stealing electricity, disrupting the delivery of electricity, and damaging the equipment). To find this metric, we created a set of interval-of-time impulse-reward variables, one for each attack step that achieves one of the three main goals. If any one of the outcomes of an attack step is selected during the course of the simulation the performance variable associated with that attack step accumulates a reward. After the simulation, we determine which attack step the adversary chose by observing which element of this set of performance variables accumulated a reward.

The second metric, the probability that the adversary will remain undetected through the end of the attack, was constructed as an instant-of-time rate-reward variable that returned the value of the *Undetected* goal variable at the end of the simulation. At the beginning of the simulation, the *Undetected* goal variable would hold a value of one. Most attack steps in the AEG had an outcome that represented the event in which an adversary was detected if the attack step was executed. If that outcome occurs at some point during the course of the simulation, one of its effects is to set the value of the *Undetected* goal variable to zero. If no outcome representing the detection of the adversary is chosen during the course of the simulation, the value of the *Undetected* goal variable remains 1. In that way we determine whether the adversary was detected during one run of the simulation. Multiple runs of the simulation show the probability that the adversary will remain undetected through the duration of the attack.

The final and perhaps most important metric, the expected monetary damage to the system in the event of an attack by an adversary, was also calculated by an instant-of-time rate-reward variable. The integer values held in the *StealElectricity*, *InterruptService*, and *DamageEquipment* goal state variables represent units of damage. We let one unit of *StealElectricity* equal $600 of damage, one unit of *InterruptService* equal $10,000 of damage, and one unit of *DamageEquipment* equal $100,000 of damage. Initially these goal state variables hold a value of 0, but the value can be increased at the successful conclusion of certain attacks.

5   Results and Discussion

The attack each adversary would attempt when faced with each possible IDS and the total monetary damage in dollars the system would sustain as result of each attack, according to our simulations, are given in Table 2. The probability that the adversary will manage to evade detection to the end of the attack is given in Fig. 3.

When we examine the results, we see that an insider adversary will attempt a major routing attack if there is no IDS or if there is a centralized IDS, but will not attempt any attack at all if the dedicated or embedded IDS is present in the system. When an insider attempts to attack the system and there is no IDS, the expected damage to the system is about one million dollars, but if a centralized

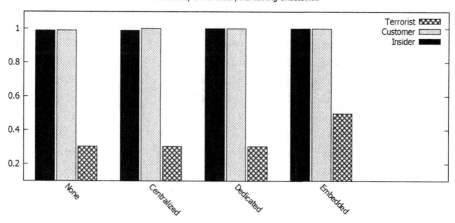

Fig. 3. Probability of remaining undetected.

IDS is present, the expected damage is halved, since the centralized IDS should be able to detect and limit the effectiveness of the routing attack. Since the insider will attempt no damaging attack when the dedicated or embedded IDS approach is used, the monetary damage to the system in this case is 0. This adversary is strongly incentivized to avoid detection, which can be seen in Fig. 3.

If there is no IDS present, the customer will attempt to jam the wireless communication between the smart meter and the rest of the network to under-report electricity consumption, causing about $600 of damage, and will successfully complete the attack without being detected in the majority of cases. However, if any of the IDS options are enabled, the customer will not attempt any attack, because the probability of obtaining the payoff is too small compared to the probability of being detected and having to pay a fine or penalty. Since no attack is attempted in these cases, the probability of remaining undetected is 1.

The terrorist is not highly incentivized to avoid detection and does not start out with many types of access, knowledge, or skills, so during our simulations the *PhysicalAttack* (which requires minimal prerequisites and causes massive damage with a high risk of detection) was chosen no matter what IDS architecture confronted the terrorist. When an adversary attempts this attack, the expected monetary damage to the system is about $5,000,000. However, there is a greater than 50 % chance that the attack will end unsuccessfully with the detection of the adversary, which we see in Fig. 3.

A utility company can use these metrics to compare intrusion detection approaches. The expected monetary loss sustained by a utility company, M, for an IDS configuration, $i \in IDS$, can be calculated with Eq. 1

$$M_i = \sum_a N_a * D_a \tag{1}$$

Table 2. Simulation results.

IDS	Adversary	Attack	Monetary damage	Error
None	Insider	Major routing	$1.07M	+/− $7.29K
	Customer	Minor jamming	$594	+/− $0.583
	Terrorist	Physical	$4.98M	+/- $50K
Centralized	Insider	Major routing	435K	+/− $2.97K
	Customer	Do nothing	$0	+/− $0
	Terrorist	Physical	$4.98M	+/− $50K
Dedicated	Insider	Do nothing	$0	+/− $0
	Customer	Do nothing	$0	+/− $0
	Terrorist	Physical	$4.98M	+/− $50K
Embedded	Insider	Do nothing	$0	+/− $0
	Customer	Do nothing	$0	+/− $0
	Terrorist	Physical	$5.02M	+/− $52.1K

where N is the expected number of attack attempts and D_a is the expected monetary damage to the system, D, per adversary, $a \in Adversaries$.

Consider a hypothetical utility that estimates 1,000 attempts by unscrupulous customers, 0.001 attempts by an insider, and 0.0001 attempts by a terrorist over a 20-year period.

Using Eq. 1 and the numbers in Table 2, we calculate the results shown in Table 3. The utility can use Table 3, along with information about installation and maintenance costs provided by vendors, to help determine the most cost-effective architecture for its system.

Table 3. Estimated monetary loss by IDS approach over a 20-year period.

IDS	Monetary damage
None	$595,568
Centralized	$933
Dedicated	$498
Embedded	$502

Space limitations force us to examine a small subset of the possible system configurations, adversaries, and attacks, but we find that the ADVISE formalism is flexible and scales well. More detail could be added by a utility company as needed. In addition, we chose to use synthetic data in our analysis as input parameters for the model. Utility companies would not have allowed us to publish unsanitized data, and it is uncertain whether any hypothetical sanitized data would have been more accurate than our educated guesses. This is not a weakness of our approach, since a utility company would already have the data needed for the input parameters for its own ADVISE model. Our synthetic data were based on the existing literature regarding security of AMI, especially [2,3].

6 Related Work

The academic community, recognizing the importance of the topic, has done prior work comparing different security approaches in AMI. The analysis in [4] uses a cost-benefit study to determine whether the added cost of RFID technology is justified given its ability to prevent electricity theft. That paper considers only energy theft, while our analysis considers attacks on the availability and integrity of the system in addition to energy theft. In [9] the authors propose an IDS for AMI and compare its security and performance with other IDSes for AMI. However, in contrast to our analysis, they explicitly do not include attacks on the meter's availability. The techniques proposed in [7] seek to compare and evaluate the security of different AMI IDSes through penetration testing and the use of archetypal and concrete attack trees. These attack trees could help a modeler create an Attack Execution Graph for an ADVISE model. In contrast to our approach, [7] does not explicitly model the attacker's attributes or motivations in detail. The authors of [2] provide a framework for evaluating the cost-effectiveness of different IDS architectures in an AMI network. However, their approach does not explicitly account for the differences in the behaviors of adversaries when attacking the system. By incorporating the adversary behavior into the model, we hope to achieve more realistic results.

7 Conclusion

In this work we showed how to use the ADVISE state-based stochastic modeling approach to calculate security metrics that are relevant in comparing different IDS architectures in an AMI network.

Unfortunately, it is often not possible to estimate many characteristics of attack steps precisely with a high degree of confidence, including the probabilities of success and the magnitude of damage given a successful attack, as well as the exact amount of protection provided by an IDS against an attack. In addition, adversary characteristics and motivations cannot usually be definitively known. Therefore, the quantitative metrics produced by the ADVISE model should not be thought of as producing exact, accurate predictions of the future. We believe, however, that these metrics can contribute to a development of a relative ranking of IDS approaches in an AMI network and provide insight into general trends of adversary behavior.

We argue that the scientific approach ADVISE offers for security evaluation is a useful complement to a common method of estimating the relative effectiveness of different security approaches: consultation of one or more security experts, who rely on intuition and experience. In contrast, the metrics calculated by ADVISE are easily auditable by other parties and assumptions are explicitly stated, which allows multiple security experts with different backgrounds to use the ADVISE formalism as a modeling language to collaboratively analyze different system designs.

Acknowledgments. The work described here was performed, in part, with funding from the Department of Homeland Security under contract HSHQDC-13-C-B0014, "Practical Metrics for Enterprise Security Engineering." The authors would also like to thank Robin Berthier, Corky Parks, Carol Muehrcke, and the anonymous reviewers of this paper for their valuable advice, as well as Jenny Applequist for her editorial assistance.

References

1. Bellman, R.: Dynamic Programming, 1st edn. Princeton University Press, Princeton (1957)
2. Cardenas, A.A., Berthier, R., Bobba, R.B., Huh, J.H., Jetcheva, J.G., Grochocki, D., Sanders, W.H.: A framework for evaluating intrusion detection architectures in advanced metering infrastructures. IEEE Trans. Smart Grid 5(2), 906–915 (2014)
3. Grochocki, D., Huh, J.H., Berthier, R., Bobba, R., Sanders, W.H., Cardenas, A.A., Jetcheva, J.G.: AMI threats, intrusion detection requirements and deployment recommendations. In: 2012 IEEE Third International Conference on Smart Grid Communications (SmartGridComm), pp. 395–400, November 2012
4. Khoo, B., Cheng, Y.: Using RFID for anti-theft in a Chinese electrical supply company: a cost-benefit analysis. In: Wireless Telecommunications Symposium (WTS), 2011, pp. 1–6 (2011)
5. LeMay, E., Ford, M.D., Keefe, K., Sanders, W.H., Muehrcke, C.: Model-based security metrics using ADversary VIew Security Evaluation (ADVISE). In: Proceedings of the 8th International Conference on Quantitative Evaluation of SysTems (QEST 2011), Aachen, Germany, 5–8 September 2011, pp. 191–200 (2011)
6. LeMay, E.: Adversary-driven state-based system security evaluation. Ph.D. thesis, University of Illinois at Urbana-Champaign, Urbana, Illinois (2011)
7. McLaughlin, S., Podkuiko, D., Miadzvezhanka, S., Delozier, A., McDaniel, P.: Multi-vendor penetration testing in the advanced metering infrastructure. In: Proceedings of the 26th Annual Computer Security Applications Conference, ACSAC 2010, pp. 107–116. ACM, New York (2010)
8. Sanders, W.H., Meyer, J.F.: A unified approach for specifying measures of performance, dependability, and performability. In: Avizienis, A., Kopetz, H., Laprie, J. (eds.) Dependable Computing for Critical Applications. Dependable Computing and Fault-Tolerant Systems, vol. 4, pp. 215–237. Springer, Heidelberg (1991)
9. Tabrizi, F.M., Pattabiraman, K.: A model-based intrusion detection system for smart meters. In: 2014 IEEE 15th International Symposium on High-Assurance Systems Engineering (HASE), pp. 17–24, January 2014
10. Möbius team: Möbius Documentation. University of Illinois at Urbana-Champaign, Urbana, IL (2014). https://www.mobius.illinois.edu/wiki/

Traffic Modeling with Phase-Type Distributions and VARMA Processes

Jan Kriege$^{(\boxtimes)}$ and Peter Buchholz

Department of Computer Science, TU Dortmund, Dortmund, Germany
{jan.kriege,peter.buchholz}@tu-dortmund.de

Abstract. In discrete event systems event times are often correlated among events of one event stream and also between different event streams. If this correlation is neglected, then resulting simulation models do not describe the real behavior in a sufficiently accurate way. In most input modeling approaches, no correlation or at most the autocorrelation of one event stream is considered, correlation between different event streams is usually neglected. In this paper we present an approach to combine multi-dimensional time series and acyclic phase type distributions as a general model for event streams in discrete event simulation models. The paper presents the basic model and methods to determine its parameters from measured traces.

Keywords: Input modeling · Stochastic simulation · Phase type distributions · Time series · Cross-correlation

1 Introduction

The adequate modeling of uncertainty is a key aspect in building realistic simulation models of real systems. The term input modeling subsumes techniques to build a mathematical model for the occurrence of events in a discrete event model [1]. In simulation historically events are characterized by distributions that are selected from a set of standard distributions like exponential, log-normal, Weibull to mention only a few examples [19]. In the recent decade also phase type distributions (PHDs), which originally have been applied in models that are solved analytically or numerically, became more popular in simulation [6,25]. PHDs allow a detailed approximation of measured densities but the parameter estimation is complex although nowadays algorithms are available that allow a fairly efficient and sufficiently exact determination of distribution parameters.

The use of distributions to model times between the occurrence of events, i.e., the inter-event times, is well established but implicitly assumes identically and independently distributed times. However, in practice, inter-event times are often correlated and the negligence of correlation results in unrealistic models and wrong results [9]. Correlations occur in various application areas like job-processing in manufacturing systems, traffic processes in computer networks or failure processes in technical systems. For modeling such dependencies time series

© Springer International Publishing Switzerland 2016
G. Agha and B. Van Houdt (Eds.): QEST 2016, LNCS 9826, pp. 295–310, 2016.
DOI: 10.1007/978-3-319-43425-4_20

[4] and Markovian arrival processes (MAPs) [23] are used. Classical time series are based on the normal distribution which is often not adequate to describe inter-event times in simulations. MAPs are a natural extension of PHDs but the adequate fitting of parameters to capture correlations is a complex and only partially solved problem such that MAPs often fail if larger correlations over various lags are present. In [7] the autoregressive to anything (ARTA) approach has been proposed which allows one to combine time series with other distributions where the distribution function can be easily inverted. This approach does not cover PHDs directly and has been extended in [17] to include a sufficiently general subclass of PHDs for modeling the distribution which is then combined with classical time series models like moving average (MA), autoregressive (AR) and autoregressive moving average (ARMA). The corresponding model is sufficiently general to model correlated inter-event times and parameter fitting can be performed efficiently.

In several applications not only the inter-event times of one event type are correlated, also inter-event times of different events show so called cross-correlation. Examples for such cross-correlations can be found in manufacturing systems where different service times are correlated [27], in internet traffic where traffic streams from different sources are correlated [26] or in grids where failure times are correlated in time and space [14]. To model such phenomena vector autoregressive to anything processes (VARTA) have been proposed in [2]. VARTA processes are based on marginal distributions of the Johnson type which allow one to fit the parameters according to the first four moments [19]. Distributions of the Johnson type are not as flexible as PHDs are.

In this paper we extend the approach from [17] to vector autoregressive processes and include also discrete time PHDs. The paper is structured as follows. In the next section we introduce the basic concepts, namely multivariate traces, PHDs, ARMA processes and the VARTA approach. Afterwards, in Sect. 3 vector correlated acyclic phase type processes are defined and the corresponding algorithms for parameter fitting are introduced. Section 4 contains some experimental results. The paper ends with the conclusions.

2   Background and Definitions

The notation we use in this paper is based on the common notation in the literature on time series [4] which is also adopted in the papers introducing the ARTA and VARTA approach [2]. Vectors and matrices are denoted by bold letters and elements are accessed by putting indices in brackets. If ε is a column vector, then ε' is the transposed row vector, $\mathbf{1}$ is the column vector where all elements are 1.

2.1   Multivariate Traces

Since we are interested in modeling several interrelated measurements simultaneously, we assume that these measurements have been recorded in form of a multivariate trace. A k-variate trace $\mathcal{T}^{(k)}$ describes k sequences of measurements (i.e. inter-event times, packet sizes etc.). We assume that the elements at position j from all sequences are associated, i.e. they are for example the inter-arrival and service time of a customer. This implies that all sequences are of equal length r. We denote the j-th entry of the i-th sequence by $t_j^{(i)}$. Statistical properties can be computed for a single sequence and the complete trace, e.g. the j-th moment of sequence i and the variance are estimated from

$$\hat{\mu}_j^{(i)} = \frac{1}{r}\sum_{m=1}^{r}(t_m^{(i)})^j \quad \text{and} \quad \hat{\sigma}_{(i)}^2 = \frac{1}{r-1}\sum_{m=1}^{r}(t_m^{(i)} - \hat{\mu}_1^{(i)})^2,$$

respectively. If we consider the complete trace the dependencies between the elements are of special interest. For a multivariate trace we can compute the auto-correlations between elements of a single sequence but also correlations between elements of different sequences resulting in correlation matrices $\hat{\rho}_h$ that contain at position (i_1, i_2) the cross-correlation between elements of sequences i_1 and i_2 that are lag h apart. Element $\hat{\rho}_h(i_1, i_2)$ is estimated as

$$\hat{\rho}_h(i_1, i_2) = \frac{1}{(r-h-1)\hat{\sigma}_{(i_1)}\hat{\sigma}_{(i_2)}} \sum_{m=1}^{r-h}(t_m^{(i_1)} - \hat{\mu}_1^{(i_1)})(t_{m+h}^{(i_2)} - \hat{\mu}_1^{(i_2)}). \tag{1}$$

In input modeling one is interested in fitting these traces, i.e. in estimating the parameters of a distribution or a process such that it resembles the characteristics of the trace. In the following we present some distributions and processes used for input modeling that are relevant for our work.

2.2   Phase-Type Distributions

PHDs describe independent and identically distributed random variables as absorption times of a finite Markov chain [23]. PHDs can be defined in continuous (CPHDs) and discrete (DPHDs) time. Most of the existing literature covers CPHDs which will be introduced first. A CPHD of order n consists of n transient and one absorbing state and is defined by an $n \times n$ matrix $\boldsymbol{D_0}$ and an initial distribution vector $\boldsymbol{\pi}$. Matrix $\boldsymbol{D_0}$ is the subgenerator of an absorbing continuous time Markov chain that contains the transition rates between transient states. It holds that $\boldsymbol{\pi}\mathbf{1} = 1$, $\boldsymbol{D_0}(i, i) < 0, \boldsymbol{D_0}(i, j) \geq 0, i \neq j$ and $\boldsymbol{D_0}\mathbf{1} \leq \boldsymbol{0}$. Events are generated whenever the absorbing state is reached and the process is restarted immediately afterwards as defined by $\boldsymbol{\pi}$. Properties of the distribution can be defined in terms of this matrix and vector, e.g. for the moments and cumulative distribution function we have

$$\mu_i = E(X^i) = i!\boldsymbol{\pi}\boldsymbol{M}^i\mathbf{1} \quad \text{and} \quad F(t) = 1 - \boldsymbol{\pi}exp(\boldsymbol{D_0}t)\mathbf{1}, \tag{2}$$

respectively, where $\boldsymbol{M} = -(\boldsymbol{D_0})^{-1}$.

DPHDs [3] are defined in a similar way. They are described by an $n \times n$ matrix $\boldsymbol{T}$ that contains the transition probabilities between transient states and an initial distribution vector τ where $T(i,j) \geq 0$, $T1 \leq 1$ and for τ the same constraints as for π apply. Then, factorial moments and distribution function are given by

$$m_i = i!\tau(I - T)^{-i}T^{i-1}1 \quad \text{and} \quad F(t) = 1 - \tau T^t 1, t \geq 0. \tag{3}$$

In general, it is difficult to obtain the moments μ_i from the factorial moments m_i. However, for the following ideas we only need mean and variance, which are given by

$$\mu_1 = m_1 \quad \text{and} \quad \sigma^2 = 2\tau(I - T)^{-2}T1 + \tau(I - T^{-1})1 - (\tau(I - T)^{-1}1)^2. \tag{4}$$

Depending on the structure of matrix D_0 (or T) and vector π (or τ) several subclasses of PHDs have been defined in the past. If the states can be ordered such that the matrix becomes upper triangular we have an acyclic CPHD or DPHD. Simpler and well known subclasses include the Exponential, Erlang and Hyper-Erlang distributions in the continuous case and the geometric and negative binomial distributions in the discrete case.

There exists a wide theory on fitting CPHDs to the empirical distribution of a trace. Many approaches are Expectation Maximization (EM) algorithms, which try to maximize the likelihood and work on the complete trace. Since EM algorithms are very slow in the general case, they have been tailored for fitting special subclasses of CPHDs like Hyper-Exponential distributions in [16] or Hyper-Erlang distributions in [28]. Faster approaches usually derive some characteristics like moments from the trace and fit the PHD according to these characteristics. An example for fitting PHDs according to empirical moments can be found in [5]. For DPHDs [3] describes a parameter estimation algorithm. A more complete list on fitting approaches can be found in [6].

2.3 (Vector) Autoregressive Moving Average Processes

Autoregressive Moving Average Processes ($ARMA(p,q)$) are well established in time series modeling (see [4]) and are defined as

$$Z_t = \alpha_1 Z_{t-1} + \alpha_2 Z_{t-2} + \ldots + \alpha_p Z_{t-p} + \beta_1 \epsilon_{t-1} + \beta_2 \epsilon_{t-2} + \ldots + \beta_q \epsilon_{t-q} + \epsilon_t \tag{5}$$

where the α_i are autoregressive coefficients, the β_i are moving average coefficients and the ϵ_t are called innovations that have normal distribution with mean zero and variance σ_ϵ^2. If the moving average terms in Eq. 5 are omitted the process becomes an $AR(p)$ and if the autoregressive terms are omitted an $MA(q)$.

A generalization of ARMA processes are Vector ARMA processes ($VARMA$ (p,q)) that can model multivariate time series [21]. If we assume that the Z_t in Eq. 5 are vectors and the coefficients are defined by matrices we immediately obtain the definition of a $VARMA(p,q)$ process:

$$Z_t = \alpha_1 Z_{t-1} + \alpha_2 Z_{t-2} + \ldots + \alpha_p Z_{t-p} + \beta_1 \epsilon_{t-1} + \beta_2 \epsilon_{t-2} + \ldots + \beta_q \epsilon_{t-q} + \epsilon_t \tag{6}$$

where the $\boldsymbol{Z_t} = (Z_t^{(1)}, Z_t^{(2)}, \ldots, Z_t^{(k)})'$ are $(k \times 1)$ vectors with the observations at time t, the $\boldsymbol{\alpha}_i$ and $\boldsymbol{\beta}_i$ are $(k \times k)$ matrices with autoregressive and moving average coefficients, respectively, and $\boldsymbol{\epsilon_t} = (\epsilon_t^{(1)}, \epsilon_t^{(2)}, \ldots, \epsilon_t^{(k)})'$ is a $(k \times 1)$ vector with innovations introducing randomness into the sequence. We will assume that the ϵ_t are Gaussian with covariance matrix $\boldsymbol{\Sigma}_\epsilon$ and $E[\epsilon_t] = 0$, $E[\epsilon_t \epsilon_t'] = \boldsymbol{\Sigma}_\epsilon$, $E[\epsilon_t \epsilon_s'] = \boldsymbol{0}, s \neq t$. Similar to the univariate case we obtain a $VAR(p)$ for $q = 0$ and a $VMA(q)$ for $p = 0$.

2.4 Stochastic Processes with an (V)ARMA Background Process

ARMA processes are very flexible in modeling autocorrelation, e.g. an $AR(p)$ can model p lags of autocorrelation exactly. However, the processes as defined in Eq. 5 result in a normal marginal distribution, which makes them not really suitable for simulation input modeling in most cases where the distribution is clearly non-normal.

A promising approach to overcome this limitation is to use ARMA processes as a background process for only modeling the autocorrelation and combine them with an arbitrary marginal distributions.

This idea has been introduced by ARTA processes [7] that combine an $AR(p)$ base process as defined in Eq. 5 with a marginal distribution F using the inversion method by setting $Y_t = F^{-1}[\Phi(Z_t)], (t = 1, 2, \ldots)$ where Φ is the standard normal cumulative distribution function. The $AR(p)$ process is constructed such that the distribution of the $\{Z_t\}$ is $N(0,1)$, resulting in $\Phi(Z_t)$ to have uniform distribution on $(0,1)$. The inverse transformation finally yields a time series Y_t with the desired marginal distribution. Since the Z_t are autocorrelated, the Y_t are as well, and it is possible to establish a relation between the two autocorrelations.

In [2] ARTA processes have been generalized to Vector ARTA (VARTA) processes that can model multivariate time series. In this case the base process is a $VAR(p)$ process as defined in Eq. 6 and the VARTA process $\boldsymbol{Y}_t = (Y_t^{(1)}, Y_t^{(2)}, \ldots, Y_t^{(k)})$ is obtained by $Y_t^{(i)} = F_{(i)}^{-1}(\Phi(Z_t^{(i)})), i = 1, \ldots, k$. [2] assumed that the $F_{(i)}$ are marginal distributions from the Johnson system of distributions though other distributions are possible as well of course.

(V)ARTA processes rely on the inversion of the distribution function and therefore are applicable for all marginal distributions F for which a closed-form expression for the inverse cdf exists or for which F^{-1} can be computed numerically in an efficient way. In practice some interesting and useful distributions cannot be used as part of ARTA processes because of this. PHDs from Sect. 2.2 have proven to be suitable distributions when it comes to modeling complicated empirical distributions from real-world observations that for example have been recorded in computer or communication networks [6]. However, as we can see from Eq. 2 the cumulative distribution function of CPHDs contains a matrix exponential which makes it in general inefficient and difficult to numerically compute the inverse of the cdf. [17] presented a different approach to combine acyclic CPHDs and ARMA processes, denoted as CAPP (Correlated Acyclic Phase-Type Process), that will be used as basis for the work in this paper.

The key idea is to interpret the CPHD as a set of paths from an initial state (i.e. states i with $\pi(i) > 0$) to the absorbing state and use a background ARMA process to choose the next path when restarting after absorption.

In the following we will extend CAPPs in two directions. First, we will integrate acyclic DPHDs into the process, which will be helpful to model measurements with discrete values. Second, we will generalize the process description to the multivariate case to allow for correlation between the different variates.

3   Vector Correlated Acyclic Phase-Type Processes

In the following we present our approach to combine k acyclic PHDs (discrete and continuous) with a VARMA background process, denoted as Vector Correlated Acyclic Phase-Type Process $VCAPP_k(n_1, \ldots, n_k, p, q)$, where $n_1, \ldots, n_k$ define the number of transient states of the PHDs and p, q give the order of the VARMA base process.

We first describe how the PHDs are represented as paths from an initial to the absorbing state (called elementary series), then show how the PHDs and the VARMA process are combined, how we can compute basic properties of the process and finally present algorithms for parameter estimation and random number generation for the process class.

3.1   Splitting Acyclic PHDs into Paths

As already mentioned we express a PHDs in terms of its elementary series, where each series describes one path from an initial state to the absorbing state. For CPHDs this concept has been introduced in [10] and it was later extended to DPHDs in [3].

For an acyclic CPHD $(\boldsymbol{\pi}, \boldsymbol{D}_0)$ the i-th series is described by a vector $\boldsymbol{\Lambda}_i$ that contains the transition rates of the states of the series and an associated probability v_i that is computed from the transition rates along the path and the initial probability of the first state of the path, i.e. v_i is the probability that this path is chosen. More formally, let $i_1, i_2, \ldots, i_k$ be the states of an elementary series. Then the probability of this series is given by

$$v_i = \boldsymbol{\pi}(i_1) \frac{\boldsymbol{D}_0(i_1, i_2)}{-\boldsymbol{D}_0(i_1, i_1)} \frac{\boldsymbol{D}_0(i_2, i_3)}{-\boldsymbol{D}_0(i_2, i_2)} \cdots \frac{\boldsymbol{d}_1(i_k)}{-\boldsymbol{D}_0(i_k, i_k)}$$

where $\boldsymbol{d}_1(i_k)$ denotes the transition rate from state i_k to the absorbing state. The vector of transition rates is given by

$$\boldsymbol{\Lambda}_i = (-\boldsymbol{D}_0(i_1, i_1), -\boldsymbol{D}_0(i_2, i_2), \ldots, -\boldsymbol{D}_0(i_k, i_k)).$$

An example for an acyclic CPHD and its elementary series is given in Fig. 1a. In the discrete case the i-th series of a PHD $(\boldsymbol{\tau}, \boldsymbol{T})$ is represented by a vector $\boldsymbol{P}_i$ with transition probabilities of the states of the series and the associated probability ϕ_i. Now we have for a series of states $i_1, i_2, \ldots, i_k$ that

$$\boldsymbol{P}_i = (\boldsymbol{T}(i_1, i_1), \ldots, \boldsymbol{T}(i_k, i_k)) \text{ and } \phi_i = \frac{\boldsymbol{\tau}(i_1)\boldsymbol{T}(i_1, i_2)}{1 - \boldsymbol{T}(i_1, i_1)} \frac{\boldsymbol{T}(i_2, i_3)}{1 - \boldsymbol{T}(i_2, i_2)} \cdots \frac{\boldsymbol{t}(i_k)}{1 - \boldsymbol{T}(i_k, i_k)}$$

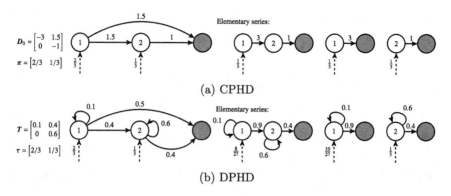

(a) CPHD

(b) DPHD

Fig. 1. Representing acyclic PHDs as elementary series

where $t(i_k)$ denotes the transition probability from state i_k to the absorbing state. An example for an acyclic DPHD and its elementary series is given in Fig. 1b.

Note, that an elementary series describes a generalized Erlang (or hypo-exponential) distribution in the continuous case and a sequence of geometric distributions in the discrete case.

3.2 Combining PHDs and VARMA Process

Now assume, that we have k PHDs each expressed in terms of their elementary series, i.e. for PHD i we have the series $j = 1, \ldots, m_i$ each consisting of an initial probability v_{i_j} and the transition rates Λ_{i_j} in the continuous case or a probability ϕ_{i_j} and vector P_{i_j} in the discrete case. To simplify notations we denote the probability of an elementary series j of PHD i with ψ_{i_j} if it is not important whether it is a series from a continuous or discrete distribution. We additionally introduce a stationary $VARMA(p, q)$ base process Z_t as defined in Eq. 6 that is used to choose between these series, thereby introducing correlation into the sequence of phase-type distributed random variables. More formally, we use the following definitions to describe VCAPPs: For the elementary series of PHD i we define

$$\underline{b}_{i_1} = 0$$
$$\bar{b}_{i_j} = \underline{b}_{i_j} + \psi_{i_j} \; j = 1, \ldots, m_i \quad \text{and} \quad \delta(U, i, j) = \begin{cases} 1, & U \in [\underline{b}_{i_j}, \bar{b}_{i_j}) \\ 0, & \text{otherwise} \end{cases} \quad (7)$$
$$\underline{b}_{i_j} = \bar{b}_{i_{j-1}} \quad j = 2, \ldots, m_i$$

for some random variable U with uniform distribution. Let $\{X_t^{(i,j)}\}$ be sequences of iid random variables that either have a generalized Erlang distribution described by the rates in Λ_{i_j} if i is a CPHD or describe a sequence of geometric distributions as defined by the probabilities in P_{i_j} if i is discrete.

Using the above definitions a $VCAPP_k(n_1, \ldots, n_k, p, q)$ describes a time-series $\boldsymbol{Y}_t = (Y_t^{(1)}, Y_t^{(2)}, \ldots, Y_t^{(k)})'$ where each $Y_t^{(i)}$ is defined as

$$Y_t^{(i)} = \sum_{j=1}^{m_i} \delta(\Phi(Z_t^{(i)}), i, j) X_t^{(i,j)}. \tag{8}$$

The $Z_t^{(i)}$ are generated by a $VARMA(p, q)$ process as defined in Eq. 6 and we require them to have standard normal marginal distribution. $\Phi(\cdot)$ is the standard normal cumulative distribution function which implies that $\Phi(Z_t^{(i)})$ have uniform distribution on $(0, 1)$ (cf. [11]). Then Eq. 8 uses the elementary series to describe a sequence of correlated random variables with the same acyclic PHD that the elementary series have been computed from (see also [17]).

In the following we relate the VARMA base process and the VCAPP autocorrelation and describe a procedure to construct a base process that results in the desired VCAPP correlation and that respects the requirements pointed out above (i.e. a stable process where the $Z_t^{(i)}$ have standard normal marginal distribution).

3.3   Computation of the VCAPP Autocorrelation

Assume that a $VCAPP_k(n_1, \ldots, n_k, p, q)$ is given. The autocorrelation and autocovariance of a $VARMA(p, q)$ process can be expressed in terms of the autocorrelation and autocovariance matrices. Note, that we required the $Z_t^{(i)}$ to have standard normal marginal distribution and thus, autocorrelation and autocovariance are identical for the base process. In the following we will sketch how to compute the autocorrelations of the $VARMA(p, q)$ base process and then relate the base process correlation with the $VCAPP_k(n_1, \ldots, n_k, p, q)$ correlation.

The theoretical background and the methods for the computation of the $VARMA(p, q)$ correlation are given in [21]. We will only summarize the basic steps. The key idea of the computation is the transformation of the $VARMA(p, q)$ into an equivalent $VAR(1)$ representation, from which the correlation can be computed easily. Let $\{\boldsymbol{Z}_t\}$ be a stable k-variate $VARMA(p, q)$ process as defined in Eq. 6. We denote the equivalent $VAR(1)$ process $\tilde{\boldsymbol{Z}}_t = \tilde{\boldsymbol{\alpha}}\tilde{\boldsymbol{Z}}_{t-1} + \tilde{\boldsymbol{\epsilon}}_t$. The construction of the matrices is summarized in [18, Sect. 2] (see also [21] for the details). For the $VAR(1)$ representation $\tilde{\boldsymbol{\Sigma}}_{\tilde{z}}(0)$, i.e. the covariance matrix at lag 0 can be easily obtained from

$$vec(\tilde{\boldsymbol{\Sigma}}_{\tilde{z}}(0)) = (\boldsymbol{I} - \tilde{\boldsymbol{\alpha}} \otimes \tilde{\boldsymbol{\alpha}})^{-1} vec(\tilde{\boldsymbol{\Sigma}}_{\epsilon}) \tag{9}$$

where the $vec()$ operator transforms a matrix into a vector by stacking the columns. $\tilde{\boldsymbol{\Sigma}}_{\tilde{z}}(0)$ now contains the desired covariance matrices $\boldsymbol{\Sigma}_Z(h)$ for $h = 0, \ldots, p - 1$ of the original $VARMA(p, q)$ process as submatrices (i.e. $\boldsymbol{\Sigma}_Z(h)$ is a block in the upper left, followed by $\boldsymbol{\Sigma}_Z(1)$ and so on) [21]. If $\boldsymbol{\Sigma}_Z(h)$ for $h = 0, \ldots, p - 1$ is known higher lags can be computed recursively from [21]

$$\boldsymbol{\Sigma}_Z(h) = \boldsymbol{\alpha_1}\boldsymbol{\Sigma}_Z(h - 1) + \cdots + \boldsymbol{\alpha_p}\boldsymbol{\Sigma}_Z(h - p). \tag{10}$$

The computation of the autocovariances requires $p > q$. Therefore, if $p \leq q$ we have to add additional matrices $\boldsymbol{\alpha}_i = 0$ before the transformation into the $VAR(1)$ process. The autocorrelations can be obtained from the autocovariances by $\boldsymbol{\rho}_Z(h) = \boldsymbol{D}^{-1}\boldsymbol{\Sigma}_Z(h)\boldsymbol{D}^{-1}$ where $\boldsymbol{D}$ is a diagonal matrix that has the square roots of the diagonal elements of $\boldsymbol{\Sigma}_Z(0)$ as elements. Recall, that this last step is not necessary in our case, because we required the $\boldsymbol{Z}_t$ to have standard normal distribution. Thus, $\boldsymbol{D}^{-1} = I$ and $\boldsymbol{\Sigma}_Z(h) = \boldsymbol{\rho}_Z(h)$.

Once the $VARMA(p,q)$ autocorrelation matrices are known, we can use them to compute the autocorrelation matrices of the VCAPP. Let $\boldsymbol{\rho}_Z(h)$ be a matrix with autocorrelation coefficients at lag h from a $VARMA(p,q)$ process computed as described above. Then, we want to determine matrix $\boldsymbol{\rho}_Y(h)$ that contains the corresponding correlation coefficients of the VCAPP. Furthermore, let $\rho_Z(i,j,h)$ and $\rho_Y(i,j,h)$ be the elements at position (i,j) of $\boldsymbol{\rho}_Z(h)$ and $\boldsymbol{\rho}_Y(h)$, respectively. Then,

$$\rho_Y(i,j,h) = Corr[Y_t^{(i)}, Y_{t+h}^{(j)}] = \frac{E[Y_t^{(i)}Y_{t+h}^{(j)}] - E[Y^{(i)}]E[Y^{(j)}]}{\sqrt{Var[Y^{(i)}]Var[Y^{(j)}]}} \tag{11}$$

Since $E[Y^{(i)}]$, $E[Y^{(j)}]$, $Var[Y^{(i)}]$ and $Var[Y^{(j)}]$ are fixed by the acyclic PHDs i and j, i.e. they can be computed using Eq. 2 or Eq. 4, the only remaining part when computing the VCAPP correlation is the joint moment $E[Y_t^{(i)}Y_{t+h}^{(j)}]$. After some substitutions and rearranging we get (cf. [17])

$$E[Y_t^{(i)}Y_{t+h}^{(j)}] = E\left[\left(\sum_{k=1}^{m_i} \delta(\Phi(Z_t^{(i)}), i, k)X_t^{(i,k)}\right)\left(\sum_{l=1}^{m_j} \delta(\Phi(Z_{t+h}^{(j)}), j, l)X_{t+h}^{(j,l)}\right)\right]$$

$$= \sum_{k,l} E\left[\delta(\Phi(Z_t^{(i)}), i, k)\delta(\Phi(Z_{t+h}^{(j)}), j, l)\right] E\left[X_t^{(i,k)}\right] E\left[X_{t+h}^{(j,l)}\right] \tag{12}$$

$$= \sum_{k,l}\left(\mu^{(i,k)}\mu^{(j,l)} \int_{-\infty}^{\infty}\int_{-\infty}^{\infty} \delta(\Phi(Z_t^{(i)}), i, k)\delta(\Phi(Z_{t+h}^{(j)}), j, l)\right.$$

$$\left. \varphi_{\rho_Z(i,j,h)}(z_t^{(i)}, z_{t+h}^{(j)})dz_t^{(i)}dz_{t+h}^{(j)}\right)$$

$$= \sum_{k,l}\left(\mu^{(i,k)}\mu^{(j,l)} \int_{\Phi^{-1}(\underline{b}_{i_k})}^{\Phi^{-1}(\bar{b}_{i_k})}\int_{\Phi^{-1}(\underline{b}_{j_l})}^{\Phi^{-1}(\bar{b}_{j_l})} \varphi_{\rho_Z(i,j,h)}(z_t^{(i)}, z_{t+h}^{(j)})dz_t^{(i)}dz_{t+h}^{(j)}\right) \tag{13}$$

where $\mu^{(i,k)}$ is the first moment of the k-th elementary series of the i-th PHD. Since the series of a PHD are either a generalized Erlang distribution or a sequence of geometric distributions, the series are, of course, simple PHDs themselves and the moments can be easily computed according to Eqs. 2 and 4. $\varphi_{\rho_Z(i,j,h)}(z_t^{(i)}, z_{t+h}^{(j)})$ is the bivariate standard normal density function with correlation $\rho_Z(i,j,h)$. Note, that Eq. 12 holds because in our process description the base process is used to determine which series of the PHDs are taken but the duration in that series is independent of the base process. Equation 13 holds because $\delta(u,i,k)$ is 1 for $u \in [\underline{b}_{i_k}, \bar{b}_{i_k})$ and 0 otherwise (cf. Eq. 7) and we can exploit this information to determine the integration bounds. For the computation of the bivariate normal integral fast numerical procedures exist [12]. Observe from Eq. 13 that for a given PHD the VCAPP correlation only depends on the VARMA correlation which appears in $\varphi_{\rho_Z(i,j,h)}(z_t^{(i)}, z_{t+h}^{(j)})$. Therefore, we may express $\rho_Y(i,j,h)$ as a function of $\rho_Z(i,j,h)$, i.e. $\rho_Y(i,j,h) = \omega(\rho_Z(i,j,h))$. In the following we assume that the elementary series of a PHD i are sorted according to their mean values $\mu^{(i,k)}$. Then $\omega(\cdot)$ has the following properties that are useful for the construction of the base process in Sect. 3.4 and that we state without proof: $\omega(\cdot)$ is a continuous and non-decreasing function. This immediately implies that the maximal and minimal possible autocorrelations $\hat{\rho}_{max}$ and $\hat{\rho}_{min}$ for a VCAPP are given by $\hat{\rho}_{max} = \omega(1)$ and $\hat{\rho}_{min} = \omega(-1)$, respectively. Furthermore we have that $\omega(0) = 0$ and $\rho_Z(i,j,h) \leq 0 \ (\geq 0) \Rightarrow \omega(\rho_Z(i,j,h)) \leq 0 \ (\geq 0)$.

3.4   An Algorithm for Fitting VCAPPs

Using the considerations from Sect. 3.3 we can sketch the algorithm from Fig. 2 for fitting VCAPPs from observations from a real system. As inputs the algorithm takes a k-variate trace $\mathcal{T}^{(k)}$ and the order p, q of the base process. In the first step the algorithm fits a PHD to each of the sequences of $\mathcal{T}^{(k)}$ using one of the available approaches mentioned in Sect. 2.2. Next the correlation matrices $\hat{\rho}_Y(i), i = 1, \ldots, H$ are estimated from the multivariate trace which describe the desired correlation the VCAPP should have. After that the algorithm determines the elementary series of the acyclic PHDs as described in Sect. 3.1 and sorts them according to their mean values. In the next step the VARMA correlation $\hat{\rho}_Z(i), i = 1, \ldots, H$ that yields a VCAPP correlation $\hat{\rho}_Y(i), i = 1, \ldots, H$ has to be determined. Observe from Eq. 13 that we can compute a $\hat{\rho}_Y(i,j,h)$ for a given $\hat{\rho}_Z(i,j,h)$ but not the other way round, which is necessary for constructing the VARMA base process. Consequently, we have to determine the base process autocorrelation numerically using Eq. 13 which can be done using a simple line search algorithm [24], since $\omega(\cdot)$ is a continuous and non-decreasing function. It should be noted that also for the original CAPP approach and the (V)ARTA approaches a numerical procedure has to be applied for computing the base process autocorrelation [2,7,17].

```
Inputs:
    k-variate trace T^(k); order of base process: p,q

1) fit k PHDs to the k sequences of T^(k)
2) compute correlation matrices ρ̂_Y(i), i = 1,...,H from T^(k) using Eq. 1
3) determine and sort elementary series of APH(n_i), i = 1,...,k
4) determine VARMA autocorrelations ρ̂_Z(i), i = 1,...,H such that VCAPP has
   correlation ρ̂_Y(i), i = 1,...,H using Eq. 13 and a search algorithm
5a) if (q==0) and (p=H): Compute VAR(p) base process according to Eq. 14
5b) otherwise: minimize Eq. 16 to find a VARMA(p,q) model for ρ̂_Z(i), i = 1,...,H
6) return VCAPP_k(n_1,...,n_k,p,q) with base VARMA(p,q) process and k PHDs
```

Fig. 2. Basic steps for parameter estimation of VCAPPs

Once the base process autocorrelations $\hat{\boldsymbol{\rho}}_Z(h)$ have been determined we have to construct a *VARMA(p,q)* base process that exhibits this structure. Depending on whether the base process is a *VARMA(p,q)* process or a *VAR(p)* process without moving average coefficients this can be done using a general purpose optimization algorithm or by solving Yule-Walker equations. Fitting a *VAR(p)* is very fast but might result in a large base process because it requires $p = H$, where H equals the number of autocorrelation lags that are considered, i.e. we need a matrix with autoregressive coefficients for each correlation matrix. Fitting a *VARMA(p,q)* process can result in a smaller model but the parameter estimation is more elaborate.

A *VAR(p)* model can be obtained by solving the the Yule-Walker equations [21], i.e. $\boldsymbol{\alpha} = (\boldsymbol{\alpha}_1, \ldots, \boldsymbol{\alpha}_p)$ is obtained from

$$\boldsymbol{\alpha} = \boldsymbol{\Sigma} \boldsymbol{\Sigma_Z}^{-1} \tag{14}$$

where $\boldsymbol{\Sigma} = (\boldsymbol{\Sigma}_Z(1), \boldsymbol{\Sigma}_Z(2), \ldots, \boldsymbol{\Sigma}_Z(p))$ and

$$\boldsymbol{\Sigma_Z} = \begin{bmatrix} \boldsymbol{\Sigma}_Z(0) & \boldsymbol{\Sigma}_Z(1) & \cdots & \boldsymbol{\Sigma}_Z(p-2) & \boldsymbol{\Sigma}_Z(p-1) \\ \boldsymbol{\Sigma}'_Z(1) & \boldsymbol{\Sigma}_Z(0) & \cdots & \boldsymbol{\Sigma}_Z(p-3) & \boldsymbol{\Sigma}_Z(p-2) \\ \vdots & \vdots & \ddots & \vdots & \vdots \\ \boldsymbol{\Sigma}'_Z(p-1) & \boldsymbol{\Sigma}'_Z(p-2) & \cdots & \boldsymbol{\Sigma}'_Z(1) & \boldsymbol{\Sigma}_Z(0) \end{bmatrix}. \tag{15}$$

Once the $\boldsymbol{\alpha}_i$ are known we compute $\boldsymbol{\Sigma}_\epsilon = \boldsymbol{\Sigma}_Z(0) - \boldsymbol{\alpha}_1 \boldsymbol{\Sigma}'_Z(1) - \cdots - \boldsymbol{\alpha}_p \boldsymbol{\Sigma}'_Z(p)$. If we set $\boldsymbol{\Sigma}_Z(i) = \hat{\boldsymbol{\rho}}_Z(i)$ in the above equations, i.e. we assume that the autocovariance equals the autocorrelation, we get a process with the desired property to have standard normal marginal distribution.

When fitting a *VARMA(p,q)* process to $\hat{\boldsymbol{\rho}}_Z(h)$ we use the Nelder-Mead algorithm from [22] to minimize

$$\underset{\boldsymbol{\alpha}_i, i=1,\ldots,p, \boldsymbol{\beta}_j, j=1,\ldots,q}{\arg \min} \sum_{l=1}^{k} \sum_{m=1}^{k} \sum_{h=1}^{H} \left(\frac{\rho_Z(l,m,h)^*}{\hat{\rho}_Z(l,m,h)} - 1 \right)^2 \tag{16}$$

where k is the number of variates, H is the number of autocorrelation lags to consider for fitting, $\hat{\rho}_Z(l,m,h)$ is the desired autocorrelation of the base process and $\rho_Z(l,m,h)^*$ is the autocorrelation of the VARMA process constructed during the fitting process. Since we are only interested in stationary base processes, non-stationary solutions have to be penalized in the fitting step. According to [21] stability of a VARMA process implies stationarity and a VARMA process is stable if all roots of the reverse characteristic polynomial $|\boldsymbol{I}_{k \times k} - \boldsymbol{\alpha}_1 z - \boldsymbol{\alpha}_2 z^2 - \cdots - \boldsymbol{\alpha}_p z^p| = 0$ lie outside the unit circle. Hence, for VARMA processes that are constructed during the fitting process and that do not fulfill this requirement we add a penalty term to Eq. 16.

Observe, that the covariance matrix of the innovations $\boldsymbol{\Sigma}_\epsilon$ is not part of the minimization in Eq. 16. Instead we set the values of $\boldsymbol{\Sigma}_\epsilon$ separately, such that the VARMA process has standard normal marginal distributions. For given AR and MA coefficient matrices we use Eq. 9 to construct a system of linear equations to determine the entries of $\boldsymbol{\Sigma}_\epsilon$. This requires $\boldsymbol{\Sigma}_\epsilon$ to be a diagonal matrix. Restricting the structure of the covariance matrix is not uncommon to cope with the complexity of VARMA processes [8]. Details on the construction of $\boldsymbol{\Sigma}_\epsilon$ can be found in [18, Sect. 3].

The minimization can be performed for different values of p and q and the best model according to Eq. 16 is selected. Of course, one has to find a compromise between a moderate size of the parameters p and q to keep the model size small and an adequate fitting of the lag h ($h = 1, \ldots, H$) correlations. Finally, the PHDs and the $VARMA(p,q)$ are combined into a $VCAPP_k(n_1, \ldots, n_k, p, q)$.

3.5 Generating Random Numbers from VCAPPs

The generation of multivariate samples from a $VCAPP_k(n_1, \ldots, n_k, p, q)$ can be efficiently performed in two steps. In the first step a multivariate sample from the $VARMA(p,q)$ base process is generated using standard theory [21]. In the second step this sample is used to generate the VCAPP random vector.

From the $VARMA(p,q)$ description the coefficient matrices $\boldsymbol{\alpha}_i, i = 1, \ldots, p$ and $\boldsymbol{\beta}_j, j = 1, \ldots, q$ and the covariance matrix $\boldsymbol{\Sigma}_\epsilon$ for the innovations are known. Furthermore, we assume that the previous observations $\boldsymbol{z}_{t-1}, \ldots, \boldsymbol{z}_{t-p}$ and the previous innovations $\boldsymbol{\epsilon}_{t-1}, \ldots, \boldsymbol{\epsilon}_{t-q}$ are known, either from previous simulation steps of the $VARMA(p,q)$ process or from an initialization step that is described in [21] and summarized in [18, Sect. 4]. Then, the next random sample can be determined recursively using Eq. 6. First, we compute the next vector with innovations $\boldsymbol{\epsilon}_t$. This is done by determining matrix $\boldsymbol{S}$ with $\boldsymbol{S}\boldsymbol{S}' = \boldsymbol{\Sigma}_\epsilon$ applying a Cholesky decomposition [13]. Now, $\boldsymbol{\epsilon}_t = \boldsymbol{S}\boldsymbol{v}_t$ where $\boldsymbol{v}_t' = (v_1, \ldots, v_k)$ contains k random numbers drawn from a standard normal distribution. Since now all values from Eq. 5 are known we can compute $\boldsymbol{z}_t$ according to that definition.

In the second step of the random number generation procedure the $\boldsymbol{z}_t$ are used to determine the VCAPP random vector $\boldsymbol{y}_t$ using Eq. 8. For each $\boldsymbol{y}_t^{(i)}$ we compute $\Phi(z_t^{(i)})$, determine the interval j for which $\Phi(z_t^{(i)}) \in [\underline{b}_{i_j}, \overline{b}_{i_j}]$. If PHD i is continuous we draw from an exponential distribution for each rate from $\boldsymbol{\Lambda}_i$. The sum of these exponentially distributed random numbers is returned as $\boldsymbol{y}_t^{(i)}$. If PHD i is discrete we draw from the geometric distributions that are defined by $\boldsymbol{P}_i$ and return the sum of the number of trials as $\boldsymbol{y}_t^{(i)}$.

4 Experimental Results

We conducted two series of experiments to show the performance of VCAPPs when fitting real traffic data. For the first experiments we used the well known trace BC-pAug89 [20] that is often used as a benchmark trace and that contains the interarrival times and packet sizes of 1 million packets in a local area network. The second more recent trace was recorded at Dartmouth College [15] and additionally includes user mobility in a wireless network.

As mentioned above BC-pAug89 is a 2-variate trace consisting of one sequence for the interarrival times of packets and another sequence with the

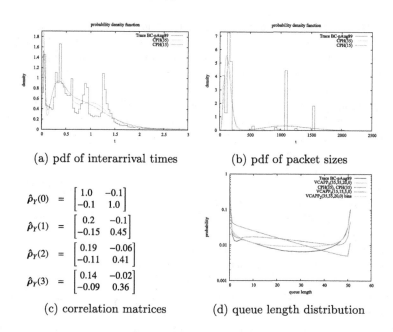

(a) pdf of interarrival times   (b) pdf of packet sizes

$$\hat{\rho}_Y(0) = \begin{bmatrix} 1.0 & -0.1 \\ -0.1 & 1.0 \end{bmatrix}$$

$$\hat{\rho}_Y(1) = \begin{bmatrix} 0.2 & -0.1 \\ -0.15 & 0.45 \end{bmatrix}$$

$$\hat{\rho}_Y(2) = \begin{bmatrix} 0.19 & -0.06 \\ -0.11 & 0.41 \end{bmatrix}$$

$$\hat{\rho}_Y(3) = \begin{bmatrix} 0.14 & -0.02 \\ -0.09 & 0.36 \end{bmatrix}$$

(c) correlation matrices   (d) queue length distribution

Fig. 3. Results for trace BC-pAug89

corresponding packet sizes. We fitted CPHDs with different number of states to both of the sequences using the approach from [28]. The results for CPHDs with 15 and 35 states are shown in Figs. 3a and b. As we can see the CPHDs provided a good approximation for the interarrival times, but only were able to capture the first peak of the packet sizes. According to the algorithm from Fig. 2 we expanded the CPHDs into two VCAPPs. The larger VCAPPs consists of the two CPHDs with 35 states and a $VAR(20)$ base process, the smaller VCAPP of the two CPHDs with 15 states and a $VAR(5)$ base process. Since we used VAR processes the VCAPPs provided an exact matching of 5 and 20 lags of correlation, respectively. The correlation matrices for the first lags are shown in Fig. 3c. As a last measure for the quality of the fitting we determined the queue length distribution of a single server queue with a buffer capacity of 50. The VCAPPs are used to obtain the interarrival times of the packets and the exponentially distributed service time depends on the packet sizes also obtained from the VCAPPs. To obtain reference values we also simulated the model with the original trace values. Figure 3d shows the queue length distribution obtained from 10 replications and approx. 1 million generated packets per replication. As we can see the large $VCAPP_2(35, 35, 20, 0)$ provides a very good approximation of the trace results, while for the smaller $VCAPP_2(15, 15, 5, 0)$ the results are not as good. For comparison we also generated results from the two CPHDs of the larger VCAPP without using the base process, i.e. the generated values are uncorrelated. The curve shows that neglecting the correlation significantly impairs the quality of the approximation. Finally, we used a modified version of the $VCAPP_2(35, 35, 20, 0)$ where we sorted the packets into bins of size 50 bytes resulting in a much smaller number of discrete values and used a DPHD for fitting. For the simulation each bin was represented by its midpoint. As shown in Fig. 3d the curves for the original $VCAPP_2(35, 35, 20, 0)$ and the variant with bins almost completely overlap.

At Dartmouth College large traces of data have been recorded from their wireless network over the past years. The traces contain movement data and tcpdump data with generated traffic. The movement data consists of a trace for every user seen in the network with dwell times and assignments to different access points. A summary of the data and collection methods can be found in [15]. Out of this data we generated a 3-variate trace consisting of the dwell times and locations from the movement data and the generated traffic from the tcpdump data. Unfortunately the movement data was recorded for a larger period of time than the tcpdump data. Moreover, tcpdump data does not exist for all access points occurring in the movement data. To obtain suitable traces for fitting we preprocessed the data in the following way: We parsed every movement trace and for each entry in the trace we checked the tcpdump data if packets have been recorded for that user in the time period specified by the movement data and summed up the amount of traffic from all packets in that period. Since traffic information is not available for all entries of the movement data we extracted consecutive sequences of at least 100 entries from the movement data for that the traffic amount could be computed. Hence, we obtained traces of the form $\mathcal{T}_i = ((t_{i,1}, l_{i,1}, p_{i,1}), (t_{i,2}, l_{i,2}, p_{i,2}), \cdots)$ describing the behavior of $i = 1, \cdots, 501$ users in the network where $t_{i,j}, l_{i,j}$ and $p_{i,j}$ are the j-th dwell time, location and amount of traffic of user i, respectively. Thus, we generated traces with realistic movement patterns, although the traces of course do not fully reflect the original situation at Dartmouth College because some original entries had to be ignored and some movements of a single user have been split into several traces. As locations we used 14 different building on the campus (i.e. all access points that lie in the same building are aggregated into a single location). The locations were fitted with a DPHD with 14 states such that we can ensure that only values from 1 to 14 can be drawn from that distribution. The CPHDs for dwell times and traffic are again fitted with the approach from [28]. For distribution fitting we can just combine all traces $\mathcal{T}_i$ under the assumptions that the users behave identically. For the computation of the correlation matrices this is not possible. Therefore, we computed correlation matrices $\hat{\boldsymbol{\rho}}_i(h)$ for each trace $\mathcal{T}_i$ and then weighted them according to the number of entries, i.e. $\hat{\boldsymbol{\rho}}(h) = \sum_i r_i/r \cdot \hat{\boldsymbol{\rho}}_i(h)$, where r_i is the number of entries in $\mathcal{T}_i$ and r is the total number of entries in all traces and fitted VCAPPs according to $\hat{\boldsymbol{\rho}}(h)$. To assess the fitting quality we set up a model with 25 users each driven by a VCAPP. For each of the locations we use a router with buffer size 50. The random samples from the VCAPP then determine what amount of traffic a user generates at a location/router and how long he stays there before moving to the next location. As reference values we again simulated the model using the trace files. Since we have more traces than users in the model each user randomly selects a trace for the simulation and we run the model with 100 replications. The queue length distributions for the

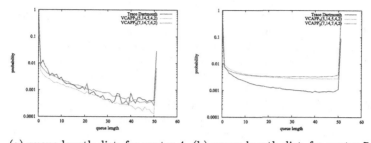

(a) queue length dist. for router 4   (b) queue length dist. for router 7

Fig. 4. Results for trace Dartmouth

traces and two different VCAPPs of two of the routers are shown in Fig. 4. As we can see we obtained a very good approximation for some routers as in Fig. 4a, while for other routers the approximation was not that good but still acceptable (Fig. 4b).

5   Conclusions

We presented an approach for parameter estimation of multivariate traces that uses PHDs for fitting the marginal distributions of the trace and a VARMA process for adding correlation. The approach divides the complex problem of fitting multivariate traces into several steps that can be performed independent of each other, i.e. the PHDs determine the correlation matrices of the VARMA process, but have no further influence when constructing the process. Our experiments with different traces suggest that the process is suitable for approximating real traffic data. For the wireless data it is probably possible to further improve the results by dividing the users into classes according to their behavior and fit a VCAPP for each of the classes, but this is subject to further research.

References

1. Biller, B., Gunes, C.: Introduction to simulation input modeling. In: Proceedings of the Winter Simulation Conference (2010)
2. Biller, B., Nelson, B.L.: Modeling and generating multivariate time-series input processes using a vector autoregressive technique. ACM Trans. Model. Comput. Simul. **13**(3), 211–237 (2003)
3. Bobbio, A., Horváth, A., Scarpa, M., Telek, M.: Acyclic discrete phase type distributions: properties and a parameter estimation algorithm. Perform. Eval. **54**(1), 1–32 (2003)
4. Box, G., Jenkins, G.: Time Series Analysis: Forecasting and Control. Holden-Day, San Francisco (1970)
5. Buchholz, P., Kriege, J.: A heuristic approach for fitting MAPs to moments and joint moments. In: Proceedings of the QEST, pp. 53–62 (2009)

6. Buchholz, P., Kriege, J., Felko, I.: Input Modeling with Phase-Type Distributions and Markov Models - Theory and Applications. Springer, Heidelberg (2014)
7. Cario, M.C., Nelson, B.L.: Autoregressive to anything: time-series input processes for simulation. Oper. Res. Lett. **19**(2), 51–58 (1996)
8. Chan, J.C., Eisenstat, E.: Gibbs samplers for VARMA and its extensions. Technical report, Australian National University, School of Economics (2013)
9. Civelek, I., Biller, B., Scheller-Wolf, A.: The impact of dependence on queueing systems. Research Showcase 8-2009, Carnegie Mellon University (2009)
10. Cumani, A.: On the canonical representation of homogeneous Markov processes modeling failure-time distributions. Micorelectron. Reliab. **22**(3), 583–602 (1982)
11. Devroye, L.: Non-Uniform Random Variate Generation. Springer, New York (1986)
12. Drezner, Z., Wesolowsky, G.O.: On the computation of the bivariate normal integral. J. Stat. Comput. Simul. **35**, 101–107 (1990)
13. Fishman, G.S.: Concepts and Methods in Discrete Event Digital Simulation. Wiley, New York (1973)
14. Fu, S., Xu, C.: Quantifying temporal and spatial correlation of failure events for proactive management. In: Proceedings of the SRDS (2007)
15. Henderson, T., Kotz, D., Abyzov, I.: The Changing usage of a mature campus-wide wireless network. In: Proceedings of the MobiCom (2004)
16. Khayari, R.E.A., Sadre, R., Haverkort, B.: Fitting world-wide web request traces with the EM-algorithm. Perform. Eval. **52**, 175–191 (2003)
17. Kriege, J., Buchholz, P.: Correlated phase-type distributed random numbers as input models for simulations. Perform. Eval. **68**(11), 1247–1260 (2011)
18. Kriege, J., Buchholz, P.: Online Companion to the Paper "Traffic Modeling with Phase-Type Distributions and VARMA Processes" (2016). http://www4.cs.tu-dortmund.de/download/kriege/publications/qest2016_online_companion.pdf
19. Law, A.M.: Simulation Modleing and Analysis. McGraw-Hill, New York (2013)
20. Leland, W., Taqqu, M., Willinger, W., Wilson, D.: On the self-similar nature of ethernet traffic. IEEE/ACM Trans. Networking **2**(1), 1–15 (1994)
21. Lütkepohl, H.: Introduction to Multiple Time Series Analysis. Springer, Heidelberg (1993)
22. Nash, J.: Compact Numerical Methods for Computers: Linear Algebra and Function Minimisation, 2nd edn. Adam Hilger, Bristol (1990)
23. Neuts, M.F.: A versatile Markovian point process. J. Appl. Prob. **16**(4), 764–779 (1979)
24. Press, W., Teukolsky, S., Vetterling, W., Flannery, B.: Numerical Recipes in C - The Art of Scientific Computing, 2nd edn. Cambridge University Press, Cambridge (1993)
25. Reinecke, P., Horváth, G.: Phase-type distributions for realistic modelling in discrete-event simulation. In: Proceedings of the SimuTools (2012)
26. Smith, R.D.: The dynamics of internet traffic: self-similarity, self-organization, and complex phenomena. Adv. Complex Syst. **14**(6), 905–949 (2011)
27. Stanfield, P., Wilson, J., King, R.: Flexible modelling for correlated operation times with application in product-reuse facilities. Int. J. Prod. Res. **42**(11), 2179–2196 (2004)
28. Thümmler, A., Buchholz, P., Telek, M.: A novel approach for phase-type fitting with the EM algorithm. IEEE Trans. Dependable Sec. Comput. **3**(3), 245–258 (2006)

An Optimal Offloading Partitioning Algorithm in Mobile Cloud Computing

Huaming Wu[1(✉)], William Knottenbelt[2], Katinka Wolter[3], and Yi Sun[3]

[1] Center for Applied Mathematics, Tianjin University, Tianjin, China
whming@tju.edu.cn

[2] Department of Computing, Imperial College London, London, UK
wjk@doc.ic.ac.uk

[3] Institut für Informatik, Freie Universität Berlin, Berlin, Germany
{katinka.wolter,yi.sun}@fu-berlin.de

Abstract. Application partitioning splits the executions into local and remote parts. Through optimal partitioning, the device can obtain the most benefit from computation offloading. Due to unstable resources at the wireless network (bandwidth fluctuation, network latency, etc.) and at the service nodes (different speed of the mobile device and cloud server, memory, etc.), static partitioning solutions in previous work with fixed bandwidth and speed assumptions are unsuitable for mobile offloading systems. In this paper, we study how to effectively and dynamically partition a given application into local and remote parts, while keeping the total cost as small as possible. We propose a novel min-cost offloading partitioning (MCOP) algorithm that aims at finding the optimal partitioning plan (determine which portions of the application to run on mobile devices and which portions on cloud servers) under different cost models and mobile environments. The simulation results show that the proposed algorithm provides a stable method with low time complexity which can significantly reduce execution time and energy consumption by optimally distributing tasks between mobile devices and cloud servers, and in the meantime, it can well adapt to environmental changes, such as network perturbation.

Keywords: Mobile cloud computing · Communication networks · Offloading · Cost graph · Application partitioning

1 Introduction

Along with the maturity of mobile cloud computing, mobile cloud offloading is becoming a promising method to reduce task execution time and prolong battery life of mobile devices. Its main idea is to improve execution by migrating heavy computation from mobile devices to resourceful cloud servers and then receiving the results from them via wireless networks. Offloading is an effective way to overcome constraints in resources and functionalities of mobile devices since it can release them from intensive processing.

© Springer International Publishing Switzerland 2016
G. Agha and B. Van Houdt (Eds.): QEST 2016, LNCS 9826, pp. 311–328, 2016.
DOI: 10.1007/978-3-319-43425-4_21

Offloading all computation components of an application to the remote cloud is not always necessary or effective. Especially for some complex applications (e.g., QR-code recognition [1], online social applications [2], health monitoring using body sensor networks [3]) that can be divided into a set of independent parts, a mobile device should judiciously determine whether to offload computation and which portion of the application should be offloaded to the cloud. Offloading decisions [4,5] must be taken for all parts, and the decision made for one part may depend on the one for other parts. As mobile computing increasingly interacts with the cloud, a number of approaches have been proposed, e.g., MAUI [6] and CloneCloud [7], aiming at offloading some parts of the mobile application execution to the cloud. To achieve good performance, they particularly focus on a specific application partitioning problem, i.e., to decide which parts of an application should be offloaded to powerful servers in a remote cloud and which parts should be executed locally on mobile devices such that the total execution cost is minimized. Through partitioning, a mobile device can benefit most from offloading. Thus, partitioning algorithms play a critical role in high-performance offloading systems.

The main costs for mobile offloading systems are the computational cost for local and remote execution, respectively, and the communication cost due to the extra communication between the mobile device and the remote cloud. Program execution can naturally be described as a graph in which vertices represent computation that are labelled with the computation costs and edges reflect the sequence of computation labelled with communication costs [8] where computation is carried out in different places. By partitioning the vertices of a graph, the calculation can be divided among processors of local mobile devices and remote cloud servers. Traditional graph partitioning algorithms (e.g., [9–11]) cannot be applied directly to the mobile offloading systems, because they only consider the weights on the edges of the graph, neglecting the weight of each node. Our research is situated in the context of resource-constrained mobile devices, in which there are often multi-objective partitioning cost functions subject to variable vertex cost, such as minimizing the total response time or energy consumption on mobile devices by offloading partial workloads to a cloud server through links with fluctuating reliability.

The problem of whether or not to offload certain parts of an application to the cloud depends on the following factors: CPU speed of mobile device, speed of the cloud server [12], network bandwidth and reliability, and transmission data size. In this paper, we improve the performance of static partitioning by taking unstable network and cloud conditions into consideration. We explore methods of how to deploy such an offloadable application in a more suitable way by dynamically and automatically determining which parts of the application should be computed on the cloud server and which parts should be left on the mobile device to achieve a particular performance and dependability target (low latency, minimization of energy consumption, low response time, in the presence of unreliable links etc.) [13]. We study how to disintegrate and distribute modules of an application between the mobile side and cloud side, and effectively utilize the cloud resources. We construct a weighted consumption graph (WCG) according to estimated computation and communication costs, and further derive a novel *min-cost offloading partitioning (MCOP)* algorithm designed especially for mobile offloading systems. The MCOP algorithm aims at finding the optimal partitioning plan that minimizes a given objective function (response time, energy consumption or the weighted sum of time and energy) and can be applied to WCGs of arbitrary topology.

The remainder of this paper is organized as follows. Section 2 introduces the partitioning models. An optimal partitioning algorithm for arbitrary topology is proposed and investigated in Sect. 3. Section 4 gives some evaluation and simulation results. Finally, the paper is summarized in Sect. 5.

2 Partitioning Models

2.1 Classification of Application Tasks

Applications in a mobile device normally consist of several tasks. Since not all the application tasks are suitable for remote execution, they need to be weighed and distinguished as:

- **Unoffloadable Tasks:** some tasks should be unconditionally executed locally on the mobile device, either because transferring relevant information would use too much time and energy or because these tasks must access local components (camera, GPS, user interface, accelerometer or other sensors etc.) [6]. Tasks that might cause security issues when executed in a different place should not be offloaded either (such as e-commerce). Local processing consumes battery power of the mobile device, but there are no communication costs or delays.
- **Offloadable Tasks:** some application components are flexible tasks that can be processed either locally on the processor of the mobile device, or remotely in a cloud infrastructure. Many tasks fall into this category, and the offloading decision depends on whether the communication costs outweigh the difference between local and remote costs or not [14].

For unoffloadable components no offloading decisions must be taken. However, as for offloadable ones, since offloading all the application tasks to the remote cloud is not necessary or effective under all circumstances, it is worth considering what should be executed locally on the mobile device and what should be offloaded to the remote cloud for execution based on available networks, response time or energy consumption. The mobile device has to take an offloading decision based on the result of a dynamic optimization problem.

2.2 Construction of Consumption Graphs

There are two types of cost in offloading systems: one is computational cost of running an application tasks locally or remotely (including memory cost, processing time cost etc.) and the other is communication cost for the application

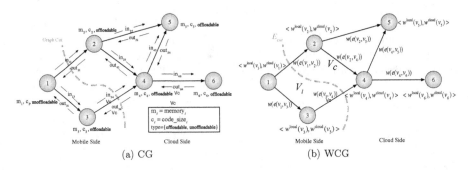

Fig. 1. Construction of Consumption Graph (CG) and Weighted Consumption Graph (WCG). (Color figure online)

tasks' interaction (associated with movement of data and requisite messages). Even the same task can have different cost on the mobile device and in the cloud in terms of execution time and energy consumption. As cloud servers usually process much faster than mobile devices having a powerful configuration, energy can be saved and performance improved when offloading part of the computation to remote servers [15]. However, when vertices are assigned to different sides, the interaction between them leads to extra communication costs. Therefore, we try to find the optimal assignment of vertices for graph partitioning and computation offloading by trading off the computational cost against the communication cost.

Call graphs are widely used to describe data dependencies within a computation, where each vertex represents a task and each edge represents the calling relationship from the caller to the callee. Figure 1(a) shows a CG example consisting of six tasks [16]. The computation costs are represented by vertices, while the communication costs are expressed by edges. We depict the dependency of application tasks and their corresponding costs as a directed acyclic graph $G = (V, E)$, where the set of vertices $V = (v_1, v_2, \cdots, v_N)$ denotes N application tasks and an edge $e(v_i, v_j) \in E$ represents the frequency of invocation and data access between nodes v_i and v_j, where vertices v_i and v_j are neighbors. Each task v_i is characterized by five parameters:

- *type*: offloadable or unoffloadable task.
- m_i: the memory consumption of v_i on a mobile device platform,
- c_i: the size of the compiled code of v_i,
- in_{ij}: the data size of input from v_i to v_j,
- out_{ji}: the data size of output from v_j to v_i.

We further construct a WCG as depicted in Fig. 1(b). Each vertex $v \in V$ is annotated with two cost weights: $\boldsymbol{w}(v) = < w^{\text{local}}(v), w^{\text{cloud}}(v) >$, where $w^{\text{local}}(v)$ and $w^{\text{cloud}}(v)$ represent the computation cost of executing the task v locally on the mobile device and remotely on the cloud, respectively. Each vertex is assigned one of the values in the tuple depending on the partitioning result of the resulting application graph [17]. The edge set $E \subset V \times V$ represents the communication cost amongst tasks. The weight of an edge $w(e(v_i, v_j))$ is denoted as:

$$w(e(v_i, v_j)) = \frac{in_{ij}}{B_{\text{upload}}} + \frac{out_{ij}}{B_{\text{download}}}, \tag{1}$$

which is the communication cost of transferring the input and return states when the tasks v_i and v_j are executed on different sides, and it closely depends on the network bandwidth (upload B_{upload} and download B_{download}) and reliability as well as the amount of transferred data.

A candidate offloading decision is described by one cut in the WCG, which separates the vertices into two disjoint sets, one representing tasks that are executed on the mobile device and the other one implying tasks that are offloaded to the remote server [18]. Hence, taking the optimal offloading decision is equivalent to partitioning the WCG into those two sets such that an objective function is minimized.

The red dotted line in Fig. 1(b) is one possible partitioning cut, indicating the partitioning of computational workload in the application between the mobile side and the cloud side. V_l and V_c are sets of vertices, where V_l is the local set in which tasks are executed locally at the mobile side and V_c is the cloud set in which tasks are directly offloaded to the cloud. We have $V_l \cap V_c = \emptyset$ and $V_l \cup V_c = V$. Further, E_{cut} is the edge set in which the graph is cut into two parts.

2.3   Cost Models

Mobile application partitioning aims at finding the optimal partitioning solution that leads to the minimum execution cost, in order to make the best tradeoff between time/energy savings and transmission costs/delay.

The optimal partitioning decision depends on user requirements/expectations, device information, network bandwidth and reliability, and the application itself. Device information includes the execution speed of the device and the workloads on it when the application is launched. For a slow device where the aim is to reduce execution time, it is better to offload more computation to the cloud. Network bandwidth and reliability affects data transmission for remote execution. If the bandwidth and reliability is very high, the cost in terms of data transmission will be low. In this case, it is better to offload more computation to the cloud.

The partitioning decision is made based on a cost estimate (computation and communication costs) before program execution. On the basis of Fig. 1(b), the overall cost can be calculated as:

$$C_{\text{total}} = \underbrace{\sum_{v \in V} I_v \cdot w^{\text{local}}(v)}_{\text{local}} + \underbrace{\sum_{v \in V} (1 - I_v) \cdot w^{\text{cloud}}(v)}_{\text{remote}} + \underbrace{\sum_{e(v_i, v_j) \in E} I_e \cdot w(e(v_i, v_j))}_{\text{communication}}, \tag{2}$$

where C_{total} is the sum of computation costs (local and remote) and communication costs of cut affected edges.

The cloud server node and the mobile device node must belong to different partitions. One possible solution for this partitioning problem will give us an arbitrary tuple of partitions from the set of vertices $< V_l, V_c >$ and the cut of edge set E_{cut} in the following way:

$$I_v = \begin{cases} 1, \text{if } v \in V_l \\ 0, \text{if } v \in V_c \end{cases} \text{ and } I_e = \begin{cases} 1, \text{if } e \in E_{\text{cut}} \\ 0, \text{if } e \notin E_{\text{cut}} \end{cases}. \tag{3}$$

We seek to find an optimal cut: $I_{\min} = \{I_v, I_e | I_v, I_e \in \{0, 1\}\}$ in the WCG such that some application tasks are executed on the mobile side and the remaining ones on the cloud side, while satisfying the general goal of a partition: $I_{\min} = \arg\min_I C_{\text{total}}(I)$. The dynamic execution configuration of an elastic application can be decided based on different saving objectives with respect to response time and energy consumption. Since the communication time and energy cost for the mobile device will vary according to the amount of data to be transmitted and the wireless network conditions. A task's offloading goals may change due to a change in environmental conditions.

Minimum Response Time. The communication cost depends on the size of data transfer and the network bandwidth, while the computation time has an impact on its cost. If the minimum response time is selected as objective function, we can calculate the total time spent due to offloading as:

$$T_{\text{total}}(I) = \underbrace{\sum_{v \in V} I_v \cdot T_v^l}_{\text{local}} + \underbrace{\sum_{v \in V} (1 - I_v) \cdot T_v^c}_{\text{remote}} + \underbrace{\sum_{e \in E} I_e \cdot T_e^{tr}}_{\text{communication}}, \tag{4}$$

where $T_v^l = F \cdot T_v^c$ is the computation time of task v on the mobile device when it is executed locally; F is the speedup factor, the ratio of the cloud server's processing speed compared to that of the mobile device. T_v^c is the computation time of task v on the cloud server when it is offloaded; $T_e^{tr} = D_e^{tr}/B$ is the communication time between the mobile device and the cloud; D_e^{tr} is the amount of data that is transmitted and received; finally, B is the current wireless bandwidth weighed with the reliability of the network.

Minimum Energy Consumption. If the minimum energy consumption is chosen as the objective function, we can calculate the total energy consumed due to offloading as:

$$E_{\text{total}}(I) = \underbrace{\sum_{v \in V} I_v \cdot E_v^l}_{\text{local}} + \underbrace{\sum_{v \in V} (1 - I_v) \cdot E_v^i}_{\text{idle}} + \underbrace{\sum_{e \in E} I_e \cdot E_e^{tr}}_{\text{communication}}, \tag{5}$$

where $E_v^l = p_m \cdot T_v^l$ is the energy consumed by task v on the mobile device when it is executed locally; $E_v^i = p_i \cdot T_v^c$ is the energy consumed by task v on the mobile device when it is offloaded to the cloud; $E_e = p_{tr} \cdot T_e^{tr}$ is the energy spent on the communication between the mobile device and the cloud including possibly necessary retransmissions; p_m, p_i and p_{tr} are the powers of the mobile device for computing, while being idle and for data transfer, respectively.

Minimum of the Weighted Sum of Time and Energy. If we combine both the response time and energy consumption, we can design a cost model for partitioning as follows [19]:

$$W_{\text{total}}(\boldsymbol{I}) = \omega \cdot \frac{T_{\text{total}}(\boldsymbol{I})}{T_{\text{local}}} + (1 - \omega) \cdot \frac{E_{\text{total}}(\boldsymbol{I})}{E_{\text{local}}}, \tag{6}$$

where $0 \leq \omega \leq 1$ is a weighting parameter used to share relative importance between the response time and energy consumption. Large ω favors response time while small ω favors energy consumption [20,21]. Performance can be traded for power consumption and vice versa [22,23], therefore we can use ω to express preferences for different applications. If $T_{\text{total}}(\boldsymbol{I})/T_{\text{local}}$ is less than 1, the partitioning will improve the application's performance. Similarly, if $E_{\text{total}}(\boldsymbol{I})/E_{\text{local}}$ is less than 1, it will reduce the energy consumption.

We only perform the partitioning when it is beneficial. Not all applications can benefit from partitioning because of application-specific properties. A precalibration of the computation cost on each device is necessary. Offloading is beneficial only if the speedup of the cloud server outweighs the extra communication cost. We compare the partitioning results with two other intuitive strategies without partitioning and, for ease of reference, we list all three kinds of offloading techniques:

- *No Offloading (Local Execution)*: all computation tasks of an application are running locally on the mobile device and there is no communication cost. This may be costly since the mobile device is limited in processing speed and battery life as compared to the powerful computing capability at the cloud side.
- *Full Offloading*: all computation tasks of mobile applications (except the unoffloadable tasks) are moved from the local mobile device to the remote cloud for execution. This may significantly reduce the implementation complexity, which makes the mobile devices lighter and smaller. However, full offloading is not always the optimal choice since different application tasks may have different characteristics that make them more or less suitable for offloading [24].
- *Partial Offloading (With Partitioning)*: with the help of the MCOP algorithm, all tasks including unoffloadable and offloadable ones are partitioned into two sets, one for local execution on the mobile device and the other for remote execution on a cloud server node. Before a task is executed, it may require a certain amount of data from other tasks. Thus, data migration via wireless networks is needed between tasks that are executed at different sides.

We define the saved cost in the partial offloading scheme compared to that in the no offloading scheme as *Offloading Gain*, which can be formulated as:

$$Offloading\,Gain = 1 - \frac{Partial\ Offloading\ Cost}{No\ Offloading\ Cost} \cdot 100\,\%. \tag{7}$$

3   Partitioning Algorithm for Offloading

In this section, we introduce the min-cost offloading partitioning (MCOP) algorithm for WCGs of arbitrary topology. The MCOP algorithm takes a WCG as input in which an application's operations/calculations are represented as the nodes and the communication between them as the edges. Each node has two costs: first the cost of performing the operation locally (e.g., on the mobile device) and second the cost of performing it elsewhere (e.g., in the cloud). The weight of the edges is the communication cost to the offloaded computation. We assume that the communication cost between tasks in the same location is negligible. The result contains information about the cost and reports which operations should be performed locally and which should be offloaded.

3.1   Steps

The MCOP algorithm can be divided into two steps as follows:

1. *Unoffloadable Vertices Merging:* An unoffloadable vertex is the one that has special features making it unable to be migrated outside of the mobile device and thus it is located only in the unoffloadable partition. Apart from this, we can choose any task to be executed locally according to our preferences or other reasons. Then all vertices that are not going to be migrated to the cloud are merged into one that is selected as the source vertex. By 'merging', we mean that these nodes are coalesced into one, whose weight is the sum of the weights of all merged nodes. Let G represent the original graph after all the unoffloadable vertices are merged.

2. *Coarse Partitioning:* The target of this step is to coarsen G to the coarsest graph $G_{|V|}$. To coarsen means to merge two nodes and reduce the node count by one. Therefore, the algorithm has $|V| - 1$ phases. In each phase i (for $1 \leq i \leq |V| - 1$), the cut value, i.e. the partitioning cost in a graph $G_i = (V_i, E_i)$ is calculated. G_{i+1} arises from G_i by merging "suitable nodes", where $G_1 = G$. The partitioning results are the minimum cut among all the cuts in an individual phase i and the corresponding group lists for local and cloud execution. Furthermore, in each phase i of the coarse partitioning we still have five steps:

   (a) Start with $A=\{a\}$, where a is usually an unoffloadable node in G_i.
   (b) Iteratively add the vertex to A that is the most tightly connected to A.
   (c) Let s, t be the last two vertices (in order) added to A.
   (d) The graph cut of the phase i is $(V_i \backslash \{t\}, \{t\})$.
   (e) G_{i+1} arises from G_i by merging vertices s and t.

3.2   Algorithmic Process

The algorithmic process is illustrated as the *MinCut* function in Algorithm 2, and in each phase i, it calls the *MinCutPhase* function as described in Algorithm 3. Since some tasks have to be executed locally, we need to merge them into one node.

The *merging* function is used to merge two vertices into one new vertex, which is implemented as in Algorithm 1. If nodes $s, t \in V$ $(s \neq t)$, then they can be merged as follows:

1. Nodes s and t are chosen.
2. Nodes s and t are replaced by a new node $x_{s,t}$. All edges that were previously incident to s or t are now incident to $x_{s,t}$ (except the edge between nodes s and t when they are connected).
3. Multiple edges are resolved by adding edge weights. The weights of the node $x_{s,t}$ are resolved by adding the weights of s and t.

For example, we can merge nodes 2 and 4 as shown in Fig. 2.

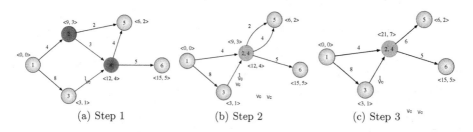

(a) Step 1   (b) Step 2   (c) Step 3

Fig. 2. An example of merging two nodes

The core of this algorithm is to make it easy to select the next vertex to be added to the local set A. We have the following definition.

Definition 1. $\exists v \in V \backslash A$, *if the potential benefit from offloading once v is offloaded:*

$$\Delta(v) = [w^{local}(v) - w^{cloud}(v)] - w(e(A, v))$$

is the minimum, then task v has the most chance to be executed locally, and the vertex v is called Most Tightly Connected Vertex (MTCV).

Further, we have the total cost from partitioning:

$$C_{cut(A-t,t)} = C^{local} - \left[w^{local}(t) - w^{cloud}(t)\right] + \sum_{v \in A \backslash t} w(e(t, v)), \qquad (8)$$

where the cut value $C_{cut(A-t,t)}$ is the partitioning cost, $C^{local} = \sum_{v \in V} w^{local}(v)$ is the total of local costs, $w^{local}(t) - w^{cloud}(t)$ is the gain of node t from offloading, and $\sum_{v \in A \backslash t} w(e(t, v))$ is the total of extra communication costs due to offloading.

Theorem 1. $cut(A - t, t)$ *is always a minimum $s - t$ cut in the current graph, where s and t are the last two vertices added in the phase, the $s - t$ cut separates nodes s and t on two different sides.*

Algorithm 1. The *Merging* function

//This function takes s and t as vertices in the given graph and merges them into one
Function: $G'=Merge(G, w, s, t)$

Input: G: the given graph, $G = (V, E)$
       w: the weights of edges and vertices
       s, t: two vertices in previous graph that are to be merged
Output: G': the new graph after merging two vertices

1: $x_{s,t} \Leftarrow s \cup t$
2: **for** all nodes $v \in V$ **do**
3:    **if** $v \neq \{s, t\}$ **then**
4:       $w(e(x_{s,t}, v)) = w(e(s, v)) + w(e(t, v))$
5:       //adding weights of edges
6:       $\left[w^{local}(x_{s,t}), w^{cloud}(x_{s,t}) \right] = \left[w^{local}(s) + w^{local}(t), w^{cloud}(s) + w^{cloud}(t) \right]$
7:       //adding weights of nodes
8:       $E \Leftarrow E \cup e(x_{s,t}, v)$ //adding edges
9:    **end if**
10:   $E' \Leftarrow E\backslash\{e(s, v), e(t, v)\}$ //deleting edges
11: **end for**
12: $V' \Leftarrow V\backslash\{s, t\} \cup x_{s,t}$
13: **return** $G' = (V', E')$

Algorithm 2. The *MinCut* function

//This function performs an optimal offloading partition algorithm
Function: $[minCut, MinCutGroupsList] \quad = \quad MinCut(G, w, SourceVertices)$

Input: G: the given graph, $G = (V, E)$
       w: the weights of edges and vertices
       SourceVertices: a list of vertices that have to be kept in one side of the cut
Output: $minCut$: the minimum sum of weights of edges and vertices among the cut
       MinCutGroupsList: two lists of vertices, one local list and one remote list

1: $w(minCut) \Leftarrow \infty$
2: **for** $i = 1 : length(SourceVertices)$ **do**
3:    //Merge all the source vertices (unoffloadable) into one
4:    $(G, w) = Merge(G, w, SourceVertices(1), SourceVertices(i))$
5: **end for**
6: **while** $|V| > 1$ **do**
7:    $[cut(A - t, t), s, t] = MinCutPhase(G, w)$
8:    **if** $w(cut(A - t, t)) < w(minCut)$ **then**
9:       $minCut \Leftarrow cut(A - t, t)$
10:   **end if**
11:   $Merge(G, w, s, t)$
12:   //Merge the last two vertices (in order) into one
13: **end while**
14: **return** $minCut$ and $MinCutGroupsList$

Algorithm 3. The *MinCutPhase* function

//This function perform one phase of the partitioning algorithm
Function: $[cut(A - t, t), s, t] = MinCutPhase(G_i, w)$

Input: G_i: the graph in Phase i, i.e., $G_i = (V_i, E_i)$
       w: the weights of edges and vertices
       SourceVertices: a list of vertices that are forced to be kept in one side of the cut
Output: s, t: the lasted two vertices that are added to A
       $cut(A - t, t)$: the cut between $\{A - t\}$ and $\{t\}$ in phase i

1: $a \Leftarrow$ arbitrary vertex of G_i
2: $A \Leftarrow \{a\}$
3: **while** $A \neq V_i$ **do**
4:    $min = -\infty$
5:    $v_{min} = null$
6:    **for** $v \in V_i$ **do**
7:      **if** $v \notin A$ **then**
8:        //Performance gain through offloading the task v to the cloud
9:        $\Delta(v) \Leftarrow [w^{local}(v) - w^{cloud}(v)] - w(e(A, v))$
10:       //Find the vertex that is the most tightly connected to A
11:       **if** $\Delta(v) < min$ **then**
12:         $min = \Delta(v)$
13:         $v_{min} = v$
14:       **end if**
15:      **end if**
16:    **end for**
17:    $A \Leftarrow A \cup \{v_{min}\}$
18:    $a \Leftarrow Merge(G, w, a, v_{min})$
19: **end while**
20: $t \Leftarrow$ the last vertex (in order) added to A
21: $s \Leftarrow$ the last second vertex (in order) added to A
22: **return** $cut(A - t, t)$

The run of each *MinCutPhase* function orders the vertices of the current graph linearly, starting with a and ending with s and t, according to the order of addition into A. We want to show that $C_{cut(A-t,t)} \leq C_{cut(H)}$ for any arbitrary $s - t$ cut H.

Lemma 1. *We define H as an arbitrary $s - t$ cut, A_v as a set of vertices added to A before v, and H_v as a cut of $A_v \cup \{v\}$ induced by H. For all active vertices v, we have $C_{cut}(A_v, v) \leq C_{cut}(H_v)$.*

Proof. As shown in Fig. 3, we use induction on the number of active vertices, k.

1. When $k = 1$, the claim is true,
2. Assume the inequality holds true up to u, that is $C_{cut}(A_u, u) \leq C_{cut}(H_u)$,
3. Suppose v is the first active vertex after u, according to the assumption that $C_{cut}(A_u, u) \leq C_{cut}(H_u)$, then we have:

$$
\begin{aligned}
C_{cut}(A_v, v) &= C_{cut}(A_u, v) + C_{cut}(A_v - A_u, v) \\
&\leq C_{cut}(A_u, u) + C_{cut}(A_v - A_u, v) \quad (u \text{ is } MTCV) \\
&\leq C_{cut}(H_u) + C_{cut}(A_v - A_u, v) \\
&\leq C_{cut}(H_v).
\end{aligned}
$$

Since t is always an active vertex with respect to H, by Lemma 1, we can conclude that $C_{cut(A-t,t)} \leq C_{cut(H)}$ which says exactly that the cost of $cut(A - t, t)$ is at most as heavy as the cost of $cut(H)$. This proves Theorem 1.

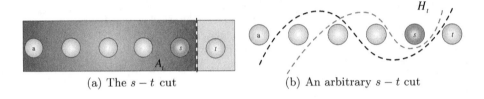

(a) The $s - t$ cut   (b) An arbitrary $s - t$ cut

Fig. 3. The illustration for the proof of Lemma 1

3.3   Computational Complexity

As the running time of the algorithm *MinCut* is essentially equal to the added running time of the $|V| - 1$ runs of *MinCutPhase*, which is called on graphs with decreasing number of vertices and edges, it suffices to show that a single *MinCutPhase* needs at most $O(|V| \log |V| + |E|)$ time yielding an overall running time. The computational complexity of the MCOP algorithm can be noted as $O(|V|^2 \log |V| + |V||E|)$.

As a comparison, Linear Programming (LP) solvers are widely used in schemes like MAUI [6] and CloneCloud [7]. An LP solver is based on branch and bound, which is an algorithm design paradigm for discrete and combinatorial optimization problems, as well as general real valued problems. The number of its optional solutions grows exponentially with the number of tasks, which means higher time complexity $O\left(2^{|V|}\right)$.

While the partitioning e.g., MAUI has exponential time complexity by using LP, our algorithm only has low-order polynomial run time in the number of tasks. Therefore, the MCOP algorithm can handler larger call graphs, which shows its advantage over simple partitioning models as used in MAUI: it can group tasks that process large amounts of data on one side, either the Cloud or the mobile, depending on the network condition.

4   Performance Evaluation

Comparing the execution time spent on the mobile device and the one on the cloud, the speedup factor F is obtained. In practice, we will first access to the cloud server to estimate the remote execution time. We use the average value, since the mobile device might assign more computation resources to a process at different moments of its execution. Therefore, during runtime of an application the link and node cost is constantly updated (the updated value will be an average of the past values and the newly obtained one).

The construction of WCG closely depends on profiling, i.e., the process of gathering the information required to make offloading decisions. Such information may consist of the computation and communication costs of the execution units (program profiler), the network status (network profiler), and the mobile device specific characteristics such as energy consumption (energy profiler). Since the focus of this paper is on the partitioning algorithm we will not enter into the details of profiling techniques, which can be found in many existing references [6, 25].

We take a face recognition application[1] as an example. By analyzing this application with Soot [26], the call graph could be built as a tree-based topology shown in Fig. 4(a). We further construct weighted consumption graph under the condition of the speedup factor $F = 2$ and the bandwidth $B = 1\,\text{MB/s}$ with reliability $= 1$, where the *main* and *checkAgainst* methods are assumed as unoffloadable nodes.

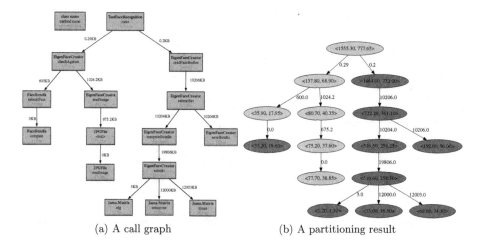

(a) A call graph   (b) A partitioning result

Fig. 4. An optimal partitioning result of the face recognition application

---

[1] The face recognition application is built using an open source code http://darnok. org/programming/face-recognition/, which implements the Eigenface face recognition algorithm.

We then implement the MCOP algorithm in Java[2]. The optimal partitioning result is depicted in Fig. 4(b), where red nodes represent the application tasks that should be offloaded to the remote cloud and the blue nodes are the tasks that are supposed to be executed locally on the mobile device.

We do a simple simulation with the WCG as predicted in Fig. 2. We have received different results under the different parameters of speedup factor F and reliable wireless bandwidth B. The partitioning results will change as B or F vary.

In Fig. 5 the speedup factor is set to $F = 3$. Since the low bandwidth results in much higher cost for data transmission, the full offloading scheme can not benefit from offloading. Given a relatively large bandwidth and stable network, the response time or energy consumption obtained by the full offloading scheme slowly approaches the partial offloading scheme because the optimal partition includes more and more tasks running on the cloud side until all offloadable tasks are offloaded to the cloud. With higher bandwidth and more stable network, they begin to coincide with each other and only decrease because all possible nodes are offloaded and the transmissions become faster. Both, response time and energy consumption have the same trend as the wireless bandwidth increases. Therefore, bandwidth and network reliability is a crucial element for offloading since the mobile system could benefit a lot from offloading in stable, high bandwidth environments, while with low bandwidth and fragile network, the *no offloading* scheme is preferred.

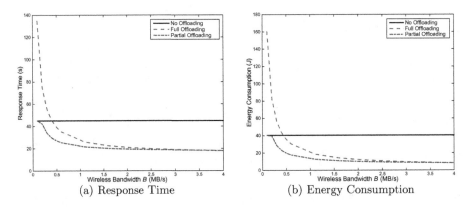

(a) Response Time   (b) Energy Consumption

Fig. 5. Comparisons of different schemes under different wireless bandwidths when the speedup factor $F = 3$

In Fig. 6 the bandwidth is fixed at $B = 3$ MB/s. It can be seen that offloading benefits from higher speedup factors. When F is very small, the full offloading

---

[2] An optimal partitioning algorithm, the code can be found in https://github.com/carlosmn/work-offload, thanks to Daniel Seidenstücker and Carlos Martín Nieto for their help.

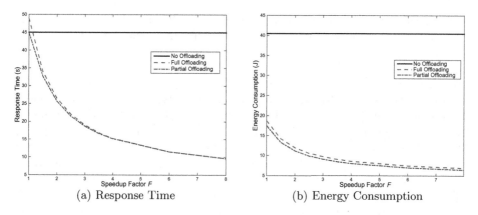

Fig. 6. Comparisons of different schemes under different speedup factors when the bandwidth $B = 3\,\mathrm{MB/s}$

scheme can reduce energy consumption of the mobile device. However, it takes much longer than without offloading. The partial offloading scheme that adopts the MCOP algorithm can effectively reduce execution time and energy consumption, while adapting to environmental changes.

From Figs. 5 and 6, we can tell that the full offloading scheme performs much better than the *no offloading* scheme under certain adequate wireless network conditions, because the execution cost of running methods on the cloud server is significantly lower than on the mobile device when the speedup factor F is high. The partial offloading scheme outperforms the *no offloading* and *full offloading* schemes and significantly improves the application performance, since it effectively avoids offloading tasks in the case of large communication cost between consecutive tasks compared to the full offloading scheme, and offloads more appropriate tasks to the cloud server. In other words, neither running all

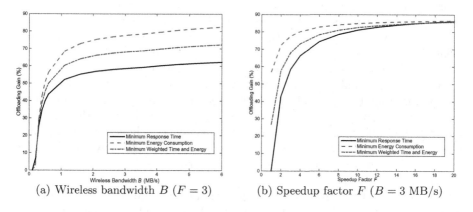

Fig. 7. Offloading gains under different environment conditions when $\omega = 0.5$

tasks locally on the mobile terminal nor always offloading their execution to a remote server can offer an efficient solution, but our partial offloading scheme can do.

In Fig. 7(a) when the bandwidth is low, the offloading gain for all three cost models is very small and almost identical. That is because more time/energy will be spent in transferring the same amount of data due to the poor network and low bandwidth, resulting in increased execution cost. As the bandwidth increases, the offloading gain first rises drastically and then the increase becomes slower. It can be concluded that the optimal partitioning plan includes more and more tasks running on the cloud side until all the tasks are offloaded to the cloud when the network condition and bandwidth increases. In Fig. 7(b) when F is small, the offloading gain for all three cost models is very low since a small value means very little computational cost reduction from remote execution. As F increases, the offloading gain first rises drastically and then approaches the same value. That is because the benefits from offloading cannot neglect the extra communication cost. From Fig. 7, the proposed MCOP algorithm is able to effectively reduce the application's energy consumption as well as its execution time. Further, it can adapt to environmental changes to some extent and avoids a sharp decline in application performance once the network deteriorates and the bandwidth decreases.

5   Conclusion

To tackle the problem of dynamic partitioning in a mobile environment, we have proposed a novel offloading partitioning algorithm (MCOP algorithm) that finds the optimal application partitioning under different cost models to arrive at the best tradeoff between saving time/energy and transmission costs/delay. Contrary to the traditional graph partitioning problem, our algorithm is not restricted to balanced partitions but takes the infrastructure heterogeneity into account.

The MCOP algorithm provides a stably quadratic runtime complexity for determining which parts of application tasks should be offloaded to the cloud server and which parts should be executed locally, in order to save energy of the mobile device and to reduce the execution time of an application. Since the reliability of wireless bandwidth can vary due to mobility and interference, it strongly affects the application's optimal partitioning result. When the network is poor, high communication cost will be incurred, and the MCOP algorithm will include more application tasks for local execution. Experimental results show that according to environmental changes (e.g., network bandwidth and cloud server performance), the proposed algorithm can effectively achieve the optimal partitioning result in terms of time and energy saving. Offloading benefits a lot from high bandwidths and large speedup factors, while low bandwidth favors the *no offloading* scheme.

The concept of optimal application partitioning under constraints generalises to many scenarios in distributed computing where should be explored further.

References

1. Yang, L., Cao, J., Yuan, Y., Li, T., Han, A., Chan, A.: A framework for partitioning and execution of data stream applications in mobile cloud computing. ACM SIGMETRICS Perform. Eval. Rev. **40**(4), 23–32 (2013)
2. Olteanu, A.-C., Ţăpuş, N.: Tools for empirical and operational analysis of mobile offloading in loop-based applications. Informatica Economica **17**(4), 5–17 (2013)
3. Wu, H., Wang, Q., Wolter, K.: Mobile healthcare systems with multi-cloud offloading. In: 2013 IEEE 14th International Conference on Mobile Data Management (MDM), vol. 2, pp. 188–193. IEEE (2013)
4. Wu, H.: Analysis of offloading decision making in mobile cloud computing. Ph.D. thesis, Freie Universität Berlin (2015)
5. Wu, H., Wang, Q., Wolter, K.: Methods of cloud-path selection for offloading in mobile cloud computing systems. In: CloudCom, pp. 443–448 (2012)
6. Cuervo, E., Balasubramanian, A., Cho, D.-K., Wolman, A., Saroiu, S., Chandra, R., Bahl, P.: Maui: making smartphones last longer with code offload. In: Proceedings of the 8th International Conference on Mobile Systems, Applications, and Services, pp. 49–62. ACM (2010)
7. Chun, B.-G., Ihm, S., Maniatis, P., Naik, M., Patti, A.: Clonecloud: elastic execution between mobile device and cloud. In: Proceedings of the Sixth Conference on Computer Systems, pp. 301–314. ACM (2011)
8. Hendrickson, B., Kolda, T.G.: Graph partitioning models for parallel computing. Parallel Comput. **26**(12), 1519–1534 (2000)
9. Stoer, M., Wagner, F.: A simple min-cut algorithm. J. ACM (JACM) **44**(4), 585–591 (1997)
10. Ali, K., Lhoták, O.: Application-only call graph construction. In: Noble, J. (ed.) ECOOP 2012. LNCS, vol. 7313, pp. 688–712. Springer, Heidelberg (2012)
11. Boykov, Y., Veksler, O., Zabih, R.: Fast approximate energy minimization via graph cuts. IEEE Trans. Pattern Anal. Mach. Intell. **23**(11), 1222–1239 (2001)
12. Liu, Y., Lee, M.J.: An effective dynamic programming offloading algorithm in mobile cloud computing system. In: Wireless Communications and Networking Conference (WCNC), 2014 IEEE, pp. 1868–1873. IEEE (2014)
13. Wu, H., Wolter, K.: Software aging in mobile devices: partial computation offloading as a solution. In: 2015 IEEE International Symposium on Software Reliability Engineering Workshops (ISSREW). IEEE (2015)
14. Kumar, K., Liu, J., Lu, Y.-H., Bhargava, B.: A survey of computation offloading for mobile systems. Mob. Netw. Appl. **18**(1), 129–140 (2013)
15. Niu, R., Song, W., Liu, Y.: An energy-efficient multisite offloading algorithm for mobile devices. Int. J. Distrib. Sens. Netw. **2013**, 1–6 (2013)
16. Giurgiu, I., Riva, O., Juric, D., Krivulev, I., Alonso, G.: Calling the cloud: enabling mobile phones as interfaces to cloud applications. In: Bacon, J.M., Cooper, B.F. (eds.) Middleware 2009. LNCS, vol. 5896, pp. 83–102. Springer, Heidelberg (2009)
17. Sinha, K., Kulkarni, M.: Techniques for fine-grained, multi-site computation offloading. In: Proceedings of the 2011 11th IEEE/ACM International Symposium on Cluster, Cloud and Grid Computing, pp. 184–194. IEEE Computer Society (2011)
18. Kao, B.Y.-H., Krishnamachari, B.: Optimizing mobile computational offloading with delay constraints. In: Proceedings of the Global Communication Conference (Globecom 14), pp. 8–12 (2014)

19. Wu, H., Wolter, K.: Tradeoff analysis for mobile cloud offloading based on an additive energy-performance metric. In: 2014 8th International Conference on Performance Evaluation Methodologies and Tools (VALUETOOLS). ACM (2014)

20. Wu, H., Sun, Y., Wolter, K.: Analysis of the energy-response time tradeoff for delayed mobile cloud offloading. ACM SIGMETRICS Perform. Eval. Rev. **43**, 33–35 (2015)

21. Wu, H., Knottenbelt, W., Wolter, K.: Analysis of the energy-response time tradeoff for mobile cloud offloading using combined metrics. In: 2015 27th International Teletraffic Congress (ITC 27), pp. 134–142. IEEE (2015)

22. Kwon, Y.-W., Tilevich, E.: Energy-efficient and fault-tolerant distributed mobile execution. In: 2012 IEEE 32nd International Conference on Distributed Computing Systems (ICDCS), pp. 586–595. IEEE (2012)

23. Wu, H., Wang, Q., Wolter, K.: Tradeoff between performance improvement and energy saving in mobile cloud offloading systems. In: 2013 IEEE International Conference on Communications Workshops (ICC), pp. 728–732. IEEE (2013)

24. Lei, L., Zhong, Z., Zheng, K., Chen, J., Meng, H.: Challenges on wireless heterogeneous networks for mobile cloud computing. In: IEEE Wireless Communications, vol. 20, no. 3 (2013)

25. Zhang, Y., Liu, H., Jiao, L., Fu, X.: To offload or not to offload: an efficient code partition algorithm for mobile cloud computing. In: 2012 IEEE 1st International Conference on Cloud Networking (CLOUDNET), pp. 80–86. IEEE (2012)

26. Soot: a framework for analyzing and transforming Java and androidapplications. http://sable.github.io/soot/

Performance Modeling

Maintenance Analysis and Optimization via Statistical Model Checking

Evaluating a Train Pneumatic Compressor

Enno Ruijters[1(✉)], Dennis Guck[1], Peter Drolenga[2], Margot Peters[2], and Mariëlle Stoelinga[1]

[1] University of Twente, EWI-FMT, P.O. Box 217,
7500 AE Enschede, The Netherlands
{e.j.j.ruijters,d.guck,m.i.a.stoelinga}@utwente.nl
[2] NedTrain Fleet Services, P.O. Box 2167, 3500 GD Utrecht, The Netherlands
{peter.drolenga,margot.peters}@nedtrain.nl

Abstract. Maintenance is crucial to ensuring and improving system dependability: By performing timely inspections, repairs, and renewals the lifespan and reliability of systems can be significantly improved. Good maintenance planning, however, has to balance these improvements against the downsides of maintenance, such as costs and planned downtime.

In this paper, we study the effect of different maintenance strategies on a pneumatic compressor used in trains. This compressor is critical to the operation of the train, and a failure can lead to a lengthy and expensive disruption. Within the rolling stock maintenance company NedTrain, we have modelled this compressor as a fault maintenance tree (FMT), i.e. a fault tree augmented with maintenance aspects. We show how this FMT naturally models complex maintenance plans including condition-based maintenance with regular inspections. The FMT is analysed using statistical model checking, which allows us to obtain several key performance indicators such as the system reliability, number of failures, and required unscheduled maintenance.

Our analysis demonstrates that FMTs can be used to model the compressor, a practical system used in industry, including its maintenance policy. We validate this model against experiences in the field, compute the importance of performing minor services at a reasonable frequency, and find that the currently scheduled overhaul may not be cost-effective.

1 Introduction

Maintenance of Critical Assets. The current trend in asset management is to use reliability-centered maintenance (RCM), with the goal of optimizing maintenance planning by maintaining critical assets more intensively than less critical ones. By focusing maintenance where it is most effective, RCM seeks to balance maintenance costs against system dependability. To achieve this balance, it is necessary to have a good understanding of the effects of a maintenance policy on the system's dependability, measured by key performance indicators like

G. Agha and B. Van Houdt (Eds.): QEST 2016, LNCS 9826, pp. 331–347, 2016.
DOI: 10.1007/978-3-319-43425-4_22

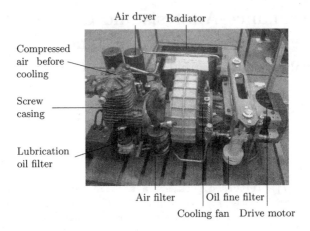

Fig. 1. *A pneumatic compressor.* Air is drawn in from the environment via the air filter and compressed by a set of screws. The compressed air is then cooled, and moisture and oil particulates are removed before the air enters the pneumatic system of the train.

availability, reliability, mean time to failure, etc. Achieving this understanding calls for an integrated analysis of system dependability and maintenance. This paper shows a method to perform such an integral analysis, namely fault maintenance trees (FMTs), and demonstrates that this method yields useful results on RCM strategies by studying a typical rolling stock asset (namely a pneumatic compressor, shown in Fig. 1) via FMTs.

Fault Trees and Fault Maintenance Trees. Fault tree analysis (FTA) [15] is a popular methodology for dependability analysis, and is commonly used in industry. A fault tree (FT) models component failures at the leaves of the tree. Then gates (like AND and OR) show how component failures lead to system failure — indeed not every single failure causes a system failure in a system with redundancy. When the failure rates of the components are known, then FTA can compute the probability for a compound event, typically a system failure.

Traditional FTA is very useful to analyse the reliability of systems when failure rates are given. In practice, however, these failure rates are strongly affected by maintenance, which is not taken into account by fault trees. Thus, FTA is not suitable to compare the performance of different maintenance policies. Moreover, many existing approaches support only exponentially distributed failure times of components.

To overcome these limitations and to determine the effect of different maintenance strategies on system reliability and costs, fault maintenance trees (FMTs) have been developed [11] combining fault trees with arbitrary failure time distributions and *maintenance models*. The latter represent the necessary elements for modelling maintenance: degradation of components, inspections, and repairs. Moreover, FMTs introduce a new gate: the rate dependency (RDEP) gate enables the failure of one component to accelerate the degradation of other components. In this paper, we find that RDEPs are necessary to accurately model the compressor. Certain failure modes, like loss of lubricating oil, severely accelerate failures of components such as motors.

FMT Analysis via Stochastic and Statistical Model Checking. FMTs are analysed by converting each element of the FMT, i.e. leaf, gate, and maintenance element, into a priced timed automaton (PTA). These automata are then composed to yield a stochastic model of the system, which is analysed using statistical model checking (SMC) [10], a Monte Carlo simulation technique [9] to obtain numerous important dependability metrics, including system reliability, availability, MTTF, expected cost, etc.

A major advantage of our approach is that, in addition to obtaining quantitative results using SMC, we can qualitatively validate the structural correctness of our model using traditional model checking techniques in the UPPAAL tool [6].

The Train-Bound Pneumatic Compressor. Many systems on modern trains, such as the brakes and automatic doors, are controlled and powered by compressed air generated by a pneumatic compressor (shown in Fig. 1). For example, when the door on a train opens, one often hears a hissing sound. This is the flow of air produced by the compressor. Since the doors and especially the brakes are safety-critical components, a loss of air pressure will leave the train stranded until it can be repaired or towed. It is thus critical to keep the compressor functioning.

The compressor generates compressed air from air in the environment. This air is filtered of dust and particulates, and pushed by motor-driven screws into a high-pressure chamber. The compressed air is then cooled and compressed moisture removed. As the screws are lubricated with oil, small droplets of oil enter the stream of air and also need to be removed. Finally, the compressed air is stored in a high-pressure reservoir to be used in the pneumatically-powered systems. Various safety elements such as pressure valves and temperature sensors ensure the compressor and the systems it powers are not damaged.

Maintenance of this compressor is required to keep it functioning correctly. The compressor contains consumable parts such as filters that need periodic replacements, and other components wear out over time. This maintenance is typical for the railway industry, with periodic replacements and inspections, and different costs for planned and unplanned maintenance. Furthermore, failure costs are high for unscheduled breakdowns during operation.

The compressor is a relevant case study for three reasons: (1) The analysis is useful for NedTrain's internal operations for logistics and maintenance engineering purposes; (2) The failure characteristics of the compressor are well documented through FMEAs, internal documentation and historical failure data; (3) Maintenance on the compressor is performed relatively independent of the rest of the train, as a defective compressor can be replaced by a functioning one from stock. This gives more freedom to optimize the maintenance program.

Modelling and Analysis. We have conducted a reliability analysis of a particular model of pneumatic compressor. We analyse the dependability of these compressors, computing the reliability, expected number of failures, and expected number of required unscheduled maintenance events. In particular, we investigate the current maintenance strategy, as well as potentially better strategies. We consider (1) variations in maintenance intervals, and (2) the usefulness of periodic overhauls.

This analysis was conducted together with NedTrain, the company that performs rolling stock maintenance for the Dutch Railways and other train operators. NedTrain is responsible for the maintenance of over 800 trains.

Our analysis finds that performing periodic servicing of the compressor has a major effect on its reliability. The periodic minor overhaul, on the other hand, does not appear to have a strong influence.

Contributions. An important contribution is the demonstration that our method can easily be extended to include system-specific constructs for modelling unusual aspects of the degradation behaviour. Specifically, we included events whose failure rate depends on the state of several other components, and maintenance actions depending on the state of components that are not relevant for the system failure.

Last but not least, we conclude that FMTs are a useful framework to investigate maintenance optimization problems from industrial practice: FMTs are a convenient model, have sufficient expressive power to capture complex maintenance aspects, and are able to produce predictive analysis results.

1.1   Related Work

A large number of analysis techniques and extensions for fault trees exist, for an overview we refer the reader to [12]. Current FTA techniques support simple repair strategies by either equipping leaves with repair times [15] or with repair boxes [3]. These techniques consider a BE to be either failed or functioning, while FMTs add support for degraded states and maintenance actions taken depending on the level of degradation.

Extending traditional fault trees, Bucci et al. [4] present a tool that can analyse FTs with non-Markovian failure distributions, which can also be used to analyse component failures due to wear over time. This method, however, does not consider maintenance to undo this wear.

An alternative extension of FTs is the Extended fault tree formalism by Buchacker et al. [5], which can model systems where some components have failure rates that depend on the status of other components. They still model failure times as exponential distributions, and do not include repairs or inspections dependent on full subtrees.

When FTA is not applicable, many techniques exist to analyse and optimize maintenance strategies without using FTA. We refer the reader to reviews such as [1] on the use of simulation techniques or [13] for techniques including analytic approximations and Bayesian reasoning.

One such approach, by Carnevali et al. [7], considers maintenance in phased systems where resources are used in a sequence of tasks, with detection and repair actions in-between these tasks.

If a system consists of a single components or a group of identical components, van Noortwijk and Frangopol [14] consider in detail two models of the effects of various maintenance choices on the reliability and cost in civil infrastructure. Neither of these models consider the failure behaviour of systems of different components.

1.2   Organization of the Paper

This paper begins with a description of the pneumatic compressor in Sect. 2 and the methodology in Sect. 3. The modelling of the compressor by FMTs is explained in Sect. 4. Then, Sect. 5 explains how the FMT is analysed, and provides the results of this analysis. Finally, we provide our conclusions in Sect. 6.

2   Case Description: The Pneumatic Compressor

Pneumatic compressors (see Fig. 1) are devices that produce compressed air. In modern trains, a pipe of compressed air runs throughout the train, and valves control the air pressure to certain installations such as the pantograph (connecting the train to the overhead power line) and automatic doors. The air pressure controls the operation of these installations, as well as providing the necessary power to operate them.

As these compressors are critical to the operation of the train, they are also a potential cause of disruptions. Various types of failures can occur, such as oil leaks and clogged filters. Inspections are performed to determine whether failures are likely to occur soon, and preventive action, such as replacing a nearly-full filter, can be taken to prevent the failure occurring in the field. Some components such as filters are also periodically replaced, since replacing them all in one service is cheaper than spreading the replacements over multiple services when inspections find a problem.

Below, we describe the operation of the compressor, its main failure modes, and the current maintenance plan.

2.1   Purpose and Operation

Pneumatic systems have long been used as a control mechanism in trains. Braking systems operated by air pressure date back to 1868 [16], and are still in use today. Although electronics are starting to replace or supplement pneumatic control, modern trains still use pneumatics for emergency brakes and other applications, such as opening and closing doors automatically and raising the pantograph to connect to the overhead electrical line.

Safety-critical pneumatic systems are designed to be fail-safe: A loss of air pressure disrupts functionality, but poses no danger. Brakes, for example, are loosed by high pressure and applied when the pressure drops. A failed compressor, therefore, does not constitute a safety risk. Nonetheless, since a failed compressor leaves the train stranded, such failures cause costly and lengthy disruptions.

To provide high-pressure air for the pneumatics, modern trains use electric compressors. In addition to generating a high pressure, the compressor also clears the air of dust and debris, and removes moisture which could cause corrosion or freezing in pipes and pneumatically-powered devices.

We examine the particular model of compressor used in Dutch VIRM 1/2/3 trains. This compressor operates using rotating screws that take air from the outside and compress it into a pipe. Before reaching the screw, the air first passes through a filter to remove any dust or debris. The screw is lubricated using oil. Due to the relatively high temperatures and airflow, micro-particles of oil are carried in the airflow through the system. To remove this oil, the air passes through two additional filters. Finally, the air is cooled and passed to the pneumatic system.

Several safety features are in place to prevent damage to the compressor or pneumatic systems: Pressure-controlled valves ensure the compressed air does not reach unsafe pressures, and a temperature switch disables the compressor if the oil temperature gets too high.

2.2 Failure Modes

Compressor failures can be divided into two categories: *Complete failures* where the compressor does not operate at all, and *degraded operation* where the compressor does not generate a sufficiently high pressure. For this paper, we consider only failures that prevent the train from operating, meaning complete failure or so much degradation that immediate repair is necessary. Other forms of degraded operation can be analyzed similarly.

Table 1 lists the types of failure that can occur, together with their failure parameters: Each failure mode is characterized by the expected time to failure assuming no maintenance is performed, and the number of degradation phases we consider in our model.

The wear of the compressor screws and the motor and bearings is complicated due to multiple causes. Particles can enter the compressor despite the filter, which causes degradation of the screws. The rate at which particles pass through the filter is significantly increased if the filter is already worn. A second mode of wear is

Table 1. *Parameters of the failure modes of the compressor.* The failure times of the components follow an Erlang distribution with the indicated number of phases and total expected time to failure (in years) assuming no maintenance is performed. The values have been scaled for anonymity. Failure mode 3 is not strictly a failure, but rather an event that is required for mode 2 to lead to failures. Also failure modes 14 and 15 are not failures, but rather indicators of degradation that are used to initiate maintenance actions, as described in Sect. 4.

Nr.	Failure mode	Nr. of phases	ETTF
1	Motor does not start when asked	3	16.6
2	De-aeration valve defective	3	200
3	Two starts in short time	2	0.001
4	Radiator obstructed	4	5.5
5	Oil thermostat defective	3	16.6
6	Low oil level	4	5.5
7	Pressure valve leakage	3	3.3
8	Air filter obstructed	2	500
9	Degraded air filter	4	5
10	Particle-induced damage	4	120
11	Oil pollution	4	5.5
12	Lubrication-induced wear	4	120
13	Motor/bearings degraded	4	120
14	Oil fine filter full	3	30
15	Degraded capacity	2	10

caused by insufficient lubrication of the screws and of the motor. This can be caused by pollution of the oil, or by insufficient oil, or a combination of both.

2.3   Maintenance

The current maintenance policy followed by NedTrain consists of some specific inspections every two days, and scheduled services every three months with a larger service every nine months. A minor overhaul is performed every three years and a major overhauls every six years (for reasons of confidentiality, these times have been scaled with the same factor as the BE failure rates).

The bi-daily inspection is mostly performed at night, while the train is prepared for service. Mechanics check the on-board diagnostic system for recorded events such as overpressure, and perform an inspection to find oil leaks or excessive noise. If this inspection finds a defect, an unscheduled service is necessary to correct it.

During the scheduled services, consumable parts such as filters are replaced, and components of the compressor are inspected for signs of wear. Some functional tests of the overall performance of the compressor are also performed, such as measuring the time needed to pressurize the pneumatic system for the entire train starting from atmospheric pressure.

Every three years, the compressor is removed from the train and shipped to NedTrain's component workshop for an overhaul. Minor and major overhauls are alternated. During an overhaul, the compressor is disassembled and all components are examined and replaced if needed. During a minor overhaul components with a small amount of wear are reused. During a major overhaul, all worn components are replaced, and the compressor is considered as good as new afterwards.

Each maintenance action can also lead to more intensive services if problems are found that cannot be corrected during the scheduled service. For example, if a minor service inspection finds that the compressor is not producing sufficient pressure but cannot find the cause, the compressor can be sent in for an overhaul.

3   Methodology

To analyse and optimize the maintenance strategy for the compressor, we have modelled the compressor in terms of fault maintenance trees. Below, we briefly describe the main ingredients of this framework: fault trees, maintenance models, analysis methods, and metrics.

3.1   Fault Trees

Fault trees (FTs) are a graphical method for performing reliability and safety analysis, widely used in industry. An FT models how component failures propagate through a system to lead to system failure, and allows a wide range of qualitative and quantitative properties to be analyzed [12,15].

FTs are directed acyclic graphs in which the leaves are called *basic events* (BEs) and describe component failures, and internal nodes are called *gates* and describe what combinations of basic events cause compound failures. The gate at the root of the tree is called the *top level event* and typically denotes a system failure or other undesired event.

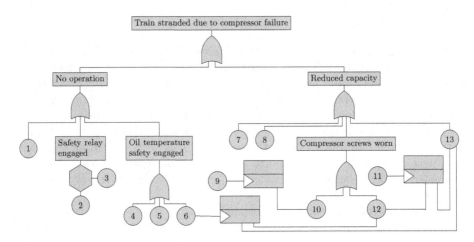

Fig. 2. *Fault Tree describing the major failure modes of the compressor.* The numbers in the basic events correspond to the numbers of the failures modes in Table 1. Failure modes 14 and 15 are not shown, as they do not contribute to the top event.

The gates of standard fault trees are AND-, OR-, and VOT(k)-gates, which fail when all, any, or at least k of their children fail, respectively. The leaves of a traditional continuous-time FT are equipped with exponential failure rates, describing the progression of failure probabilities over time.

Classic fault tree analysis includes techniques to compute the reliability and availability of the system, to find the biggest contributors to system unreliability, and to compute the sensitivity of these metrics to the parameters of the BEs [12].

3.2 Fault Maintenance Trees

Fault maintenance trees (FMTs) [11] are an extension of FTs that can model several additional contributors to system reliability, such as maintenance through inspections and repairs, degradation of components over time, and situations where one failure causes accelerated wear of another component. The FMT modelling the compressor is shown in Fig. 2.

Extended Basic Events. The BEs in an FMT are more expressive than in standard BEs: Standard BEs generally model only exponential or Weibull distributions of failure times, while FMTs support failures that occur when a component gradually wears out, and where the effect of this wear can be reversed by maintenance actions.

BEs represent the components' failure behaviour over time. A BE can be equipped with multiple phases, representing different stages of degradation. A threshold specifies at which phase an inspection should trigger a maintenance action. The transition time into a next phase can be described by an arbitrary probability distribution, but usually follows an exponential distribution, in which case the total failure behaviour of a BE is described by an Erlang distribution.

RDEP Gates. FMTs support all the gates of static and dynamic FTs [8]. Additionally, they include a rate dependency (RDEP) gate, representing dependencies between components leading to accelerated wear. This gate has one input event, and one or more dependent children. When the input event occurs, the failure behaviours of the dependent children are all accelerated by a factor γ, independently specified for each child. When the input is repaired, degradation of the children returns to their normal rates.

Repair and Inspection Modules. Standard FTs can support relatively simple repair policies using distributions over repair times, or via repair boxes [3]. FMTs enable more advanced maintenance policies via *repair modules* (RM) and *inspection modules* (IM).

An IM describes at what frequency components are inspected as well as the so-called repair threshold. The latter is the (minimal) degradation phase where repairs will be performed. When the repair threshold is reached, the next inspection will trigger a repair and send a repair request to the RM associated with the IM.

The RM listens for repair requests of specific IMs and initiates the repair or partial replacement of a specific set of BEs. When the RM is invoked, the BEs change their phases to a less degraded phase. Moreover, the RM can invoke a periodic renewal of components, e.g. the replacement of a tire after four years.

IMs and RMs can be combined to model more complex policies, such as periodic replacements or simultaneous repair of a group of components when one fails.

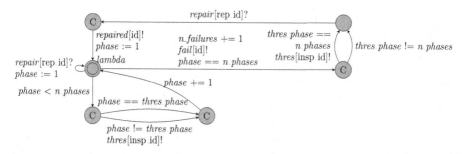

Fig. 3. *PTA of a basic event* with failure time given by an Erlang distribution with *n_phases* phases and a repair threshold at *thres_phase*. From the initial state, the PTA waits an exponentially distributed time with mean *lambda*, and moves downward if it has not yet reached the last phase in the Erlang distribution, or rightward if it has. If it is not in the final phase, it advances by one phase, and it may emit a signal *thres*[insp_id] to a listening inspection module. The BE may also receive a signal *repair*[rep_id] and return the initial phase. Upon completing the final phase, the failure counter is incremented and a signal *fail*[id] is emitted. A threshold signal may be sent, and then the BE waits to receive a *repair*[rep_id] signal. After receiving this signal, the failed BE emits a signal *repaired*[id], and returns to the initial phase and state.

3.3 Analysis of FMTs by Statistical Model Checking

Technically, FMTs are analysed using statistical model checking of priced timed automata (PTAs) [2]. That is, we first convert the FMTs into a network of PTAs and use the statistical model checker UPPAAL [6] to compute the requested metrics.

PTAs are an extension of timed automata with costs on locations and actions. PTAs are transition systems using real-valued clocks to specify deadlines, with enabling conditions for actions. Costs can be incurred at a fixed amount when taking a transition, or proportional to the time spent in a certain location.

Each element of the FMT (i.e. each BE, IM, RM, and gate) is assigned a unique ID, and a template of the appropriate PTA is instantiated with the specific parameters for the element. The PTAs for the basic event, repair module and the inspection module are shown in Figs. 3, 4, and 5, respectively. The IDs are used to instantiate the synchronization signals.

The PTA is then analyzed using the UPPAAL model checker. This approach has the advantage of allowing both quantitative analysis of the metrics described in Sect. 3.4 using statistical model checking, and qualitative analysis and validation of the structural correctness of the model using traditional model checking. The latter enables us to check properties of the model such as that every BE can be repaired, that every gate can fail, etc.

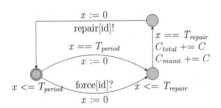

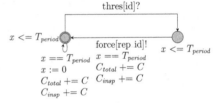

Fig. 4. *PTA for a repair module.* The PTA begins in the leftmost state with clock x initially zero. It waits until either the waiting time for a periodic repair (T_{period}) elapses, or a repair request signal (*force[id]*) is received. In either case, the module waits some time T_{repair}, incurs the C for a repair, sends a signal (*repair[id]*) to any BEs repaired by this module, and resets the timer.

Fig. 5. *PTA for an inspection module.* The PTA begins in the leftmost state, and waits until either the time until the inspection interval (T_{period}) elapses, or a threshold signal (*thres[id]*) is received from a BE. If the time elapses before a signal is received, the inspection cost is incurred and the timer resets. If a threshold signal is received, the module waits for the scheduled inspection time, then signals its associated repair module to begin a repair (*force*[rep_id]), and resets the timer.

Qualitative checks require state-space exploration of the model, which leads to exponential time-complexity as the number of FMT elements increases. Fortunately, the statistical model checker does not need to generate the full state-space, and thus its computation time is relatively independent of the number of elements, but rather grows with the desired accuracy of the result.

3.4 Metrics

We analyse several aspects of the dependability of the compressor, namely the reliability, expected number of failures, and expected number of unplanned maintenance activities. These can be used to compare different maintenance policies and help in deciding which policy is better, as well as to check that the compressor population will meet its performance requirements under a given policy.

Reliability. The probability of experiencing no system failures within a given time period. We compute the probability that within a certain period, there is never a time where a set of BEs is in a failed state leading to the occurrence of the top level event of the FMT. In LTL, we annotate the state corresponding to the failure of the top-level event as *failed*, and express the reliability within time t as $P(\diamond^{\leq t} failed)$.

The term *unreliability* denotes the probability that at least one failure occurs in the time of interest.

Expected Number of Failures. We compute the expected number of occurrences of the top event within a given time window. Since the compressor can always be repaired after a failure, there can be multiple failures over time. We can also compute the number of failures of individual components or subtrees of the FMT.

Expected Number of Unplanned Maintenance Activities. We compute the expected number of times that an inspection finds a defect that is not corrected during the normal maintenance procedure. In the case of the compressor, this occurs for any failure found during the bi-daily inspection, or at certain levels of degradation of components during servicing. These cases require a repair to be scheduled in the maintenance depot or overhaul facility. During an overhaul, all repairs are considered part of planned maintenance.

4 Modelling of the Compressor

The fault tree and maintenance plan described in Sect. 2 were constructed by the research department of NedTrain.

Based on documentation of failure characteristics and expert opinions of system engineers and mechanics, the structure of the FMT was constructed. The resulting FMT is displayed in Fig. 2. As described in Sect. 2.2, compressor failures are divided into complete failures and reduced capacity. This division helps validate the model, since these categories of failures are easy to distinguish in a practical fault condition.

Table 2. *Specification of the acceleration factor of BEs 12 and 13, depending on the states of BEs 6 and 11. The non-degraded state is state 0, the failed state is state 3.*

State BE 11	State BE 6			
	0	1	2	3
0	1	2	4	6
1	2	4	6	10
2	4	6	10	15
3	6	10	15	30

While modeling the compressor, it was noted that several failure modes are related to each other, such as degradation of the air filter leading to increased wear of the screws. While it is possible to model these independently as 'particle-induced wear under normal condition' and 'particle-induced wear with ruptured filter' (since a degraded air filter is not by itself a cause of failure), this leads to difficulty when describing the maintenance policy. The RDEP gates offer a much more natural description of a single BE with degradation that is accelerated by another BE. For BEs 12 and 13, special variants are used that capture the simultaneous but non-linear accelerating effects of BEs 6 and 11. Table 2 specified how much the affected BEs are accelerated depending on the states of the triggering BEs.

Quantitative parameters on degradation patterns and parameters were estimated based on interviews with maintenance engineers responsible for the maintenance plan and system engineers specialized in pneumatics, as well as experiment reports of a simulation environment where compressors can be tested.

While FMTs support arbitrary failure time distributions, determining the exact distribution of each BE was beyond the scope of this case study. Instead, we have modeled the BEs as exponential distributions or Erlang distributions with few phases, as these overestimate the number of failures in the relatively short times between maintenance actions. Due to the very high cost of failure compared to maintenance, relying on a conservative model and performing more maintenance than required is preferable to using an optimistic models and experiencing more failures in the field.

While describing the maintenance policy, we found two properties of the system that are used in maintenance scheduling (BEs 14 and 15), which are in fact complex properties influenced by the degradation of most basic events. Since the exact effect is too complex to include in the model, we instead treat these as basic events that do not contribute to the top level event but are included in the maintenance policy.

Table 3. *Maintenance description for the compressor.* Given a BE, a phase of degradation, and a maintenance action, the table lists the effect of that action on the degradation of the BE. I.e. the last column lists the phase to which the BE moves when the given action is performed while the BE is in the listed phase. If the top event occurs, and after some maintenance actions denoted with result 'O2', a large overhaul is immediately performed resetting all components to their undegraded state.

BE	Phase	Maintenance action	Result phase
1	2	M1	1
1	2	O1	1
2	2	O1	1
3	2	Any	1
4	3	M1	2
4	Any	O1	1
5	2	M1	O2
5	2	O1	1
6	Any	M1	1
6	Any	O1	1
7	2	I1	1
7	2	M1	1
8	Any	M1	1
8	Any	O1	1
9	Any	M1	1
9	Any	O1	1
11	3 or 4	M1	1
11	Any	M2	1
11	Any	O1	1
13	2 or 3	M1	1
13	2 or 3	O1	1
14	2	M1	1
14	3	M1	O2
14	Any	O1	1
15	2	M1	O2
15	Any	O1	1
	Legend		
I1	Bi-daily inspection		
M1	Three-monthly maintenance		
M2	Nine-monthly maintenance		
O1	Minor overhaul		
O2	Major overhaul		

Another behaviour that was not included in the model is the low oil level, which can be accelerated by oil leaks in several components. Since it is unlikely that multiple such leaks occur at the same time, we instead chose to model the oil pressure as a single BE.

The parameters of the BEs are listed in Table 1. The failure rates were obtained by consultation with experts within NedTrain, specifically system engineers and mechanics, to include both theoretical estimates and practical information. The estimates were further informed by experiments conducted at the overhaul facility operating a compressor in a simulated environment.

4.1 Maintenance Modelling

We compare the dependability and costs of compressors subject to different maintenance policies. This allows us both to validate the model against actual recorded failures, and to offer suggestions for improvements in the policy that lead to cost savings or increased dependability.

NedTrain has specified the current maintenance policy, which is based on a balance between performance, risks, and costs. The specification of this policy consists of the frequency with which each maintenance action must be performed, and for each BE and degradation level the effect of the action.

In the FMT, inspection modules describe the inspection rates and the threshold at which corrective action is performed. Different BEs have different thresholds, depending on the visibility of the degradation of a component and the importance of correction.

Most maintenance actions return various components to the undegraded state if they are found in a certain degraded state. This is modelled using separate inspection and repair modules for the different BEs. For example, as shown in Table 3, an inspection module inspects BE 11 every month checking whether it has reached phase 3 and if so, repairs it. Some repair actions, in particular the major overhaul, are initiated when other maintenance actions find excess wear. In this case, the BE is modified to have multiple inspection thresholds for different inspection modules.

The current model makes a few assumptions: First, we assume that all maintenance is carried out exactly on schedule. In practice, maintenance actions with scheduled intervals greater than one month are sometimes performed in the last 10–20 % of the interval, to optimise allocation of resources. Since the fluctuations in inspection times are small compared to the inspection interval and do not occur often, we expect this assumption not to significantly distort the results.

We also assume that inspections are perfect, i.e. an inspection always leads to a repair if the degradation level is past the threshold. While this may seem questionable, we argue that the actual inspections are performed well enough that this is not a significant source of error in the model. Moreover, we assume that repairs occur instantly. Since the degradation rates already factor in that the compressor is not in use all the time, we consider it reasonable to also factor in the relatively short time spent in repair.

5   Analysis and Results

In this section we describe the results of several experiments we conducted on the FMT of the compressor. As a first step, we have validated the FMT using the current maintenance policy against observations from the field. Therefore, we used the model as constructed, i.e. we analysed the compressor under the current policy. Since we concluded that the model is in line with our expectations based on failure data, we continued with finding possible improvements of the current policy. Therefore, the maintenance strategy within the FMT was modified by changing inspection frequencies and replacements. This led to a description of how an optimal maintenance strategy of the compressor can be constructed.

Note that the results in this section are averages of 40,000 simulation runs each. The variance between the simulation runs is low enough that a 95 % confidence interval around the mean results has a width of less than 5 % of the indicated value, both with the original and the anonymized values. The analysis required approx. 6 CPU-hours per model on an Intel Opteron 4386.

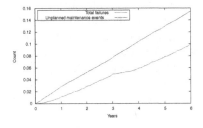

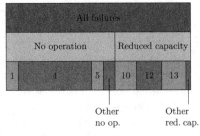

(a) Number of compressor failures and unplanned maintenance events over time. Note that each failure also causes unplanned maintenance, but these are not included here.

(b) Breakdown of the failures of the compressor by cause. The numbers in the bottom row correspond to the failure causes in Table 1.

Fig. 6. Results for the compressor under the current maintenance policy. (Color figure online)

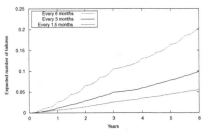

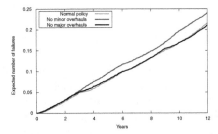

(a) Effect of different frequencies of the small service.

(b) Effect of the minor and major overhauls.

Fig. 7. Expected number of failures for variations on the maintenance policy. (Color figure online)

First, we estimate the total failure rate of the compressor over time. We consider NedTrain's fleet of 239 compressors since this model of train began operation in 1994 until 2015. Although a direct comparison with the model is not possible due to long periods of time when compressors are kept unused in a warehouse, the model's prediction is in agreement with NedTrain's estimate of the operational failure rate to within 50 %.

A graph of the cumulative number of failures over time is shown in Fig. 6a.

Table 4. *Listing of the expected failure rates of different causes of compressor failures.* Values are yearly occurrences in a population of 233 compressors.

Failure cause	Failure rate
Motor does not start when asked	0.41
De-aeration valve defective	0.025
Radiator obstructed	2.48
Oil thermostat defective	0.40
Low oil level	0.34
Pressure valve leakage	0.22
Air filter obstructed	0
Particle-induced rupture	0.71
Lubrication-induced wear	0.86
Motor/bearings degraded	0.82

We observe that the unplanned maintenance events increase almost linearly with time, as they are mostly caused by failures that are not wear-related, and thus occur with exponentially distributed failure times. We only consider the interval between two major overhauls, since the compressor is expected to be as good as new after a major overhaul.

Other Maintenance Policies. To examine the leading causes of failures, the expected number of occurrences of each failure mode per year was estimated. Table 4 shows the annual number of expected failures, averaged over the six-year period between major overhauls. A graphical breakdown of the causes of failures is displayed in Fig. 6b. We see that the failure mode 'radiator obstructed' is by far the leading cause of failure. The current maintenance policy for the radiator is to remove large obstructions when found during visual inspections, and more thoroughly clean it during larger maintenance operations. Our analysis suggests that more frequent cleaning may cheaply reduce failures, although we note that these failures are also usually quickly and cheaply resolved when they do occur.

Next, we consider two possible variations to the maintenance policy: Fig. 7a shows the number of failures over time for different frequencies of the minor service. We find that this service has a significant effect on the expected failure rate. It is therefore useful to carefully examine the costs associated with this service, to find an optimal balance between servicing and failure costs.

We also consider the possibility of omitting the minor overhaul after three years, and of omitting the major overhaul after twelve years (instead performing a minor overhaul at this time). The effects of which are graphed in Fig. 7b. After six years, the minor overhaul has prevented approx. 0.02 failures per compressor. This suggests that the overhaul may not be cost-effective, although this depends strongly on the relative costs of the overhaul and the failure. Furthermore, the effects of replacing the major overhaul by a minor one are too small to be measured by our approach, offering a further possibility for cost savings. We do note, that although we have no indications that the degradation behaviour will be noticeably different after six years, we do not have the data to prove that nonlinear effects such as metal fatigue will not cause more unexpected failures.

6 Conclusion

We have modelled and analysed several maintenance policies for the compressor via fault maintenance trees. We conclude that FMTs are a useful tool for maintenance analysis and optimization. In particular, the modelling process is not too difficult, and the analysis provides useful insights. Obtaining correct failure rates and degradation data from the field required additional effort, but was also feasible in practice.

We obtain dependability estimates for the compressor, which maintenance planners can use in combination with known costs of maintenance and effects of failures to determine which plan results in the lowest cost with optimal effectiveness.

Future work includes the extension of FMTs with continuous degradation phases, models that take into account specific conditions and usage scenarios that influence degradation. We would further like to explore how to convert the per-compressor failure estimates into per-train estimates, given that compressors are commonly swapped between trains or left in storage for extended periods of time. Finally, we would like to extend FMTs to include imperfect maintenance, such as inspections that have some probability of not detecting a degraded component.

Acknowledgements. This work has been supported by STW and ProRail under the project ArRangeer (122238), the EU FP7 project TREsPASS (318003), and the NWO project BEAT (612.001.303).

References

1. Alrabghi, A., Tiwari, A.: State of the art in simulation-based optimisation for maintenance systems. Comput. Ind. Eng. **82**, 167–182 (2015)
2. Behrmann, G., Larsen, K.G., Rasmussen, J.I.: Priced timed automata: algorithms and applications. In: de Boer, F.S., Bonsangue, M.M., Graf, S., de Roever, W.-P. (eds.) FMCO 2004. LNCS, vol. 3657, pp. 162–182. Springer, Heidelberg (2005)
3. Bobbio, A., Codetta-Raiteri, D.: Parametric fault trees with dynamic gates and repair boxes. In: Proceedings of the Reliability and Maintainability Symposium, pp. 459–465 (2004)
4. Bucci, G., Carnevali, L., Vicario, E.: A tool supporting evaluation of non-markovian fault trees. In: Proceedings of the 5th International Conference on Quantitative Evaluation of Systems (QEST), pp. 115–116, September 2008
5. Buchacker, K.: Modeling with extended fault trees. In: Proceedings of the 5th IEEE International Symposium on High Assurance Systems Engineering (HASE), pp. 238–246 (2000)
6. Bulychev, P., David, A., Larsen, K.G., Mikučionis, M., Poulsen, D.B., Legay, A., Wang, Z.: UPPAAL-SMC: statistical model checking for priced timed automata. In: Proceedings of 10th Workshop on Quantitative Aspects of Programming Languages (2012)
7. Carnevali, L., Paolieri, M., Tadano, K., Vicario, E.: Towards the quantitative evaluation of phased maintenance procedures using non-Markovian regenerative analysis. In: Balsamo, M.S., Knottenbelt, W.J., Marin, A. (eds.) EPEW 2013. LNCS, vol. 8168, pp. 176–190. Springer, Heidelberg (2013)

8. Dugan, J.B., Bavuso, S.J., Boyd, M.A.: Fault trees and sequence dependencies. In: Proceedings of Reliability and Maintainability Symposium, pp. 286–293 (1990)

9. Fishman, G.: Monte Carlo: Concepts, Algorithms, and Applications. Springer, Heidelberg (1996)

10. Legay, A., Delahaye, B., Bensalem, S.: Statistical model checking: an overview. In: Barringer, H., Falcone, Y., Finkbeiner, B., Havelund, K., Lee, I., Pace, G., Roşu, G., Sokolsky, O., Tillmann, N. (eds.) RV 2010. LNCS, vol. 6418, pp. 122–135. Springer, Heidelberg (2010)

11. Ruijters, E., Guck, D., Drolenga, P., Stoelinga, M.: Fault maintenance trees: reliability centered maintenance via statistical model checking. In: Proceedings of Reliability and Maintainability Symposium, January 2016

12. Ruijters, E., Stoelinga, M.: Fault tree analysis: a survey of the state-of-the-art in modeling, analysis and tools. Comput. Sci. Rev. **15–16**, 29–62 (2015)

13. Sharma, A., Yadava, G.S., Deshmukh, S.G.: A literature review and future perspectives on maintenance optimization. J. Qual. Maintenance Eng. **17**(1), 5–25 (2011)

14. van Noortwijk, J.M., Frangopol, D.M.: Two probabilistic life-cycle maintenance models for deteriorating civil infrastructures. Probab. Eng. Mech. **19**(4), 345–359 (2004)

15. Vesely, W.E., Goldberg, F.F., Roberts, N.H., Haasl, D.F.: Fault Tree Handbook. Office of Nuclear Regulatory Reasearch, U.S. Nuclear Regulatory Commision, Washington, DC (1981)

16. Westinghouse, G.: Improvement in steam-power-brake devices. US Patent 88,929 (1869)

Performance Evaluation of Train Moving-Block Control

Giovanni Neglia[1]([✉]), Sara Alouf[1], Abdulhalim Dandoush[2], Sebastien Simoens[4], Pierre Dersin[3], Alina Tuholukova[1], Jérôme Billion[3], and Pascal Derouet[4]

[1] Université Côte d'Azur, Inria, Sophia Antipolis, France
{giovanni.neglia,sara.alouf,alina.tuholukova}@inria.fr
[2] ESME Sudria, Paris, France
dandoush@esme.fr
[3] Alstom Transport, Saint-Ouen, France
[4] Alstom Transport, Villeurbanne, France
{sebastien.simoens,pierre.dersin,jerome.billion,
pascal.derouet}@transport.alstom.com

Abstract. In moving block systems for railway transportation a central controller periodically communicates to the train how far it can safely advance. On-board automatic protection mechanisms stop the train if no message is received during a given time window.

In this paper we consider as reference a typical implementation of moving-block control for metro and quantify the rate of spurious Emergency Brakes (EBs), i.e. of train stops due to communication losses and not to an actual risk of collision. Such unexpected EBs can happen at any point on the track and are a major service disturbance.

Our general formula for the EB rate requires a probabilistic characterization of losses and delays. Calculations are surprisingly simple in the case of homogeneous and independent packet losses. Our approach is computationally efficient even when emergency brakes are very rare (as they should be) and can no longer be estimated via discrete-event simulations.

Keywords: Emergency brakes · Communication Based Train Control (CBTC) · European Train Control System (ETCS)

1 Introduction

In order to avoid collisions between consecutive trains traveling on the same track, the track is traditionally divided in fixed sections—called blocks—and only one train at a time is allowed to be in a given block.

The increasing demand for efficient mass transit transport requires to utilize railway infrastructure more efficiently. The improvements of train-sidetrack wireless communications, on board processing and actuators have made possible the introduction in the last 15 years of moving block systems, where blocks are dynamically calculated. Figure 1 schematically illustrates the two different approaches. The moving-block control can reduce the headway taking into

© Springer International Publishing Switzerland 2016
G. Agha and B. Van Houdt (Eds.): QEST 2016, LNCS 9826, pp. 348–363, 2016.
DOI: 10.1007/978-3-319-43425-4_23

account the actual distance between the trains as well as their speeds. It is being deployed as Communication-Based Train Control (CBTC) for urban mass transit system and is under consideration for next generation of European Train Control System (ETCS). This is referred as ETCS level 3 and is currently under standardization.

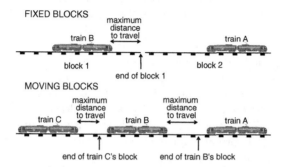

Fig. 1. Fixed-block and moving-block operation.

Moving-block systems require a continuous information exchange (detailed in Sect. 2) between an on board local controller, called the Carborne Controller (CC) and an external ground controller, called the Zone Controller (ZC) because it monitors all the trains in a given zone. Safety-critical messages are exchanged using standard or proprietary radio technologies. If no message is received during a given interval then the CC will no longer have valid guarantees that train movement is still safe and will trigger an Emergency Brake (EB). It is clearly desirable to limit the frequency of spurious emergency brakes, i.e. emergency brakes that are simply due to losses on the wireless channel and not to a potential collision risk. Indeed spurious emergency brakes can be themselves a cause of danger, with trains potentially blocked in tunnels, risks of passengers disembarking on the tracks, etc. Moreover, a spurious EB can generate legitimate EBs on the following trains on the track, causing in this way major service disturbance. For this reason, the so-called performance based contracts can bind rail transport companies to specify the maximum number of spurious emergency brakes over a given period of time.

In spite of their criticality, the estimation of the rate of spurious EBs is mostly based on historical operational data. This approach strongly limits the possibility to evaluate ahead of time the performance when significant changes are deployed and in particular when new lines based on new technologies are built. It is often required to experimentally adapt different system parameters (e.g. transmission power levels, timer values, ...) after the deployment of the line, and sometimes even to deploy additional trackside equipment (e.g. radio transmitters). These difficulties are often considered one of the reasons for the delay in the standardization of ETCS level 3. For example [8] shows that the official quality of service specifications for the different subcomponents of the ETCS level 3 system can lead to a ridiculously high rate of spurious EBs (one every 30 min).

A model-based analysis can then play a fundamental role for a preliminary evaluation of the real performance of moving block control. Some work has been done in this direction following [8], and then considering its abstraction from ETCS level 3 specifications mostly using Stochastic Petri Nets (SPNs) [1–3,5,9]. In particular the approach proposed in [8] to numerically solve the SPN works only under the so-called enabling restriction, i.e. only one transition can be generally distributed and all the others should be exponential random variables. In the more realistic cases, the authors rely then on Monte Carlo simulations of the SPN. The naive simulation approach presented in [8] cannot manage to quantify EB rate smaller than 2 EBs per hour. Importance splitting techniques used in [9] allow to estimate much smaller rates (about 10^{-10} per hour). It is not clear if the computational cost of this numerical approach is insensitive to the packet loss probability p. References [5,7] show how UML descriptions can be used to describe the moving block control in ETCS level 3 and can be automatically translated to MoDeST formal language (a process algebra-based formalism) and to SPNs, but they do not solve the problem of quantitative evaluation of such rates when losses are rare. In the very recent paper [2] Carnevali et al. use the tool ORIS to solve numerically the SPN proposed in [8,9], without the need to rely on Monte Carlo simulations. The tool indeed overcomes the limit of the enabling restriction thanks to recent advancements based on the method of stochastic state classes [6]. Moreover, it allows for a transient analysis of the system. As a case study, the authors consider a toy-example similar to that in [8] leading to very high EB rates. From a preliminary analysis using their tool, it is not clear if more realistic scenarios can be solved in a reasonable amount of time.

Our approach differs from the related literature in three main aspects. First, rather than moving from the current proposals for ETCS level 3, we consider as reference an actual implementation of the moving-block system for metro by Alstom, one of the world largest company in the domain of rail transport and signaling. Looking at an actual implementation has led us to identify the importance of the time-slotted operation of the two controllers (the CC and the ZC). Indeed, the most important delay component in the messages' exchange between the CC and the ZC is due to the waiting time for the next clock tick at which the controller can process the message. This waiting time can be equal to hundreds of milliseconds versus the tens of milliseconds due to network delays. This aspect was ignored in the previous literature and we show that it has to be addressed to correctly evaluate the system performance. In particular, a consequence of the time-slotted operation is that the EB rate exhibits non-trivial discontinuity as the timer value changes. A second (methodological) difference in comparison to the direction of [8] and follow-ups is that we push as further as possible the probabilistic analysis to derive closed-formula expressions. We derive a general formula for the rate of spurious EBs under general loss and delay processes, and a simple formula for the case of independent and homogeneous packet losses. The analysis allows to better understand the role of the different system parameters. On the contrary, the existing literature only relies on simulations or (in the case of [2]) on the numerical solution of a SPN. In both cases the dependence on the system parameters is hidden. Finally, from the algorithmic point of view, it is not clear if the numerical approaches proposed until now can be practically used to estimate EB rates as low as in this paper. Our guess is that this is probably not the case but, perhaps, for [2,9]. Indeed our approach does not need to simulate rare sequences of packet losses and is then practically implementable.

The paper is organized as follows. In Sect. 2 we describe our assumptions about the train scenario and the details of the moving-block control including typical values for system parameters. Then in Sect. 3 we describe our general approach to study the system, we show that a worst case analysis is of limited utility (Sect. 3.1) and then move to derive a general formula for the EB rate (Sect. 3.1) that requires to characterize system delays (Sect. 3.2) and losses. The case of independent and homogeneous packet losses is considered in Sect. 3.3. Some numerical experiments are in Sect. 4. Section 5 concludes the paper and discusses how to extend our approach to more general loss scenarios. The most frequently used acronyms are listed in Table 1. Due to space constraints some of the results are in the companion technical report [4].

Table 1. List of acronyms

CBTC	Communication Based Train Control
CC	Carborne Controller
DCS	Data Communication Sub-System
EB	Emergency Brake
EOA	End-Of-Authority
ETCS	European Train Control System
LOC	Location report
TM	validity duration Timer of a LOC
ZC	Zone Controller

2   Scenario

Here we describe the specific railway scenario we consider. In our description we will refer to transmission technologies and parameters typical of a urban rail network (and then of a CBTC system), but our following analysis does not depend on these specific implementation details. What is instead required is that the random variables (r.v.s) defined below (train speed, distances between access points, etc.) have bounded support and are lower bounded by a positive constant. For a given r.v. α, we denote by $\alpha_{min} > 0$ its lower bound and by $\alpha_{max} < \infty$ its upper bound.[1]

We consider a train moving on an infinitely long track. The train has two WiFi On Board Modems (OBMs) with directional antennas: one is located at

---

[1] Throughout the paper Greek letters always denote random variables, while capital letters usually denote system parameters.

the front of the train, the other at the back. We refer to them respectively as the blue and the red OBMs. Along the track there are pairs of closely-located WiFi Access Points (APs), using the same channel. The pair is called a Trackside Radio Equipment (TRE). Each AP in a TRE is devoted to communicate with one of the two OBMs and is connected to an independent wired network through which the Zone Controller (ZC) can be reached. We also label the APs, the wireless channels and the wired networks blue or red as the corresponding OBM. Hence communications between the train and the ZC are possible through separate paths, each with a single wireless link.

2.1   Train Moving-Block Control

In this section we describe the detailed operation of a moving block system considering as reference the specific CBTC implementation by Alstom.[2]

Figure 2 shows a messages exchange between the on board controller (the CC) and the ground controller (the ZC). Observe that both the controllers operate in discrete time on the basis of clock periods of hundreds of milliseconds. This is due to the fact that they are actually e-out-of-f voting systems where different processors perform in parallel the same calculations and a time-slotted operation simplifies the synchronism of the processors. The clock periods at the ZC and at the CC (respectively T_{ZC} and T_{CC}) are in general different because the subsystems are provided by different vendors and also because they have different computational loads during one period.

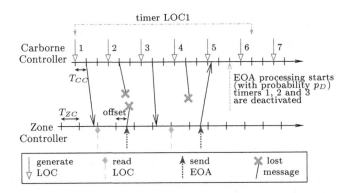

Fig. 2. Illustration of LOC-EOA exchanges.

The most important CBTC messages are location reports (LOC) and end-of-authority ones (EOA). A LOC is a message periodically transmitted from the on board CC through the Data Communication Sub-System (DCS) to the ground ZC. The message is actually sent twice through the blue and the red networks. The first LOC arriving at the ZC is processed. Each LOC is acknowledged by an EOA

---

[2] The parameters' values have been slightly changed and some specific implementation details are hidden to protect Alstom industrial know-how.

message in the reverse direction (again sent through the two networks). The EOA communicates to the CC how far the train can advance. The LOC has a validity duration TM and a timer with such duration is activated at the generation of the LOC. An EOA is said to be valid if the timer of the corresponding LOC has not expired yet. The CC-ZC-CC exchange works as follows.

1. A LOC is generated at the CC every T_{LOC}, multiple of the CC clock period T_{CC}.
2. The LOC (say LOC k) is ready to be emitted and passed to the DCS after a processing delay equal to T_{CC}.
3. The delivery delay introduced by the DCS is a random variable χ_1 with support in $[T_{DCS,\min}, T_{DCS,\max}]$.
4. At the ZC the LOC is available for computing at the next tick of the clock.
5. The computing time at the ZC required to process the LOCs from all the trains in the zone and generate the corresponding EOAs is T_{ZC}.
6. The EOA k is emitted within the next cycle of the ZC at an offset O depending on the train.
7. The EOA is delivered to the CC after a random delay χ_2, distributed as χ_1, but independent from it.
8. At the CC the EOA gets in a processing queue, at the next tick of the CC clock the most recent EOA present in the queue is processed unless there are higher priority tasks arrived during the same CC clock period (which happens with probability p_D). In any case an EOA processing is not delayed more than an additional CC period.
9. The EOA k is actually processed only if it remains valid until the end of the current CC clock. Once processing starts, all the pending timers for older LOCs (i.e. LOC h for $h \leq k$) are deactivated.
10. If the timer of a LOC is not deactivated before its expiration, the EB procedure is triggered.

In what follows we refer to the k-th LOC and its corresponding EOA as the k-th LOC-EOA exchange, but note that any later EOA can deactivate the timer of the k-th LOC. We say that a LOC-EOA exchange is lost if either the LOC or the EOA does not arrive to destination.

3   Analysis

In this paper we consider that the system is described by a stationary stochastic process and calculate the steady-state rate at which emergency brakes occur (as common to all the related literature but [2]). In particular we consider that the train is moving according to some stationary mobility model and the algorithm described above is running all the time, even after the occurrence of an emergency brake. Ignoring the train stopping time after an EB is a reasonable approximation because we are estimating rare events.

We denote by $\mathcal{L}_k$ the event that the exchange k is lost, $\mathcal{T}_k$ the event that the k-th LOC experiences a timeout and $\bar{A}$ the complement of set A. The k-th

Table 2. Notation and typical values for the variables. In the paper some of the variables appear with subscripts. A subscript b (r) denotes that the variable refers to the blue (red) OBM or network. A subscript L (E) denotes that it refers to a LOC (an EOA).

Symbol	Quantity	Value
T_{ZC}	ZC clock period	378 ms
T_{CC}	CC clock period	225 ms
T_{LOC}	LOC generation period	$3T_{CC}$
TM	validity duration of a LOC	5.5 s
T_{DCS}	transmission delay	$[10, 50]$ ms
τ	positive random component of T_{DCS}	$[0, 40]$ ms
ϕ	positive random component of T_{DCS} for first message to arrive	$[0, 40]$ ms
O	EOA transmission offset	$[0, T_{ZC}]$
ω_{CC}	number of CC ticks an EOA waits until CC processes it	$\{0, 1\}$
p_D	probability that ω_{CC} is 1	0.01
ω_{ZC}	time interval between LOC arrival at ZC and next ZC tick	
σ	time interval between earliest arrival time of a LOC at ZC and next ZC tick	
q_{EB}	emergency brake probability	
r_{EB}	emergency brake rate	
p	packet loss	
$\tilde{p}$	probability to lose a LOC-EOA exchange	
T_k	arrival time of k-th EOA	
γ_k	tick at which k-th EOA is processed	
$\mathcal{D}_k$	event that k-th EOA is late to deactivate the timer of LOC 1	
$\mathcal{T}_k$	event that k-th LOC experiences a timeout	
$\mathcal{L}_k$	event of k-th LOC-EOA exchange loss	

LOC experiences a timeout if the k-th exchange is lost and the later EOAs do not arrive or arrive too late, then $\mathcal{T}_k \subset \mathcal{L}_k$[3]. We observe that a sequence of consecutive timeouts generates a single EB and then a timeout for a given LOC, say LOC 1, is counted as an EB only if the previous LOC 0 does not experience a timeout. The probability q_{EB} that a random LOC experiences an emergency break is then $q_{EB} = \Pr(\bar{\mathcal{T}}_0 \cap \mathcal{T}_1)$ that does not depend on the specific pair of LOCs considered because the process is stationary. Moreover, under the condition that LOC 1 experiences a timeout, LOC 0 experiences a timeout if and only if the corresponding exchange is lost, because later EOAs are not able to block the timer of LOC 1 and a fortiori the timer of LOC 0. Then $\bar{\mathcal{T}}_0 \cap \mathcal{T}_1 = \bar{\mathcal{L}}_0 \cap \mathcal{T}_1$ and the rate of emergency brakes is

$$r_{EB} = \frac{q_{EB}}{T_{LOC}} = \frac{\Pr(\bar{\mathcal{L}}_0 \cap \mathcal{T}_1)}{T_{LOC}}. \tag{1}$$

---

[3] In this paper $A \subset B$ denotes that A is a subset of B, not necessarily proper.

3.1   EB Probability

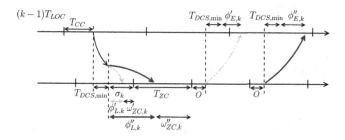

Fig. 3. Different delay components of the k-th LOC-EOA exchange for two different values of the LOC transmission delay $\phi'_{L,k}$ and $\phi''_{L,k}$.

In this section we first derive some simple bounds for q_{EB}. The bounds will reveal to be too loose to be practically used, but they are nevertheless useful for the subsequent analysis. We conclude the section with a general formula for the EB rate, whose terms will be calculated in the following sections. We report numerical values corresponding to the typical scenario presented in Sect. 2.

Minimum and Maximum LOC-EOA Round Trip Times. We calculate the minimum and the maximum time between the generation of a LOC and the instant T when the corresponding EOA is available for computation at the CC. Consider a LOC generated at time 0. Its EOA arrives at the CC at time (see also Fig. 3):

$$T = T_{\min} + \phi_L + \phi_E + \omega_{ZC} + O, \tag{2}$$

where $T_{min} = T_{CC} + 2T_{DCS,\min} + T_{ZC} = 623$ ms, ω_{ZC} is the time interval between the arrival of the LOC at the ZC and the next ZC tick and ϕ_L and ϕ_E are the random components of the transmission delays respectively for the first LOC and the first EOA to arrive at destination.

The earliest arrival time $T_{\min} + O$ occurs when the LOC and the EOA experience the minimum travel times on the DCS (i.e. $\phi_L = \phi_E = 0$) and the LOC is available for computing at the ZC immediately before a ZC tick (i.e. $\omega_{ZC} = 0$).

The latest arrival time $T_{\max} + O$ occurs when the LOC and the EOA experience the maximum travel time on the DCS (i.e. $\phi_L = \phi_E = T_{DCS,\max} - T_{DCS,\min}$) and the LOC is available for computing at the ZC immediately after a ZC tick. In this case the LOC will wait an additional T_{ZC} before being processed (i.e. $\omega_{ZC} = T_{ZC}$). Hence $T_{\max} = T_{CC} + T_{DCS,\max} + T_{ZC} + T_{ZC} + T_{DCS,\max} = 1081$ ms.

Number of Potential LOC-EOA Exchanges Before a TimeOut. Even if a LOC or an EOA is lost, the EOAs corresponding to following LOCs could still deactivate its timer and then the emergency brake would be prevented. In this section we calculate how many LOC-EOA exchanges can happen between the generation of a LOC and the expiration of the corresponding timer, i.e. how many other EOAs can have a chance to block the timer.

Let us consider that the first LOC is generated at time $t = 0$, then its timer would expire at time $t = TM$. The maximum number $n_{\max}$ of LOC-EOA exchanges can be calculated considering that i) the last potentially useful EOA arrives in the shortest time possible and ii) it is immediately processed by the following CC tick, which is the last one before the timer expires.

The last potential useful EOA arrives at $(n_{\max}-1)T_{LOC}+T_{\min}+O$ and it can then be processed at $T_{CC}\lceil((n_{\max}-1)T_{LOC}+T_{\min}+O)/T_{CC}\rceil$. The CC tick just before the timer expires occurs at time $T_{CC}\lfloor TM/T_{CC}\rfloor$. We determine $n_{\max}$ by imposing that $\left\lceil\frac{(n_{\max}-1)T_{LOC}+T_{\min}+O}{T_{CC}}\right\rceil = \left\lfloor\frac{TM}{T_{CC}}\right\rfloor$,[4] and we can manipulate this equality as in [4], to obtain:

$$n_{\max} = 1 + \left\lfloor \frac{TM - \left\lceil\frac{T_{\min}+O}{T_{CC}}\right\rceil T_{CC}}{T_{LOC}} \right\rfloor . \tag{3}$$

Similarly the minimum number $n_{\min}$ of LOC-EOA exchanges can be calculated considering that (i) the last potentially useful EOA arrives in the longest time possible and (ii) it is processed 2 CC ticks later in correspondence of the last tick before the timer expires. Then we determine $n_{\min}$ by imposing that $\left\lceil\frac{(n_{\min}-1)T_{LOC}+T_{\max}+O}{T_{CC}}\right\rceil = \left\lfloor\frac{TM}{T_{CC}}\right\rfloor - 1$, and proceeding as above we obtain:

$$n_{\min} = 1 + \left\lfloor \frac{TM - \left(\left\lceil\frac{T_{\max}+O}{T_{CC}}\right\rceil + 1\right) T_{CC}}{T_{LOC}} \right\rfloor . \tag{4}$$

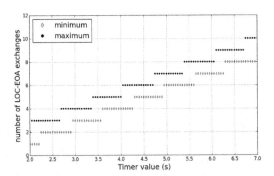

Fig. 4. Minimum and maximum number of LOC-EOA exchanges for $O = 50$ ms, calculated through Eqs. (4) and (3).

---

[4] This assumes $n_{max} > 1$. The first EOA needs to be valid until the end of the CC clock during which it is processed and then its processing time should start the latest at the tick number $\left\lfloor\frac{TM-T_{CC}}{T_{CC}}\right\rfloor$.

The difference between $n_{\max}$ and $n_{\min}$ depends on the timer TM and also on the offset. For the typical values in Table 2 they differ by at most 2 exchanges, i.e. $n_{\max} \leq n_{\min} + 2$. Figure 4 shows $n_{\min}$ and $n_{\max}$ for different values of the timer TM and an offset $O = 50$ ms. It also shows that the difference of two exchanges is achieved for some values of TM.

The two values $n_{\min}$ and $n_{\max}$ allow us to provide respectively upper and lower bounds for the EB probability and then for the EB rate, but these bounds can be too loose for practical uses. We are going to show it in the simple case when packet losses on the two wireless blue and red channels are independent Bernoulli random variables with parameter p. In this case a LOC or an EOA message is received with probability $1 - p^2$ and the probability $\tilde{p}$ to lose a LOC-EOA exchange is then $\tilde{p} = 1 - (1 - p^2)^2$. An emergency brake requires that the exchange 0 is not lost. Moreover the EB will necessarily occur if the $n_{\max}$ following LOC-EOA exchanges are lost (even if the $(n_{\max} + 1)$-st EOA arrives, it will be after the timer expiration) and cannot occur unless $n_{\min}$ exchanges are lost (the first $n_{\min}$ EOA cannot arrive late even in the worst case). It follows that

$$(1 - \tilde{p})\tilde{p}^{n_{\max}} \leq q_{EB} \leq (1 - \tilde{p})\tilde{p}^{n_{\min}}. \tag{5}$$

With the values in Table 2 the upper bound can be up to $\tilde{p}^{-2}$ times larger than the lower bound. A typical value for the packet loss probability is $p = 5\,\%$, and then $\tilde{p} \approx 0.5\,\%$ and the ratio of the two bounds is almost 4×10^4. In this case, as we are going to show later, the upper bound can be too pessimistic and practically of no utility to set the parameter TM. For this reason a more refined analysis is required.

Exact Formula. LOC 1 is generated at time $t = 0$ and then the k-th LOC is generated at $(k - 1)T_{LOC}$. The k-th EOA is the EOA corresponding to the k-th LOC. The timer of LOC 1 would expire at time $t = TM$. Remember that $\mathcal{L}_k$ denotes the event that the k-th LOC-EOA exchange is lost. Let $\mathcal{D}_k$ denote the event that the k-th EOA arrives too late to deactivate the timer of LOC 1. The two events are disjoint, i.e. $\mathcal{L}_k \cap \mathcal{D}_k = \emptyset$. LOC 1 experiences a timeout if and only if all the following exchanges are lost or their EOAs arrive too late, i.e.

$$\mathcal{T}_1 = \bigcap_{k=1}^{\infty} (\mathcal{L}_k \cup \mathcal{D}_k) = \bigcap_{k=1}^{n_{\max}} (\mathcal{L}_k \cup \mathcal{D}_k), \tag{6}$$

where the last equality follows from the fact that only the first $n_{\max}$ exchanges have a possibility to stop the timer ($\Pr(\mathcal{L}_k \cup \mathcal{D}_k) = 1$ for $k > n_{\max}$).

Due to timing constraints EOAs cannot arrive out of order. A consequence is that if the k-th EOA arrives too late to deactivate the timer of LOC 1, no later EOA will be able to deactivate it. In particular later EOAs will be lost or will arrive too late, i.e. $\mathcal{D}_k \subset \mathcal{D}_{k'} \cup \mathcal{L}_{k'}$ for all $k' \geq k$. This simple relation allows us to conclude [4] that for any m

$$\bigcap_{k=1}^{m} (\mathcal{L}_k \cup \mathcal{D}_k) = \bigcup_{k=1}^{m} \left(\mathcal{D}_k \cap \left(\bigcap_{h=1}^{k-1} \mathcal{L}_h \right) \right) \cup \left(\bigcap_{h=1}^{m} \mathcal{L}_h \right). \tag{7}$$

We can now move to calculate q_{EB}. From Eqs. (6) and (7), it follows that

$$q_{EB} = \Pr\left(\bar{\mathcal{L}}_0 \cap \mathcal{T}_1\right) = \Pr\left(\bar{\mathcal{L}}_0 \cap \overset{n_{\max}}{\underset{k=1}{\cap}}(\mathcal{L}_k \cup \mathcal{D}_k)\right)$$

$$= \Pr\left(\bar{\mathcal{L}}_0 \cap \left(\overset{n_{\max}}{\underset{k=1}{\cup}}\left(\mathcal{D}_k \cap \left(\overset{k-1}{\underset{h=1}{\cap}}\mathcal{L}_h\right)\right)\cup\left(\overset{n_{\max}}{\underset{h=1}{\cap}}\mathcal{L}_h\right)\right)\right). \qquad (8)$$

This expression can be simplified observing that the first $n_{\min} - 1$ EOAs cannot arrive late ($\Pr(\mathcal{D}_k) = 0$ for $k \leq n_{\min}$)

$$q_{EB} = \Pr\left(\bar{\mathcal{L}}_0 \cap \left(\overset{n_{\max}}{\underset{k=n_{\min}+1}{\cup}}\left(\mathcal{D}_k \cap \left(\overset{k-1}{\underset{h=1}{\cap}}\mathcal{L}_h\right)\right)\cup\left(\overset{n_{\max}}{\underset{h=1}{\cap}}\mathcal{L}_h\right)\right)\right). \qquad (9)$$

Equation (9) can be read as follows: a timeout occurs if there is a sequence of $n_{\min}$, $n_{\min} + 1$ up to … $n_{\max} - 1$ exchanges lost and the following EOA arrives late or if all the $n_{\max}$ exchanges are lost. These events are disjoint, because $\mathcal{D}_k \cap \mathcal{L}_k = \emptyset$, and then we can conclude:

$$q_{EB} = \sum_{k=n_{\min}+1}^{n_{\max}} \Pr\left(\mathcal{D}_k \cap \left(\bar{\mathcal{L}}_0 \cap \overset{k-1}{\underset{h=1}{\cap}}\mathcal{L}_h\right)\right) + \Pr\left(\bar{\mathcal{L}}_0 \cap \overset{n_{\max}}{\underset{h=1}{\cap}}\mathcal{L}_h\right) \qquad (10)$$

$$= \sum_{k=n_{\min}+1}^{n_{\max}} \Pr\left(\mathcal{D}_k \,\middle|\, \bar{\mathcal{L}}_0 \cap \overset{k-1}{\underset{h=1}{\cap}}\mathcal{L}_h \cap \bar{\mathcal{L}}_k\right) \Pr\left(\bar{\mathcal{L}}_0 \cap \overset{k-1}{\underset{h=1}{\cap}}\mathcal{L}_h \cap \bar{\mathcal{L}}_k\right)$$

$$+ \Pr\left(\bar{\mathcal{L}}_0 \cap \overset{n_{\max}}{\underset{h=1}{\cap}}\mathcal{L}_h\right). \qquad (11)$$

The last equality holds because $\mathcal{D}_k = \mathcal{D}_k \cap \bar{\mathcal{L}}_k$. The reason why we introduce the additional set $\bar{\mathcal{L}}_k$ will be clear in the following sections, where we will move to characterize delays and losses in order to compute the terms appearing in Eq. (11). We denote this sequence of loss events as $S_{\mathcal{L},k} \triangleq \bar{\mathcal{L}}_0 \cap \cap_{h=1}^{k-1} \mathcal{L}_h \cap \bar{\mathcal{L}}_k$.

As observed, for the typical values in Table 2 it is $n_{\max} \leq n_{\min} + 2$ and then there are at most 3 terms in Eq. (11).

3.2   Delay

In this section we characterize the event $\mathcal{D}_k$. In particular, we are interested to evaluate the probabilities $\Pr(\mathcal{D}_k \mid S_{\mathcal{L},k})$ appearing in Eq. (11). To this purpose we will study in detail the different components that determine if the k-th EOA arrives before or after the expiration of the timer of the first LOC.

Again, assume that LOC 1 is generated at time 0. If the k-th exchange LOC-EOA is not lost, then the arrival time of the k-th EOA is

$$T_k = T_{\min,k} + \phi_{L,k} + \phi_{E,k} + \omega_{ZC,k} \qquad (12)$$

where $T_{\min,k} = T_{CC} + 2T_{DCS,\min} + T_{ZC} + (k-1)T_{LOC} + O$ and the random variables $\omega_{ZC,k}$, $\phi_{L,k}$, $\phi_{E,k}$ represent the same quantities as those in Eq. (2), but are referred to the k-th exchange rather than to the first one. The EOA is processed at the tick

$$\gamma_k \triangleq \left\lceil \frac{T_k}{T_{CC}} \right\rceil + \omega_{CC,k}, \tag{13}$$

where $\omega_{CC,k}$ represents the processing delay at the CC expressed in number of ticks. According to the description in Sect. 2.1 $\omega_{CC,k}$ can assume value 0, if the EOA is going to be processed at the first CC tick after T_k, or value 1, if it is going to be processed at the following tick. We are going to characterize the Bernoulli r.v. $\omega_{CC,k}$ soon, for the moment we observe that the EOA arrives too late if $\gamma_k > \frac{TM}{T_{CC}}$ i.e. the EOA starts being processed after the expiration of the timeout. Then, the event $\mathcal{D}_k$ can be expressed as $\mathcal{D}_k = \bar{\mathcal{L}}_k \cap \left\{ \gamma_k > \frac{TM}{T_{CC}} \right\}$, and

$$\Pr\left(\mathcal{D}_k \mid S_{\mathcal{L},k} \right) = \Pr\left(\gamma_k > \frac{TM}{T_{CC}} \mid S_{\mathcal{L},k} \right), \tag{14}$$

because $\bar{\mathcal{L}}_k \subset S_{\mathcal{L},k}$. In order to calculate this probability we now move to consider each source of randomness in γ_k.

Processing Delay at the CC. Observe that $\omega_{CC,k}$ is independent of the arrival time of the k-th EOA T_k, as well as on arrival of any other EOA. In fact the queuing delay for the k-th EOA depends only on higher-priority traffic and not on the previous EOAs (that may or not being present in the processing queue), because only the most recent EOA is processed. It follows that $\omega_{CC,k}$ is independent of the event $\cap_{h=1}^{k-1}\mathcal{L}_h$ and its conditional distribution is equal to the a priori distribution provided in Sect. 2.1, i.e. $\omega_{CC,k}$ in Eq. (14) is a Bernoulli random variable with parameter p_D. While $\omega_{CC,k}$ as introduced is defined only when the k-th exchange is not lost, we can define it for any k as an independent Bernoulli random variable with parameter p_D. It can then be interpreted as the processing delay experienced by a hypothetical EOA arriving at a given time. The distribution of $\omega_{CC,k}$ does not depend on k and is independent of $S_{\mathcal{L},k}$.

Processing Delay at the ZC. Going back to Eq. (12), the random variable $\omega_{ZC,k}$ is dependent on the relative position of the ticks of the two clocks but also on the value of $\phi_{L,k}$. In fact the later the LOC arrives at the ZC (the larger $\phi_{L,k}$) the less the LOC has to wait until the next ZC tick (the smaller $\omega_{ZC,k}$), unless the LOC arrives so late that it misses the first available ZC tick and needs to wait for the next one. While we cannot get rid completely of this dependence, it is simpler to reverse it. With reference to Fig. 3, we express T_k with this equivalent expression:

$$T_k = T_{\min,k} + \sigma_k + \mathbb{1}_{\phi_{L,k} > \sigma_k} T_{ZC} + \phi_{E,k} \tag{15}$$

where σ_k denotes the time interval between the earliest possible instant at which the k-th LOC could be received at the ZC and the next ZC tick and $\mathbb{1}_{\phi_{L,k} > \sigma_k}$

is a Bernoulli random variable indicating if the random component of the communication delay will cause the LOC to miss this ZC tick and then to wait for the following one. It can be easily verified that σ_k depends on the specific LOC we are considering because the two clock periods are different. Then coherently with the idea that, in order to evaluate q_{EB}, the first LOC is chosen at random, σ_k is a random variable. Observe that the variable σ_k is independent of the loss processes and in particular of $S_{\mathcal{L},k}$. Moreover, it is independent of communication delays (i.e. of the variables $\phi_{L,k}$, $\phi_{E,k}$) and of processing delay at the ZC (i.e. of $\omega_{CC,k}$). Our next task is to determine σ_k's distribution.

Given the value $\sigma_1 = s_1$ for the first LOC, the values of the other r.v.s σ_k for $k > 1$ are uniquely determined, let $\sigma_k = s_k$. Assuming that T_{ZC} and T_{LOC} are commensurable numbers and choosing an opportune unit so that their values can be expressed as integers, in [4] we show that the possible values for s_k are the values s in $[0, T_{ZC})$ for which the following Diophantine equation in m and n admits integer solutions:

$$mT_{ZC} - nT_{LOC} = s - s_1. \tag{16}$$

The study of this equation in [4] leads to the conclusions that s_k assumes all and only the values in the set $S = \{\tilde{s} + iM, \ i = 0, 1, \ldots q_{ZC} - 1\}$ where M is the greatest common divisor of T_{ZC} and T_{LOC}, $T_{ZC} = q_{ZC}M$ and $\tilde{s} = s_1 \% M$. For example for the typical values we consider ($T_{ZC} = 378$ ms, $T_{LOC} = 675$ ms) it is $M = 27$, $q_{ZC} = 14$. Moreover, the sequence s_n is periodic with period q_{ZC} and then assumes the q_{ZC} values in S only once during each period. When we consider that the first LOC is a LOC selected at random, we conclude then that the variable σ_k is a uniform random variable over the set $S = \{\tilde{s} + kM, k = 0, 1, \ldots q_{ZC} - 1\}$.[5]

Communication Delays. In order to completely characterize the probability in Eq. (14), we need to discuss the two random variables $\phi_{L,k}$ and $\phi_{E,k}$. Remember that $\phi_{L,k}$ is the delay experienced by the "fastest" of the two LOC packets conditional on one of them arriving at the ZC. Let $\tau_{r,L}$ denote the random component of the delay experienced by the k-th LOC packet transmitted on the red network if it is not lost (we omit for simplicity the dependence on k). We can similarly introduce $\tau_{b,L}$, $\tau_{r,E}$ and $\tau_{b,E}$. These delays are independent and identically distributed random variables with Cumulative Distribution Function (CDF) $F_\tau(t)$. In particular, under the typical values in Sect. 2.1 they have support $[0, 40]$ ms.

3.3 Independent Losses

As an application of Eq. (11) we consider the case when packet losses are independent and homogeneous and Eq. (11) reduces to an easy-to-calculate exact formula. The independence allows to write:

---

[5] The analysis can be easily adapted to take into account the effect of clocks' frequency-shift [4].

$$\Pr\left(\mathcal{D}_k \mid S_{\mathcal{L},k}\right) = \Pr\left(\mathcal{D}_k \mid \bar{\mathcal{L}}_k\right) = \Pr\left(\gamma_k > \frac{TM}{T_{CC}}\right) \triangleq d(k), \qquad (17)$$

where γ_k is a function of the independent r.v.s $\omega_{CC,k}$, σ_k (already characterized in the previous section) and $\phi_{L,k}$ and $\phi_{E,k}$, whose CDF $F_\phi(t)$ can be easily derived by conditioning on the number of packets arriving at the ZC/CC:

$$F_\phi(t) = \frac{(1-p)^2}{1-p^2}\left(1-(1-F_\tau(t))^2\right) + \frac{2(1-p)p}{1-p^2}F_\tau(t) = \frac{F_\tau(t)\left(2-F_\tau(t)(1-p)\right)}{1+p}.$$

Our definition of $d(k)$ stresses that $\Pr(\gamma_k > TM/T_{CC})$ is a function of k, but this happens because of the constant $T_{\min,k}$, while the distributions of the r.v.s $\omega_{ZC,k}$, $\sigma_{CC,k}$, $\phi_{L,k}$ and $\phi_{E,k}$ do not depend on k.

Finally, by developing the terms $\Pr\left(S_{\mathcal{L},k}\right)$ in Eq. (11), we obtain

$$q_{EB} = \sum_{k=n_{\min}+1}^{n_{\max}} d(k)\tilde{p}^{k-1}(1-\tilde{p})^2 + \tilde{p}^{n_{\max}}(1-\tilde{p}), \qquad (18)$$

where $\tilde{p} = 1 - (1-p^2)^2$ is the probability that an exchange is lost.

4   Numerical Experiments

In this section we validate Eq. (18) through discrete-event simulations of the system, for which we have developed an ad-hoc Python simulator. The scenario tested by discrete-event simulations matches that described in Sect. 2 and considered in our analysis. For constant system parameters and the support of random variables, we have considered the typical values indicated in Table 2.

Figure 5 shows the EB rate versus different values of the packet loss probability p for $TM = 5.5$ s. The red solid curve is obtained through Eq. (18). Simulation results obtained by the Python simulator for selected values of p are reported as 95% confidence intervals in blue. About the computational time, Eq. (18) requires a few seconds on a current commodity PC. On the same machine the

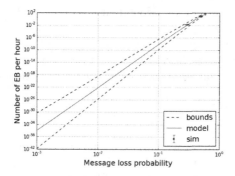

Fig. 5. Number of emergency brakes per hour when $TM = 5.5$ s. (Color figure online)

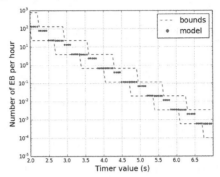

Fig. 6. Rate of emergency brakes when $O = 50$ ms and $p = 0.3$. (Color figure online)

Python simulator is able to simulate roughly 10^4 hours of train operation in one hour. It follows a rate of the order of 10^{-4} EBs per hour requires roughly 100 hours to be estimated with a precision of 1 % through the Python simulator. It is clear that lower EB rates are out of reach for the Python simulator.

Figure 5 also shows the black dashed curves that plot the functions $(1 - \tilde{p})\tilde{p}^{n_{\min}}/T_{LOC}$ and $(1 - \tilde{p})\tilde{p}^{n_{\max}}/T_{LOC}$ and that correspond to the upper and lower bound in Eq. (5) in presence of independent Bernoulli packet losses with probability p. We observe that the produced bounds are very loose.

As a final application of our methodology, Fig. 6 shows the expected number of emergency brakes per hour for different values of the timer TM, $O = 50$ ms and packet loss probability $p = 0.3$. The theoretical values calculated from Eqs. (18) and (1) (red dots) are compared with the bounds (black dashed lines). The figure shows that the simple upper bound can be orders of magnitude larger than the actual value. We now discuss the discontinuities appearing in the EB rate curve. From Eq. (18) we observe that the EB probability exhibits discontinuities only if $n_{\min}$, $n_{\max}$ or the functions $d(k)$ do. The small gaps of the EB rate correspond indeed to changes in the values $n_{\min}$ or $n_{\max}$ as it is revealed by the corresponding jumps of the bounds. The other gaps correspond to changes of the functions $d(k)$. We remember that $d(k) = \Pr(\gamma_k > TM/T_{CC})$, where γ_k is an integer. Then $d(k)$ does not depend on TM as far as $h \leq TM/T_{CC} < h+1$ for some integer h. Indeed, it can be checked that the other discontinuities in the curve (when neither $n_{\min}$ nor $n_{\max}$ change) correspond to integer values of TM/T_{CC}. This high sensitivity to the timer value is not only easily revealed by our numerical method, but well explained by our theoretical analysis.

5   Conclusion

In this paper we study the moving block control to quantify the rate of spurious EBs. Differently from existing literature, our starting point is not the current recommendation for the future ETCS level 3, but an actual implementation for metro. Equation (11) characterizes the EB rate in a general stationary setting, but it requires to compute the probability to observe specific patterns of packet losses, that can be a difficult task in general. Nevertheless, in the simple case of independent and homogeneous packet losses, the equation reduces to a simple analytical formula whose computational cost does not depend on the loss probability value. The formula can then be used to quantify extremely rare events (as emergency brakes should be). We are currently working to study more general loss scenarios, where losses are strongly correlated and time-variant. Our current results are in [4] and rely on a Monte Carlo approach to efficiently sample from the stationary distribution of the system.

This work is partially funded by the Inria-Alstom virtual lab.

References

1. Babczyński, T., Magott, J.: Dependability and safety analysis of ETCS communication for ERTMS level 3 using performance statecharts and analytic estimation. In: Zamojski, W., Mazurkiewicz, J., Sugier, J., Walkowiak, T., Kacprzyk, J. (eds.) DepCoS-RELCOMEX. AISC, vol. 286, pp. 37–46. Springer, Heidelberg (2014)
2. Carnevali, L., Flammini, F., Paolieri, M., Vicario, E.: Non-Markovian performability evaluation of ERTMS/ETCS level 3. In: Beltrán, M., Knottenbelt, W., Bradley, J. (eds.) EPEW 2015. LNCS, pp. 47–62. Springer, Heidelberg (2015)
3. Flammini, F., Marrone, S., Iacono, M., Mazzocca, N., Vittorini, V.: A multiformalism modular approach to ERTMS/ETCS failure modeling. Int. J. Reliab. Qual. Saf. Eng. 21(1), 1450001 (2014). (29 pages)
4. Neglia, G., et al.: Performance evaluation of train moving-block control. Research Report RR-8917. Inria, May 2016
5. Hermanns, H., Jansen, D.N., Usenko, Y.S.: From StoCharts to MoDeST: A comparative reliability analysis of train radio communications. In: Proceedings of WOSP 2005, pp. 13–23 (2005)
6. Horváth, A., Paolieri, M., Ridi, L., Vicario, E.: Transient analysis of non-Markovian models using stochastic state classes. Perform. Eval. 69(7–8), 315–335 (2012)
7. Trowitzsch, J., Zimmermann, A.: Using UML state machines and petri nets for the quantitative investigation of ETCS. In: Proceedings of Valuetools 2006 (2006)
8. Zimmermann, A., Hommel, G.: A train control system case study in model-based real time system design. In: Proceedings of IPDPS 2003 (2003)
9. Zimmermann, A., Hommel, G.: Towards modeling and evaluation of ETCS real-time communication and operation. J. Syst. Softw. 77(1), 47–54 (2005)

Decoupling Passenger Flows for Improved Load Prediction

Stefan Haar and Simon Theissing[(✉)]

MExICo Team, INRIA and LSV, CNRS and ENS de Cachan, Cachan, France
simon.theissing@inria.fr

Abstract. This paper continues our work on perturbation analysis of multimodal transportation networks (TNs) by means of a stochastic hybrid automaton (SHA) model. We focus here on the approximate computation, in particular on the major bottleneck consisting in the high dimensionality of systems of stochastic differential balance equations (SDEs) that define the continuous passenger-flow dynamics in the different modes of the SHA model. In fact, for every pair of a mode and a station, one system of coupled SDEs relates the passenger loads of all discrete points such as platforms considered in this station, and all vehicles docked to it, to the passenger flows in between. In general, such an SDE system has many dimensions, which makes its numerical computation and thus the approximate computation of the SHA model intractable. We show how these systems can be canonically replaced by lower-dimensional ones, by decoupling the passenger flows inside every mode from one another. We prove that the resulting approximating passenger-flow dynamics converges to the original one, if the replacing set of balance equations set up for all decoupled passenger flows communicate their results among each other in vanishing time intervals.

Keywords: Stochastic hybrid automata · Transportation networks · Fluid Petri nets · Stochastic differential equations modelling

1 Introduction

Apart from some exceptions, the different modes and lines in modern multimodal transportation networks do not share infrastructure elements, but are loosely connected through passenger transfers. Understanding how these passenger transfers connect their modes and lines is thus crucial if one wants to analyse how perturbations spread across such TNs. In this context, the present work is a contribution to our SHA model from [6] that we have developed for the computation of passenger load forecasts in multimodal TNs; given (i) estimations for all uncertain initial passenger loads (platforms, vehicles, etc.) and uncertain *continuous* passenger arrival flows, and (ii) the possibility to track individual vehicles so as to study the impact of well-directed interventions to their operation such as early departures.

© Springer International Publishing Switzerland 2016
G. Agha and B. Van Houdt (Eds.): QEST 2016, LNCS 9826, pp. 364–379, 2016.
DOI: 10.1007/978-3-319-43425-4_24

Our SHA Model. Our SHA model from [6] extends a previous *deterministic* hybrid automaton (DHA) model from [4]. In this, a finite set of vehicles is operated, and every vehicle is confined to a particular mode or line which does not share infrastructure elements with any other modes or lines. Passengers are grouped into a finite set of *trip profiles*, which define routes in the TN at hand, together with preferences for choosing different vehicle missions. Every mode of the DHA model corresponds to a particular configuration of the vehicles' discrete positions and discrete operational states. With these parameters, every mode defines which passenger flows between stations and vehicles are possible. In this way, a system of coupled ordinary differential equations (ODEs), one equation per station, is associated to every mode. This system relates the passenger loads of all stations and of all *stopped* vehicles docked to these stations, to the passenger flows such as boarding and alighting in between. Transitions *between* modes are triggered by (i) vehicles that *must depart*, i.e. whose elapsed driving and dwell times exceed some deterministic thresholds fixed by operation rules, and (ii) by passenger load trajectories hitting some pre-defined regions and thus triggering the departure of some vehicle (examples: boarding a train must stop if the train is full, or if the number of passengers on the platform is small and the train is scheduled to leave, etc.).

Now a TN is everything but deterministic: The influx of passengers into the system is a random process (from a macroscopic point of view, in fact a very continuous and measurable random process as compared to e.g. single passenger incidents), and the distribution of the passengers over the different possible trip profiles - is also unknown and can only be given statistically. This motivated the stochastic hybrid automaton (SHA) model that we introduced in [6]: Compared to our above DHA model, we replaced all systems of ODEs by systems of (Itô-) stochastic differential equations (SDEs), so as to be able to (i) start our analyses with uncertain initial passenger loads, and (ii) include uncertain passenger arrival flows into the model's continuous time dynamics. The mechanism of triggering mode transitions via thresholds remains the same; however, these hitting times are not deterministic, isolated points in time any more, but rather random variables with a continuous range of values.

Our SHA model thus does not fully cover the dynamical spectrum of the stochastic hybrid system (SHS) from [8], but only implements a particular realization thereof: In our SHA model, there are no mode transitions which are exponentially distributed w.r.t. time. In this context, also note that the SHS from [8] is an abstract mathematical model for a system with a mixed discrete and continuous dynamics; no more no less. The definition of e.g. all vector fields or possible mode transitions therein might be non-trivial and often cannot be done by pen and paper. That is why, we employ artefacts from the Petri nets formalism so as to e.g. derive all differential balance equations in a canonical way; which was proposed in many papers such as [10] before.

Problem Formulation. In [5], we introduced a strategy for the approximate computation of our above SHA model: We let the automaton change its mode only at equidistantly-spaced discrete points in time. Several challenges then arise. On the one hand, we are confronted with an explosion of the SHA model's timed mode graph, that - as its name suggests - captures the evolution of the SHA model's mode in discrete time steps; but we do not consider this combinatorial problem here, it will be treated in another work. Rather, our present paper focusses on another major bottleneck, namely the high dimensionality of the SDE systems defining the passenger flow dynamics in any given mode. The dimension of theses systems of coupled SDEs that we set up for every pair (mode, station) in the SHA model from [6] corresponds to the number of the passengers' different trip profiles, multiplied by the number of different discrete positions for the passengers within this station and the vehicles docked to it. Our major concern with this high dimensionality then is the fact that all algorithms that we have found so far are prone to what is known as the curse of dimensionality.

Simulation of SDEs. Monte Carlo simulations [9] require to sample realizations of the uncertain initial states of the considered random variables. For one-dimensional RVs subjected to one-dimensional SDEs this sampling might be trivial e.g. by employing the inverse transform sampling. However, it seems that sampling the uncertain initial state of multidimensional RVs is a non-trivial task that is active and still an open problem. Among the algorithms proposed thus far, we mention the Metropolis-Hastlings and the Gibbs sampler, which can be integrated into what is called a Markov Chain Monte Carlo simulation [1]. Other more exotic sampling techniques might involve e.g. neural networks [7].

Analytic Methods. Instead of sampling as above, another approach that we shall study elsewhere is to numerically integrate a multivariate Fokker-Planck equation. Such a system of partial *ordinary* differential equations is derived from the original multidimensional SDE, and describes the time evolution of an initial probability density function (PDF) under the system's dynamics; here, it concerns the passenger load vector's density function, giving the distribution of the number of passengers in the different trip profiles. However, many computational drawbacks also come along with this method, or more specifically with the numerical integrations required. First, not all numerical integration schemes can ensure the conservation of the probability flux in their basic set up; with the Finite Volume method [2] being one exception. Second, those schemes which can ensure the conservation of the probability flux are not easily extendible from common two or three dimensional applications to higher-dimensional problems.

Alternative Approaches. Alternatives to the computation or simulation of high-dimensional SDEs might involve their discrete approximation, which we do not pursue here. The technique studied here aims at decoupling the dynamics in the SDEs, so as to produce an alternative set of lower-dimensional SDEs that reproduces, or at least approximates, the original model dynamics. For instance, the authors of [3] mention the *local* specification of flows in a fluid stochastic Petri net model as a means for the decoupling. However, in contrast to our approach, they look at *scalar* rather than vectorial (passenger) flows.

In the rest of this paper, we shortly review our SHA model from [6] in Sect. 2 together with the discrete time computation of its state space from [5]. We also discuss the set up of all high-dimensional SDEs for the passenger flow dynamics in the SHA model's different modes. We then explain in Sect. 3 how the passenger flows in all modes can be systematically decoupled so as to replace the original systems of SDEs by approximating lower-dimensional ones. In this context, we also proof asymptotic convergence of the dynamics produced by the lower-dimensional SDEs w.r.t. the original dynamics. Last but not least, we summarize the contribution of our approach, and give a brief outlook on future work in Sect. 4.

2 Our SHA Model

2.1 Model Structure

Infrastructure. Basic modelling blocks of the SHA model are place/transition nets (= Petri nets with the token flow left out), which capture the structure of a finite set of stations S and a finite set of transportation grids G (TGs).

Every station $s \in S$ is made up of a finite set P_s of gathering points $p \in P_s$ (= places; represented by double circles) that can accommodate a limited number of passengers, and a finite set T_s of corridors $t \in T_s$ (= transitions; represented by double boxes) connecting (i) GPs to other GPs, or (ii) GPs to the station's exterior (cf. Fig. 1 below). Here, connected means "possibility of a passenger flow" in the direction of the edges that connect the corridors with the GPs.

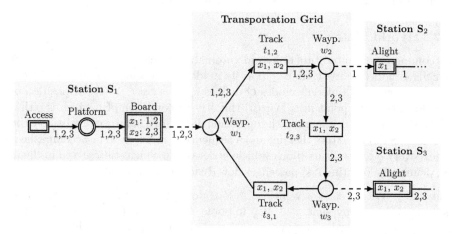

Fig. 1. Representation of the infrastructure of a sample TN in our SHA model, together with (i) the paths of two different vehicle missions x_1 and x_2, and (ii) an indication of the stops along these paths for the specification of three different trip profiles (TPs).

Every TG $g \in G$ captures the structure of a particular mode or line; and in doing so, all possible vehicle movements between its finite set W_g of discrete waypoints $w \in W_g$ (= places; represented by simple circles) which accommodate the vehicle tokens (at maximum one vehicle per waypoint) via tracks (= transitions; represented by simple boxes).

A finite set of tuples $(a, b) \in \mathcal{I}$, with $\mathcal{I} \subseteq (T \times W) \cup (W \times T)$, $T :=$ $\bigcup_{s \in \mathcal{S}} T_s$ and $W := \bigcup_{g \in \mathcal{G}} W_g$, composed of a transition in a station and a way-point in a TG, defines the interface between the stations and the TGs (repre-sented by dashed arcs in Fig. 1 above): Every tuple $(a, b) \in \mathcal{I}$ either connects some GP in a station $s \in \mathcal{S}$ to a waypoint in a TG $g \in \mathcal{G}$, in which case $a \in P_s$ and $b \in W_g$; or vice versa. In this way, every tuple defines which passenger flow between a vehicle stopped at a waypoint in a TG and a GP (= platform) in a station is possible for the purpose of boarding & alighting; see below.

Vehicle Operation. At the heart of the operation of a finite set $\mathcal{V}$ of all vehicle tokens $v \in \mathcal{V}$ considered in the SHA model are missions: Every mission defines a path in a particular transportation grid, together with (i) a sequence of stops at the waypoints along that path; (ii) deterministic-timed (minimum & maximum dwell times) and passenger load-dependent departure conditions from the stops which might state that a vehicle cannot depart from a stop as long as some passengers still want to alight from or board it; and (iii) driving times between all waypoints which might be functions of the positions of all vehicle tokens.

Passenger Routing. We group all passengers into a finite set $\mathcal{Y} := \{1, 2, \ldots, n\}$ of $n \in \mathbb{N}$ different trip profiles (TPs): Every $y \in \mathcal{Y}$ defines a particular path in TN's infrastructure, together with the passengers' preferences for the different vehicle missions (cf. Fig. 1 above). However, this does not mean that the passengers cannot change their TPs as we will highlight in a short (see Sect. 2.3).

2.2 Hybrid State

As common in the literature of hybrid automata, we refer to the discrete state of our SHA model at any time $\tau \geq 0$ as its mode: A particular mode $q \in Q$ from the finite set of all different modes Q defines for every $v \in \mathcal{V}$ (i) the position of v in form of a waypoint in a TG; (ii) the driving condition of v which is either parked, stopped or driving; (iii) the operational state of v in form of a mission to be executed, a discrete state therein, and a sequence of missions to be accomplished. Thus, every $q \in Q$ tells us which vehicle is docked to which station; and in doing so, defines the (continuous) passenger flow dynamics in TN.

Remark 1. We say that a vehicle $v \in \mathcal{V}$ is docked to a station $s \in \mathcal{S}$ iff (i) v is stopped at a waypoint $w \in W_g$ in some TG $g \in \mathcal{G}$; (ii) acc. to $\mathcal{I}$, either passengers can board v stopped at w from some GP in s, or alight from it to some GP in s. Moreover, we denote by $\mathcal{V}(s, q) \subseteq \mathcal{V}$ the subset of all vehicles that are docked to s in q.

Remark 2. If k is a row (column) vector, then we denote by $k[i]$ the element in its i-th column (row).

The continuous state of the SHA model at any $\tau \geq 0$, defines (i) the elapsed dwell times of all stopped vehicles, (ii) the elapsed driving times of all moving vehicles, and (iii) the passenger load $M(b, \tau)$, with $M : (b, \tau) \in (P \cup \mathcal{V}) \times \mathbb{R}_{\geq 0} \rightarrow$ $\mathcal{M}(b)$ and

$$\mathcal{M}(b) := \left\{ k \in (\mathbb{R}_{\geq 0})^{|\mathcal{Y}|} : \sum_{i=1}^{|\mathcal{Y}|} k[i] \leq c(b) \right\},\tag{1}$$

for every vehicle $b \in \mathcal{V}$ and every GP in a station $b \in P$. Therein, $P := \bigcup_{s \in \mathcal{S}} P_s$, $\mathrm{M}(b, \tau)[i]$ gives the number of passenger at/on-board b, who travel acc. to the TP $i \in \mathcal{Y}$, and $c(b)$, with $c : P \cup \mathcal{V} \to \mathbb{R}_{>0}$, gives the maximum number of passengers b can accommodate at the same time.

2.3 Balance Equations

For any $q \in Q$, we adapt the notation ${}^{\bullet}b(q)$ for the preset and $b^{\bullet}(q)$ for the postset of any $b \in P \cup \mathcal{V}'(q)$, with $\mathcal{V}'(q) := \bigcup_{s \in \mathcal{S}} \mathcal{V}(s, q)$, from the Petri nets literature for our purposes: ${}^{\bullet}b(q)$ denotes the set of all corridors in the stations that are connected by an arc pointing towards b. Accordingly, $b^{\bullet}(q)$ denotes the set of all corridors in the stations that are connected by an arc pointing away from b. For $b \in \mathcal{V}'(q)$, those arcs (dashed arcs in Fig. 1 above) point towards/away from the waypoint which accommodates b.

Note that all corridors in the stations of our SHA model are connected in a special way to the rest of the modelled infrastructure (GPs in the stations and waypoints in the TGs).

Remark 3. For any $t \in T$, we denote by ${}^{\star}t(q) := b$ the single GP in a station or vehicle docked to a station $b \in P \cup \mathcal{V}'(q)$ which is connected to t in q by an arc pointing towards t iff $t \in b^{\bullet}(q)$. Accordingly, we denote by $t^{\star}(q) := a$, for any $t \in T$, the single GP or vehicle docked to a station $a \in P \cup \mathcal{V}'(q)$ which is connected to t in q by an arc pointing away from t iff $t \in {}^{\bullet}a(q)$.

This special structure allows us to decompose all corridors in $q \in Q$ into three disjoint sets; implementing inflows, transfer flows, and outflows: Inflows model the arrival processes of the passengers who join the SHA model from TN's exterior.

Definition 1 (Inflow). *An inflow is a passenger flow assigned to any $t \in T_1$, with*

$$T_1 := \big\{ t \in T : \exists p \in P \text{ s.t. } t \in {}^{\bullet}p \,\wedge \\ \nexists p' \in P \text{ s.t. } t \in p^{\bullet} \,\wedge \nexists w \in W \text{ s.t. } (w, t) \in \mathcal{I} \big\}.\tag{2}$$

Transfer flows model passenger flows within the SHA model; including passenger transfers between the GPs in the stations, as well as passenger transfers between GPs in the stations and vehicles docked to the stations.

Definition 2 (Transfer Flow). *A transfer flow in $q \in Q$ is a passenger flow assigned to any $t \in T_2(q)$, with*

$$T_2(q) := \big\{ t \in T : \exists b \in P \cup \mathcal{V}(q) \text{ s.t. } t \in {}^{\bullet}b \,\wedge \\ \exists b' \in P \cup \mathcal{V}'(q) \text{ s.t. } t \in (b')^{\bullet} \big\}.\tag{3}$$

Finally, outflows model the SHA model's drain of passengers to TN's exterior.

Definition 3 (Outflow). *An outflow is a passenger flow assigned to any $t \in T_3$, with*

$$T_3 := \{t \in T : \exists p \in P \text{ s.t. } t \in p^\bullet \wedge$$
$$\not\exists p' \in P \text{ s.t. } t \in p^\bullet \wedge \not\exists w \in W \text{ s.t. } (t, w) \in \mathcal{I}\}. \tag{4}$$

With that said, we denote by $T'(q)$, with $T'(q) := T_1 \cup T_2(q) \cup T_3$, the set of all corridors active in $q \in Q$; and by $\gamma(\tau)$, with $\gamma : \mathbb{R}_{\geq 0} \to Q$, the mode of our SHA model at time $\tau \geq 0$.

$$\mathrm{dM}(b, \tau) := \sum_{t \in {}^\bullet b \cap T'(\gamma(\tau))} \mathrm{R}(t) \overbrace{[\phi(t, \tau) \, \mathrm{d}\tau + \delta(t) \, \mathrm{d}\mathcal{W}(\tau)]}^{\text{Passenger flow into b}} -$$

$$\sum_{t \in b^\bullet \cap T'(\gamma(\tau))} \underbrace{[\phi(t, \tau) \, \mathrm{d}\tau + \delta(t) \, \mathrm{d}\mathcal{W}(\tau)]}_{\text{Passenger flow leaving b}} \tag{5}$$

then defines the time evolution of the passenger load of every GP in a station and of every vehicle docked to a station $b \in P \cup \mathcal{V}'(q)$ at any time $\tau \geq 0$ when the SHA model is in $q \in Q$. This balance equation relates $\mathrm{M}(b, \tau)$ to all passenger flows into b and leaving it: We capture the routing of all passengers along the different TPs as well as their local re-routing among these TPs in so-called routing matrices.

Remark 4. We denote by $\Psi^{d_1 \times d_2}$, for some $d_1, d_2 \in \mathbb{N}_{>0}$ and any set Ψ, the set of all matrices with d_1 rows and d_2 columns, whose elements are from Ψ. In the case that $d_2 = 1$, we drop d_2 in $\Psi^{d_1 \times d_2}$ and write Ψ^{d_1} instead.

The i-th row and the j-th column of a particular routing matrix $\mathrm{R}(t)$ assigned to $t \in T$, with

$$\mathrm{R} : T \to \left\{ K \in (\mathbb{R}_{\geq 0})^{|\mathcal{Y}| \times |\mathcal{Y}|} : \sum_{i=1}^{|\mathcal{Y}|} K[i, j] = 1, \forall j = \mathcal{Y} \right\},$$

defines the relative amount of the flow of passengers who join t acc. to the TP $j \in \mathcal{Y}$, and who leave t acc. to the TP $i \in \mathcal{Y}$; and the fact that every column of $\mathrm{R}(t)$ must either sum up to one or to zero, implies that all passenger flows are conserved.

Remark 5. Time could be included in the domain of the routing matrices above so that they might change values during mode transitions of the SHA model depending on the hybrid state; so as to account e.g. for loudspeaker announcements.

We next write down the passenger flow assigned to every corridor $t \in T(q)$ in q acc. to its impact on $\mathrm{M}(p, \tau)$ as the sum of a drift term $\phi(t, \tau)$, with

$$\phi : (t, \tau) \in \bigcup_{q \in Q} T'(q) \times \mathbb{R}_{\geq 0} \to \left\{ v \in (\mathbb{R}_{\geq 0})^{|\mathcal{Y}|} : \sum_{i=1}^{|\mathcal{Y}|} v[i] \leq \phi_{\max}(q, t) \right\},$$

and a constant diagonal diffusion term

$$\delta : \bigcup_{q \in Q} T'(q) \to \left\{ K \in \mathbb{R}^{|\mathcal{Y}| \times |\mathcal{Y}|} : K[i,j] = 0, \forall i \neq j \right\}.$$

Therein, $\phi_{\max}(q,t)$, with $\phi_{\max} : q \in Q \times T'(q) \to \mathbb{R}_{\geq 0}$, is the maximum passenger throughput of the corridor $t \in T'(q)$, when the SHA model is in $q \in Q$.

Remark 6. Let X be a continuous RV. Then, $\mathrm{pdf}(X)$ denotes its PDF; $\sigma(X)$ denotes its state space; and $\mathrm{pdf}(X,x)$ denotes the evaluation of $\mathrm{pdf}(X)$ at x for some $x \in \sigma(X)$.

We discuss the specification of $\phi(\cdot)$ and $\delta(\cdot)$ in more detail in the rest of this paper. Here, only note that the drift term of a flow into some $b \in P \cup \mathcal{V}'(q)$ shifts the density of $\mathrm{M}(b,\tau)$ in its domain. The flow's diffusion term narrows or broadens the density of $\mathrm{M}(b,\tau)$.

2.4   Grouping of Balance Equations

In principle, the passenger flows in (5) can be defined as any functions of the SHA model's complete hybrid state as long as they are capacity- and demand-sensitive; crucial properties that we assume for all passenger flows in our SHA model: We say that some passenger flow is capacity-sensitive iff its drift does not cause the passenger load of some GP or vehicle to exceed the capacity limit of that GP or vehicle.

Definition 4 (Capacity-Sensitive Flow). *A passenger flow assigned to some* $t \in T'(q)$ *in* $q \in Q$ *is capacity-sensitive iff* $t \in T_3$ *or*

$$\sum_{i=1}^{|\mathcal{Y}|} \mathrm{M}(t^\star, \tau)[i] \to c(t^\star)$$

implies that $\phi(t,\tau) \to 0$ *for any* $\tau \geq 0$.

Additionally, we say that a passenger flow is demand-sensitive iff its drift does not cause any passenger load to become negative.

Definition 5 (Demand-Sensitive Flow). *A passenger flow assigned to some* $t \in T'(q)$ *in* $q \in Q$ *is demand-sensitive iff* $t \in T_1$ *or*

$$\mathrm{M}({}^\star t, \tau)[j] \sum_{i=1}^{|\mathcal{Y}|} \mathrm{R}(t)[i,j] \to 0$$

implies that $\phi(t,\tau)[j] \to 0$ *for all* $j \in \mathcal{Y}$ *and for any* $\tau \geq 0$.

Remark 7. Definitions 4 and 5 taken alone cannot ensure the non-negativity and capacity limits of the passenger loads assuming non-zero diffusion terms in (5). Instead both properties must be explicitly ensured during the computation or simulation of (5) in form of reflecting boundary conditions. See e.g. [6], where we derive reflecting boundary conditions for the numerical integration of a multivariate Fokker-Planck equation obtained from (5).

For our purposes however, we do not need this kind of global inclusion of the SHA model's complete hybrid state into the specification of the passenger flows: We restrict the domains of their drift terms to the passenger loads in their presets and postsets.

Definition 6 (Local Flow). *A passenger flow assigned to some* $t \in T'(q)$ *in* $q \in Q$ *is local iff for any* $\tau \geq 0$,

- $t \in T_1$, *and the flow's drift term only depends on* $M(t^\star, \tau)$, *or*
- $t \in T_2(q)$, *and the flow's drift term only depends on* $M(^\star t, \tau)$ *and* $M(t^\star, \tau)$, *or*
- $t \in T_3$, *and the flow's drift term only depends on* $M(^\star t, \tau)$.

This *local specification* of all passenger flows produces a natural decomposition of all SDEs set up for any $q \in Q$: The balance equations in form of (5) set up for the passenger loads of all GPs $p \in P_s$ and vehicles $v \in V(s, q)$, for some station $s \in S$, are independent from the passenger loads of all GPs outside s and vehicles not docked to s. We can thus group them into one common system of coupled SDEs of dimension $k := (|P_s| + |V(s, q)|) |Y|$, which latter system is decoupled from those systems set up for all other stations.

Remark 8. In practice, we do only have to consider all those TPs in the domain specification for the passenger load of a particular GP or vehicle, whose paths cover this GP or vehicle. Thus, k as defined above only defines an upper bound for the dimension of the system of SDEs set up for s in q.

2.5   Mode Transitions

We assume that at the initial simulation time $\tau = 0$, with $\tau \geq 0$, our SHA model is in one particular mode with marginal probability one, and we know the elapsed driving &dwell times of all vehicles. We then let our SHA model transition between its discrete modes only at discrete time steps $\tau = i \, _\Delta\tau$, with $i \in \mathbb{N}_{>0}$, of fixed length $_\Delta\tau > 0$. In this context, we also let the elapsed driving &dwell times of all vehicles only evolve at $\tau = i \, _\Delta\tau$ by $_\Delta\tau$. A directed acyclic graph (DAG) then captures the time evolution of our SHA model's vehicle load (= particular mode and particular realization of all elapsed discrete driving &dwell times). We do not go into details of its computation here, but only stress some important points. Refer to [5] for more information: Every node, say m, in this DAG, say G, represents a particular vehicle load for our SHA model in the half-closed time interval $[h_m \, _\Delta\tau, (h_m + 1) \, _\Delta\tau)$ iff $h_m \in \mathbb{N}_{\geq 0}$ is the height of m in G. Thus, two nodes with the same height $h' \in \mathbb{N}_{\geq 0}$ in G represent two alternatives for our SHA model's vehicle load in $[h' \, _\Delta\tau, (h' + 1) \, _\Delta\tau)$. Two or more branches away from m indicate the possibility of mode transitions; with one branch for every alternative mode transition, and one additional branch for the continuation of m-th mode. Several nodes with the same height in G can have the same mode and thus the same passenger flow dynamics in common.

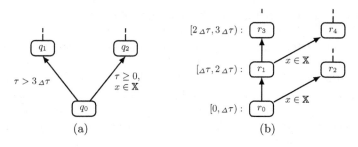

Fig. 2. Schematic comparison of a (classical) mode graph (a) and a timed mode graph (b) for our SHA model: $\mathbb{X}$ denotes a compact region in the SHA model's complete passenger load space as entrance condition for a not further specified passenger load-driven mode transition, and $_{\Delta}\tau > 0$ is the fixed time step that separates every pair of two consecutive time layers when the SHA model can change its mode

2.6 Propagation of Passenger Loads

At any simulation time $\tau = i_{\Delta}\tau$, with $i \in \mathbb{N}_{\geq 0}$ and $_{\Delta}\tau > 0$, one single marginal joint PDF, say pdf (i), defines the passenger loads of all GPs in the stations and of all vehicles. For $i = 0$, we assume that pdf (i) is known with marginal probability one. Then, starting from $i = 0$, all passenger loads have to be propagated forward in time from one time layer in the SHA model's DAG to the next: For the computation of pdf $(i + 1)$, for some $i \in \mathbb{N}_{\geq 0}$, all high-dimensional systems of SDEs defined by our SHA model's different modes in the time layer $\left[i_{\Delta}\tau, (i + 1)_{\Delta}\tau \right)$ of the DAG, must be computed from $\tau = i_{\Delta}\tau$ to $\tau = (i + 1)_{\Delta}\tau$ with pdf (i) as common initial PDF. Depending on the particular use case at hand so as to e.g. forecast the risk of overcrowded platforms, this forward propagation is normally terminated once the simulation time exceeds some constant threshold. Refer to [5] for more details.

3 The Decoupling of All Passenger Flows

3.1 Overview

Our decoupling approach is perhaps best described by the following sequence of images: We assume that every GP in a station and every vehicle $b \in P \cup V$ has the shape of a circular area, say A_b. We next assume that the passenger load of b is equally distributed on A_b at any simulation time step $\tau = i_{\Delta}\tau$, with $\tau \geq 0$, $i \in \mathbb{N}_{\geq 0}$, and $_{\Delta}\tau > 0$; in which $_{\Delta}\tau$ is the fixed time step that separates every pair of two consecutive time layers confining all mode transitions.

Remark 9. We denote by $\Gamma(\tau)$, with $\Gamma : \mathbb{R}_{\geq 0} \to 2^Q \setminus \emptyset$, the subset of all modes our SHA model can be in at time $\tau \in \mathbb{R}_{\geq 0}$.

For any time $\tau \in \mathcal{H}_i$, from the time interval $\mathcal{H}_i := \left[i_{\Delta}\tau, (i + 1)_{\Delta}\tau \right)$, any mode $q \in \Gamma(\tau)$, and any $b \in P \cup V'(q)$, we divide A_b into $|(^\bullet b \cup b^\bullet) \cap T'(q)|$ non-overlapping slices (cf. Fig. 3 below); in which one slice is attributed to every passenger flow into or leaving b, i.e., the passenger flow assigned to every corridor $t \in (^{\bullet}b \cup b^{\bullet}) \cap T'(q)$. Our assumptions above then imply that at $\tau = i \, _{\Delta}\tau$ (i) the surface area of a particular slice defines how many passengers it accommodates at τ, and (ii) the distribution of this latter number of passengers w.r.t. the passengers' different TPs is identical to the distribution of the total number of passengers at b and τ w.r.t. the different TPs. We moreover assume that a retractable wall is installed along every frontier separating two neighbouring slices (dashed lines in Fig. 3 below). These walls prevent the equidistant re-distribution of the slices' passenger loads at any $\tau \in \mathcal{H}_i$, which *diffusion* is restricted to the discrete time step $\tau = (i+1) \, _{\Delta}\tau$ when all walls are removed.

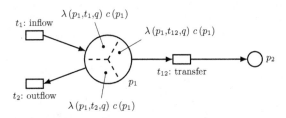

Fig. 3. Schematic representation of our decoupling approach: all GPs and vehicles docked to the stations in a particular mode, say q, of the SHA model are divided into slices, with impenetrable walls separating neighbouring slices until the next discrete point in time, say τ, arrives when the SHA model can change its mode. As long as the SHA model stays in q, all passengers flow into or out of the slices. They do not flow into or out of the original GPs and vehicles. A re-distribution of the slices' passenger loads occurs at τ.

So in our physically-touched model above, the slices' passenger loads are decoupled at any $\tau \in \mathcal{H}_i$, which implies that they might be filled and emptied at different rates if we assume that the passengers flow into and leave the slices of b; instead of flowing into and leaving b itself. For the specification of the slices' surface areas, we use the maximum passenger throughputs assigned to the corridors for the different modes; see below.

3.2 Decoupled Balance Equations

General Structure. The system of SDEs that we will set up for the decoupled passenger flow assigned to every $t \in T'(q)$ for any $q \in Q$ next, defines how this flow manipulates the passenger load $M_{q,t}(^{\star}t, \tau)$ of the isolated slice from $^{\star}t$ attributed to t in q and/or the passenger load $M_{q,t}(t^{\star}, \tau)$ of the isolated slice from $t^{\star}$ attributed to t in q; when our SHA model is in q. We write it down in the very general form of

$$dX_{q,t}(\tau) := A_{q,t}(X_{q,t}(\tau)) \, d\tau + B_{q,t}(X_{q,t}(\tau)) \, dW(\tau), \qquad (6)$$

with the state vector $X_{q,t}$, the drift vector $A_{q,t}$, the diffusion matrix $B_{q,t}$, and the vector of $|\mathcal{Y}|$ uncorrelated Wiener processes $\mathcal{W}$.

Remark 10. We write the tuple of a mode $q \in Q$ and a transition $t \in T'(q)$ in form of subscript separating both in the given order by a comma next to a variable or constant iff we refer to the projection of that variable or constant in (6) set up for the decoupled passenger flow assigned to t in q.

Projection of Passenger Loads and Flows. As outlined in the figurative overview of our decoupling approach above, we project $M(b, \tau)$, for any $b \in P \cup V'(q)$ and $q \in Q$, to $M_{q,t}(t, \tau)$, with $M_{q,t} : T'(q) \times \mathbb{R}_{\geq 0} \to \mathcal{M}_{q,t}(b)$ and

$$\mathcal{M}_{q,t}(b) := \left\{ k \in (\mathbb{R}_{\geq 0})^{|\mathcal{Y}|} : \sum_{i=1}^{|\mathcal{Y}|} k[i] \leq \lambda(b, t, q) \, c(b) \right\},$$

at $\tau = i_{\Delta\tau}$, with $i \in \mathbb{N}_{\geq 0}$, acc. to

$$M_{q,t}(b, i_{\Delta\tau}) := \lambda(b, t, q) \, M(b, i_{\Delta\tau}) \tag{7}$$

iff our SHA model is in mode q at $\tau = i_{\Delta\tau}$. Therein, $\lambda(b, t, q)$, with

$$\lambda(b, t, q) := \frac{\phi_{\max}(q, t)}{\sum_{t' \in (\bullet b \cup b^\bullet) \cap T'(q)} \phi_{\max}(q, t')}, \tag{8}$$

defines the maximum number $\lambda(b, t, q) \, c(b)$ of passengers the isolated slice from $b \in P \cup V'(q)$ assigned to $t \in (\bullet b \cup b^\bullet) \cap T'(q)$ in q can accommodate (cf. Fig. 3 above). This simple projection implies

$$\mathrm{pdf}(M_{q,t}(b, i_{\Delta\tau}) = \lambda(b, t) \, k) = \mathrm{pdf}(M(b, i_{\Delta\tau}) = k), \forall k \in \mathcal{M}(b), \tag{9}$$

with $\mathcal{M}(b)$ from (1). We also use (8) to project $\phi(t, \tau)$ - which we assume to be local, demand- &capacity sensitive - to $\phi_{q,t}(t, \tau)$ acc. to Table 1 below, which implies that all qualitative properties of $\phi(t, \tau)$ such as demand-sensitiveness are adopted by $\phi_{q,t}(t, \tau)$.

Table 1. Specification of $\phi_{q,t}(t, \tau)$ assigned to $t \in T'(q)$ in $q \in Q$

Inflow:	$\phi\left(\lambda^{-1}(t^*, t, q) \, M_{q,t}(t^*, \tau)\right)$
Transfer Flow:	$\phi\left(\lambda^{-1}(\bullet t, t, q) \, M_{q,t}(\bullet t, \tau), \lambda^{-1}(t^*, t, q) \, M_{q,t}(t^*, \tau)\right)$
Outflow:	$\phi\left(\lambda^{-1}(\bullet t, t, q) \, M_{q,t}(\bullet t, \tau)\right)$

Inflows. In general, we neither know the passengers' exact arrival times, nor the TPs of the new arriving passengers. However, in most situations we know some reference values, and we can estimate quite reasonably fluctuations around them (e.g. from statistical considerations); which latter knowledge we can then map to the systems of SDEs set up for all decoupled inflows. More specifically, we set up for every $t \in T_1$ a balance equation in form of (5), which defines the impact of the inflow assigned to t, to the passenger load of t^*; and integrate this balance equation into (6). Table 2 lists the corresponding ingredients.

Transfer Flows. Once having joined the SHA model, we assume that the passenger transfer dynamics regarded in isolation within the SHA model in a particular mode is deterministic; which implies zero diffusion terms for the specification of all decoupled passenger transfer flows: For every $t \in T_2(q)$ in $q \in Q$, we set up two balance equations in form of (5). The first balance equation defines the impact of the transfer flow assigned to t, to the passenger load of $^*t$. Accordingly, the second balance equation relates the passenger load of t^* to the same decoupled transfer flow. We then integrate both balance equations into (6) acc. to Table 2.

Table 2. Specification of the system of SDEs set up for the decoupled inflow, transfer flow, or outflow assigned to $t \in T'(q)$ in mode $q \in Q$ of our SHA model

	Inflow	Transfer Flow	Outflow
Schematic structure	$t\ \square\!\!\rightarrow\!\!\bigcirc\ t^*$	$^*t\ \bigcirc\!\!\rightarrow\!\!\square\!\!\rightarrow\!\!\bigcirc\ t^*$	$^*t\ \bigcirc\!\!\rightarrow\!\!\square\ t$
$X_{q,t}(\tau)$	$M_{q,t}(t^*, \tau)$	$\begin{bmatrix} M_{q,t}(^*t, \tau) \\ M_{q,t}(t^*, \tau) \end{bmatrix}$	$M_{q,t}(^*t, \tau)$
$A_{q,t}(\tau)$	$R(t)\,\phi_{q,t}(t, \tau)$	$\begin{bmatrix} -\phi_{q,t}(t, \tau) \\ R(t)\,\phi_{q,t}(t, \tau) \end{bmatrix}$	$-\phi_{q,t}(t, \tau)$
$B_{q,t}$	$\delta(t)$	0	0

Outflows. Similar to the specification of all transfer flows above, we demand zero diffusion terms for all passenger outflows: For every $t \in T_3$, we set up a balance equation in form of (5) and integrate it into (6). This balance equation relates the passenger load of $^*t$, to the outflow assigned to t (cf. Table 2).

3.3 Correctness of Our Decoupling Approach

Assume that our SHA model is in mode $q \in Q$ at time $\tau = i\,_\Delta\tau$, for some $i \in \mathbb{N}_{\geq 0}$; in which $_\Delta\tau > 0$ is the fixed time step that separates every pair of two consecutive time layers confining all mode transitions. Moreover, assume that we like to compute the probability of a particular mode transition of the SHA model at time $\tau = (i+1)\,_\Delta\tau$; which is triggered by the passenger load trajectory of some GP in a station or vehicle docked to a station $b \in P \cup V'(q)$ taking a value from $k \in K$, with $K \subseteq M(b)$ and $M(b)$ from (1). More formally speaking, we thus like to compute the probability

$$\mathbb{P}(M(b, (i+1)\,_\Delta\tau) \in K) := \int_K \mathrm{pdf}(M(b, (i+1)\,_\Delta\tau) = k)\,\mathrm{d}k \qquad (10)$$

with $M(b, \tau)$ specified at $\tau = i\,_\Delta\tau$ by $\mathrm{pdf}(M(b, i\,_\Delta\tau))$ acc. to (9).

Remark 11. Let $X_1, X_2, \ldots, X_n$ be a vector of $n \in \mathbb{N}_{>0}$ continuous RVs. Then, $\mathrm{pdf}(X_j; j \in \{1, 2, \ldots, n\})$ denotes the joint PDF of $X_1, X_2, \ldots, X_n$; $\mathrm{pdf}(X_j = x_j; j \in \{1, 2, \ldots, n\})$ denotes the evaluation of $\mathrm{pdf}(X_j; j \in \{1, 2, \ldots, n\})$ at $(x_1, x_2, \ldots, x_n)$, with $x_j \in \sigma(X_j)$, $\forall j \in \{1, 2, \ldots, n\}$.

Look at

$$\mathbb{P}\left(\sum_{t\in(^\bullet b\cup b^\bullet)\cap T'(q)} M_{q,t}\left(b,(i+1)\,_{\Delta}\tau\right)\in K\right) =$$

$$\int_K \mathrm{pdf}\left(\sum_{t\in(^\bullet b\cup b^\bullet)\cap T'(q)} M_{q,t}\left(b,(i+1)\,_{\Delta}\tau\right)=k\right)\,\mathrm{d}k \qquad (11)$$

instead, which is the probability that the sum of the decoupled passenger loads of the different isolated slices from b (isolated in q) takes a value from K at $\tau = (i+1)\,_{\Delta}\tau$. Let

$$l := |(^\bullet b\cup b^\bullet)\cap T'(q)|, \qquad (12)$$

and introduce the set $\overline{\mathcal{M}}(b,k)$, with

$$\overline{\mathcal{M}}(b,k) := \left\{(k_1,k_2,\dots,k_l)\in(\mathcal{M}(b))^l : \sum_{j=1}^l k_j = k\right\} \qquad (13)$$

Moreover, let $\{t_1,t_2,\dots,t_l\} := (^\bullet b\cup b^\bullet)\cap T'(q)$. Then, write down (11) in form of

$$\mathbb{P}\left(\sum_{t\in(^\bullet b\cup b^\bullet)\cap T'(q)} M_{q,t}\left(b,(i+1)\,_{\Delta}\tau\right)\in K\right) =$$

$$\int_K \int_{\overline{\mathcal{M}}(b,k)} \mathrm{pdf}\left(M_{q,t_j}\left(b,(i+1)\,\tau\right)=k_j; j\in\{1,2,\dots,l\}\right)\,\mathrm{d}(k_1,k_2,\dots,k_l)\,\mathrm{d}k \qquad (14)$$

Therein, note that $M_{q,t_1}\left(b,(i+1)\,_{\Delta}\tau\right),\dots,M_{q,t_l}\left(b,(i+1)\,_{\Delta}\tau\right)$ are independent RVs. Thus, (14) simplifies to

$$\mathbb{P}\left(\sum_{t\in(^\bullet b\cup b^\bullet)\cap T'(q)} M_t\left(b,(i+1)\,_{\Delta}\tau\right)\in K\right) =$$

$$\int_K \int_{\overline{\mathcal{M}}(b,k)} \prod_{t_j\in(^\bullet b\cup b^\bullet)\cap T'(q)} \mathrm{pdf}\left(M_{t_j}\left(b,(i+1)\,\tau\right)=k_j\right)\,\mathrm{d}(k_1,k_2,\dots,k_l)\,\mathrm{d}k \qquad (15)$$

Theorem 1. *For any $q\in Q$, $b\in P\cup\mathcal{V}'(q)$, and $k\in\mathcal{M}(b)$, the integral*

$$\int_{\overline{\mathcal{M}}(b,k)} \prod_{t_j\in(^\bullet b\cup b^\bullet)\cap T'(q)} pdf(M_{t_i}\left(b,(i+1)\,\tau\right)=k_i)\,\mathrm{d}(k_1,k_2,\dots,k_l)$$

from (15) converges to $pdf(M\left(b,(i+1)\,_{\Delta}\tau\right)=k)$ from (10) for $_{\Delta}\tau\xrightarrow{_{\Delta}\tau\ge 0}0$.

Note that Theorem 1 implies that our above decoupling approach produces a set of SDEs (one for every decoupled flow) for the different modes of our SHA model; this set approximates the original coupled passenger flow dynamics in the limiting case of vanishing discrete simulation time steps, when we let the decoupled slices communicate their results.

Proof of Theorem 1. Common Initial State: From (7), note that

$$
\sum_{t\in(\bullet b\cup b\bullet)\cap T'(q)} \mathrm{M}_{q,t}\left(b, i\,_{\Delta}\tau\right) = \sum_{t\in(\bullet b\cup b\bullet)\cap T'(q)} \lambda\left(b,t,q\right)\mathrm{M}\left(b, i\,_{\Delta}\tau\right)
$$
$$
= \mathrm{M}\left(b, i\,_{\Delta}\tau\right) \sum_{t\in(\bullet b\cup b\bullet)\cap T'(q)} \lambda\left(b,t,q\right). \tag{16}
$$

From (9) follows

$$
\sum_{t\in(\bullet b\cup b\bullet)\cap T'(q)} \lambda\left(b,t,q\right) = 1, \tag{17}
$$

which in turn implies

$$
\sum_{t\in(\bullet b\cup b\bullet)\cap T'(q)} \mathrm{M}_{q,t}\left(b, i\,_{\Delta}\tau\right) = \mathrm{M}\left(b, i\,_{\Delta}\tau\right). \tag{18}
$$

Common Differential Dynamics: The continuous time evolution of

$$
\sum_{t\in(\bullet b\cup b\bullet)\cap T'(q)} \mathrm{M}_{q,t}\left(b, \tau\right)
$$

in the time interval $\tau \in \left[i\,_{\Delta}\tau, (i+1)\,_{\Delta}\tau\right)$ is defined by

$$
\mathrm{d}\left(\sum_{t\in(\bullet b\cup b\bullet)\cap T'(q)} \mathrm{M}_{q,t}\left(b, \tau\right)\right) = \sum_{t\in(\bullet b\cup b\bullet)\cap T'(q)} \mathrm{dM}_{q,t}\left(b, \tau\right), \tag{19}
$$

with initial state

$$
\mathrm{M}_{q,t}\left(b, i\,_{\Delta}\tau\right),
$$

for some $i \in \mathbb{N}_{\geq 0}$ and $\tau_{\Delta}\tau > 0$, which is identical to (5) for $_{\Delta}\tau \to 0$ given the specification of (6) acc. to Tables 1 and 2, q.e.d.

3.4   Consequence of Our Decoupling Approach

In the original approximate computation of our SHA model's state space, we were confronted with one system of coupled SDEs for every station $s \in \mathcal{S}$ in every mode. The dimension of this system is $n := (n_{s,1} + n_{s,2})\,n_y$ iff $n_{s,1}$ corresponds to the number of different gathering points in s, $n_{s,2}$ corresponds to the number of vehicles docked to s, and $n_y := |\mathcal{Y}|$ corresponds to the number of the passengers' different trip profiles in the TN at hand. Now our decoupling approach replaces this n-dimensional system of SDEs by a set of probably much smaller systems of ODEs (with uncertain initial states) and SDEs: Every of this new/replacing system of equations has $2\,n_y$ dimensions if it captures a transfer flow, and n_y dimensions otherwise.

4 Summary and Outlook

In this paper, we have considered one major bottleneck that may arise in the approximate computation of our SHA model from [5]: the numerical computation of the many high-dimensional SDEs, which define the passenger flow dynamics in its different modes. More specifically, we have shown how all passenger flows can be systematically decoupled in the different modes of our SHA model, which produces a set of lower-dimensional ODEs and SDEs replacing the original SDEs. We proved correctness of this decoupling approach. Numerical experiments are under way. We want to share our insights obtained from them in future publications, where we also intend to (i) discuss improvements targeting the computation of the SHA model's discrete state, and (ii) show how our model and algorithms for its approximate computation can be applied to the perturbation analysis of a multimodal TN.

Acknowledgement. This research work has been carried out under the leadership of the Technological Research Institute SystemX, and therefore granted with public funds within the scope of the French Program "Investissements d'Avenir".

References

1. Brooks, S., et al.: Handbook of Markov Chain Monte Carlo. Chapman & Hall/CRC Handbooks of Modern Statistical Methods. Chapman and Hall/CRC, Boca Raton (2011)
2. Causon, D.M., Mingham, C.G.: Introductory Finite Volume Methods for PDEs. Ventus Publishing ApS, Frederiksberg (2011)
3. Ciardo, G., Nicol, D., Trivedi, K.S.: Discrete-event simulation of fluid stochastic Petri nets. IEEE Trans. Softw. Eng. **25**, 207–217 (1997)
4. Haar, S., Theissing, S.: A hybrid-dynamical model for passenger-flow in transportation systems. In: 5th IFAC Conference on Analysis and Design of Hybrid Systems (2015)
5. Haar, S., Theissing, S.: Forecasting Passenger Loads in Transportation Networks (2016). https://hal.inria.fr/hal-01259585 (working paper)
6. Haar, S., Theissing, S.: Predicting traffic load in public transportation networks (2016). https://hal.archives-ouvertes.fr/hal-01286476 (working paper)
7. Hoogerheide, L., Kaashoek, J., van Dijk, H.: Functional approximations to posterior densities: a neural network approach to efficient sampling, December 2002. http://hdl.handle.net/1765/1727
8. Hu, J., Lygeros, J., Sastry, S.S.: Towards a theory of stochastic hybrid systems. In: Lynch, N.A., Krogh, B.H. (eds.) HSCC 2000. LNCS, vol. 1790, p. 160. Springer, Heidelberg (2000)
9. MacKay, D.J.C.: Introduction to Monte Carlo methods. In: Proceedings of the NATO Advanced Study Institute on Learning in Graphical Models (1998)
10. Wolter, K.: Modelling hybrid systems with fluid stochastic Petri nets. In: Proceedings of the 4th International Conference on Automation of Mixed Processes: Hybrid Dynamic Systems (2000)

Author Index

Printed in the United States
By Bookmasters